ACSM's
Exercise Management for Persons With Chronic Diseases and Disabilities

FOURTH EDITION

Geoffrey E. Moore, MD, FACSM
Healthy Living and Exercise Medicine Associates

J. Larry Durstine, PhD, FACSM
University of South Carolina

Patricia L. Painter, PhD, FACSM
University of Utah

Human Kinetics

Library of Congress Cataloging-in-Publication Data

Names: Moore, Geoffrey E., 1957- , editor. | Durstine, J. Larry, editor. |
    Painter, Patricia Lynn, editor. | American College of Sports Medicine.
Title: ACSM's exercise management for persons with chronic diseases and
    disabilities / Geoffrey E. Moore, J. Larry Durstine, Patricia L. Painter,
    editors.
Other titles: Exercise management for persons with chronic diseases and
    disabilities
Description: Fourth edition. | Champaign, IL : Human Kinetics, [2016] |
    Includes bibliographical references and index.
Identifiers: LCCN 2015026995 | ISBN 9781450434140 (print)
Subjects: | MESH: Exercise Therapy–standards–Practice Guideline. | Chronic
    Disease–rehabilitation–Practice Guideline. | Disabled
    Persons–rehabilitation–Practice Guideline. | Exercise
    Test–methods–Practice Guideline.
Classification: LCC RM725 | NLM WB 541 | DDC 615.8/2–dc23 LC record available at http://lccn.loc.
gov/2015026995

ISBN: 978-1-4504-3414-0 (print)

The web addresses cited in this text were current as of December 7, 2015, unless otherwise noted.

Senior Acquisitions Editor: Amy T. Tocco; **Developmental Editor:** Melissa J. Zavala; **Managing Editor:** B. Rego; **Copyeditor:** Joyce Sexton; **Proofreader:** Red Inc.; **Indexer:** Bobbi Swanson; **Permissions Manager:** Dalene Reeder; **Senior Graphic Designer:** Fred Starbird; **Graphic Designer:** Dawn Sills; **Cover Designer:** Keith Blomberg; **Art Manager:** Kelly Hendren; **Associate Art Manager:** Alan L. Wilborn; **Illustrations:** © Human Kinetics, unless otherwise noted; **Printer:** Walsworth

Printed in the United States of America   10 9 8 7 6 5 4 3 2 1

The paper in this book was manufactured using responsible forestry methods.

Human Kinetics
Website: www.HumanKinetics.com

United States: Human Kinetics
P.O. Box 5076
Champaign, IL 61825-5076
800-747-4457
e-mail: info@hkusa.com

Canada: Human Kinetics
475 Devonshire Road Unit 100
Windsor, ON N8Y 2L5
800-465-7301 (in Canada only)
e-mail: info@hkcanada.com

Europe: Human Kinetics
107 Bradford Road
Stanningley
Leeds LS28 6AT, United Kingdom
+44 (0) 113 255 5665
e-mail: hk@hkeurope.com

Australia: Human Kinetics
57A Price Avenue
Lower Mitcham, South Australia 5062
08 8372 0999
e-mail: info@hkaustralia.com

New Zealand: Human Kinetics
P.O. Box 80
Mitcham Shopping Centre, South Australia 5062
0800 222 062
e-mail: info@hknewzealand.com

E5762

Contents

Part I **Foundations of Exercise in Chronic Disease and Disability** **1**

Geoffrey E. Moore, J. Larry Durstine, and Patricia L. Painter

Chapter 1 **Exercise Is Medicine in Chronic Care** **3**

Robert Sallis and Geoffrey E. Moore

Chapter 2 **Basic Physical Activity and Exercise Recommendations for Persons With Chronic Conditions** **15**

Benjamin T. Gordon, J. Larry Durstine, Patricia L. Painter, and Geoffrey E. Moore

Chapter 3 **Art of Clinical Exercise Programming** **33**

Patricia L. Painter and Geoffrey E. Moore

Part VIII **Case Studies** **301**

Geoffrey E. Moore

Foreword

I f you are a primary care doctor, buy this book and keep it readily available in your clinic. It will become very dog-eared in short order.

If you are an allied health care provider working with patients who have chronic conditions but you and your coworkers aren't knowledgeable about exercise in chronic disease, then you too should buy this book.

If you are a clinical exercise physiologist or a physical therapist, or if you are studying to become an exercise professional working with people who have a chronic disease or disability, you certainly need this book.

If you already have an earlier edition, take a close look at this one because it is more than a simple update. Rather, it is a substantial rewriting designed to help forge your profession in tomorrow's health care environment.

Abundant epidemiological and clinical trial data prove that physical inactivity and lack of physical fitness are strong, independent risk factors for many chronic conditions, for disability, and for all-cause mortality (notably from cardiovascular disease and cancer). Exercise programs and regular physical activity are known to

- counteract metabolic states that cause cardiovascular disease,
- reduce disability,
- improve quality of life,
- help maintain physical independence,
- help maintain cognitive ability,
- delay loss of independence, and
- in some circumstances, increase longevity.

These are the issues that patients with chronic conditions and their families really care about—maintaining the vitality, physical functioning, and independence that they sense are slipping away.

The evidence for the role of exercise in maintaining health, well-being, and the physical functioning required to maintain independent living is overwhelming. While not all of these outcomes can be realized for all conditions discussed in this book, no other single health prescription has the potential to achieve all of these benefits. Medical schools and postgraduate training of physicians across the globe must begin to address exercise in the contemporary training of physicians, and health care systems must begin to incorporate exercise into chronic care management. If that describes a health care system you want to create, no other resource is so thoroughly focused on helping you achieve that transformation.

Geoffrey E. Moore, MD, FACSM
Healthy Living & Exercise Medicine Associates

Preface

The first edition of *ACSM's Exercise Management for Persons with Chronic Diseases and Disabilities* (affectionately referred to by the editors as *CDD*) was an effort to encourage people working in cardiopulmonary rehabilitation to expand their skills and knowledge to other populations. Most of the recommendations in the first and subsequent editions came from professional experience and approached each chronic health problem in terms of a "special and unique population." They provided minimal guidance on how to consider exercise for individuals with more than one condition. One major advance in the fourth edition is that it conceptually addresses how exercise can be managed in persons with various combinations of chronic conditions.

People with multiple chronic conditions are numerous, and the medical literature does not provide much help in thinking about how to address the problem of exercise management in such people. Despite the fact that many people present to health and exercise professionals with multiple chronic conditions, this situation is rarely studied because of the scientific complexity of interacting pathophysiologies that result in quite heterogeneous exercise responses. Such individuals also have a high rate of intercurrent illness and thus are more likely to miss training sessions for extended periods of time. These issues make obtaining and interpreting research data very difficult, because technically these subjects may not follow the study protocol (even if they eventually complete all phases). As a result, scientists shy away from studying these complex health problems for good reasons: (a) It is difficult to obtain funding; (b) these types of studies are very difficult to design; and (c) study outcomes are not always publishable in peer-reviewed journals because of between-subject heterogeneity in the intervention. Regardless of these difficulties, much more evidence is presently available to support exercise in chronic conditions than was available 20 years ago when we developed the first edition of *CDD*.

A second major advance in this fourth edition of *CDD* is the refocusing of goals of exercise beyond primary and secondary prevention of cardiometabolic disease, toward the goal of keeping patients and clients physically active in order to optimize their physical functioning and full participation in life activities. This approach parallels the modern practice of gerontology, in which a major goal is to preserve cognitive function and independent living. The accumulating evidence (from both research and anecdotal information) demonstrates that the most important benefit of exercise in people with chronic conditions is the ability to maintain or improve physical functioning and independence. In this perspective, exercise intolerance may not improve very much, nor will the pathology be cured, but preventing decline in cognitive and physical functioning helps maintain quality of life. Regardless of the condition, physical functioning—the ability to do activities of daily living for oneself and to participate in recreational activities—is highly predictive of longevity and the ability to live independently. *Regular physical activity and exercise is the only prescription that can preserve these personal freedoms.*

The third major advance in this edition of *CDD,* and perhaps the most important, is the drafting of the book as a key resource for primary care providers. The American College of Sports Medicine has spearheaded the Exercise Is Medicine global initiative, with a goal of having doctors everywhere prescribing exercise to their patients. Most of what the world's people need is exercise for primary prevention; but secondary prevention—exercise as a medicine in treating persons with chronic disease or disability—is an additional goal of the Exercise Is Medicine effort. Primary care medicine is changing dramatically, especially in the United States, but also around the world. Two major innovations, still evolving, are (1) the chronic care model and (2) the concept of the medical home. To some degree, these two concepts go together, with the medical home serving to help patients coordinate all aspects of their care and most especially chronic disease management. There remains a tendency, however, not to use exercise in patients who really would benefit from exercise training. Most physicians and even trained epidemiologists are underinformed on the power of exercise and physical fitness in the promotion of health, well-being, quality of life, and longevity. Patients need their physicians to advise them to be more active, and our aim is to help physicians know how to do this for their patients with complex chronic health problems.

Accordingly, we have added a primer on exercise—why it's important, how to follow proper exercise prescription protocols, and some guidance on knowing when and how to improvise. This is not simple for many, if not most, patients with multiple chronic conditions, whose needs for exercise programming aren't to be found in a peer-reviewed guideline. Providers must have a sense of how to improvise. Readers who need information on how to prescribe exercise will find this in chapters 2 through 4. These chapters cover the basics of exercise prescription, explain when someone needs to have diagnostic exercise testing, discuss how to use existing exercise resources in the community, and provide a brief overview of counseling methods commonly in use by exercise professionals.

A fourth major advance is that in chapter 2 we are putting forth a new *Basic CDD4 Recommendation,* which is based on consensus statements and streamlined to better suit a chronic disease population. The application of these recommendations is discussed for a number of very common chronic conditions, as well as the common combinations of chronic conditions. These recommendations are based in part on expert panel statements and opinions (and sometimes on the expert opinions of the authors and editors of this edition). The "art" of exercise counseling is presented to provide insight on how and when the "rules" presented in established guidelines might be adjusted to reduce barriers and to encourage people to adopt physical activity as part of the medical management of their condition. Finally, a variety of less common conditions are discussed, leaving out no condition that we've addressed in the past.

The fifth major advance is in how we have chosen to present the case studies. In the past, case studies were presented in conjunction with the relevant chapter. This is convenient, but the format also has a tendency to convey the notion "*Here's how to manage someone with ____*". In this edition, we have collected the case studies into the new part VIII, called "Case Studies." We also think of these as "exercise rounds." Health professionals use meetings called rounds to share knowledge about a patient case both for ongoing care and to discuss the management of such situations. The purpose of the case studies, as in all clinical rounds conducted by health and exercise providers everywhere, is partly to discuss the "how to" of a case. But a more elegant function stems from the fact that real life isn't neat and that every single person—*everyone*—has a unique story that reveals the challenges the individual faces and also the failings of medical knowledge. The most important function of part VIII is to help the reader see the gaps in care (and research findings) to help guide individualization and improvisation in care.

From a liability perspective, readers must be aware that exercise management of persons with chronic conditions requires clinical training. Only a handful of patients with chronic conditions are "apparently healthy," and while exercise is generally safe, there is risk in working with such persons. Exercise and health professionals who work independently with these individuals need to understand both the pathophysiology (which is only briefly covered in this book), the medical management, and the risks of exercise—content far beyond the expertise of exercise professionals without clinical training. Exercise professionals without clinical training should not work independently with this population, but can provide exercise services under the supervision of an appropriately qualified clinician.

From the perspective of the editors, the most important barrier our health care system needs to overcome is the lack of interaction between the medical and exercise professions. Few physicians are well trained in the use of medically directed exercise, and they could refer many more of their patients than they do. As a result, programs like cardiac rehabilitation, diabetes education, and intervention are underused, as are exercise programs for persons with arthritis, cancer survivors, and people with many other conditions. Many fitness facilities offer such programs, but almost none provide a full range of services covering all conditions discussed in this book. Thus, perhaps the most important goal of this book is to reveal common ground and to create an approach involving more collaboration and teamwork between medical and exercise professionals.

We therefore have strived to make *CDD4* helpful for primary care physicians and staff working in a patient-centered medical home while retaining the same user-friendliness for allied health and exercise professionals who liked prior editions of this book. We hope this text will provide guidance for primary  care practices to incorporate exercise specialists into their practice team and as a critical part of their referral network. We also hope this text will help exercise professionals see how to extend themselves to primary care practices that need their expertise. Beyond more communications, for medical and exercise professionals to become a "team," some learning about each other's professional culture will need to occur.

If our work is to be adopted across the world, exercise management must become an embedded operation in primary care practices, specialty practices, and health and fitness businesses. To those on the medical team (doctors, nurses, physical therapists, clinical exercise physiologists, occupational therapists, counselors, and so on), the people this book is about are called patients; but after these patients go to a medical exercise program, these same people need to join a gym or go to a personal trainer who thinks of them as clients. Only when this pathway is heavily traveled will people with chronic conditions have the hope of optimizing their physical function so they can remain independent and have a reasonable quality of life.

Acknowledgments

On behalf of Larry Durstine, Trish Painter, and myself, I would like to thank ACSM's Publications Committee for investing in this series of textbooks. Special thanks to that committee's chairs—Larry Kenney, Jeff Roitman, and Walt Thompson—for all their support. Thanks to Katie Feltman (ACSM) and Amy Tocco (Human Kinetics) for their help and patience in the remaking of this edition, as I know the changes we came up with put us far behind the original time line. Thanks to Deb Riebe for collaborating with us to help blend this book with the 10th edition of ACSM's Guidelines—this blending is such an important element and much more complicated than it sounds. Larry Durstine and I are particularly grateful for the inspiration and support of Loarn Robertson, former acquisitions editor at Human Kinetics, who guided us through the first three editions of this book and provided valuable senior leadership in the design of this series. On behalf of clinicians and exercise researchers everywhere, we thank the patients and research subjects for the priceless gift of allowing us to learn from them—there is nothing so generous as putting one's trust in someone who is leading you into the unknown. Without you, this book couldn't have been written.

On a personal note, I thank John Rudd, CEO of Cayuga Medical Center at Ithaca, for allowing me to spend some office time working on this textbook. Lastly, I'd like to thank my best of friends, Trish Painter and Larry Durstine, for 25 years of devotion to the cause of creating this series of books—Trish for providing the inspiration and Larry for making it happen. I've learned so much from both of you.

Geoffrey E. Moore, MD, FACSM

Notice and Disclaimer

Care has been taken to confirm the accuracy of the information presented and to describe generally accepted practices. However, the authors, editors, and publisher are not responsible for errors or omissions or for any consequences from application of the information in this book and make no warranty, expressed or implied, with respect to the currency, completeness, or accuracy of the contents of the publication.

Application of this information in a particular situation remains the professional responsibility of the practitioner; the clinical treatments described and recommended may not be considered absolute and universal recommendations.

The authors, editors, and publisher have exerted every effort to ensure that drug selection and dosage set forth in this text are in accordance with the current recommendations and practice at the time of publication. However, in view of ongoing research, changes in government regulations, and the constant flow of information relating to drug therapy and drug reactions, the reader is urged to check the package insert for each drug for any change in indications and dosage and for added warnings and precautions. This is particularly important when the recommended agent is a new or infrequently employed drug.

Some drugs and medical devices presented in this publication have Food and Drug Administration (FDA) clearance for limited use in restricted research settings. It is the responsibility of the health care provider to ascertain the FDA status of each drug or device planned for use in their clinical practice.

FOUNDATIONS OF EXERCISE IN CHRONIC DISEASE AND DISABILITY

Part I of this textbook is a primer in exercise management, primarily for health care professionals who have little or no formal training in exercise physiology. But it's also designed to help exercise specialists on the health care team collaborate with clinical staff in the process of exercise programming in persons with chronic conditions. The goal is for the entire health care team to function at "the top of their pay grade" with regard to exercise management.

Physicians and primary care or medical home staff mainly serve to provide motivation, as well as to diagnose and resolve any exercise-related problems. Exercise specialists, such as physical therapists, clinical exercise physiologists, occupational therapists, personal trainers, and fitness professionals usually provide the vast majority of exercise intervention and counseling. Everyone needs to be comfortable working with complex patients and know when to confer with clinical staff or a physician to solve a problem.

It is important that everyone working with a patient proceed with the same expectations, so one purpose of this section is to create common understanding of when exercise specialists should confer with physicians about additional exercise testing and diagnostic evaluation. What physicians usually want is to know when things are not going according to expectations, because that is a situation in which the physician needs to figure out why the patient is not responding as expected. It could be that the patient is not going to have a great response to exercise training, because many people with a severe burden of chronic disease show limited adaptation to training. But it could also be that there is a problem not fully diagnosed or not yet adequately managed medically, and the physician's job is to figure that out. Some physicians want to be more involved in the day-to-day progress, but most want to be problem solvers, and problem solving is what they're best trained to do.

Another issue is the multidisciplinary nature of exercise management. Experts who have worked in this field a long time know that physicians and allied health care staff have very different professional cultures and training and don't think in the same fashion. At first, it's a little shocking to learn this, because one might think that all of health care would be based on the same "textbook." But, in fact, that is often not the case, and it's common to find various health care staff having very different takes on a situation. Some easy examples, not often mentioned in this book, are massage, mechanical modes of therapy, and prosthetic shoe inserts. The various health professions often have very strong differences of opinion on these treatment modalities. This section seeks to provide a foundation that will facilitate communication and help all members of the exercise management team.

Lastly, primarily for physicians to get a deeper understanding of what happens in an exercise program, we have provided an elementary introduction to commonly used counseling techniques. This section may also be helpful to students who are new to dealing with patients, but it is only a superficial review intended to illustrate, not teach, how to be a good counselor.

.

Exercise Is Medicine in Chronic Care

The ability to do physical activity, be it physical labor or leisure-time activity or recreation or sport, is one of the most central aspects of being a person. Early in youth, all of us have a natural urge to want to do things by ourselves, for ourselves, as we develop our own sense of autonomy and independence. Late in life, or in people of any age who suffer with the burden of a chronic condition, the decline in ability to do physical activity takes on increasing importance as patients draw closer to not being able to do things for themselves. The threatened loss of autonomy and independence is emotionally traumatic and is one of nearly everyone's greatest fears in life.

Vulnerability to this threat is closely related to the loss of ability to do activities of daily living, and thus is about exercise and physical functioning. Whether a patient has a disease of metabolism or of an organ or from a physical disability, the ability to preserve physical functioning sufficient to maintain independence is central to the human psyche.

Beyond this spiritual element of life, abundant data now show that physical activity and physical fitness are immensely powerful in their ability to both lengthen life and enhance quality of life. Research over the last several decades has accumulated sufficiently to allow us to state confidently that there is no body tissue or system that does not benefit from regular physical activity. Further, there are extremely few chronic conditions in which the burden of the chronic condition, comorbidities related to the chronic condition, or the disease-related quality of life are not made better with some kind of exercise program.

In some cases, exercise prescription can be seen as secondary or tertiary prevention—averting coronary artery disease in a person who has hypertension, or preventing a second myocardial infarction. In some cases, exercise may be mainly helpful at retarding the rate of decline in functional capacity or cognitive decline. Thus, all physicians, especially primary care physicians, should help all

Benefits of Exercise Training With Chronic Conditions

- Increases longevity and mitigates disability in some conditions
- Increases the length of disability-free life
- Improves metabolic function, shifting away from diabetes and cardiovascular disease
- Improves physical functioning and quality of life

patients with a chronic condition optimize their program of physical activity or exercise. From the perspective of helping the patient maintain vitality, it's one of the most important jobs a physician can do.

On these bases, it is clear that exercise functions like a medicine and has a far broader spectrum of application than any single medication. *There is no other prescription with such pluripotent potential.*

Exercise Is Medicine

A key purpose of this fourth edition of *ACSM's Exercise Management for Persons With Chronic Diseases and Disabilities,* known to the editors as *CDD4,* is to advance the goal of helping physicians use exercise as easily as they use medications. Here are three barriers to that goal:

- It's easy to prescribe a pill.
- It's difficult to counsel patients on lifestyle.
- Many societies don't pay health care professionals for exercise management.

For physicians, this situation creates a reliance on pills and causes an unintended consequence of transferring responsibility for health away from patient behavior and onto molecular chemistry. This often results in patients being less active because of their reliance on medications! The same can be said of nutrition and dietary supplements, intended to meet nutritional needs that aren't being met by one's diet, when most nutritionists feel it would be better to just eat well.

There is debate about whether physicians should try to change the physical activity habits of their patients. The U.S. Preventive Services Task Force (USPSTF) found insufficient evidence to conclude that having physicians prescribe exercise is effective in actually getting patients to do more exercise. In contrast, other groups found evidence of benefit but were unable to conclude what system design best supports the patients and practices in the goal of increasing medically advised physical activity.

We hold, however, that some physician behaviors should be driven by principle and moral imperative, not by an evidence base. This is not to say that evidence should be ignored, but rather that in some domains of health care, notably areas of healthy behaviors, physicians should always adopt an affirmation model. The ethical origins of this principle lie in a deontological rule that physicians should advise patients toward behavior choices that carry beneficence. In the case of physicians prescribing exercise, the issue is more how physicians should operationalize practice protocols to express this moral imperative on the value of exercise than whether or not they should express their advice to exercise. Physician encouragement is the number one reason that people cite for what prompted them to quit smoking. If physicians are able to persuade patients to take insulin or cholesterol-lowering or antihypertensive drugs and to quit smoking, shouldn't physicians also advise patients to exercise? Brief counseling and pedometer programs significantly increase physical activity.

Exercise Is Medicine in the Medical Home Care Model

The Exercise Is Medicine (EIM) initiative was established "to make physical activity assessment and exercise prescription a standard part of the disease prevention and treatment paradigm for all patients." This initiative was started in November 2007 by the American College of Sports Medicine (ACSM) in conjunction with the American Medical Association at a national launch held in Washing-

ton, DC, attended by acting U.S. Surgeon General Dr. Steve Galson, along with the directors of the President's Council on Physical Fitness and Sports and the California Governor's Council on Physical Fitness and Sports. In May 2008, the first World Congress on EIM was held in conjunction with the ACSM annual meeting to announce the global launch of this program. Exercise Is Medicine has been adopted in over 50 countries, including six regional centers around the world in North America, Europe, Latin America, Asia, Africa, and Australia, showing the worldwide acceptance of the basic tenets of EIM, including recommendations for weekly physical activity to improve health.

Exercise Is Medicine was not conceived for patients with chronic diseases and disabilities, but more for apparently healthy individuals who are able to safely do activities such as walking. As such, the EIM recommendations for exercise are geared more to promote population health than to specifically address the exercise needs of persons with a chronic condition. Many people with a chronic condition have mild manifestations that are well controlled—for example, hypertension managed with a DASH (Dietary Approaches to Stop Hypertension) diet and a diuretic to achieve resting blood pressures of 126/84. Such an individual likely does not require any special accommodations for the condition in an exercise prescription, and for such a patient ACSM recommends following the ninth edition of *ACSM's Guidelines for Exercise Testing and Prescription.*

In contrast, *CDD4* is for persons who do require some accommodation for their chronic condition. One important need, then, is to help physicians and health care professionals blend the skills needed for EIM with the skills needed for *CDD4.*

Annual Wellness Visit

Primary care physicians have long followed the practice of having an annual visit with a physical exam. Health care system redesign is increasingly moving toward viewing this encounter as an *annual wellness visit,* where the intent and focus are on creating or updating the patient's plan for health promotion and disease prevention. With regard to promoting physical activity among the apparently healthy population, primary care physicians should look at this visit as a variation on the preparticipation physical exam, where the objective is to clear patients for the regular physical activity needed to stay healthy, and provide a prescription that will help the patient adopt and maintain an active lifestyle.

In this visit, rather than clearing patients to participate in sport, physicians are either clearing and guiding them to participate in regular exercise, or referring them to a program that will help them make the transition.

These are examples of programs that can help the transition:

- Physical and occupational therapy
- Cardiac and pulmonary rehabilitation
- A medically supervised exercise program (e.g., aquacise for patients who have arthritis or are obese)
- A carefully prescribed and monitored independent home program for those with stable disease

The goal of such programs is to help patients increase their physical activity and improve physical functioning to the point that they can do their own self-directed program, or do so with the aid of an exercise specialist or personal trainer. For patients who have a chronic condition, physicians should consider exercise an essential part of the treatment plan and devise a plan to help the patient adopt and maintain a regular exercise routine. By looking at each patient as you would at an athlete, you can be much more effective in helping him achieve the physical activity he needs to stay healthy.

Exercise Vital Sign and Health Risk Assessment

A basic tenet of the EIM initiative is that physical activity should be regarded as a vital sign according to which every patient has her exercise habits assessed so that a proper physical activity or exercise prescription can be provided. The exercise vital sign (EVS) is a simple way to do this and to also get the topic of exercise into the exam room with every patient. The EVS can be administered by the medical assistant as part of the assessment of the traditional vital signs of blood pressure, pulse, respirations, and temperature.

In part based on ACSM recommendations, the health informatics group at the National Institutes of Health recommends that all medical records include two simple questions (see "Exercise Vital Sign Questions").

Multiplying the two responses together gives the number of minutes per week of self-reported moderate to vigorous physical activity (MVPA) done each week by that patient. An electronic medical record can automatically display this

Exercise Vital Sign Questions

- On average, how many days per week do you engage in at least moderate to vigorous physical activity like a brisk walk?
- (Response range: 0-7 days)
- On those days, for how many minutes do you engage in physical activity at this level?
- (Response range: 10, 20, 30, 40, 60, >60 min)

value, and adults doing less than 150 min per week can be flagged with an alert. Practices that don't have an electronic medical record should have the medical assistant flag such patients. Kaiser Permanente installed an EVS in their electronic medical records in 2009, and over 90% of adult patients had an EVS recorded on their chart after 3 years of use. In patients over 65, 96% had an EVS on their chart.

Senior citizens, as the population most at risk and burdened with chronic conditions and disabilities, are patients who can especially benefit from doing regular exercise.

Another popular tool is a health risk assessment (HRA) questionnaire; in the United States these are mostly in use by wellness programs sponsored by employers or health insurance plans. Many HRAs have the EVS questions built in to the questionnaire and thus can provide the MVPA score as a part of the report. Health risk assessment tools go beyond physical activity and also ask the patient questions about diet, tobacco, stress, and other lifestyle-related risk domains. For purposes of lifestyle-related health promotion, an annual (or periodic) wellness visit coupled with an HRA that includes a MVPA score is an excellent model to implement as a standard operating procedure for a primary care or medical home practice.

Role of the Physician

What can busy physicians do during a short office visit to encourage their patients to be physically active? Well, first off, they should insist that the practice implement annual or periodic wellness visits that employ an HRA with an EVS as a standard operating procedure, with participation in these visits a mandatory requirement for patients in the practice. Then, not only has the patient provided the necessary lifestyle information (including information on physical activity), but the physician (or designee) has an entire visit that focuses on health promotion and the provider has an opportunity to discuss exercise with the patient. If the health care system doesn't compensate physicians for doing wellness visits, the practice can develop an employee with the necessary skills to do these visits and designate the role to that employee. Such a visit at least obtains the patient's data and shows patients that their doctor is willing to devote special resources to assessing their lifestyle-related needs (including exercise).

Outside of a wellness visit protocol, physicians often feel they have to squeeze a discussion about exercise into a visit scheduled for either acute or chronic care. "Tips on Bringing Up Physical Activity in a Clinic Visit" can help guide physicians regarding how to integrate a discussion about exercise into a disease care visit.

Table 1.1 includes some ideas for starting a discussion about exercise with an apparently healthy person or someone whose chronic condition does not constrain her ability to be active and exercise.

For apparently healthy patients in whom the main goal is risk reduction, provide an exercise prescription (see chapter 2) or suggest useful resources:

Tips on Bringing Up Physical Activity in a Clinic Visit

If the visit is for acute care (assuming it is not an injury):

Consider skipping any discussion about physical activity.

Maybe the patient should even take a few days off.

Exercise training is a chronic phenomenon that presupposes a stable medical status.

If the visit is for chronic care:

It's a perfect opportunity to discuss the role of exercise in the patient's health!

- Buying a pedometer to measure walking steps (daily goal of 8,000-10,000 steps)
- Joining an exercise class or getting an exercise video game or DVD
- Seeking a community resource such as a YMCA
- Getting advice from a local exercise professional or wellness coach

If the physician has 5 min or more for brief counseling, or more appropriately chooses to spend substantial time on the subject of exercise for purposes of treating the patient's chronic condition, then he can assess the patient's readiness for change regarding exercise habits. He might ask questions such as these:

- What would the patient want to do to be more active?
- What barriers are preventing this from happening?

Consider brainstorming with patients on how to get around these barriers, or explaining in detail how exercise can affect the conditions they have or may be at risk for and how they can go about incorporating physical activity into their daily life.

Exercise Prescription

Physicians should write exercise prescriptions because a physician's prescription carries with it the moral weight of the physician's judgments designed to help the patient—it elevates exercise to the same stature as all the other recommendations the physician has for the patient. There are two simple ways to write an exercise prescription; this isn't really any different than for any other prescription. These are discussed in detail in chapter 2 and are illustrated here in table 1.1 and "Two Styles of Exercise Prescription." In brief, they use a mnemonic called FITT. The idea is just to think of exercise like any other medication: The type or kind of exercise (e.g., walking) is like the type of drug; the dose is how many minutes (duration) and how hard to go (intensity); and the frequency of dosing is every day (or however many days a week; maybe even twice a day if the patient has difficulty sustaining exercise). For more details on how to write an exercise prescription, including activities most appropriate for persons with chronic conditions and low physical functioning, see chapter 2.

Table 1.1   Sample Discussions With Patients on Physical Activity

Physician time available	Ideas to start a discussion on exercise
No time or running late	Skip it! At least the EVS was measured, sending a message on the importance of exercise to the patient's health.
1 min	If MVPA score is >150: "Great job. You are getting the exercise you need to be healthy!" *[Encouragement]* If MVPA score is <150: "Have you ever thought about how physical activity affects your health? When we have you back for follow-up, I'd like to spend more time talking about how we can help you get more active." *[Assesses for stage of change (see chapter 3) and prompts awareness, prompts decisional balance in preparation for next visit]*
2 min	If MVPA score is >150: "Great job on the exercise! Is there a question you have about exercise that I can help you with, or help you understand how exercise affects your health conditions?" *[Reinforces the role of exercise in patient's own personal situation]* If MVPA score is <150: See 1 min example. OR "I see you aren't able to get the recommended amount of exercise. What do you think is preventing you from being able to do so?" *[Assesses for stage of change (chapter 3) and probes for the patient's self-perceived barriers (e.g., time, money, motivation, stress)]*
3-5 min (not as an add-on message, but as part of the visit)	If MVPA score is >150: "Tell me about your exercise program and what you like about it." *[Implies that you want to know what is important to the patient. Also, gathers info on what is useful to the patient (may help other patients if you know about it!)]* If MVPA score is <150: "I can't find any of the causes of high blood pressure that are due to other problems, so we need to address treating your high blood pressure. I'm going to recommend this diet [handing out materials], but studies show that 30 minutes of brisk walking, every day, will lower your blood pressure about 10 mmHg. Why don't you try this diet and a walking program, 30 minutes at least 5 days a week, and let's see if we can avoid putting you on medication." *[Example for an initial visit for high blood pressure]*

EVS = exercise vital sign

MVPA = moderate to vigorous physical activity

Two Styles of Exercise Prescription

Here are two styles to consider when writing an exercise prescription.

FITT Style

Frequency—how often to do activities in number of days per week

Intensity—the subjectively perceived or objectively quantified level of exertion

Time—the duration for which each activity should be pursued on each day

Type—the specific nature of the activity or activities to be performed

Pharmacy Style

Indication—the chronic condition for which exercise training is prescribed

Specific exercises—the specific nature of the activity or activities to be performed

Dose as measured by

- duration (time) or number of repetitions and
- intensity of subjectively perceived or objectively quantified level of exertion.

Progression and renewals—the time course over which the program should continue

Adverse effects—chronic conditions or exercise-specific risks worthy of caution

Power of Walking

Walking should be the default form of physical activity for anyone who can walk, just as pushing oneself in a wheelchair should be the default for those who use a wheelchair. Walking is extremely accessible for all ages and fitness levels and can be done alone or in groups. Walking is low cost and doesn't require a gym or specialized equipment. It is also easy to measure walking by using a pedometer or a watch. Walking has generally good long-term adherence and has been proven to benefit health. Finally, use of walking as a form of commuting (and also cycling) saves not only in terms of health care costs but also in fossil fuel consumption as an environmentally friendly way to commute.

Exercise Is Medicine in the Chronic Care Model

We have briefly reviewed the basis for EIM in the primary care or medical home environment, mainly aimed at apparently healthy persons or those with one or more chronic condition(s) that do not substantially alter the person's ability to exercise safely. This material is a foundation for the far more challenging problem of exercise management in persons with a chronic disease or disability in whom the condition substantially affects their ability to exercise.

The degree to which a patient's condition affects her ability to exercise safely is a complex function of both the specific condition or combination of conditions and the relative burden of severity. Today there is no simple algorithm to quantify this, so physicians and exercise professionals who assist such patients must use their best judgment based on their knowledge of the condition, the patient's physical exam, and any useful laboratory data (such as measures of physical functioning or exercise tolerance testing). This assessment of the patient's *disease burden* then must be factored in together with the individual's goals, socioecological status, and available resources to help the patient become more active (with a long-term goal of meeting physical activity guidelines). This coordination of exercise or physical activity with socioecological factors and available resources is a logical function of the medical home as part of the chronic care model.

The chronic care model (figure 1.1) has risen to favor among health care system designers, layered onto what has become known as the medical

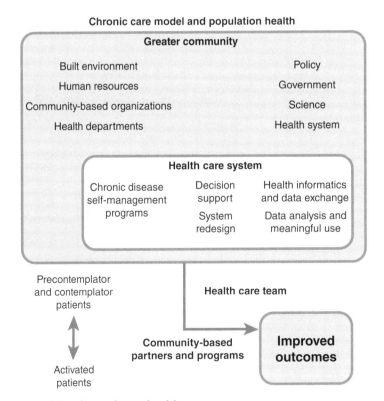

Figure 1.1 Chronic care model and population health.
Adapted from V. Barr and the McColl Institute for Healthcare Innovation.

home or patient-centered medical home (PCMH) model of primary care. Most readers are familiar with the basic concept that one key function of the PCMH is to serve as a coordinating center for all needs related to a patient's health care. Examples include helping improve patient health literacy, assisting patients with obtaining social or mental health services, facilitating transportation to and from health care visits, arranging for home health aides, and performing many other complex care-coordination tasks. Modern health care systems and patient care plans are extremely complex, especially in situations that involve multiple chronic conditions or disability, and the task of making sure all aspects of care are well coordinated is often too difficult for patients and their families to accomplish.

The chronic care model is a *generalized conceptual model* rather than a *care protocol,* and objective evidence supporting the chronic care model is generally positive, albeit modest. Most chronic conditions involve far too much variability in socioecological situation to be formulaic, as almost every patient's specific situation has an enormous number of socioecological variables beyond the influence of the PCMH. There are too many locally defined elements to make a simple case for the chronic care model. Lack of a firm evidence base that supports including exercise as a component of the chronic care model might cause one to have reservations on the best way to implement exercise management. But clinicians have the difficult task of turning a clinical guideline into operations that deliver better health care, and patients have immediate needs that cannot wait for outcomes research.

The chronic care model has weaknesses, most of which are flawed assumptions or dilemmas that the model in and of itself cannot resolve. One example is the internal challenge (i.e., within the health care system itself) of using electronic medical records to share patient data, embed decision-making support with the system, and use advanced analytical techniques to improve population health. Meeting this challenge is not possible until sufficient technological investments have created the necessary data management infrastructure. Other challenges are more external to the health care system, such as the creation of a built environment that facilitates an active lifestyle, or one in which high-quality produce is abundant and inexpensive, thus facilitating healthy diets. These socioecological aspects of healthy living, especially an environment that

promotes a healthy diet and exercise or physical activity, are things the health care community itself cannot provide but are essential elements if lifestyle interventions are to be successful.

The chronic care model appropriately notes that these are essential linkages, but building them into a sustainable system is very difficult without a unifying economic force. Perhaps the biggest flaw in the chronic care model is that there is often no business model to sustain the linkages that figure 1.1 portrays. For example, delivery of exercise services can be particularly difficult when patients are aging and burdened with multiple chronic conditions that have led to disability. Such patients often have limited financial resources and are intimidated about starting an exercise program, but supervision of exercise services is not compensated by insurance. The patient needs guidance to get started; the primary care physician can provide an EIM type of prescription, but the patient can't afford an exercise specialist, and insurance won't cover the cost of the exercise specialist. In the United States this is *a very common scenario,* far too common to allow reliance on charity care without destabilizing the revenues that support exercise specialists.

Unfortunately, this is not the only financial design flaw in the chronic care model. Another assumption in the model is that individual patients and communities will become "activated" to work toward enhanced health and well-being (see figure 1.1). With about one-fourth of the population totally sedentary and two-thirds of the population not meeting the recommended level of physical activity, getting (and keeping) these people "activated" about exercise is a formidable challenge. Behavior change specialists (especially in the area of exercise) know that the majority of people don't sustain their efforts, so the concept of activation is far easier to diagram than to make happen in real life! This is another area in which redesign of the health promotion system, to provide economic incentives for everyone to be physically active, is an essential element missing from the PCMH-based chronic care model.

Proponents of the chronic care model advocate disease self-management interventions, though implementation of such programs is easier in some environments and in some aspects of health than in others. In areas where low health literacy, lack of education, and low socioeconomic status are prevalent, such barriers can be difficult to overcome as a sustainable business. Mental health issues, often superimposed on low literacy, undermine cognitive behavioral techniques that are the core of lifestyle interventions. Add in cultural issues and the lack of stable business models in certain sectors of our societies, and key parts of the chronic care model (especially efforts to promote exercise) may not work well in environments burdened with disparity.

The ACSM and its partner organization, the Clinical Exercise Physiology Association (CEPA), are actively striving to promote the inclusion of clinical exercise physiologists as full-fledged members of the care management team whose services are a benefit covered by health insurance. In many health care systems, especially in the United States, the services of clinical exercise specialists are not compensated by health care plans. If health care systems are to make use of therapeutic exercise in the care of chronic disease, this deficiency must be corrected. Primary care physicians need the support of clinical exercise physiologists as a referral resource if they are to implement wide-scale adoption of exercise prescriptions.

Primary care physicians, more than any other group, need to call for inclusion of clinical exercise physiologists in the PCMH team and for insurance plans to compensate exercise physiologists for their services. Much of what this book seeks to help accomplish—getting patients with chronic conditions more active—will be a real challenge in societies where services of exercise professionals are not valued in the health care system. This is a most notable barrier in the United States, where the health care system undervalues physical activity (though it is arguably one of the best buys, if not *the* best buy, in all of medicine).

Rationale for Including Physical Activity in Chronic Care Management

In persons with a stable chronic condition or combination of chronic conditions, this advice should almost always be the first and most important aspect of their ongoing management:

Incorporate a physically active lifestyle.

The primary reason to emphasize a physically active lifestyle is to avoid what has been termed the *disuse syndrome* or, alternatively, the *downward spiral of chronic disease* (figure 1.2).

Increasingly, studies show that the special challenge for persons who develop a chronic condition or disability is that they are vulnerable to becoming increasingly sedentary, which has a cascade of adverse effects:

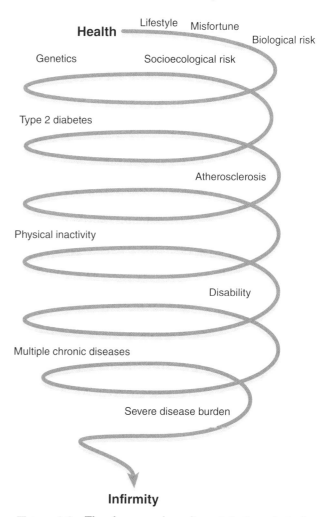

The downward cardiometabolic spiral of chronic disease and disability

Health — Lifestyle   Misfortune
Genetics   Biological risk
     Socioecological risk
Type 2 diabetes
          Atherosclerosis
Physical inactivity
          Disability
Multiple chronic diseases
     Severe disease burden

Infirmity

Figure 1.2   The downward cardiometabolic spiral of chronic disease and disability.

- Low functional capacity, which predicts poor outcomes and mortality
- Reduced gait speed and lower-extremity function associated with loss of independence
- Loss of independence, which has a negative impact on quality of life
- Increased risk of excessive weight gain
- Skeletal muscle insulin resistance or frank type 2 diabetes, with subsequent cardiovascular disease
- A gradual deterioration toward being disabled

These findings explain the long-standing observation that in persons with disability, the most common cause of death is heart disease.

Exercise Prescription, Counseling, and the Common Barriers to Exercise

In the first three editions of this book, each condition had its own recommendations for testing and prescription. But in most cases these were very similar; and since we wrote *CDD3*, a number of national and global governments and nongovernmental societies have issued guidelines on physical activity. In general, all these recommendations are similar. Moreover, if the *CDD* series followed the condition-by-condition path of logic, one would eventually end up with thousands of special case recommendations, the vast majority of which would be similar. Accordingly, for *CDD4*, the editors elected to converge onto one basic recommendation (referred to in this book as the Basic *CDD4* Recommendation), which is based on expert opinion of the editors and several authors who have worked on all the prior editions. This Basic *CDD4* Recommendation is consistent with current Department of Health and Human Services (HHS) recommendations and ACSM Guidelines, though slightly different in the following ways:

- It advises 150 min/week of MVPA (lower limit recommended by HHS/ACSM).
- It advises 150 min/week of light-intensity activity for those who can't do MVPA.
- It adds sit to stand, step-ups, and arm curls as the recommended strength training exercises.

Thus, for the majority of patients with a chronic disease or disability, the Basic *CDD4* Recommendation is not different from the guidelines for apparently healthy persons or persons with chronic conditions that have minimal impact. For individuals who are older and have more compromised functioning, the modified HHS Guidelines presented by the American Heart Association (AHA) for older individuals are appropriate. For deeper insight and understanding, the reader should see parts II through V, as well as the parts of *CDD4* that are relevant to the chronic condition of interest.

Counseling and behavioral intervention recommendations for persons with chronic conditions are mostly similar to those for the apparently healthy population (see chapter 4). What tends to be a somewhat more formidable obstacle for persons with chronic conditions are the barriers to becoming more active. This is because chronic conditions that pose a more severe burden on the patient tend to be associated with disability and subsequent disparities in resources. The chronic care model might prove a useful tool in addressing these challenges, but this remains to be proven.

Very often patients relate a variety of barriers that prevent them from doing exercise. The most common barrier is a lack of time due to competing demands. Patients often describe having a job (or more than one job) that requires them to work long hours, often with a long commute and family duties that leave them with little free time. Such competing demands frequently leave the patient too tired and without the time needed to do regular physical activity. Patients with a chronic condition often complain of a physical limitation that prevents them from doing regular exercise, such as lower-extremity arthritis or a severe medical condition that makes physical functioning a challenge. Some patients find that exercise is simply too boring and unenjoyable.

An important job for the physician, exercise professional, and chronic care management team is to help break down these barriers and to counsel the patient on how to get to the point where exercise is a habit and not an option. One simple recipe for getting apparently healthy patients to exercise is the following:

- Park their car farther away from their worksite each morning, then take a brisk 10-min walk to their work station
- At lunchtime, walk for 5 min and return before eating
- After work, take the same 10-min walk back to their car

This gives them 30 min for the day, so if they do this on 5 days each week, they will be assured of getting 150 min of moderate exercise each week.

Apparently healthy people and patients who have mild chronic conditions that don't impair their physical functioning, and who simply cannot get their exercise done during the week, can opt to do 75 min of moderate exercise on Saturday and another 75 min on Sunday, achieving all 150 min on the weekend. It may help if these patients change their mindset from the weekend as a time to rest to a time to be physically active for at least part of both days. This "weekend warrior" approach is less appropriate and useful for patients who have lower physical functioning because it is likely too exhausting to attempt 75 min on 2 consecutive days.

Other ideas to help minimize the time constraint include increasing intensity, especially for those who are apparently healthy or who don't have much limitation in physical functioning. Another helpful idea is to find an exercise partner; each partner holds the other accountable and improves adherence to the program. Owning a dog can be very helpful because the dog becomes the exercise partner, and dogs tend to insist on going out for a walk. As a result, dog owners tend to do more walking.

Take-Home Message

Exercise is a very powerful tool to treat and prevent chronic disease, mitigate the harmful effects of obesity, reduce mortality rates, and improve physical functioning and quality of life. In effect, Exercise is Medicine. Unfortunately, physical inactivity is one of the major public health problems of our time. For these reasons, physicians, especially primary care physicians and the staff of the medical home, have a responsibility to assess physical activity habits in their patients, inform them of the risk of being inactive, and provide a proper exercise prescription.

- Physician practices should use the exercise vital sign to assess MVPA.

- Physicians should support patients who are meeting physical activity guidelines.

- Physicians should encourage those who don't meet these guidelines.

- Patients who have a chronic condition that does not impair physical functioning or increase risk can be advised to walk or to do the same kinds of activities recommended for the apparently healthy population.

- Patients who have a more severe burden of chronic disease or disability often benefit from referral to resources in their community, such as physical therapy, occupational therapy, cardiopulmonary rehabilitation, or other medically supervised programs.

- Patients who have recently been hospitalized are often more willing to invest in physical activity, particularly if they obtain guidance, so transitions of care are important moments when the physician should be alert to encouraging more physical activity.

Facilitating regular exercise in a physical activity program is one of the most important functions of health care providers using the chronic care model.

Suggested Readings

Anderson LH, Martinson BC, Crain AL. Health care charges associated with physical inactivity, overweight, and obesity. *Prev Chronic Dis.* 2005;2:A09.

Bindman AB, Blum JD, Kronick R. Medicare payment for chronic care delivered in a patient-centered medical home. *JAMA.* Published online August 8, 2013 [Accessed August 9, 2013]. doi: 10.1001/jama.2013.276525.

Blair SN. Physical inactivity: the biggest public health problem of the 21st century. *Br J Sports Med.* 2009;43:1-2.

Bravata DM, Smith-Spangler C, Sundaram V, Gienger AL, Lin N, Lewis R, Stave CD, Olkin I, Sirard JR. Using pedometers to increase physical activity and improve health: a systematic review. *JAMA.* 2007;298(19):2296-2304.

Chin MH. Quality improvement implementation and disparities: the case of the health disparities collaboratives. *Med Care.* 2010 August;48(8):668-675. doi: 10.1097/MLR.0b013e3181e3585c.

Coleman KJ, Ngor E, Reynolds K, Quinn VP, Koebnick C, Young DR, Sternfeld B, Sallis RE. Initial validation of an exercise "vital sign" in electronic medical records. *Med Sci Sports Exerc.* 2012;44(11):2071-2076.

Foster C, Porcari JP, Anderson J, Paulson M, Smaczny D, Webber H, Doberstein S, Udermann B. The talk test as a marker of exercise training intensity. *J Cardiopulm Rehabil Prev.* 2008;28(1):24-30.

Frølich A. Identifying organisational principles and management practices important to the quality of health care services for chronic conditions. *Dan Med J.* 2012;58(2):B4387.

Fromer L. Implementing chronic care for COPD: planned visits, care coordination, and patient empowerment for improved outcomes. *Int J Chron Obstruct Pulmon Dis.* 2011;6:605-614.

Haskell WL, Lee IM, Pate RR, Powell KE, Blair SN, Franklin BA, Macera CA, Heath GW, Thompson PD, Bauman A. Physical activity and public health: updated recommendation for adults from the American College of Sports Medicine and the American Heart Association. *Circulation.* 2007;116:1081-1093. Originally published online August 1, 2007. doi: 10.1161/CIRCULATIONAHA.107.185649.

Hatzakis Jr MJ, Allen C, Haselkorn M, Anderson SM, Nichol P, Lai C, Haselkorn JK. Use of medical informatics for management of multiple sclerosis using a chronic-care model. *J Rehabil Res Dev.* 2006;43(1):1-16.

Joy EL, Blair SN, McBride P, Sallis RE. Physical activity counseling in sports medicine: a call to action. *Br J Sports Med.* 2013;47:49-53.

Lots to Lose: How America's Health and Obesity Crisis Threatens our Economic Future. Bipartisan Policy Center. 2012.

McCorkle R, Ercolano E, Lazenby M, Schulman-Green D, Schilling LS, Lorig K, Wagner EH. Self-management: enabling and empowering patients living with cancer as a chronic illness. *CA Cancer J Clin.* 2011;61(1):50-62.

Minkman M, Ahaus K, Huijsman R. Performance improvement based on integrated quality management models: what evidence do we have? A systematic literature review. *Int J Qual Health Care.* 2007;19(2):90-104.

Physical activity. *Lancet.* July 18, 2012. www.thelancet.com/series/physical-activity [Accessed July 27, 2012].

Pollock ML, Wenger NK. Physical activity and exercise training in the elderly: a position paper from the Society of Geriatric Cardiology. *Am J Geriatr Cardiol.* 1998;7:45-46.

Pratt M, Macert CA, Wang G. Higher direct medical costs associated with physical inactivity. *Phys Sportsmed.* 2000;28(10):1-11.

Sabaté E, ed. *Adherence to Long-Term Therapies: Evidence for Action.* Geneva: World Health Organization; 2003.

Sui X, LaMonte MJ, Laditka JN, Hardin JW, Chase N, Hooker, SP, Blair SN. Cardiorespiratory fitness and adiposity as mortality predictors in older adults. *JAMA.* 2007;298(21):2507-2516.

U.S. Department of Health and Human Services. *Physical Activity and Health: A Report of the Surgeon General.* Atlanta: U.S. Department of Health and Human Services, Centers for Disease Control and Prevention, National Center for Chronic Disease Prevention and Health Promotion; 1996. www.cdc.gov/nccdphp/sgr/index.htm [Accessed November 9, 2015].

Basic Physical Activity and Exercise Recommendations for Persons With Chronic Conditions

The American College of Sports Medicine (ACSM) published the first edition of *The Guidelines for Graded Exercise Testing and Exercise Prescription* (the ACSM Guidelines) in 1975. Since then, the Guidelines have evolved and become the standard in exercise management. The principles for exercise prescription established in the ninth edition of the ACSM Guidelines provide the foundation for the recommendations in *ACSM's Exercise Management for Persons With Chronic Diseases and Disabilities, Fourth Edition (CDD4),* but the Basic *CDD4* Recommendation is slightly more conservative because the clinical population of interest in *CDD4* is persons who have a more severe impact on their ability to do exercise.

The Guidelines primarily focus on exercise prescription for normal adults (so-called apparently healthy adults) but also include recommendations on testing and prescription for persons with some chronic conditions. Most of the chronic conditions addressed in the Guidelines relate to conditions that increase cardiometabolic risk, and in many of those conditions the impact of the condition on exercise is small. Hypertension is a good example, in that well-controlled hypertension doesn't appreciably alter physical functioning, and in such cases the *CDD4* recommends using the Guidelines.

The Basic *CDD4* Recommendations, however, are mainly intended for clinical populations in whom the exercise response and exercise capacity are dramatically diminished because of the condition. While *CDD4* also addresses hypertension, for example (see chapter 6), hypertension is not the sole entity that affects exercise capacity; rather, it is hypertension in combination with other comor-bid chronic conditions that has the net effect of reducing exercise capacity. The role of hypertension in this situation is more complicated than for hypertension as a sole entity. Some patients demonstrate a brisk hypertensive response to exercise, but others have blood pressure that is pharmacologically well controlled, with normal exercise responses. Yet others have one or more additional comorbid conditions that substantially impair their capacity to do physical work; and, in addition, there are people who are unable to exert themselves hard enough to show much of a pressor response to exercise. This book is mainly about the latter two groups—those individuals with multiple comorbidities that complicate the standard guidelines for exercise. The editors of *CDD4* expect the reader to be familiar with the Guidelines and to be able to judge when the Basic *CDD4* Recommendations can apply to a particular patient. If there is uncertainty, the reader is advised to be conservative. As the awareness of the benefits of regular physical activity for people with chronic conditions increases, there will be more referrals for exercise in more complicated individuals for whom exercise may previously have been considered inappropriate.

Definitions Used in This Book

It is important for everyone on the health care team to have a common language for physical activity and exercise, but many clinicians have little or no formal training in exercise science. For this reason, we provide definitions of terms and phrases commonly used in exercise science.

Physical Activity and Exercise

The terms *physical activity* and *exercise* are widely used in reference to many different activities, but their meanings are often vague and they are sometimes used interchangeably. Before delving into the Basic *CDD4* Recommendations, it is prudent to clarify the terms used in this book. The World Health Organization (WHO) defines physical activity as "bodily movement produced by skeletal muscles that requires energy expenditure." Technically, then, *exercise* is a form of physical activity. *Exercise training,* however, is *regular* physical activity and can be defined as "planned, structured, and repetitive physical activity for the purpose of developing physical fitness."

Participation in *exercise training,* as a repeated series of individual exercise sessions over a period of days, weeks, and months, is usually done to improve physical fitness or functioning but can be done with other objectives in mind. One reason for confusion is that some sources refer to exercise training just as *exercise.*

Light, Moderate, and Vigorous Physical Activity

Most national and International guidelines categorize physical activity into three general levels of exercise intensity: light, moderate, and vigorous. Additionally, the reintroduction of research in the area of high-intensity exercise makes it important to define this fourth level. These are the four levels (see also table 2.1):

1. Light: An aerobic activity that causes a barely noticeable change in breathing, involving increased depth (volume of each breath) more than rate; can usually be sustained by an untrained individual for 60 min or more

2. Moderate: An aerobic activity that can be performed at a ventilatory demand that allows carrying on a conversation (also known as talk test); can usually be sustained by an untrained individual for 30 to 60 min

3. Vigorous: An aerobic activity that is sufficiently demanding of ventilation that talking cannot be maintained during the activity; can usually be sustained by an untrained individual for only 20 to 30 min

4. High: An aerobic or combination activity (i.e., a combined exertion of muscle contractions at or near their maximal strength for an extended number of repetitions, such as circuit weight training); can be sustained only briefly, typically <10 min

It perhaps should be noted that any exertion less than light in intensity would be considered sedentary and that these categories apply mainly to untrained individuals; athletic people are likely to have higher thresholds.

As an alternative way to estimate perceived exertion, a 0-10 scale, as shown in table 2.2, may be appropriate. This scale is mainly validated for use in pulmonary disease but can be helpful in other chronic conditions. In this particular 0-10

Table 2.1   Intensity of Physical Activity and Breathing

Intensity	Ventilatory symptomatology
Light: 1.5-3 METs	Barely detectable, increased depth
Moderate: 3-6 METs	Can pass the talk test
Vigorous: 6-9 METs	Can't pass the talk test
High: >9 METs	Heavy breathing

1 MET = metabolic energy expenditure at rest, approximately 3.5 mL O_2 / kg / min

Table 2.2   0-10 Point RPE Scale

Rating of perceived exertion	Subjective description
0	Nothing (rest)
0.5	Very, very light
1	Very light
2	Light
3	Moderate
4	Somewhat hard
5	Hard
6	
7	Very hard
8	
9	
10	Very, very hard

Glossary of Important Terms in Exercise Science

The following terms are important for everyone on the health care team to be familiar and conversant with, because care management is likely to involve discussions on these issues.

exercise training: Has historically meant a program of recurrent and regular physical work that results in increased fitness. Such work and programs could be physical labor on the job; recreational activities such as hiking or gardening; or sporting activities and competition such as ball games, competitive running, or bicycle touring. Organized "sports" are a subset of these, but exercise training doesn't have to involve a sport.

physical activity: Refers to all-encompassing acts that involve physical work. Leisure-time physical activity (discussed in the section so titled) is intended to denote physical activity done when not on the job. Leisure-time physical activity is a very commonly used research measure. From a caloric expenditure perspective, physical work done on the job and LTPA can be equivalent, and public health recommendations of 150 min/week of moderate to vigorous physical activity on the job or in LTPA are considered interchangeable.

exercise capacity: Refers to the ability of an individual to perform aerobic work. Even though the terms are not synonymous, exercise capacity is often referred to as maximal exercise capacity. Exercise capacity commonly means the ability to consume oxygen as measured by a large-muscle, whole-body form of exercise such as walking-running or cycling. Nomograms may be used to estimate exercise capacity, or a direct measurement of work in the form of watts or oxygen uptake (or both) may be used.

physical functioning: Refers to the practical application using whole-body integration of exercise capacity. Examples include carrying bags of groceries into the house, which involves strength to lift the bags; coordination to get up the stairs; and the aerobic ability to walk (while carrying the bags). The term *functional capacity* blends the terms *physical functioning* and *exercise capacity*.

maximal exercise: Usually refers to maximal aerobic capacity as measured using a treadmill or cycle ergometer. Other more specific tests, such as swimming or skating or skiing, can be done (if the exercise lab has the proper highly specialized equipment); but these are more commonly used in sport physiology laboratories than for persons with chronic conditions. One exception is that some exercise labs have the ability to do wheelchair exercise testing.

peak exercise: Closely related to maximal exercise; worth noting because many persons with chronic conditions cannot achieve a so-called *true maximum*. In a "true max," the measured oxygen uptake increases in a linear relationship to physical work until oxygen uptake levels off and does not increase any further with increasing work. Technically, this *plateau* in oxygen uptake must not increase more than 250 mL/min despite increasing work rate, and most exercise laboratories would also prefer to measure a respiratory exchange ratio (RER) of >1.15 (where RER is the ratio of CO_2 produced to O_2 taken up, or $\dot{V}CO_2/\dot{V}O_2$).

six-minute walk: A test commonly used in cardiovascular studies of pharmaceutical agents because it is very inexpensive and is sensitive for folks who have low physical functioning. The basic format of this test is to have a measured walkway; the patient walks at his own pace with the objective of going as far as he can in 6 min.

sit to stand: There are a number of variations of this simple test; one basic version measures how long it takes to rise from a sitting position to a standing position 10 consecutive times, trying to complete the test as fast as possible. Performance is assessed by time. One variation for those who cannot do 10 consecutive repetitions includes a timed test measuring how many repetitions are completed in 30 s.

get-up-and-go: This is another simple test, requiring the individual to get up out of a chair and walk 12 ft [3.7 m], turn around, and return to sit back down. Performance is assessed by time, with the goal of completing the test as fast as possible.

sit and reach: The sit and reach is a functional mobility test in which the patient sits in a chair against a wall with one leg extended, then bends forward to reach as far as she can. There are variations of this test to accommodate persons with back discomfort. Performance is measured by the distance the individual can reach forward from the seated position.

scale, one would generally expect level 5 (hard) to be very near the ventilatory threshold. In persons with neuromuscular disease or conditions, however, where fatigue is a main symptom, level 5 may occur well below the ventilatory threshold.

Leisure-Time Physical Activity

Physical activity is achieved during leisure time (accumulated outside the work environment) or occupationally (accumulated in the work environment). Because modern societies have experienced a significant reduction in physically demanding jobs to the point that most jobs are at most light-intensity exertion (or even sedentary), *leisure-time physical activity* (LTPA) is now generally considered an indicator of the overall physical activity (leisure-time + occupational physical activities, combined).

Having patients incorporate LTPA and exercise into their daily routine will add to the total daily physical activity. The sample recommendation in chapter 1, of brisk walking for 10 min between parking the car and getting to the work station, 10 min more at lunchtime, and another 10 min after leaving the work station for the car, is a good example of LTPA. Depending on individual goals and the clinical severity of the chronic disease, the medical management plan should provide guidance on how to increase LTPA, provide a well-designed exercise prescription for exercise training, or do both.

Physical Activity Guidelines

Over the last 40 years, the recommendations for the optimal type and volume of physical activity or exercise have evolved as research evidence on this subject has emerged. In 2008, the U.S. Department of Health and Human Services (HHS) published evidence-based guidelines on physical activity based on an exhaustive expert review of the research literature, with overwhelming findings pointing to the following recommendation:

- Everyone should be physically active, defined as accumulating a minimum weekly total of 150 min of moderate physical activity or, alternatively, 75 min of vigorous physical activity.
- Adults should participate in 2 or more days of muscle strengthening activities that involve all major muscle groups.

- Individuals at risk for falls should incorporate activities to improve balance.

This recommendation has essentially become a worldwide consensus. Since 2002, the WHO has recommended 30 min of moderate-intensity physical activity on at least 5 days a week, or at least 20 min of vigorous-intensity exercise on 3 days a week. The European Union (EU)-approved guidelines, published in 2008 shortly after the HHS Guidelines, also recommend 60 min of daily physical activity for children and adolescents, as well as strength or balance conditioning (or both) for seniors over the age of 65. The EU Guidelines are more broad-reaching in that they extensively address the process of implementation of their recommendations. See table 2.3 for summary information on a variety of current guidelines on physical activity.

Guidelines from ACSM and the American Heart Association (AHA) for older adults seem more appropriate for many individuals who have a chronic health condition. These guidelines differ from the HHS Guidelines in that the recommendations for intensity of aerobic activity take into account the current fitness level and focus on using perceived exertion to guide exercise intensity. They also include recommendations on maintaining flexibility and for balance training in individuals who are at risk for falls.

The only national guidelines that specifically address chronic conditions are the Swedish National Institute of Public Health's *Physical Activity in the Prevention and Treatment of Disease*. This document presents recommendations for most chronic diseases that are affected by physical activity and exercise. Unlike the U.S. Physical Activity Guidelines, these recommendations outline physical activity and exercise for specific diseases with the intent of better accommodating the varying pathophysiological effects (as well as treatments) of chronic disease.

Basic *CDD4* Recommendations for Physical Activity or Exercise in Chronic Conditions

After four editions of the *CDD* series, with many decades of clinical experience on the part of the contributors to *CDD4,* the main working group of

Table 2.3   Comparison of Exercise Guidelines From Leading Organizations

Guidelines	World Health Organization (WHO) Physical Activity Guidelines	U.S. National Human and Health Services Physical Activity Guidelines	European Union Physical Activity Guidelines	American Heart Association Physical Activity Guidelines	American College of Sports Medicine Position Stand for Exercise Prescription in Older Adults	Swedish *NIPH Physical Activity in the Prevention and Treatment of Disease*
Frequency?	Yes, most days of the week recommended	Yes, most days of the week recommended	Yes, most days of the week recommended	Yes, 3-5 days a week	Yes, 3-5 days a week for aerobic, 2-3 days for resistance training	Yes, but varies for each chronic condition
Specific intensity?	No, only general intensities given (e.g., vigorous)	No, only general intensities given (e.g., vigorous)	No, only general intensities given (e.g., vigorous)	No, only general intensities given (e.g., vigorous)	Yes, but varies on the volume of exercise used	Yes, but varies for each chronic condition
Multiple modes of exercise?	Yes, aerobic and muscle strengthening	Yes, aerobic and muscle strengthening	Yes, aerobic and muscle strengthening	No, only aerobic addressed in AHA general guidelines	Yes, aerobic, muscle strengthening, and flexibility	Yes, aerobic, muscle strengthening, and flexibility
Duration?	Yes, 150-300 min weekly	Yes, 150-300 min weekly	Yes, 150-300 min weekly	Yes, 30 min per session	Yes, varies on volume of exercise used	Yes, varies for each chronic condition
Are the guidelines designed specifically for chronic disease patients?	No, guidelines specifically designed for healthy adults	No, guidelines specifically designed for healthy adults	No, guidelines specifically designed for healthy adults	No, guidelines specifically designed for healthy adults	No, guidelines specifically designed for healthy adults	Yes

authors concluded that it is confusing and unnecessary to sustain disease-specific recommendations, for these reasons:

- There are thousands of chronic conditions and causes of disability.
- The vast majority of recommendations seem similar for most chronic conditions.
- Ultimately, exercise is fairly simple and needs to be seen as elegantly powerful.
- The complexities and nuances are matters of clinical judgment for safety's sake.
- The main concern in the chronic conditions in *CDD4* is loss of independent living, which is primarily a function of light-intensity physical activities.

For these reasons, the Basic *CDD4* Recommendations are consistent with but differ very slightly from the Guidelines and from the various physical activity guidelines discussed in the preceding section, because the *CDD4* editors also want to make certain that sufficient attention is paid to light-intensity physical activity, especially the ability to perform instrumental activities of daily living with the goal of having patients remain independent. With these considerations in mind, the Basic *CDD4* Recommendations are as follows:

- Every person with a chronic condition should be physically active, accumulating a minimum weekly total of
  - 150 min of preferably moderate-intensity physical activity or, if that is too difficult, then
  - 150 min of light-intensity physical activity may be substituted.

- At least 2 days per week of flexibility and muscle strengthening activities that should *minimally* involve
  - chair sit-and-reach stretches on left and right,
  - at least eight consecutive sit-to-stand exercises,
  - at least 10 step-ups (or a flight of steps), leading with each foot, and
  - at least eight consecutive arm curls with a minimum of 2 kg held in the hand; 4 kg is recommended.
- Individuals at risk for falls should be evaluated for causes of falls. Not all falls are caused by a condition that can be treated with exercise training. If the diagnosis of the causes suggests that exercise training can reduce the likelihood of a fall, then activities to improve balance should be incorporated into individuals' exercise regimens, under the supervision of an exercise therapist trained in fall prevention.

The higher the aerobic intensity, muscle forces, and required range of motion, the greater the likelihood of an adverse event.

These Basic *CDD4* Recommendations are summarized in table 2.4. Readers should always bear in mind the goals behind the Basic *CDD4* Recommendations, because these goals are helpful in drafting an individualized program to meet the unique needs of each patient. Everyone should be physically active to an extent sufficient to maintain independent living:

- Let no barrier block someone from doing light-intensity physical activity.
- Independent living requires a minimum ability to perform activities involving (or demanding)
  - light-intensity aerobic work (or exertion), combined with
  - strength, flexibility, and balance and coordination.

Adverse events from exercise cannot be completely eliminated, but there are two main categories to consider:

1. Activity-dependent risks (due to the nature of the activity)
2. Disease-dependent risks (those that relate to the pathophysiology)

The best way to minimize activity-dependent risks is to encourage the patient to practice safety precautions. If there is concern that the individual cannot do this independently, then he needs a supervised exercise program, at least to get started.

One major concern is whether or not the advice to do physical activity exposes the patient to the possibility of a disease-dependent risk. Such risks are associated with the intensity of exercise. Accordingly, if the recommendation is to complete vigorous- or high-intensity physical activities, it is prudent to follow the Guidelines on exercise testing and prescription. Most people with a chronic condition can safely participate in moderate-intensity physical activity, and if there is any concern that a particular individual cannot do so, she should either undergo some disease-specific diagnostic exercise testing or be referred to a supervised exercise program (or both).

There are few data beyond anecdotal cases to support the concern that light-intensity physical activities are likely to precipitate a disease-dependent adverse event (especially sudden death or myocardial infarction). Discerning the epidemiological role of light activities in precipitating such events would be exquisitely difficult, because participation in light-intensity activities is so ubiquitous in daily life. If someone is medically unstable to the point that activities of daily living threaten injury or death, then the Basic *CDD4* Recommendations do not apply because the individual is not able to maintain independence and belongs in either a hospital or a nursing home. Indeed, it is likely that many people end up in a nursing facility sooner than they need to because no one recommended that they do light-intensity physical activity.

These nuances of safety are a key reason why an exercise professional is a necessary member of the chronic care team, because these staff are trained and have the experience to make good judgments regarding when exercise is safe and when it is not safe. See chapter 3 for a more in-depth discussion on how these judgments are made, which often involves more art than science.

Table 2.4 **Basic *CDD4* Recommendations**

Mode	Frequency	Duration	Intensity	Progression
Aerobic • Large-muscle easily accessible activities such as walking as the basic program • Aquatics recommended for those with musculoskeletal problems during weight-bearing activity • Other fun-to-do large-muscle activities such as cycling or gardening are alternatives	4-5 days/week	Start at any duration, as tolerated Goal of 40 min/session 20 min if a combined session with strength training exercises	Start at self-selected walking speed, at an intensity meeting the talk test Gradually increase to an RPE of 3-5/10	From self-selected pace, over 4 weeks gradually increase time to 40 min each session, increasing intensity as tolerated. Persons interested in higher-intensity exercise should obtain guidance before doing it.
Strength • Functional gravity-based exercises recommended as the basic program • Weight training an alternative for those who are interested and motivated to do it	2-3 days/week	For body weight exercises: Functional exercises (see chapter 4), one set during TV commercials For weights: 1 set of 8-12 reps to fatigue	• Sit to stand: 8 reps • Alternative: stair steps (standard is 10 steps) • Arm curls: 8 reps with ~4 kg (can use plastic milk jug filled with water) • 50-70% of 1RM	Build gradually to as many sets a day as tolerated. For curls and weight training: Increase to 2 sets over ~8 weeks.
Flexibility Hips, knees, shoulders, and neck	3 days/week	20 s/stretch	Maintain stretch below discomfort point	Discomfort point should occur at a ROM that does not cause instability. This discomfort point will vary between people and with different joints in each person.
Warm-up and cool-down	Before and after each session	10-15 min	Easy RPE <3/10	Should be maintained as transition phase, especially for those doing higher-intensity physical activity.

RPE = rating of perceived exertion

1 RM = 1 repetition maximum, the amount of weight that can be lifted just 1 time

ROM = range of motion

How to Prescribe Physical Activity or Exercise in Chronic Care

It is important for everyone on the health care team to understand an exercise prescription, including primary care providers and allied health care staff who may not be familiar with exercise prescriptions. There are two basic methods, one that is traditionally used by exercise physiologists, which is known by the acronym FITT. The other method, which may seem more natural to some physicians, is to use their standard method for prescribing medications. These are outlined in the following sections.

FITT Model of Exercise Prescription

The traditional method of prescribing exercise used by many clinical exercise physiologists is the FITT method: frequency, intensity, time, and type (of exercise).

Frequency, *or how many days per week of a particular exercise:*

For aerobic activities, the recommendation is for all persons to participate in activities requiring aerobic exertion on 4 or 5 days per week.

For strengthening or functional exercises, the recommendation is that all persons do functional activities that require muscular strength two to three times each week.

Intensity, *or how hard to exercise,* which depends on the kind of exercise being performed:

The recommendation is that the intensity of exercise be based on perceived exertion in persons with chronic conditions. The reasons for this are multiple:

Many patients are on medications that alter the heart rate response to exercise (e.g., β-blockers).

Persons with a disability or those who require a prosthetic are often markedly less efficient than people without disabilities and thus have dramatically less efficient exercise economy.

Many health conditions alter exercise heart rates.

Time, *or duration or how long to exercise during each session:*

For aerobic activities, the recommendation is to work up to a duration of 30 to 40 min per session and accumulate a minimum of 150 min of moderate-intensity aerobic exercise each week.

For strengthening–functional activities, all persons should complete the following (or an equivalent): a minimum of two sets of 10 repetitions of arm curls, two sets of 10 repetitions of sit to stands, and two repetitions of a 10-step stair climb (or step-ups).

Type, *or what kind(s) of specific exercise to perform:*

For aerobic exercise, the recommendation is walking as the primary type of physical ac-

tivity. The reason is that walking is the basic form of locomotion for humans and is essential for independent living and maintaining quality of life.

- Other activities with a similar amount of energy expenditure (cycling, swimming, and so on) are acceptable substitutes for walking and are preferred in situations in which weight-bearing activity is a problem.

- For individuals with disabilities who cannot walk, pushing a wheelchair (i.e., propelling themselves) is their walking equivalent.

For strengthening exercises:

- Work up to a total of 20 arm curls with a minimum of a 2 kg (~4 lb) object held in each hand (e.g., 2 L milk jugs filled with water).

- Handgrip exercises are an alternative if the person is not sufficiently strong to do all 20 arm curls with the 2 kg weight (weaker persons can use foam squeeze sponges, stronger individuals can use tennis balls or spring-loaded handgrip devices).

- When the patient can do all 20 arm curls on each side with 2 kg, she can repeat the process with a 4 L jug filled with 3 L water and then a full 4 L.

For stretching, the recommendation is that all patients be able to reach their toes on the left and right sides in a chair sit and reach.

For basic physical functioning exercises, recommendations are as follows:

- Do sit-to-stand drills (which can be done at home with nothing more than a chair).

- Do stair climbing or step-ups (using a handrail in the case of fall risk).

Many different strengthening, functional and patient-appropriate exercises can be invented. Those suggested here are a minimum to help maintain quality of life and independence. There is almost no limit to functional types of exercises. If an invented form of exercise is practical and efficient with regard to what an individual needs to live a better quality of life and is safe, then it may be added to the exercise prescription.

FITT Style of Exercise Rx

Frequency:

Be physically active 5 days a week.

Do aerobic or mixed activities 3 or 4 days a week and strengthening activities 1 or 2 days a week.

Intensity:

During aerobic activities, go at a pace at which you can still manage to talk comfortably.

During strengthening activities, aim to do exercises such that you can do only 8 to 12 repetitions before you can't go on and have to rest.

Time:

For aerobic activities, do 30 to 40 min a day and accumulate 150 min for the week.

For strengthening activities, build up to doing two full sets of the exercise (doing each one twice).

Type:

Aerobic activities such as walking, jogging, cycling, swimming, rowing, and cross-country skiing.

Strengthening exercises such as lifting weights, push-ups, sit-ups, pull-ups, squats, and digging.

Mixed activities are things like gardening, digging, and snow removal (with a shovel) that are prolonged and hard enough to be aerobic but that also require some strength.

Pharmacologic Model of Physical Activity or Exercise Prescription

Exercise prescription closely parallels the way a pharmacologic prescription is written; some physicians and health care providers may find it more natural to prescribe exercise in this way. The following example provides the same information as in the FITT example but is organized in the same way as a prescription for a medication.

Specific activities or exercises to perform are analogous to a specific drug. Every drug has an indication, a diagnosis that makes the drug appropriate, so the indication is a specific chronic condition to be treated with exercise (e.g., hypertension or diabetes).

For aerobic exercise, the recommendation is walking because it is the basic form of locomotion for humans and has benefit in the specified condition.

- Other activities with a similar amount of energy expenditure (cycling, swimming, and so on) are acceptable substitutes for walking.
- For individuals who have disabilities and cannot walk and who transport themselves in a wheelchair, pushing their wheelchair is their walking equivalent.

For strengthening and physical functioning exercises, the recommendation is as follows:

- Arm curls using objects held in the hand (e.g., bottles filled with water)
- Sit-to-stand drills (which can be done at home using a chair)

Other useful strength and physical functioning exercises include these:

- Stair climbing or single step-ups (assuming that the person is not at risk of falling)
- Handgrip (using foam squeeze sponges or spring-loaded devices)

For stretching, the recommendation is chair sit-and-reach exercises.

Dose of exercise (the product of how intensely and how long an activity is performed):

In persons with chronic conditions, the intensity of physical activity or exercise is usually based on subjective perception (see table 2.1) rather than on target heart rates or some other objective measure. The reasons for this are multiple:

- Many patients are on medications that alter the heart rate response to exercise (e.g., β-blockers).
- Persons with a disability or a prosthetic can be markedly less efficient than people without disabilities and thus have dramatically less efficient exercise economy.
- Many health conditions can markedly alter exercise heart rates.

Time or duration of each session of activity:

- For aerobic exercise, work up to a duration of 30 min per session (longer or shorter bouts are useful and acceptable)
- For strengthening–functional exercises, a minimum of two sets of eight repetitions of arm curls, handgrips, sit to stand, and stair climb
- For stretching, holding the position for 30 s on each side

Frequency is similar for the two methods and relates to how many days per week:

For aerobic activities, 4 or 5 days per week, accumulating a minimum of 150 min of moderate-intensity exercise each week (or light intensity if moderate is too difficult)

For strengthening–functional exercises, at least two times each week

Stretching before starting and after finishing exercise—helps promote mind–body awareness

Lastly, note that some patients may need to do (or be able to do) only part of the prescription as written here, in which case only the parts that the individual should complete are written.

Judging Exercise Intensity

Patients must learn to judge intensity level as part of monitoring their exercise, and exercise monitoring is done in terms of absolute or relative intensity. Measures of absolute intensity are based on the *rate of energy expenditure*. Energy expenditure is classically measured in oxygen uptake (mL O_2 · kg body weight^{-1} · min^{-1}) and is normalized in a unit known as a metabolic equivalent (MET). In normal human biology, this is the resting rate of oxygen uptake:

$$\sim 3.5 \text{ mL } O_2 \cdot \text{ kg body weight}^{-1} \cdot \text{min}^{-1} = 1 \text{ MET}$$

Reconsidering the physical activity guidelines, one can convert a subjective respiratory response to exercise into METs. See Suggested Readings (Ainsworth et al., 2013) for a useful compendium of energy expenditures that covers a wide variety of physical activities.

In addition to ventilatory symptomatology, one can objectively rate effort or intensity by measuring the exertional heart rate (HR) or rate it subjectively by using a psychophysical rating of perceived exertion (RPE). Ventilatory symptom-

Medical Style of Exercise Rx

SIG:

For _____ (specific condition; e.g., type 2 diabetes)

Do aerobic activities 3 or 4 days a week working up to 30 to 40 min a day, including activities such as walking, jogging, cycling, swimming, rowing, and cross-country skiing.

Do strengthening activities 1 or 2 days a week, such as weightlifting, wrestling, push-ups, sit-ups, pull-ups, squats, and digging.

On aerobic days, you can substitute mixed activities like gardening, digging, and snow removal (with a shovel) that are prolonged and hard enough to be aerobic but also require some strength.

Intensity:

During aerobic activities, go at a pace at which you can still manage to talk comfortably.

During strengthening activities, aim to do exercises such that you can do only 8 to 12 repetitions before you can't go on and have to rest.

Time:

For aerobic or mixed activities, work up to accumulate 150 min of activity each week.

For strengthening, build up to doing two to four full sets of these activities each week.

---

atology, HR, and RPE are physiologically related, so any of these measurements is useful in monitoring exercise intensity. Historically, HR is the traditional measure and is the most objective exercise intensity measure. But measuring for some persons is technically more difficult; in addition, other things besides exercise can influence HR, and various chronic conditions or medications can alter the HR response. For this reason, in persons with chronic conditions, RPE is the preferred measure and is commonly calibrated by each individual's ventilatory abilities through use of the talk test.

There are two RPE scales that are used widely, one ranging from 6 to 20 and the other ranging from 0 to 10. Exercise professionals who are familiar and comfortable with these scales may wish to use either one; we discuss them in some detail for readers who are less familiar with them. With the 6 to 20 scale, a value of 6 represents rest while a value of 20 represents maximal effort. With the 0 to 10 scale, 0 represents rest while 10 represents maximal effort.

At first glance to someone not trained in exercise physiology, these scales may appear interchangeable. A real physiological difference does exist between the scales: The 6 to 20 scale is based on the linear relationship between exertion and HR, and the 0 to 10 scale is based on the nonlinear relationship between exertion and ventilation. The range of 6 to 20 may seem odd, but this scale stems from the exercise response in healthy young college students in whom the nominal resting HR is 60 and maximal HR is about 200 (truncating the zero yields 6 and 20). The 0 to 10 scale has a more intuitive range, but it is modeled on perceived exertion and ventilation and is nonlinear at high levels of exertion.

In people who have a normal cardiopulmonary response to exercise, the 6- to 20-point scale is recommended because this scale is perceived by patients as easier when making fine adjustments in work rate at the higher ends of the scale. This scale is particularly helpful for higher-functioning individuals who seek to do high-intensity interval training. The 0- to 10-point scale may be particularly well suited, however, for persons with chronic pulmonary conditions (for whom this scale was originally developed).

There are several other well-validated methods people can use to judge exercise intensity, mostly based on HR or on percentage of maximal oxygen uptake. Any of these methods can be used in patients with chronic conditions, but in 20 years and after four editions of the *CDD* series, the editors cannot recall a single author who did not prefer to use perceived exertion as the method of judging exercise intensity. Remember that with subjective scales like RPE, persons who are less fit will have a higher rating of effort at any given absolute level of exertion when compared to the rating of effort among those who are more physically fit. Use of RPE also allows for variability in symptoms such as fatigue that are common yet often variable in people with chronic conditions.

Graded Exercise Testing

All guidelines listed in table 2.3 advise that persons with chronic disease consult their physician before beginning an exercise program. Note that

this recommendation for a preparticipation evaluation has been the standard advice for several decades, despite a widespread lack of physician training in exercise management. The uncomfortable truth is that most primary care doctors have never been trained in proper preparticipation evaluation and, lacking better guidance, either give clearance, order a graded exercise test with 12-lead electrocardiogram (ECG), or refer the patient to a cardiologist. Very shortly prior to the publication of *CDD4,* ACSM published new recommendations on preparticipation evaluation, which postdated the drafting of this textbook. While not identical, *CDD4* and these new recommendations have the same general approach. In this section we address when individuals can just be started on a physical activity program and when they need to be more carefully evaluated.

When to Order an Exercise Test

Exercise testing has long been an integral part of the preparticipation health status evaluation. However, almost no data are available to suggest that a preparticipation evaluation prevents adverse events during exercise in persons with chronic conditions. A key question for the physician is this:

When do I need to order an exercise test? And what kind of test?

The value of an exercise test for secondary disease treatment is multifaceted, with benefits ranging from better exercise prescriptions to usable diagnostic information. In a recent position stand by the AHA, graded exercise testing was recognized for the following uses:

- Detection of coronary artery disease (CAD) in people with chest pain syndromes or potentially equivalent symptoms
- Evaluation of the anatomic and functional severity of CAD
- Prediction of cardiovascular events and all-cause death
- Evaluation of physical capacity and effort tolerance
- Evaluation of exercise-related symptoms
- Assessment of chronotropic competence, arrhythmias, and response to implanted device therapy
- Assessment of the response to medical interventions

Exercise tests for such purposes are generally completed with a 12-lead ECG, but valuable information from an exercise test can be learned without an ECG.

These uses of the exercise test are extremely valuable in order to optimize the prescription of physical activity and exercise training, but many patients may not be good candidates for an exercise test. The Guidelines specify the following attributes as reasons for not performing an exercise test:

- Extreme deconditioning
- Orthopedic limitations
- Left ventricular dysfunction that limits exertion by shortness of breath
- Known coronary anatomy
- Recent successful revascularization
- Recent uncomplicated or stable myocardial infarction
- Recent pharmacologic stress test

The following is the rationale behind these reasons for not ordering a graded exercise test with 12-lead ECG:

- Deconditioning, orthopedic limitations, and exertional dyspnea are likely to prevent the individual from achieving 85% of age-predicted maximal heart rate and thus are likely to lead to a nondiagnostic study, or
- The cardiac diagnosis is already known.

__D__oes the result of the test alter medical management or preclude exercise and physical activity (or both) until after the results are known?

It is most important for the physician doing a preparticipation evaluation to consider what exercise testing will achieve for an individual with a particular chronic condition or set of chronic conditions. If the individual's physical functioning does not seem substantially limited by his chronic conditions and he is interested in starting a vigorous-intensity program or high-intensity interval training, obtaining an exercise test is prudent. This would be particularly important if the individual was not regularly active and had many risk factors for cardiovascular disease. Exercise testing would be less important if the individual was already very active and had no risk factors and just wanted to increase his fitness.

If the patient's physical functioning is highly limited by chronic conditions, as is very often the case, with the exercise test likely to be nondiagnostic, a prudent course is to advise starting with light-intensity physical activities, perhaps supervised by an exercise specialist, and wait to see if symptoms develop that merit ordering a diagnostic exercise test. This strategy can be conceived as

__S__tart low, progress slowly, and be alert (for symptoms).

Most systems in the United States will not cover the cost of a test just for the purposes of prescribing exercise. Such nonreimbursement issues are unfortunate, because exercise tests yield valuable data for creating an exercise prescription. Without knowledge of someone's exercise responses (e.g., peak exercise capacity, peak heart rate, blood pressure response, presence or absence of symptoms of cardiac ischemia or other conditions, reason for stopping), it will take longer to optimize an exercise program because the exercise specialist has to develop this knowledge through observation of many exercise sessions. In such circumstances, a number of exercise tests exist that are more geared toward measuring physical functioning and are likely more applicable to many patients with chronic conditions. Prudent judgment must always prevail when one is determining whether an individual with a chronic disease needs an exercise test or not, and what type of test to perform.

Alternative Forms of Exercise Tests

Many people are not able achieve an intensity of exertion to reach 85% of their predicted maximal heart rate, and for them a more logical path is to consider tests of physical functioning more than cardiopulmonary and vascular function. These individuals tend to have neuromuscular disabilities with cardiometabolic or pulmonary conditions, or multiple chronic conditions, and to be frail or elderly or both.

Particularly in geriatrics and neuromuscular conditions, strong rationale exists for using the Short Physical Performance Battery (SPPB), which is well validated for many populations. In situations in which this form of testing is the best path to follow, *CDD4* authors have referred to this recommendation in their chapter. In addition, it is not unreasonable for physicians and exercise specialists to use the SPPB in any population when useful. *CDD4* stops short of recommending the SPPB for all chronic conditions, however, because insufficient data exist to justify such testing (especially the balance tests in persons not at risk for falls). Rather, the Basic *CDD4* Recommendation is a minimum of physical functioning tests because the editors and contributors had the opinion that low performance of key factors in physical functioning put the patient at greatly increased risk of losing independence. Presumably, forthcoming research will guide subsequent editions of this book as to the best set of physical functioning tests to use. Another consideration is whether or not to develop a universal test battery (such as the SPPB) or whether there are optimal tests to evaluate physical functioning for particular chronic conditions.

The Basic *CDD4* Recommendation for evaluation of physical functioning is that all patients, regardless of chronic condition(s), should be able to do the following physical function tests *at a minimum:*

- 6 or 8 m gait speed >0.6 m/s
- Eight sit-to-stand repetitions in 30 s
- Eight arm curls with a 4 kg mass
- Ascending a flight of 10 steps in under 30 s
- Chair sit and reach to the toes (0 in.) on both sides

> *M*uscular strength and endurance are important requirements for independent living, so it is important to note to patients how quickly they've improved from their program.

Minimum Exercise Recommendations When an Exercise Test Is Not Available

There are many situations in which one must start an exercise program without having the benefit of formal exercise test results. Everyone on the health care team needs to be comfortable with how to do this safely. Many experienced exercise professionals are quite comfortable with this approach and actually prefer it. Within just a few sessions, skilled professionals will learn everything they would have used from an exercise test, and they will have seen how the person's body responds and recovers from repeated sessions.

Aerobic Exercise Training

The guidelines for prescribing exercise without an exercise test are extremely limited and are generally based on an exercise specialist's experience with a particular chronic condition. Here are some general suggestions for initial intensity:

- Activities in the range of 2 to 4 METs
- Exercise HR = resting HR + 20 contractions/min
- An RPE no higher than 11 to 14 (6-20 scale)
- Start low, progress slowly, and be alert (for symptoms)

In many persons with chronic conditions who have been sedentary before the program, starting at these levels will produce fatigue fairly quickly (10-20 min). Surprisingly often, such individuals rapidly gain endurance and after just a few sessions are able to last more than 30 min at this same work rate.

At this point, patients with a chronic condition can advance and attempt some moderate-intensity intervals of 1 to 5 min with the work rate increased to a RPE more in the 13 to 17 range (6-20 scale). It is very useful to have an exercise specialist help patients learn this progression and guide them in gradually increasing the moderate-intensity interval duration, with the goal of eventually achieving the recommended 30-40 continuous minutes. With this approach to starting an exercise program, it is worth repeating one last time: Start low, progress slowly, and be alert.

Strength Training

A key rationale for the physical functioning tests recommended in the preceding section is that these tests can convert to strengthening exercises for people to do at home without having to buy any exercise equipment. The only needs are a sturdy chair, a place to walk, a flight of steps with a handrail for safety, and a 2 L (half-gallon) jug to fill with water for the arm curls. The following is a minimal strength conditioning program designed to supplement the recommendations for aerobic exercise presented earlier:

- Perform two sets of 30 s of sit-to-stand repetitions.
- Perform two sets of eight arm curls with a 4 kg mass (e.g., a milk jug filled with water).
- Perform 10 step-ups two times, leading with each leg and building up to ascending a flight of 10 steps.
- Perform chair sit and reaches holding for 30 s two times on both sides.

High-Intensity Interval Training

The interval training concept has long been a part of exercise training theory. A walk–jog program is nothing more than a form of interval training. Childhood play often follows a form of high-intensity exertion with rest, because children have no sense of pace and by nature use interval training. Applying an interval training concept to persons with a chronic condition is an obvious idea.

High-intensity interval training has traditionally been used as a method to train athletes, but this concept is relatively new to people with chronic conditions (though similar things have been tried in the past). The simplest definition for interval training is an exercise session that consists of alternating periods of high-intensity and low-intensity exertion or even periods of rest. Early in the 20th century, German physiologists reported tremendous cardiovascular system gains in competitive athletes who repeated brief intervals of high exertion with rest intervals in between. Using these higher-intensity intervals short in duration, participants are able to complete greater exercise volumes at higher intensity levels than during a sustained, continuous effort. High-intensity interval training provides an effective alternative program that can produce greater changes in various physiological, performance, and health-related factors.

An additional form of high-intensity intervals is low-volume interval training. Because low-volume interval training is only just receiving attention, little scientific information pertaining to health benefits is known. Current information suggests that low-volume interval training can stimulate physiological changes that are comparable to those with sustained, continuous training, despite reduced total exercise volume. For people who have difficulty finding or making time to exercise, this may prove beneficial.

In interval training, the duration and intensity of the exercise interval are adjusted to alter the exercise volume, the exercise intensity, or both. The way in which duration and intensity and rest are manipulated depends on the individual's goals and fitness level. Interval training involves essentially four variables that can be used to individualize an exercise session (note that the first entry on the following list includes two variables):

- Volume of each exercise interval
  - Duration
  - Intensity
- The length of recovery between exercise intervals (minutes)
- The number of exercise intervals performed

Intensity may or may not be the same in all the intervals, depending on the goal of the exercise session. Variable-intensity, variable-duration intervals, known as *fartlek* in running (Swedish term for speed play), are the most sophisticated form of intervals and are recommended only in highly advanced and exercise-trained patients. In general, constant-intensity, constant-duration intervals are recommended, at least until the individual has developed a sense of how to do interval training and how much is well tolerated.

Interval training potentially offers a great deal for individuals who are inactive or who have a chronic disease and are starting an exercise program. In the easiest application, alternating fast walking with slower walking, or stationary cycling at high work output with lower work output, is a form of interval training.

Emerging research suggests that many, perhaps most, persons with a chronic health condition such as cardiovascular, metabolic, or pulmonary diseases can benefit from interval training. Accordingly, a major consideration in developing an exercise prescription is the selection of an appropriate training strategy, then tailoring this strategy to the cardiovascular, metabolic, pulmonary, and peripheral muscle limitations in order to maximize training benefits.

Clinically Supervised Exercise Programming

The professionals involved in the care of individuals with chronic diseases may vary depending on the specific needs of each person and can require different facilities or clinics to support the care. The following section is devoted to various clinic and facility options for people with chronic diseases.

Physical Therapy

Specialty-trained profession typically focuses on neuromuscular disorders and musculoskeletal injuries.

Care is typically provided for persons with diseases such as osteoarthritis and rheumatoid arthritis, Parkinson's disease, late phase of multiple sclerosis, and stroke.

These professionals are routinely involved with inpatient care and continued follow-up outpatient care, depending on the condition.

Patients or clients are often individuals who are older with chronic conditions that cause severe exercise intolerance, but they vary and can include persons of all ages with limitations in physical functioning.

Occupational Therapy

Demographic is similar to that for physical therapy programs, though usually the target population has severely impaired physical functioning.

Work is commonly performed in hospitals and in community settings with a focus on activities specific to returning to work or tasks related to activities of daily living.

Occupational therapy is less commonly prescribed for persons with chronic disease than physical therapy (though occupational therapy is likely underused by physicians).

Cardiac Rehabilitation

Multiphase programming is specific for persons with particular cardiac diagnoses, though many of these individuals typically have multiple chronic conditions.

Multifactorial risk reduction is the primary objective in these programs, with an emphasis on stress management, nutrition, smoking cessation, or exercise depending on the individual.

Pulmonary Rehabilitation

Comprehensive, multidisciplinary interventions are individualized, focusing attention on pulmonary outcomes for performing normal daily activities.

Because of the irreversible effects of most pulmonary diseases, symptom management and psychological interventions are important for this population.

Speech Pathology

Speech programs are primarily for people with stroke or autism.

Programs attempt to improve speech, articulation, dysphagia or swallowing difficulties, stammering and stuttering.

Depending on the severity of the condition, speech pathologists typically work closely with outpatient therapists and physical therapists.

Exercise Programs Supervised by Exercise Specialists

The risk level and complexity of persons accepted into these programs depend on experience and certification level of the exercise specialist.

Individuals admitted to such programs typically are ambulatory and do not have severe musculoskeletal or orthopedic limitations that would require the expertise of physical or occupational therapists.

These programs may be a continuation of other programs such as cardiac rehabilitation.

Community-Based Programs

Low- to moderate-risk persons with chronic disease are appropriate referrals.

People who have functional limitations or high cardiovascular risk factors should be considered for some type of supervised programming.

Some community-based programs offer supervised programming, but the experience and education of the exercise specialist are variable and unregulated.

Commercial Services (Fitness Centers, Personal Trainers)

These types of programs are applicable to low- to moderate-risk persons with chronic disease.

Individuals who have functional limitations or serious high-risk factors should be considered for some type of supervised programming.

Some fitness centers offer supervised programming, but the experience and education of the exercise specialist are variable and unregulated.

ACSM's Exercise Personnel Certifications

Many professional organizations offer certification programs that facilitate the entrance of professionals into specific work areas. In this regard ACSM has taken a leadership role since 1975 in training and certifying exercise professionals. To develop these certification programs, a job task analysis (JTA) was conducted with a primary purpose of defining the major areas of professional practice, describing tasks performed, and identifying the knowledge and skills necessary to work with specific populations. The populations that were considered were healthy adults, adults who are older, and individuals with chronic disease or disability. The ACSM has several levels of certifications that entitle individuals to work with a wide range of populations from healthy adults to persons with chronic diseases. The following lists the five current primary ACSM certifications that prepare individuals to work in exercise programming.

Exercise Is Medicine Credential (Level 1)
- National Commission for Certifying Agencies (NCCA)–accredited fitness professional certification
- Successful completion of the Exercise Is Medicine (EIM) credential training course and examination

    Note: The EIM Level 1 is mentioned for completeness, but staff with this credential are not qualified to work with persons who have chronic conditions. (Levels 2 and 3 generally qualify through other ACSM credentials, listed next.)

Registered Clinical Exercise Physiologist (RCEP)
- Required to sit for the exam: a graduate degree in clinical exercise physiology containing coursework in exercise testing, exercise prescription, and exercise training, along with 600 hours of clinical experience (external to classroom or laboratory) working with individuals with chronic disease
- Highest ACSM exercise professional certification
- Able to conduct testing and training of individuals with a cardiac, pulmonary, or metabolic (e.g., diabetes, chronic kidney disease) disorder

Certified Clinical Exercise Physiologist (CEP)
- Required to sit for the exam: a bachelor's degree in exercise physiology, exercise science, kinesiology, or an exercise science–based field, along with 400 hours of clinical experience if from a Commission on Accreditation of Allied Health Education Programs (CAAHEP)-accredited program or 500 hours if from a nonaccredited program
- Typically works in a hospital- or university-based setting and is qualified to conduct testing and training of persons with cardiac, pulmonary, or metabolic (e.g., diabetes, chronic kidney disease) disorders

Certified Exercise Physiologist (EP-C)
- Bachelor's degree in health-related field required to take exam
- May assist low- to moderate-risk individuals with medically controlled diseases and health conditions, as well as apparently healthy clients
- Requires more guidance from the medical management team and other health care professionals than RCEP and CES professionals

Certified Personal Trainer (CPT)
- High school diploma required to sit for the exam
- Typically works with apparently healthy individuals and those with low-risk health challenges who are able to exercise independently
- Requires more guidance from the medical management team and other health care professionals relative to RCEP and CES professionals

Certified Cancer Exercise Trainer (CCET)
- Must have ACSM- or NCCA-accredited health–fitness-related certification and a bachelor's degree in a health-related field with 500 hours of clinical experience
- Designs and administers fitness assessments and exercise programs specific to an individual's cancer diagnosis, treatment, and current recovery status
- May have the ability to work with other moderate-risk persons who have chronic disease but specializes in those with cancer

These certifications offer a variety of options for chronic disease care, with each certification playing a different role. Other organizations such as the National Strength and Conditioning Association and the National Academy of Sports Medicine offer certifications, but their certifications are not focused on the care of people with chronic disease.

Referring physicians and medical care teams should carefully consider the level of expertise and certification of exercise specialist team members based on the level of complications of the populations involved. Individuals with interest in and enthusiasm about exercise and those who have worked at a club or as a personal trainer are not qualified to work with medically complicated persons without supervision. Such an individual can be employed as an aide to an adequately trained exercise specialist.

Suggested Readings

Ainsworth BE, Haskell WL, Herrmann SD, Meckes N, Bassett DR Jr, Tudor-Locke C, Greer JL, Vezina J, Whitt-Glover MC, Leon AS. 2011 Compendium of Physical Activities: a second update of codes and MET values. *Med Sci Sports Exerc.* 2011;3:1575-1581.

Artinian NT, Fletcher GF, Mozaffarian D, Kris-Etherton P, Van Horn L, Lichtenstein AH, Kumanyika S, Kraus WE, Fleg JL, Redeker NS, Meininger JC, Banks J, Stuart-Shor EM, Fletcher BJ, Miller TD, Hughes S, Braun LT, Kopin LA, Berra K, Hayman LL, Ewing LJ, Ades PA, Durstine JL, Houston-Miller N, Burke LE. Interventions to promote physical activity and dietary lifestyle changes for cardiovascular risk factor reduction in adults: a scientific statement from the American Heart Association. *Circulation.* 2010;122(4):406-441.

Balady GJ, Ades PA, Comoss P, Limacher M, Pina IL, Southard D, Williams MA, Bazzarre T. Core components of cardiac rehabilitation/ secondary prevention programs: a statement for healthcare professionals from the American Heart Association and the American Association of Cardiovascular and Pulmonary Rehabilitation Writing Group. *Circulation.* 2000;102:1069-1073.

Blair SN, Kohl III HW, Paffenbarger RS Jr, Clark DG, Cooper KH, Gibbons LW. Physical fitness and all-cause mortality. A prospective study of healthy men and women. *JAMA.* 1989;262:2395-2401.

Celli BR. Pulmonary rehabilitation in patients with COPD. *Am J Respir Crit Care Med.* 1995;152:861-864.

Cress ME, Buchner DM, Prohaska T, Rimmer J, Brown M, Macera C, Dipietro L, Chodzko-Zajko W. Best practices for physical activity programs and behavior counseling in older adult populations. *J Aging Phys Act.* 2005;13:61-74.

Gordon NF, Gulanick M, Costa F, Fletcher G, Franklin BA, Roth EJ, Shephard T. Physical activity and exercise recommendations for stroke survivors: an American Heart Association scientific statement from the Council on Clinical Cardiology, Subcommittee on Exercise, Cardiac Rehabilitation, and Prevention; the Council on Cardiovascular Nursing; the Council on Nutrition, Physical Activity, and Metabolism; and the Stroke Council. *Circulation.* 2004;109:2031-2041.

Hand C, Law M, McColl MA. Occupational therapy interventions for chronic diseases: a scoping review. *Am J Occup Ther.* 2011;65:428-436.

Heran BS, Chen JM, Ebrahim S, Moxham T, Oldridge N, Rees K, Thompson DR, Taylor RS. Exercise-based cardiac rehabilitation for coronary heart disease. *Cochrane Database Syst Rev.* 2011;7:CD001800.

Hung WW, Ross JS, Boockvar KS, Siu AL. Recent trends in chronic disease, impairment and disability among older adults in the United States. *BMC Geriatr.* 2011;11:47.

Lacasse Y, Goldstein R, Lasserson TJ, Martin S. Pulmonary rehabilitation for chronic obstructive pulmonary disease. *Cochrane Database Syst Rev.* 2006;CD003793.

McDonald CM. Physical activity, health impairments, and disability in neuromuscular disease. *Am J Phys Med Rehabil.* 2002;81:S108-S120.

Norton K, Norton L, Sadgrove D. Position statement on physical activity and exercise intensity terminology. *J Sci Med Sport.* 2010;13:496-502.

Pescatello LS, ed. *ACSM's Guidelines for Exercise Testing and Prescription.* Philadelphia: Lippincott Williams & Wilkins; 2014:42-46.

Riebe D, Franklin B, Thompson PD, Garber CE, Whitfield G, Magal M, Pescatello L. Updating ACSM's Recommendations for Exercise Participation Health Screening. Med Sci Sports Exerc. 2015;11:2473-2479.

Ries AL, Kaplan RM, Limberg TM, Prewitt LM. Effects of pulmonary rehabilitation on physiologic and psychosocial outcomes in patients with chronic obstructive pulmonary disease. *Ann Intern Med.* 1995;122:823-832.

Rijken PM, Dekker J. Clinical experience of rehabilitation therapists with chronic diseases: a quantitative approach. *Clin Rehabil.* 1998;12:143-150.

Sawatzky R., Liu-Ambrose T, Miller WC, Marra CA. Physical activity as a mediator of the impact of chronic conditions on quality of life in older adults. *Health Qual Life Outcomes.* 2007;5:68.

Art of Clinical Exercise Programming

Chapter 1 outlines general principles for recommending physical activity in healthy individuals, and chapter 2 outlines commonly used principles for more specific exercise prescription. Practitioners can follow either the FITT (frequency, intensity, time, type) model or the medical type of prescription, whichever seems simplest or more natural to their way of thinking. Chapter 2 also reviews the general physical activity recommendations from several organizations, reviews some basic recommendations from *ACSM's Guidelines for Exercise Testing and Prescription, 9th edition* (the Guidelines), and contrasts the Guidelines with the slightly more conservative Basic *CDD4* Recommendations presented in this book. It would be wonderful if that were enough to guide exercise programming in all patients with chronic conditions, because then the book that is *CDD4* could simply be a position statement! Unfortunately, a great many patients with chronic conditions don't clearly fall within a well-developed evidence base, and the more severely affected they are by their chronic conditions, the more difficult it is for them to adhere to an exercise prescription. Also, it is common for such patients to become medically unstable, ranging from just not doing well on a particular day, to developing an acute exacerbation of their conditions, to developing an acute illness or even a new and potentially life-threatening condition. Thus an exercise specialist must be able to improvise, often on the spot during a session, to help patients keep on track as best as possible. With the seemingly infinite number of medical and socioecological challenges that can arise, it is not possible to easily reduce such improvisation into an algorithm. The chronic care team, and especially the clinical exercise specialist, must develop the ability to guide exercise management in chronic conditions with artful skill—the *art of exercise management.*

For individuals who are unaccustomed to physical activity or who have never exercised as a part of their lifestyle, adopting an exercise program can be a challenge. Individuals newly diagnosed with a chronic condition are often advised to begin a program, but side effects of many conditions or the treatments (or both) can cause fatigue, weakness, depression, pain, shortness of breath, or other discomfort, and may present significant challenges to getting started and sustaining a program. Therefore, in many situations, getting the patient to stick with the program will require adjustments in the exercise prescription, and the ability to be flexible in meeting the specific needs of each individual client is critically important.

A successful prescription is one that facilitates incorporation of regular physical activity into the patient's lifestyle and is based on the individual's clinical status and goals. As with the nuances in other aspects of clinical practice, there is little (if any) evidence base to guide the art of exercise management. The suggestions in this chapter come from the shared experiences of exercise professionals who have extensive experience working with individuals who have chronic conditions. What follows are essential steps that should be integrated into the process of exercise programming in individuals with chronic conditions, in the hope of facilitating every patient's adoption of an active lifestyle.

Step 1: Assess Current Health Status

The exercise professional must have a clear understanding of the clinical status, including the history of the disease, current symptoms, physical and cardiovascular restrictions, and treatment, as well as any side effects of the condition or treatment. It is important to understand how the patient is coping with the disease, especially in terms of participation in life activities and physical activity. The individual's history of physical activity and the extent to which the disease or treatment or both have affected participation in regular exercise, leisure-time activities, activities of daily living (ADLs), and instrumental activities of daily living (IADLs) will help determine starting levels and goals. People's perception of their physical ability may be determined by limitations they have actually experienced or by messages they have received from health care providers, family, or friends about what they are able to do. They may have specific concerns about participation in physical activity or exercise based on their own previous experiences, the experiences of others, or things they have read. This assessment should also identify current and past barriers to participating in physical activity, as well as the understanding of the benefits of physical activity or exercise in general and how they relate it to their condition. The following should be identified:

- Any absolute and relative contraindications to exercise due to the condition
- Any clinical or patient concerns related to the safety of participating
- What clinical aspects of the condition might be improved with physical activity
- What motivates and sparks the patient's interest in starting and sustaining participation

Step 2: Assess Current Level of Physical Activity

Assessing the patient's current level of physical activity can range from simple questions to more formalized questionnaires to logbooks or the use of accelerometers (commonly called step counters). A good place to start is with a simple question, such as one of these:

How much do you walk each day?

How often do you climb stairs?

What do you do for exercise?

In contrast to specific and pointed questions in the exercise vital sign, during the preparticipation evaluation it's better to start with a more open-ended and broader question. For example, one key insight that will be helpful is how the patient

Be Curious About Follow-Up Details

A skilled counselor is going to be curious about how patients perform the activities and what this means to them, which will give each patient a sense that the exercise specialist or counselor really is trying to figure out how to help him achieve his goals.

himself defines exercise. Depending on his answer, it may be worthwhile to ask some follow-up questions that more clearly delineate his ideas about physical exertion. One could have three different patients, all with psoriatic arthritis and fatigue, with very different responses (and very different expectations):

- I like to run and do about three marathons a year, but I'm worried if I can do that now.
- I like to ride my Harley (motorcycle) in the summer and go snowmobiling in the winter.
- I like to play horseshoes, and I go to tournaments on most weekends.

The follow-up questions should become more focused, dealing with the frequency and duration of the activities and any symptoms the client has experienced. Another approach is to have individuals describe their activities over the course of a typical day or week. In this approach, practitioners must have a good working concept about the physical demands of many kinds of activities and in their mind are assessing the patient's responses for

- indicators of strength, aerobic ability, and endurance, as well as
- evidence of problems with balance, coordination, and flexibility.

In some instances, counselors might need to look up activities in order to more fully characterize the patient's responses and ask additional questions to fully understand the patient's limitations. For example, they may want to probe further with questions like these:

- How many rounds does your usual horseshoe tournament last?
- Do you get tired enough that it affects your play?

Step 3: Identify Exertional Symptoms That Limit Physical Activity

The practitioner should try to assess any symptoms that limit physical activity. It is helpful if the limiting symptom is both assigned an objective value and described qualitatively, as in these examples:

I can walk for 10 minutes before I have to stop because my legs cramp up.

I can only fold laundry for 5 minutes before I get short of breath and my shoulders start to hurt.

It is important to get an idea of how long the limiting symptoms last after the patient stops to recover. This *recovery time* is another important measure of physical function that may have implications in diagnosis, in prognosis, and in developing the exercise prescription. An individual who has to stop walking after 10 min due to leg cramping, but who can resume walking after a standing rest for 2 min, will probably respond differently to exercise than one who has to sit and rest for 10 min before she has recovered. The person who needs a lot of recovery is likely going to require alternative modes (or types) of exercise, at least at the start of the program. In some cases one would avoid activities that cause certain symptoms, but in other cases one might actually want to use exercises that provoke symptoms, all depending on the nature and cause of symptoms. For instance, in a patient who has a myopathy, one might avoid activities that lead to cramps; but in a patient who has intermittent claudication, part of the intent of the exercise is to cause a level of muscle pain that is tolerable but just shy of causing cramps.

Intermittent claudication is not the only example of a condition for which one objective of exercise is to induce some mild symptoms, but in general that is not a good approach. For the most part, if an individual experiences any type of discomfort with physical activity, the chances of his exercising regularly are slim. So the general objective is to avoid symptoms, not cause them! If a patient experiences symptoms during physical activity, exercise training, ADLs, or IADLs, this should be discussed among the chronic care team members. Keep an eye toward determining if a symptom is a feature of the chronic condition or is related to treatment (i.e., aggressive blood pressure management that results in postural or exertional hypotension, or anemia resulting from chemotherapy). If the cause is unclear, it needs to be evaluated by the physician.

Symptoms that may be related to general deconditioning and weakness may indicate the need to start with a focus on strengthening with very gradual progression to aerobic training to ensure successful adaptation. Joint discomfort and stiffness that are associated with exercise may indicate the need for specific physical therapy interventions to enhance range of motion, reduce joint pain, or both. By and large, exertional symptoms are an indication that

- a supervised program would be the best approach, so
- the exercise specialist can provide exercises to mitigate the symptoms, and so
- the exercise specialist can guide the patient to the most appropriate mode of exercise.

Step 4: Evaluate Physical Function and Performance

The preparticipation evaluation by the physician or midlevel practitioner should, at a minimum, characterize the patient into one of four basic categories:

1. Mildly impaired to normal
2. Moderately impaired, low functioning
3. Severely impaired, very low functioning
4. Needs aid, debilitated

These categories are roughly similar to the New York Heart Association (NYHA) classes I through IV for the ability to do physical activity. The comparisons in relation to external work are shown in table 3.1; note especially the impact on ADLs.

Patients who have chronic conditions but have mild to no impairment in physical functioning should generally have their exercise prescription and programming in accordance with the ACSM Guidelines. Such patients may still need a referral to a formal exercise or rehabilitation program because of cardiovascular or neuromusculoskeletal issues, but enrollment is usually brief, and most such patients can rapidly progress to home-based self-directed activities or community-based programs.

Table 3.1   Exercise Intolerance and Physical Functioning

Impairment in physical functioning	Nearest NYHA class	Approximate METs	Approximate range of aerobic capacity	Approximate range of physical functioning
Mild to none	I	≥ 6	≥ 20 mL $O_2 \cdot kg^{-1} \cdot min^{-1}$	Walk-jog or better
Moderate	II	4-5	14-20 mL $O_2 \cdot kg^{-1} \cdot min^{-1}$	Can walk briskly, but at high RPE and not for long
Severe	III	3-4	10-14 mL $O_2 \cdot kg^{-1} \cdot min^{-1}$	Can walk and do all ADLs but must rest often
Needs aid	IV	≤ 3	<10 mL $O_2 \cdot kg^{-1} \cdot min^{-1}$	Struggles to do ADLs

1 MET = metabolic energy expenditure at rest, approximately 3.5 mL O_2 / kg / min
RPE = rating of perceived exertion
ADL = activities of daily living
NYHA = New York Heart Association

Patients who have chronic conditions but have moderate impairment in physical functioning or worse should generally be referred to an exercise specialist for more in-depth evaluation of physical functioning. Some physicians are able to do this as well, especially those who have been trained in physical medicine, sports medicine, neurology, and geriatrics. This more thorough examination should identify the ability to perform a variety of physical functions, ranging from simple observation of basic movements to the more complex physical functioning measures in the Basic *CDD4* Recommendations. Examples of basic movements that should arouse clinical suspicion include

- Reaching with arms, tremors, coordination
- Inability to bend and stoop
- Festinating gait, waddling, wide-base stance, pectoral–pelvic dyssynchrony
- Any other signs of neurological impairment

Physicians usually detect such problems, but they may be subtle and overlooked. If the exercise specialist finds such abnormalities and the referring physician has not previously noted these findings, the physician should be contacted to confer on the need for a neurology consultation.

The vast majority of patients with chronic conditions, especially the cardiometabolic conditions, do not have gross neurological signs, and many have little or no impairment in physical functioning. For such patients, the in-depth functional assessments are not necessary, and the exercise specialist may choose to observe higher-level activities such as brisk walking and running. But

Special Considerations

Keep these points in mind when working with people who have diabetes or those with greater risk of falling:

Individuals With Type 1 or Type 2 Diabetes

Be especially alert for subtle signs of neurological impairment in patients with diabetes, because these are easy to miss and suggest the presence of diabetic neuropathy.

Individuals With Fall Risk

At this time, the appropriate evaluation and treatment for balance issues is sufficiently complex and multifactorial that there is no simple solution with regard to exercise programming.

Signs of instability on the balance assessment should thus be referred to a therapist trained specifically in balance problems.

in people who have a more severe burden of their condition, it is useful to obtain some or all of the following measures:

- Chair sit and reach
- Gross ability to rise from a chair, sit-to-stand performance
- Gait speed and ability to climb stairs
- Upper-body strength (lifting, handgrip, and so on)

Step 5: Selecting Physical Performance Assessments

Physical functioning measures have primarily been developed for use in geriatrics; these consist of standardized tasks that test physical performance limitations for specific movements such as walking, getting up from a chair, and other tasks encountered in daily life. The many advantages to using these measures include the following:

- Ease of use in the clinic (no special equipment)
- Reproducibility of protocol
- Efficient use of clinic time
- High cost-effectiveness (no costs for special equipment)
- Low preparation, aftereffects, and time or cost burdens for the patient

A major advantage of these measures is that many of them are widely used and have well-established norms (including some condition-specific populations). These tests are highly predictive of disability, nursing home admission, and health care utilization in older individuals.

There are a plethora of these types of measures, each developed to address a specific question or situation; thus choice of a specific test should be made based on the reported limitations and goals of the client and the program to be prescribed. Unfortunately, there is no set algorithm for choosing a test; however, these types of tests are well suited for very deconditioned or frail individuals, regardless of age. Use in other, higher-functioning individuals runs the risk of a ceiling effect—that is, an individual can walk 4 m only so fast, so any improvement resulting from an intervention may not be observed. Thus a higher-level test, such as a 400 m walk or a shuttle walk, may be indicated to assess mobility and changes with an intervention.

Activities of Daily Living and Instrumental Activities of Daily Living

Assessment of ADL and IADL limitations should be performed for individuals who are older and those who are severely compromised by their disease (regardless of age). This can be done using a standardized questionnaire such as the Katz ADL questionnaire, which identifies limitations in six ADLs (bathing, dressing, grooming, eating, toileting, and transferring). Instrumental ADL limitation can be assessed using the Lawton questionnaire, which identifies limitations in more complex activities such as preparing meals, shopping, handling money/bills, and housework. In individuals who are older, limitations in ADLs or IADLs are predictive of poor outcomes and predict admission to assisted living.

In some chronic conditions, physical or occupational therapy or both can be useful because limitations in ADLs and IADLs are highly prevalent. For example, among patients with end-stage renal disease treated with dialysis, 41% have at least one ADL limitation. In both younger and older patients on dialysis, ADL limitation is associated with a 3.4 times higher risk of mortality than in those without ADL limitations.

Assess limitations in ADLs when working with patients who have significant physical deconditioning or multiple comorbidities.

Commonly Used Tests of Physical Functioning

This section describes several of the more commonly used tests, including the tests in the Basic *CDD4* Recommendations. Many additional performance-based test procedures, with scoring and other details, can be found at http://web.missouri.edu/~proste/tool.

Walking Tests

Walking is the most important mode of physical activity, because an individual who cannot move about independently on foot needs either aid with locomotion or assistive devices such as a wheelchair. Accordingly, there are multiple methods to assess someone's ability to walk. None of these methods is clearly superior, so a health or exercise professional needs to individualize the choice of walking test based on the patient's or individual's unique situation.

Six-Minute Walk Test

Various walking tests have been reported and are most commonly used in the cardiopulmonary domains, including 2-, 6-, and 12-minute walk tests, which determine the distance travelled in the time allotted. Of these, the 6-minute walk test (6MW) is the most commonly used functional test in cardiopulmonary research and practice. Patients choose their own intensity of walking and are allowed to stop and rest during the test. Maximal exercise capacity is not achieved, nor does the test measure cardiorespiratory fitness or peak oxygen uptake. The 6MW is probably an indicator of integrated responses of the systems involved during exercise and has the main advantage of being very inexpensive. It is not a replacement for cardiorespiratory exercise testing, however, and cannot be used to determine the mechanisms of exercise intolerance. The correlation coefficient between the 6MW test and peak oxygen uptake in patients with cardiac or pulmonary disease is highly variable, from $r = 0.51$ to 0.90, and is much lower in people who are elderly and have mild to moderate mobility limitations. A decline of 53 to 71 m in 6MW performance can identify clinically meaningful deterioration in cardiopulmonary populations. Thus, the 6MW test has clinical utility but is relatively limited, and most clinical exercise physiologists do not consider it that useful. Accordingly, it can be a valuable tool but is not included as a Basic *CDD4* Recommendation.

Intermittent Shuttle Walk Test

The intermittent shuttle walk test (ISWT) is a graded walking test that has more defined end points than the 6MW. It uses recorded voice signals to direct the pace of the patient, who walks back and forth across the floor over a 10 m distance. The walking speed increases every minute until the subject is unable to reach the turnaround point (10 m away) within the required time. It has good reproducibility and is more highly correlated with $\dot{V}O_2$peak than the 6MW. The structure of the test allows for assessment of total walking distance as well as peak gait speed.

Gait Speed Test

Gait speed has become one of the most widely used physical functioning tests, with leading researchers advocating that gait speed become a vital sign for older individuals. Of all the individual measures of functional limitation, gait speed is the best predictor of nursing home admissions, morbidity, and mortality. Figure 3.1 shows gait speed requirements for various activities. Gait speed is typically measured as the subject's *usual pace* over short distances (8 to 20 ft [2.4 to 6 m]). A usual gait speed of <0.6 m/sec is associated with poor

Basic CDD4 Recommended Physical Functioning Tests

- 4-meter gait speed
- 30-second sit to stand
- 30-second step test
- 30-second arm curls, performed while holding a 2 kg mass

outcomes, as is a decline in gait speed of more than 0.10 m/sec.

Variations in this test include the so-called fast gait speed (instruction to walk as fast as comfortably possible), which may have a sensitivity advantage in detecting morbidity but does not improve the association with survival. Note that there is probably a ceiling effect on gait speed in younger or more fit individuals. This phenomenon resulted in the development of a 400 m walk test for use in healthier or more fit individuals.

In assessing mobility, it may be just as easy to ask the patient how far she can comfortably walk. If a walking program is to be recommended, this *comfortable walk distance* could be the starting point for the program, gradually increasing to progressively longer walks. Any of the aforementioned walk tests could be used at baseline and later in follow-up to objectively assess an improvement.

Core and Lower-Extremity Functional Tests

The ability to lift one's own weight against gravity is paramount to independent living, and can be easily assessed with simple functional tests that are known to correlate with health outcomes. The ability to rise from a seated to a standing position is immensely important and perhaps more basic than walking, since one must rise to a bipedal stance in order to start walking. A closely related function is stair climbing, because navigating through the inside of buildings and landscaped grounds and on sidewalks involves ascending and descending stairs. In an individual who, upon observation and inspection, does not easily move about, one or more of the following measures of physical functioning should be assessed.

Chair-to-Stand Tests

Moving from a seated to a standing position requires lower-extremity strength and is an indica-

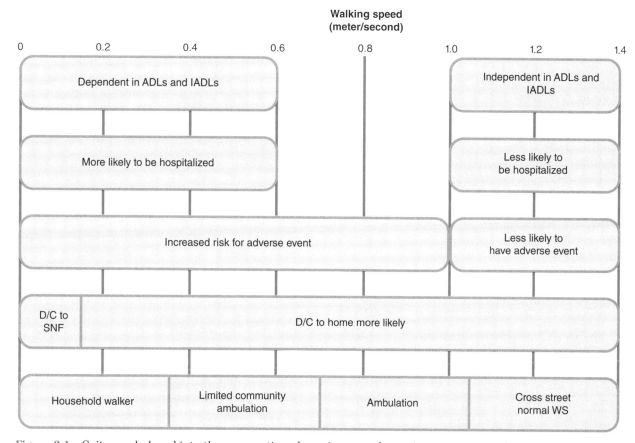

Figure 3.1   Gait speed placed into the perspective of requirements for various activities and outcomes (D/C = discharge; SNF = skilled nursing facility).

Adapted, by permission, from S. Fritz and M. Lusardi, 2009, "White paper: Walking speed—the sixth vital sign," *Journal of Geriatric Physical Therapy* 32(3): 2-5.

tor of lower-extremity function, specifically muscle power. In chair stand tests, patients are asked to rise unassisted from a chair that is a standard height of 43 cm (17 in.). *Unassisted* means that the patient must fold his arms across his chest so he cannot use his arms to push off. There are a variety of chair stand protocols:

- The time it takes to stand up and sit down 5 times (STS5)
- The time it takes to stand up and sit down 10 times (STS10)
- The number of sit-to-stand cycles completed in 30 s
- The number of sit-to-stand cycles completed in 60 s

An advantage of the STS5 is that it is a predictor of falls in individuals who are elderly; an advantage of the 30- or 60-s sit-to-stand test is that it is an indicator of muscle endurance or fatigability.

*T*he Basic *CDD4 Recommendation includes the 30-s sit-to-stand test because*

- *fatigue is an important factor in physical functioning and independent living, and*
- *fewer than eight repetitions has been associated with poor outcomes.*

Many patients with chronic conditions struggle with the sit-to-stand test and become concerned when they're unable to complete more than a few repetitions. A surprising number of patients cannot do even one repetition, though they are able to live independently by using their arms and other aids to rise to a standing position. For these reasons, *CDD4* recommends the 30-s protocol to improve ability to detect low endurance while minimizing the embarrassment and anxiety in patients who don't do well on this test. Such patients may feel as though they "failed" the test, a response that should be refuted, though it is useful to know how much they're relying on their arms and thus help motivate them to do strength training exercises.

Stair Climb Power Test

Climbing stairs is an important basic activity that can be easily assessed as a part of a physical functioning evaluation. Stair climbing tests can be used to discover

- whether or not people are able to climb stairs,
- any coping methods used (alternate steps, use of the handrail), and
- the speed at which they are able to ascend.

The stair climb power test can identify mobility limitations; lower-extremity power (force × velocity) can be determined from a test in which the subject climbs one flight of stairs as fast as safely possible. Stair climb time and vertical height of the stairs are used to calculate velocity, with body mass and gravity used to calculate the generation of downward force.

Timed Up-and-Go Test

The timed up-and-go is a commonly used test of the time in seconds it takes for a subject to rise from a standard-height armchair, walk to a line on the floor 3 m away, turn around, return and sit back down. Reliability is well documented; the test highly correlates with several other tests of mobility, and it can discriminate patient health status including independent living status and history of falls. Normative reference values have been determined through meta-analysis.

Test Batteries

Combining single tests into a battery of tests has been suggested as a way to more accurately assess overall physical functioning. Some of these focus on balance, strength, and power of the lower extremities, while others incorporate upper-extremity tasks and simple tests of dexterity for upper-extremity function.

The Short Physical Performance Battery (SPPB) is a well-validated test battery that is a measure of overall lower-extremity function. It is reliable and valid for predicting disability, institutionalization, hospital admission, and mortality in older individuals. Extensive data are available for comparison in older individuals, and the battery has been used in several large studies including the Established Population for Epidemiologic Studies in the Elderly (EPESE) cohort. The SPPB consists of three different tests of lower-extremity function:

- 4 m gait speed
- Five-repetition chair stand (STS5)
- Balance tested in three different standing positions

In the SPPB, values from these three performance measures are assigned a score ranging from 0 to 4, with 4 indicating the highest level of performance and 0 the inability to complete the task. A summary score ranging from 0 (worst performers) to 12 (best performers) is calculated. Many experts support the use of the SPPB broadly, and this has some surface validity, but the SPPB was not included in the Basic *CDD4* Recommendation because of a lack of sufficient condition-specific data for many conditions in this book.

Upper-Extremity Tests

While most of the ability to live independently is determined by the ability to rise and walk without falling, upper-extremity function is also extremely important. This is because people must not only move through the environment but must also have the ability to open and close doors and to carry objects. A person who can go to the grocery store and ascend a flight of steps, but cannot carry their bag of groceries up the steps, will need help.

The relationship of upper-extremity functioning to health outcomes is not as well characterized as core and lower-extremity physical functioning. Grip strength is most commonly used in studies, and low grip strength has been associated with poor health outcomes. The expert opinion of the editors and authors of relevant sections in *CDD4*, however, is that grip strength is not adequate as a measure of physical functioning. The ability to perform isometric contractions in a small muscle group has little to do with overall physical functioning or the performance of ADLs, such as carrying groceries and living independently.

The upper extremity of humans is a remarkably versatile organ, and complete characterization of upper-extremity function is very complex. In a desire to characterize minimal function that is necessary for fully independent living, the editors and authors of *CDD4* sought to make a simple test that is quick and easy to use in a clinical setting. We considered that while some objects can be carried with the elbow fully extended and the arm hanging at one's side, many objects are more securely carried via flexing the elbow and nestling the object next to the torso. Such an activity involves both grip and flexion of the elbow and elevation with internal rotation of the humerus-shoulder against resistance (the weight of the object). Thus, the movement that best measures this is the arm curl.

Arm Curl Test

Until more science is available, the Basic *CDD4* Recommendation for assessing upper-extremity physical functioning is the number of arm curls the person can perform in 30 s while holding a 2 kg mass. This test has the advantage of being similar in format to the 30-s sit to stand, can be administered by a wide variety of clinic staff, mirrors the exercises the patient will be advised to do in physical or occupational therapy, and does not require costly materials (e.g., a 2 L or half-gallon jug of water will suffice). We anticipate that this recommendation will be scientifically examined and revised accordingly in future editions of this textbook.

Step 6: Considerations for Formal Exercise Tolerance Testing

As noted in chapter 2, these are the primary purposes of clinical exercise testing:

- Diagnostic assessment of symptoms of ischemic heart disease
- Assessment of abnormal symptoms associated with exertion such as light-headedness or dizziness, irregular heart rhythm or racing pulse, excessive shortness of breath
- Assessment of blood pressure management

Although results from diagnostic exercise testing can be used for exercise prescription and to track improvements in fitness from exercise training, in many health systems this is regarded as a more secondary use of exercise testing. If diagnostic exercise testing is to be used in patients who have a severe burden in physical functioning, the laboratory must be notified to use a low-level or ramp protocol. The treadmill protocols most commonly used for diagnostic purposes (e.g., the Bruce and Modified Bruce protocols) have relatively large increases in metabolic equivalents (METs)/stage, and such protocols typically are not gradual enough for many patients with chronic conditions. The protocol rapidly exceeds their functional capacity and thus is not very helpful for developing an exercise prescription.

If a diagnostic exercise test is ordered, the provider ordering the test should inform the laboratory performing the study to use one of the following methods:

- Balke or Modified Naughton
- Low-level constant-increment protocol
- Continuous low-level ramping protocol
- Branching low-level protocol

One useful approach to customizing a low-level constant-increment protocol is to have patients walk on the treadmill at 0% grade, and increase the speed until they feel as if it's *their idea of a brisk walk*. Most patients with a severe burden of chronic disease have biomechanical difficulties going faster than that. So use this "brisk walking speed" for the entire test, and increase the grade by 1% to 2% every 1-2 min. If the patient seems fairly high functioning and would seem to have a peak aerobic capacity above $20 \text{ mL O}_2 \cdot \text{kg}^{-1} \cdot \text{min}^{-1}$, one should use 2% increments. If she is lower functioning, then one uses 1% increments. With such an approach, few patients with limited physical functioning will last longer than 8 to 10 min.

With use of a cycle ergometer, a good incremental protocol is to increase work rate by 25 W every 1-2 min. Again, few patients will exceed 150 W, and thus most will be done in 8 to 10 min. In higher-functioning patients, 50-W increments may be a better choice, but very few patients with chronic conditions will exceed 200 W in peak power output.

It is important to reinforce the reasons why Bruce and Modified Bruce protocols are not helpful in most patients with chronic conditions. The relatively large increments in work rates in these tests usually lead to the rapid onset of fatigue, because the jump between stages is too big to resolve low levels of exercise tolerance and overwhelms the individual's capability. Comparatively speaking, this would be like giving an athlete a treadmill test that progresses from a slow walk at 2 mph (3.2 kph) to jogging at 8 mph (12.9 kph) to a full sprint at 25 mph (40.2 kph); such a test doesn't provide useful information about what they can do. Since many individuals with chronic conditions have severely affected physical functioning or are severely deconditioned, their baseline exercise tolerance will be 5 METs or less. A test using the Bruce protocol will not go three stages, most likely because of leg fatigue. Moreover, many patients with chronic conditions have difficulty with the large increase in elevation and stop the test because of biomechanical discomfort. Although the ACSM Guidelines recommend that protocols with large increments/stage are best suited for younger or active individuals, these protocols remain the most commonly used protocols in most diagnostic exercise laboratories. The medical team must seriously consider the cost/benefit analysis for ordering a diagnostic exercise test before starting an exercise program, and the exercise laboratory used for such patients must be practiced in the use of low-level protocols.

Other considerations are whether or not an order for an exercise test will result in an expensive but ultimately not useful series of diagnostic studies that serve only to delay initiation of a low-level walking program. There are many conditions in which diagnostic testing will produce equivocal results due to chronic electrocardiogram (ECG) changes such as left ventricular hypertrophy, conduction blocks, electrolyte abnormalities, medication effects, and sometimes prior myocardial infarctions. In these cases, the start of exercise training will likely be delayed, pending whatever additional diagnostic testing is deemed necessary following an equivocal stress test. This often results in a missed opportunity, with patients falling through the cracks and never getting started on their program.

Conditions in which the prevalence of cardiac disease is high (e.g., in those with chronic kidney disease or survivors of stroke), but risk of mortality is associated more strongly with physical inactivity than comorbidity risk, highlight the need for some form of physical activity regardless of any diagnosis of cardiac disease. In patients who have survived a stroke, forthcoming guidelines (as of the writing of *CDD4*) from the American Heart Association will no longer call for diagnostic stress testing before exercise training in survivors of stroke. The rationale for this change is that the immediate poststroke period is immensely important in neural retraining, and the patients cannot afford to lose precious time awaiting an appointment for an exercise test before they can get started.

In situations like these, in which the testing is not likely to be of high predictive value or the test risks a delay in starting a low-level program that the patient really should begin as soon as possible, the care team should consider using a functional exercise trial. In an extended trial, the initial sessions start at a low level; the responses to a given session are assessed by the exercise specialist; and gradual increases in intensity or duration are attempted and again are assessed with each successive session. Thus an appropriate individualized progression and program can be developed without formalized exercise testing. These sessions can be monitored using ECG, blood pressure, and symptoms, and thereby provide even better information on the individual responses and related symptoms than is obtained in a formalized exercise test.

The extended trial has the benefit of focusing on process and effort, allowing clients to experience success and learn and *feel* how their body responds to activity (chapter 4), thus setting up a positive feedback situation, alleviating fears that activity may harm them or cause excessive fatigue. The extended trial also acknowledges clients' abilities and breaks down the barriers related to their perceived ability to participate safely and to benefit from physical activity.

In addition to providing the physiological measures of heart rate and blood pressure, the extended trial is valuable to the exercise specialist in that it provides the opportunity to get to know the client in a more real-world setting, giving the exercise specialist unique insights among the care team. The specialist should be careful to listen to comments about the activity and watch clients' body language as they gradually increase intensity. This is an ideal situation for providing positive feedback and coaching, as well as education that will be reinforced by muscle memory. It is one thing to sit down in an office and explain the ratings of perceived exertion (RPE) scale; it is a totally different thing to have the client experience those ratings. The number of sessions can vary, depending on the client and the perceived confidence of his or her ability to proceed to less supervised or even unsupervised exercise. The frequency of supervised sessions can be gradually reduced and increasingly used just for a check-in to monitor progress and evaluate any new symptoms.

Step 7: Considerations for Program Referral

Whether an individual needs to participate in a supervised program or is able to adopt exercise independently at home varies between individuals and changes over time in the same individual. The level of program supervision falls on a continuum ranging from continuous hemodynamic monitoring in a clinical setting to occasional phone follow-up by the health care team to assess independent participation. Although availability, convenience, and insurance coverage may be the real determinants of participation in a clinically supervised program, ideally, the factors that determine referral to a formal program should be:

- Specific limitations in physical functioning
- Clinical condition and safety of exercise (requiring ECG or other monitoring)
- Patient preference (usually out of insecurity with independent exercise)
- Location that encourages attendance
- Patient's understanding of symptoms

Staffing levels and qualifications at the referral facility should also be considered. A facility that focuses on higher-intensity exercise training or sport–competition training, with staff not familiar with clinical issues in patients who have multiple chronic conditions, may be intimidating and inappropriate for an individual starting at lower levels or someone who has clinical concerns and needs to progress gradually.

Physical therapy is indicated for specific impairments or deficits in range of motion, strength, or mobility. Occupational therapy is indicated to assist with specific impairments in basic self-care and ADLs. Physical and occupational therapy both can be continued as long as there is demonstration of progress toward predetermined goals of treatment. In the United States, insurance coverage requires documentation of progress. Unfortunately, physical and occupational therapy in the United States typically do not include general conditioning, which would be nice given that many patients with chronic conditions present in a severely deconditioned state. Patients who need a reconditioning program often have to use their own funds or obtain financial support to participate in a medically supervised program.

Since the goal is to educate, motivate, and facilitate incorporation of regular physical activity or exercise training (or both) into the lifestyle, for most individuals the program should avoid creating a dependence on clinical supervision of exercise. Thus, depending on the clinical condition and safety considerations, supervision should be gradually reduced over time, usually over a few weeks to a few months (depending on the patient's specific conditions). The objective is for patients to understand their own responses to exercise, how to manage any symptoms they may have, how to know when their symptoms indicate a need to be evaluated, and how to progress and sustain their program independently in whatever setting that facilitates their ongoing participation. Depending on the needs of the individual, regular follow-up visits might include a supervised exercise session from time to time (it helps exercise specialists immensely if they can see the patient exercising) or only a phone call check-in by the management team. Assessment of participation in physical activity or exercise training should be a part of the routine medical care assessment, and deficits in participation or identified problems associated with participation should result in a reassessment by the exercise specialist on the team. Important medical information can come from monitoring and lead to changes in medical management.

Step 8: Develop a Strategy for Monitoring Progress

Sustaining participation in a regular program of exercise is a challenge, and many strategies for enhancing adherence can be used depending on individual preferences, style, and needs. Factors that are monitored will differ depending on the safety concerns related to the clinical condition.

Monitoring participation is critical, obviously, because there is no chance of benefits if there is no participation. Strategies for monitoring participation range from regular phone follow-up to diaries to online websites that can be set up for the individual and the team members to monitor. Many online programs can link individuals to social groups for peer support and motivation, allowing tracking of progress and encouragement from others. Monitoring using step counters is an option for walking prescriptions. When participation drops off, the participant should be contacted to identify the reason for reduced participation. Changes in life situations such as changes in work schedules or additional domestic responsibilities should be assessed, and problem solving with the individual is needed to find time for participation. Participation can drop off with changes in clinical status or new onset of depression; any of these should be referred for evaluation. Any new-onset pain or discomfort could contribute to reduced participation and should be assessed, as well as any new symptoms as already described.

Monitoring of clinical responses may include blood glucose, heart rates, pre- and postexercise blood pressures, and any symptoms that may be experienced. Thus, any reporting form (i.e., diary, phone interview, phone reporting) should include those clinical data that will provide the team important monitoring not only of the exercise responses but also of the clinical status of the individual. Symptoms that could be monitored include rating of perceived exertion, angina-type symptoms, levels of dyspnea, claudication or cramping, muscle fatigue, and overall fatigue associated with exercise. Changes in symptoms experienced during or after a routine bout of exercise should be evaluated by the medical care team for assessment of a change in clinical status.

The frequency of monitoring again should be individually determined. Check-in by either phone contact or a monitored session should be done frequently as the program begins; as regular participation and a stable program have been achieved, monitoring frequency can be reduced. If any abnormal responses are reported, reassessment of the program and monitoring frequency would be indicated.

Monitoring of fitness should be done regularly. If there were functional limitations at the initiation of the program, those limitations should be reassessed regularly (i.e., quarterly or biannually, depending on the extent of limitations). Activities of daily living and IADL limitations would be reassessed using self-report, whereas basic movement limitations and performance limitations would be assessed using objective measures described previously (e.g., stair climbing, chair stand tests). Objective measures of fitness can be obtained using a monitored exercise session (i.e., documenting heart rate, blood pressure, and RPEs at standard submaximal levels) or repeat physical performance tests such as the timed walking tests or stair climbs. Participants could be instructed on how to do their own fitness checks at home using a timed walking test to assess their progress. For example, have them walk around a track once or twice and time how long it takes, and have them rate their exertion level and record it. Then have them repeat this monthly to document changes in time as well as their effort ratings. Such checks can serve as motivation as well as monitoring strategies.

Important Exercise Reminders

Remind patients and clients of these important considerations:

When to Reduce the Intensity of Exercise

- If your effort level feels "very hard" or "very, very hard"
- If you are breathing too hard to be able to talk
- If you have muscle soreness that prevents exercise the next day
- If you do not feel fully recovered within 1 h of stopping exercise
- If you are excessively fatigued after an exercise session

When Not to Exercise on a Given Day

- If you have a fever (over 101 °F)
- If you have a new illness that has not been treated

- If ambient temperature and humidity are excessive
- If exercise causes pain
- If you feel ill in any way

When to Stop Exercise and Ask for Guidance

- If you feel chest discomfort or inappropriate shortness of breath
- If you unexpectedly have an irregular heart rhythm
- If you feel dizziness or light-headedness during exercise, or you have dizziness or light-headedness that does not resolve after you stop exercising
- If you experience leg cramps that persist after stopping exercise
- If your vision is blurry
- If you have excessive fatigue

It is critical to educate participants on how to monitor their own responses to exercise. This includes hemodynamic responses (specifically heart rate and heart rhythm), exertion ratings, and symptoms. The following are abnormal responses that every patient needs to know:

- Excessive shortness of breath
  - Such that a conversation could not be carried out (the talk test)
  - That does not normalize with reducing intensity or stopping exercise
- Excessive heart rate
  - When the heart rate is higher than prescribed or than what is usual for a given level of exercise
  - When there is the feeling that the heart is racing
  - When the heart rate remains >100 beats/min (tachycardic) 30 min after stopping the exercise
- Muscle or joint pain that prevents continuation of the exercise despite reduction in intensity or that persists after stopping exercise
- Nausea
- Headache
- Dizziness or light-headedness
- Chest pain

Patients also need to know when they should not exercise, should reduce intensity or duration, and should stop an exercise session, and when to call the health care team or go to the emergency room. They must also be aware of conditions and situations in which exercise must be deferred until they are reevaluated and "cleared" for exercise by the health care team.

In follow-up visits, it is important to assess for changes in activity since the last visit, as a decline may be associated with new-onset symptoms (e.g., a new episode of low back pain), a change in clinical status (e.g., gradual development of anemia), or recent changes in medications (e.g., a change in a statin causing musculoskeletal symptoms that lead the patient to exercise less). Decreases in activity level can be concerning, as they often indicate a new or worsening health problem that needs medical evaluation. For example, development of exertional dyspnea may indicate a new or worsening cardiac problem. It is also possible that this new onset of shortness of breath could be the result of the individual's attempt to increase exercise intensity more than recommended, which should be addressed through education in awareness of symptoms of excessive effort.

Take-Home Message

This list presents some caveats to consider.

- Exercise should "energize" or "invigorate"; if it causes fatigue, the intensity may be too high.
- Exercise that causes pain sufficient to prevent participation the next day should be reduced in intensity, or the mode of activity should be changed (i.e., from weight bearing to nonweight bearing). Also, if the pain seems biomechanical in origin, consider referral to physical therapy, an assessment of footwear, referral for a biomechanical assessment, or more than one of these.
- Gradual progression in intensity and duration of activity will enhance adaptation, minimize injury, and possibly be more sustainable.
- Some individuals with high comorbid burden may not demonstrate objective improvement in fitness or performance. In such patients, maintaining their level of physical function is a positive outcome (or they would likely be deteriorating over time).
- There are many situations in which heart rate monitoring is not recommended, due to treatments or conditions (e.g., changes in fluid status with renal failure) or medications that alter the chronotropic response to exercise. The *CDD4* recommendation is that the individual focus on subjective ratings of exertion and symptoms as a guide for intensity monitoring. Don't make monitoring "too scientific" or "too technical," which may seem burdensome for people who are older or those with significant disease burden. If such a patient asks for target heart rates, provide them, but the default approach should be subjective ratings of exertion.
- In the vast majority of patients who are sedentary, maybe older, and burdened with multiple comorbidities, getting them to "just move" is success. Developing and negotiating strategies to get started should be the focus. Once patients have established a habit of getting up and moving

daily, regardless of the intensity or duration, they may have some receptiveness to guidance for progressing in intensity or duration and discussing other aspects of fitness.

- Every effort should be made to incorporate monitoring and encouraging participation in the overall health care plan, such that everyone who sees the client assesses his or her level of participation in physical activity, providing encouragement and ensuring that the importance of participation is transmitted from all team members to the patient and family or caregivers.

Suggested Readings

Abellan van Kan G, Rolland Y, Andrieu S, et al. Gait speed at usual pace as a predictor of adverse outcomes in community-dwelling older people: International Academy on Nutrition and Aging (IANA) Task Force. *J Nutr Health Aging.* 2009;13(10):881-889.

American Thoracic Society. ATS statement: guidelines for the six-minute walk test. *Am J Respir Crit Care Med.* 2002;166(1):111-117.

Bean JF, Kiely DK, LaRose S, Alian J, Frontera WR. Is stair climb power a clinically relevant measure of leg power impairments in at-risk older adults? *Arch Phys Med Rehabil.* 2007;88(5):604-609.

Fritz S, Lusardi M. Gait speed placed into perspective of requirements for various activities and outcomes. White paper: walking speed - the sixth vital sign. *J Geriatr Phys Ther.* 2009;32(2):2-5.

Singh SJ, Morgan MD, Scott S, Walters D, Hardman AE. Development of a shuttle walking test of disability in patients with chronic airways obstruction. *Thorax.* 1992;47:1019-1024.

Solway S, Brooks D, Lacasse Y, Thomas S. A qualitative systematic overview of the measurement properties of functional walk tests used in the cardiorespiratory domain. *Chest.* 2001;119(1):256-270.

Studenski S. Bradypedia: is gait speed ready for clinical use? *J Nutr Health Aging.* 2009;13(10):878-880.

Studenski S, Perera S, Wallace D, Chandler JM, Duncan PW, Rooney E, Fox M, Guralnik JM. Physical performance measures in the clinical setting. *J Am Geriatr Soc.* 2003;51(3):314-322.

Additional Resources

A compendium of questionnaires is accessible on the interactive Physical Activity Resource Center for Public Health website, www.parcph.org [Accessed October 30, 2015].

Short Physical Performance Battery training video plus literature on development of the SPPB, www.grc.nia.nih.gov/branches/ledb/sppb [Accessed October 30, 2015].

For performance-based test procedures, with scoring and other details, http://wcb.missouri.edu/~proste/tool [Accessed October 30, 2015].

Examples of comprehensive patient education for starting exercise for older individuals or those with chronic conditions:

Free booklet: *Exercise: A Guide for People on Dialysis.* Available at: http://lifeoptions.org/catalog [Accessed October 30, 2015].

Exercise & Physical Activity: Your Everyday Guide from the National Institute on Aging. Available from the NIH/NIA website, http://www.nia.nih.gov/health/publication/exercise-physical-activity [Accessed October 30, 2015].

Art of Exercise Medicine: Counseling and Socioecological Factors

Chapter 1 has a brief discussion on an approach a physician can take to provide brief counseling for the purposes of promoting physical activity and exercise. The discussion in chapter 1 assumes that the physician is not the primary counselor on exercise but has more of a "head coach" role, in which the job is mainly to provide the moral imperative on being active and to establish the importance of physical activity in promoting health. The examples provide some simple suggestions depending on how much time the physician has for this purpose. For patients who don't have a chronic condition, the practice may only wish to recommend some exercise resources. Patients who do have a chronic condition, however, are likely to need substantially more counseling on exercise. Physicians can do this if they choose, but most physicians are probably not well prepared to do exercise counseling and will want to refer to an exercise professional.

CDD4 is the first edition in this series to address counseling, sometimes called coaching, to improve physical activity. A complete discussion of counseling is well beyond the scope of *CDD4,* but it is important to introduce basic concepts since one of the key roles of the entire medical home team—especially the primary care physician—is to help patients be motivated to sustain their exercise program. While it is the health coaches and exercise professionals who do all the real work described in this book, patients look to the primary care physician to establish the essential elements of managing their health. Primary care physicians thus need to be more aware of counseling methods and must be prepared to refer patients to appropriate exercise professionals to individualize and implement a care plan. Earlier editions of this textbook overlooked this essential teamwork and did not adequately address how

physicians can collaborate with the exercise community. Physicians may or may not elect to engage in counseling services, depending on their business model. But they do need to be aware of how to address physical activity and exercise, how to collaborate with exercise professionals, and what counseling methods the rest of the health care team is using. Counseling services for lifestyle intervention are evolving in response to today's health care needs; but sadly, at this point, lifestyle counseling has been disappointing and unable to facilitate sustained behavior change, even when it is included as part of patient care.

Anyone who does lifestyle intervention is wise to expand his skills and seek his genuine voice as a counselor. Patients have a finely honed ability to detect when they are being given a recipe or formula, and only a few patients actually want that approach. Most patients prefer a counselor who is trying to hear and understand their unique problems and feelings about their situation. Lifestyle counselors who primarily focus on what they want their patient to do often express frustration and experience burnout, as they see only that their patients are all making the same "wrong choices." Counselors should be encouraged to uncover each patient's unique life path as an intriguing story and be captivated by the challenges each patient faces (even though many details will sound similar). Good counseling is less about the number of steps or minutes or calories per day of MVPA (moderate to vigorous physical activity), or time at a target heart rate, than it is about helping patients overcome the barriers they face, many of which are deeply embedded in their way of life and sense of self. Good counselors who truly care about their patients respond to each patient's unique story with an honest and humble reaction, with a gentle but firm sense of their own voice.

An exercise prescription is merely what you advise as a path to better health.

- Focus on the exercise prescription, and your counseling will be boring and ineffective.
- Focus on the tale that is the patient's life, and every encounter will be provocative.

Common Behavioral Techniques Used in Exercise Counseling

There are very many counseling techniques, some of which have been more thoroughly explored with regard to promoting physical activity. The discussion below is not exhaustive but briefly reviews the more commonly used methods for exercise counseling.

Self-Efficacy

In the 1960s, Albert Bandura developed a social–cognitive theory of behavior, commonly called *self-efficacy*. Self-efficacy is easy to use in a clinical context because it employs a 10-point Likert-type or visual analog scale to assess how confident the patient is that she can execute a particular behavior under hypothetical contexts. Self-efficacy measures are especially revealing in social situations that make a choice difficult. For example,

How confident are you that you will walk every day for 30 minutes?

0 -------------------- 5 -------------------- 10

| Certain I | 50/50 chance | Certain I |
| couldn't do it | I could do it | could do it |

How confident are you that you can go outside for a 30-minute walk when it is cold and raining?

0 -------------------- 5 -------------------- 10

| Certain I | 50/50 chance | Certain I |
| couldn't do it | I could do it | could do it |

In this example, the individual might respond 7 or 8 to the first question but only 2 on the second question. As another example, one common self-efficacy question for tobacco cessation is how confident the smoker is that he could resist having a cigarette when others are smoking around him.

Self-efficacy is a helpful tool to detect insecurity and lack of resilience. Counselors can then work to build patients up to a stronger level of self-efficacy through encouragement and the use of small successes as achievements that demonstrate to patients that they do have the ability to succeed. Building on such successes, the counselor's objective is to help patients improve their self-efficacy with experience, using the self-efficacy scale to mark progress.

Stages of Change

In the early 1980s, Prochaska and DiClemente developed the *transtheoretical model* of behavior, more commonly known as *stages of change*. In contrast to self-efficacy, stages of change is more a theory than a tool. The strengths of this theory are that it's easy to generalize to a wide variety of lifestyle behaviors (it was originally developed for tobacco cessation) and provides a framework for approaching any situation. The main weakness of this theory is that it is highly nonspecific and therefore does not provide the counselor with detailed direction on what to say or do in any given circumstance. Also, this theory gives the impression that change is linear and sequential (which is not necessarily so) while overemphasizing the cognitive and underemphasizing the emotional or spiritual aspects of change. Stages of change theory is very helpful in structuring the counselor's perspective of the patient's state of mind at any given moment.

Stages of change theory categorizes the change process into five basic categories:

1. Precontemplation—not yet aware of the need to change
2. Contemplation—aware of the need to change and actively pondering it
3. Preparation—getting emotionally ready and gathering mechanisms to support change
4. Action—actively involved in trying to change behaviors
5. Maintenance—having established a new behavior, working to maintain it long-term

Stages of change theory can be extremely helpful at providing a rough notion of the patient's mindset, thus giving the counselor clues on how to facilitate progress. Consider three common chief complaints of people who have been referred to cardiac rehabilitation:

1. "Well, I just got out of the hospital. My doctor said I had a heart attack, and he made me come see you, but I don't really know what for. He told me to eat more vegetables, but I hate them."

2. "Well, I just got out of the hospital. My doctor said I had a heart attack, and he told me to go to cardiac rehabilitation. He said you can help me avoid having another heart attack."

3. "Well, I just got out of the hospital because I had a heart attack. I haven't had a cigarette since then, I bought a book about the Mediterranean diet, and I've been walking 30 minutes a day."

Obviously, each of these perspectives involves a very different mental state! The first person is a *precontemplator* and is not likely to do any exercise prescription you give her. The second is in *preparation* and likely to be receptive to your recommendations. The third is leading the *action* of his own changes and very likely to enthusiastically adopt your suggestions. So the dialog between the counselor and these patients should be very different, each discussion individualized with consideration given to the patient's current stage.

Stages of change becomes complex in that smoking behavior and exercise and diet can be highly compartmentalized in completely different stages. While the five stages of change appear to be a linear progression, transitions can be rapid or slow and can involve recidivism in which the person reverts to a less progressed stage, vacillating back and forth between higher and lower stages. It would not be uncommon, for example, for the second and third of these patients, after a year, to have stopped exercising, gone back to smoking cigarettes (because their whole family continues to smoke), and to be relying on medications to lower their cholesterol and blood pressure—while the first patient could have changed her diet, lost weight, and is exercising for an hour three times a week!

Stages of change theory can be very helpful for understanding the patient's mindset regarding each domain of behavior (tobacco, exercise, diet, mental health), guiding the counselor on what kind of approach to take for each of the lifestyle domains. Stages of change is thus a useful construct, applicable to many situations, but limited in that the counselor needs to develop many counseling skills above and beyond categorization in each of the stages.

Self-Determination Theory

In the 1970s, Edward Deci and Richard Ryan advanced a form of *self-determination theory* that is well suited as a motivational tool to promote adherence to exercise and physical activity. The self-determination concept relies less on an external motivating force (e.g., "Exercise is good for me, will improve my quality of life and my longevity, so therefore I should dance") than on the intrinsic rewards (e.g., "I will dance because it's fun").

Self-determination theory seems particularly well suited for exercise counseling. Compared to tobacco cessation and changing to a heart-healthy diet (which many people perceive as self-deprivation), exercise typically gives participants an immediate sense of satisfaction and fulfillment—if not during the exercise, then very soon afterward. Self-determination counseling leverages this property of exercise to increase the patient's self-efficacy and motivation to maintain an active lifestyle.

As with many other behavioral interventions, self-determination theory studies have applied the 7A's model of counseling, though it is often executed in the simpler 5A's model (in bold).

1. Address—the topic of exercise, physical activity, physical functioning

2. **Ask**—whether the patient is regularly physically active and satisfied with her physical functioning

3. **Advise**—the patient to pursue a more active lifestyle to maintain physical functioning

4. **Assess**—the patient's readiness to change, especially noting barriers to change

5. Agree—in a motivational interview style (see next section), on a patient-selected plan

6. **Assist**—in an active way to help find solutions to problems and overcome barriers

7. **Arrange**—for follow-up to ensure that the patient succeeds in becoming more active

The 5A's tool, originally developed for tobacco cessation, is helpful in promoting physical activity. Self-determination can theoretically be combined with a self-efficacy tool and the counseling aspect of the 5A's driven by stages of change theory, with individual advice being aimed at helping the patient advance to a more progressed stage of change (and avoid recidivism). Good exercise counselors likely employ all of these methods with an ease and alacrity that typically takes a few years of clinical experience to master.

Motivational Interviewing

In the 1990s, William Miller and Stephen Rollnick developed what is actually more a style of counseling, known as *motivational interviewing*. Good counselors, whether trained in motivational interviewing or not, almost always develop a style consistent with the tenets of motivational interviewing. This style of counseling emphasizes patients' autonomy by respecting their personal sensitivities and perspectives. Like many behavioral techniques, it was originally developed for alcohol and tobacco cessation programs and subsequently broadened to other areas of lifestyle intervention.

While motivational interviewing has become popular in recent years, it is no panacea with respect to helping people make and sustain change. In one study with African American adolescents, motivational interviewing was actually inferior in comparison to attention control. Nonetheless, motivational interviewing is usually a helpful tool. The most important benefit of motivational interviewing is that it helps a counselor develop voicing that expresses respect for the patient's views, thereby establishing that the counselor hears and understands the patient's unique situation.

One nuance of motivational interviewing is that it can sound insincere when poorly executed, it can have an artificial feeling, not seem like a genuine voicing of concern and thereby undermine the entire intent of the motivational interviewing technique! Consider this example:

> Mrs. Jones is a 60-year-old mother with disability and with obesity, type 2 diabetes, heart disease, and arthritis whose daughter is a single mother with three children (ages 3, 9, and 14). The daughter makes a living as a maid, working two jobs—one for a motel and another as a per diem worker for a housekeeping service. The patient's monthly disability check is not enough to cover her expenses and medications, so she has moved in with her daughter, in part for herself and in part to help her daughter save on child care. The patient finds herself wanting to enroll in an aquatic exercise program at a gym, but her child care duties interfere with the time of the aquacise sessions, and the cost of the program isn't within her budget. She's upset and fretting over whether or not she can get financial aid from a local charity to help pay for the program and is finding the application form difficult to complete.

This fictitious example represents a surprisingly common scenario that counselors face. The following are two responses to hearing Mrs. Jones's story, both compatible with motivational interviewing theory. If you were Mrs. Jones, which response sounds more genuine and caring?

> Well, Mrs. Jones, I hear you saying you have some tough choices to make. Perhaps my staff can help you with the application form.

> Wow. You're in a difficult spot. Do you want some help with the form?

Motivational interviewing can be difficult for technologically trained clinicians to master, perhaps especially so for physicians, whose perspectives and biases differ from the patient's preferences and desires. Years of schooling in the highly technical fields of health care can lead many health professionals to feel as if they know what is "right," and it can be hard to accept a patient's desire to make a choice based more on emotions and feelings than on scientific findings and technological data. But it is the rare patient in whom a paternalistic approach yields sustained change. Education about "the facts" alone is a poor agent of change, and motivational interviewing usually facilitates progress down a path of least resistance to thereby foster a higher degree of self-efficacy. There are moments—generally uncommon—when the dynamic partnership between a provider and a patient is best served by the counselor's asserting his recommendations in a paternalistic manner. One hallmark of superb counselors is their ability to blend all the counseling styles together, including a strong paternalism when that is the voice the patient responds best to in the interaction. Perceiving the right time and place for such moments is an artful application of counseling skill.

> Motivation is achieved through feelings, not by acquisition of knowledge.

The irony for lifestyle counselors is that their patients' successes are determined not so much by the facts the counselors learn in school—details such as target heart rates, body composition assessment, exercise techniques and program design, or calories and carbohydrate counting and mindful eating. Rather, success is determined more by how well the counselor connects emotionally with the patient and is able to facilitate the patient's being in touch with her own feelings and the things she does to hang in there when the going gets tough.

Goal Setting

Goal setting has long been thought to be important in counseling, including motivational interviewing in which the counselor facilitates patients in setting their own goals. For example, one goal is to do the minimum amount of recommended exercise because those who meet it reduce their cardiometabolic risk. For "big" goals, such as losing 20 lb or achieving 150 min of MVPA each week, a stepped-goal approach divides the big goal into smaller intermediate goals that are more achievable in the short term (even as short as the rest of today, or the next hour!). The inability of most people to succeed, even with the stepped-goal approach, brings into question the applicability of goal setting for all individuals.

Of course, many patients come preloaded with a goal that is highly likely to end in failure because it's unrealistic (e.g., lose 50 lb before their daughter's wedding next month). Good counseling in that type of situation would be advice adapted to suit the patients' psychological state while also helping them better see what they might realistically be able to do (e.g., lose 10 lb before the wedding).

Other Aspects of Exercise Counseling

The previous sections briefly discuss the most thoroughly researched counseling approaches used in exercise management, but by no means do these constitute an exhaustive list. New theories on human behavior and learning continue to arise; some have been expressly tested in exercise research, and some have not but have obvious implications. Because the aforementioned methods and tools have shown disappointing long-term success, this section discusses some nascent ideas on counseling and behavior change.

Self-Theory

Self-theory is a model of learning and motivation, primarily developed by Carol Dweck, wherein the impact of a failed attempt to learn a new skill emerges in subsequent attempts. The person's prior experiences influence how she perceives herself (her self-theory) and very strongly influence how well she does in subsequent learning attempts and in the long term. The self-theory model dichotomizes people into two groups based on self-perception of having

- *fixed potential,* determined by their intrinsic and innate abilities, or
- *unlimited potential,* with an ability to master new and difficult challenges.

Self-theory arose primarily in child psychology but also has been demonstrated in adolescents and adults. Self-theory has not been studied as a specific application to exercise in chronic conditions, and it runs somewhat counter to traditional exercise specialist training, where setting a goal is an important aspect of an exercise plan (e.g., it is common for patients to set a goal such as walking 10,000 steps a day or participating in a 5 km run or walkathon). Most health care providers are predisposed (if not specifically trained) to be goal oriented and to set goals. For this reason, the findings coming from self-theory are potentially very important in exercise management because of how self-theory interacts with goal setting.

Some people think of themselves as having an intrinsically fixed potential, as if they are what they are, with no ability to rise above their innate potential. For such people, their prior failures are seen as evidence that they don't have as much natural ability as others:

I'm only a small skinny kid, so it's not surprising that I can't make the team.

Other people think of themselves as more of a blank slate, such that they can overcome difficulties and master difficult skills that aren't easy to learn. In people like this, prior failures have no relevance on their perceived ability to succeed in the future:

Geez, I didn't make the team. I'll have to start practicing more.

Ultimately, all of us happen upon something that is too difficult for us to do, and we either don't succeed easily or just fail. How each person perceives this failure affects how he responds to future challenges. Those who weren't good at sporting games as a child, for example, who came to see themselves as nonathletic, often develop a negative attitude toward their own capacity to participate in and enjoy physical activity. It turns out that about half of people are in the fixed potential group and the other half are in the unlimited potential group. Given this, one wonders how our society's approach to games and sports affects children's self-theories with regard to exercise and whether or not it is related to the high prevalence of physical inactivity in today's adults.

Recent studies, primarily in children, adolescents, and young adults, suggest that focusing on *process* and *effort* may be a better way of guiding many people through incremental change and advancement. Focusing on process or effort has the advantage of reducing self-judgment, which can lead to negative self-thoughts. Some people, in response to difficulty or to failure, begin to question and undermine their belief in their own capabilities. Such negative thinking is very hard to overcome, and an emphasis on goals may set such people up to be vulnerable to negative thinking should they fail to achieve the goal. Process- and effort-oriented thinking more intrinsically acknowledges that some things are hard to do, and may even be harder for some people than for others, and thus is a more optimistic mindset suggesting that everyone can eventually succeed.

Self-theory interacts adversely with *goal setting.* If people see themselves as having fixed potential, then goal setting is risky because failure to achieve a goal reinforces their notion that they have limited potential. If, on the other hand, people focus on the process of learning new things (rather than on achieving specific goals), they have better results in all areas of life—school, sports, social life, marriage, and work. The process-oriented perspective fosters a self-theory of unlimited potential and helps avoid failure in a person with a self-theory of fixed potential (because there is no concrete goal to fail to achieve). Self-theory has not been as rigorously tested with regard to exercise, but the strength of findings in other aspects of complex behaviors suggests it has potential to improve adherence to regular physical activity.

The very important lesson from self-theory research, though not yet proven to apply to exercise in chronic conditions, is to emphasize the process of being active and maintaining physical functioning, rather than aiming for a specific goal.

Note that this runs somewhat counter to exercise guidelines, even the Basic *CDD4* Recommendation of 150 min a week of moderate to vigorous physical activity! All of ACSM's recommendations, including those in *CDD4,* are goal oriented, and goal setting remains the standard recommendation. But if a patient is poorly adherent and isn't doing well with her exercise program, think about trying self-determination theory to evoke intrinsic rewards and emphasizing the *process* of becoming active rather than achieving a specific exercise goal.

Since self-theory is likely to be counterintuitive to most health care and exercise professionals, who generally are biased toward the goal-setting, high-achieving, and unlimited potential mindset, counselors who recognize this in themselves will need to resist projecting it onto their patients and work to encourage goals or process as best suited to each individual patient.

Motivation and Grit

Motivation is almost always a perplexing issue in lifestyle counseling and is seemingly easier to undermine than to build up. Patients who are intrinsically motivated often don't seem to need much help from a counselor, but patients who are depressed, self-deprecating, and mired in negative thinking can seem nearly impossible to help get motivated. In general, the more patients *feel* a positive connection to the efforts they are making, the more likely they are to maintain motivation. Knowing what one should do doesn't provide the emotional drive that powers the engine of behavior change.

> Motivation is primarily an emotive, not cognitive, property.

Grit, a trait that has become a popular topic in recent years, refers to a resilience and ability to persevere and demonstrate discipline during difficult challenges. Self-love, that is, doing hard emotional work for one's own benefit, is a necessary element of lifestyle change when it's not easy and when change doesn't come naturally to the patient. Grit and self-love are closely related traits that some people have in reserve and some don't exhibit much at all. Successful and sustained behavior change is closely allied to these traits, and good lifestyle intervention counseling nurtures these characteristics. Unfortunately, there is no known and surefire way to prompt motivation or grit or self-love, so the best one can say at this point is that these aspects of counseling must be highly individualized to suit the patient. This is an area in which patients have a highly refined sense of when someone cares about them or is simply reciting recipes, so counselors must develop their own unique voicing to convey true caring.

A Difficult Cycle to Escape

Severe chronic disease and disability have a tendency to undermine a person's ability to make a living, causing quality of life to take a downward spiral because he or she lacks personal and financial resources to invest time, money, and energy in their health. This only leads to a cycle of worsening health status, less ability to make a living, and higher burden of chronic disease.

- Chronic conditions adversely affect socioeconomic position.
- Life course socioeconomic position sets the stage for chronic conditions.

Explore Motivation and Negotiate Objectives

Motivational interviewing theory suggests that all individuals have unique objectives for participating in exercise, and generally any goals they have should initially supersede those of the practitioner—or at least there should be a negotiated balance of the two. Goals can be an emotionally difficult challenge, however, and while exercise goals are important in many patients, it may be better to focus on process and accepting the challenge.

One complexity about goals is that they can vary widely, because needs are heavily influenced by socioecological circumstances that are independent of the chronic condition. Imagine two patients with congestive heart failure (CHF) that is medically similar.

- An older retiree who wants to maintain independence in activities of daily living and instrumental activities of daily living to remain in his home and delay having to move to assisted living
- A middle-aged high school teacher who wants to maintain employment and coach her kids in sports and thus wants to achieve as high a level of physical fitness as possible

Obviously the exercise prescriptions need to be very different. For the retiree, 150 min a week at a somewhat-hard to hard rating of perceived exertion may not be necessary, and he may not want to work that hard. For such a person, maintaining strength, range of motion, and general mobility is often a good focus for the program. For a younger patient, the ACSM Guidelines may be more appropriate, so higher-intensity interval training may be a good choice.

Many clients haven't thought about physical activity because they have received discouraging messages (or no message at all) about "overexertion" from family members and health care providers. All too often, patients are told, "Take it easy" or "Don't overdo it." Unfortunately, significant deconditioning can occur as a result, especially if extended hospitalization is involved. Once patients are stabilized on treatment, there is an opportunity to present realistic possibilities for activities that they can look forward to and work toward and to help get them motivated.

One very common desire, seen in almost every patient, is the possibility of reducing the need for medication. The following are common conditions in which exercise (especially when combined with a heart-healthy diet and good stress management training) presents a real possibility of reducing medications:

- Hypertension and hypercholesterolemia
- Prediabetes, diabetes
- Musculoskeletal pain
- Depression

All members of the management team, the patient, and the family must be clear on what exercise can and can't accomplish in a given condition. Exercise is great therapy, but it is no panacea.

Only a very small number of patients have a realistic chance to get off all medications, but a realistic chance at reducing need for some medications can be a big motivator.

One idea that can help patients is the notion that exercise can be an antidote to the side effects of some medications. A good example is immunosuppression following organ transplantation, as exercise can attenuate or eliminate some of the side effects of these medications (i.e., muscle weakness and weight gain with chronic prednisone therapy). It is important to portray exercise as a treatment that can make medications more tolerable with less negative effects on quality of life.

Fear Avoidance

Patients' success in adopting an exercise program can also be determined by their coping style, which in many patients is passive and not very effective. Patients can be influenced by fear, anxiety, or depression, which warrant particular attention. Motivation to participate in an exercise program can be heavily influenced by health literacy, locus of control, affective state, and fear. People who do not understand their disease may believe that there is nothing they can do to improve their quality of life. It is common for patients with heart conditions to be fearful that activity and exercise are unsafe or will make their condition worse. The exercise professional must therefore assess health literacy, coping style, self-efficacy, fear-avoidance beliefs, and catastrophic ideation about the situation.

Fear avoidance is the concept that one should avoid activities due to the belief that those activities will cause pain or further injury. Fear avoidance has mostly been studied in painful neuromusculoskeletal conditions but has a strong parallel in chronic disease. Catastrophic ideation is a maladaptive cognitive–affective response to illness (best studied in chronic pain) in which the individual tends to magnify the threat, to feel helpless, and to be unable to inhibit negative thoughts (about pain, disability, or death). Fear avoidance is common in people with chronically painful conditions such as low back pain and fibromyalgia. An analogy in cardiometabolic situations is the so-called cardiac cripple (who is incapacitated by fear of cardiac death). The classic model of fear avoidance is initiated by a clinical event, causing an increase in event-related fear, catastrophic ideation, hypervigilance (regarding the feared condition), and a response of withdrawal and avoidance, followed by sedentary behavior and deconditioning, affective decline (flattening of mood and affect with an increase in depression and anxiety), and a resulting loss of functioning and resilience (see figure 4.1).

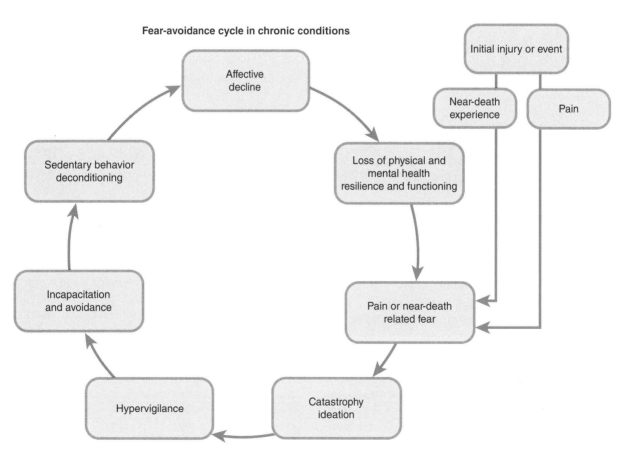

Figure 4.1   Fear-avoidance cycle in chronic conditions.

The fear-avoidance cycle influences the level of disability, and while a certain degree of fear avoidance is normal and healthy, excessive fear avoidance can undermine the patient's self-efficacy for participation in exercise. Excessive fear avoidance must therefore be addressed, with all members of the team reassuring the individual of the benefits and safety of the program. In patients, exercise testing and especially a maximal exercise test is likely to cause extreme fatigue, symptoms, or pain (due to the intensity), which will reinforce their fears. While there are many questionnaire tools to characterize fear avoidance, they may not be that useful outside the research environment. The *CDD4* recommendation is that the clinician use subjective interviewing to collect information regarding health literacy, coping style, fear-avoidance beliefs, and any catastrophic ideation. The clinician should do this by directly asking patients what they think their condition means in terms of impact on lifestyle, what they think is safe and unsafe activity, and what they can do to have a positive effect on their condition.

Motivation, understanding of one's condition, self-efficacy, and fear are all modifiable. Controlled graded exposure to exercise in a supervised setting may be the best method to address fear avoidance, using the functional exercise trial approach (see chapter 3). Such an approach provides an opportunity to demonstrate the patient's usual response to physical activity, including various symptoms such as pain, fatigue, and shortness of breath. The individual can gradually learn that symptoms are signals intended to protect the body yet are not necessarily indicators of danger. Symptoms should not be ignored but should be gauged and managed with appropriate guidelines, not feared or completely avoided. By providing techniques for the individual to manage these symptoms as they arise, the clinician can encourage a more active coping style and improve self-efficacy.

Prolonged Frequent Contact

With a genuine and caring connection between the counselor and the patient having been established, sustained involvement leads to a higher likelihood of achieving better outcomes. This is best summarized as the concept of *prolonged frequent contact,* which may be nothing more than a marker of the process- and effort-oriented approach to maintaining motivation and gradual change. Prolonged frequent contact may thus be a good benchmark for a counselor to aim for, and interventions should probably be designed to foster prolonged frequent contact. Harkening back to an era when counseling was most of what physicians had to offer, the words of Dr. Francis Peabody get at the heart of counseling—caring about that individual. Making patients feel that they are cared for is an essential element of prolonged frequent contact.

Socioecological Disparities and Exercise in Chronic Conditions

Counseling for physical activity and exercise is difficult with apparently healthy persons. The preceding discussion on counseling and goal setting demonstrates that counseling is often substantially more difficult with persons with a significant burden or disability from a chronic condition. Experienced health care providers and exercise professionals will also be quick to note that socioecological disparities have an enormous impact on lifestyle change and can present a substantial barrier to a physical activity plan of care. An in-depth discussion of disparities is well beyond the scope of *CDD4,* and this discussion only introduces the topic. An effective lifestyle intervention plan for a person with one or more chronic conditions needs to consider disparity, because many patients with chronic conditions have a disability and end up with disparities as a major factor to consider.

The socioecological model is traditionally a child development framework, but in exercise management it is also a useful construct for adults and seniors. The socioecological model posits that individuals are strongly influenced by their family and peer environment, their physical environment, the cumulative impact of school and media and communication inputs they receive, and their resources (physical, capital, human, and others). Any person who provides individual or group counseling soon discovers that—for some individuals—these socioecological factors (and any disparities) can present a major barrier to implementation of an exercise–physical activity program. This is particularly evident in people who need more than just an exercise program, who also need a full multifactorial lifestyle risk-reduction intervention.

Patients' socioecological situation has a profound interaction with their ability to adhere to an exercise or physical activity program. Persons who grew up in or spent much of their life in a disadvantaged socioecological situation are more likely to develop a chronic condition and become disabled, and they have as much as a 17-year reduction in disability-free life expectancy. Many people with chronic conditions end up in poverty or near poverty, in part because of the costs of health care and in part because of the lost income due to being disabled. Health professionals involved in exercise management for persons with chronic conditions need to have a strong understanding of each patient's socioecological situation. Failure to do so will almost surely result in a low success rate with regard to achieving a sustained active lifestyle.

Psychosocial Aspects

Exercise can be an extremely important component of a lifestyle intervention program, because many patients see exercise as having intrinsic rewards (see self-determination theory, previously discussed); conversely, dieting is often perceived as deprivation rather than reward. Moreover, group exercise, while not effective for every patient, is on the whole better for increasing program adherence. Some people, though, prefer to exercise alone or in a setting where they can engage more with the environment than with other humans (e.g., hiking trails). These preferences are influenced by the socioecology of their life, including any adverse social experiences they may have had (separately as well as in the context of exercise). One should bear such issues in mind when assisting people in configuring their exercise plan.

Economics of Exercise as Medical Treatment

Exercise has long been known to be a very efficient use of health care dollars and as one of the most cost-effective health care interventions. Despite this, exercise remains undervalued and underappreciated as a medical treatment, with few physicians having any formal training in exercise-related topics. Counseling a patient who needs exercise often raises economic issues that must be explored and resolved, noting that patients will make an emotional decision based on how they feel about investing their time, money, and energy in any exercise program.

- *Who is spending, and who is saving?*
- *Low tech + low cost = perception of low value?*

Historically, health care systems have presumed that an exercise program is a personal investment that patients must make for themselves. For goal-oriented plan administrators who have the means or for whom exercise is a personal preference, this seems a simple decision, and many health plan administrators do not understand beneficiaries who won't make an investment in exercise. But for patients who have been sedentary all their lives or who are in an adverse socioecological position, exercise may seem like a poor investment of their limited resources. In recent years, more health plans have recognized the need to couple good incentives with health promotion literacy, a pairing that will potentially lead to more support for exercise programming.

Education, Literacy, and Health Literacy

At the level of population health, educational attainment has an enormous impact on reducing the risk of developing a chronic condition or disability and also significantly increases the likelihood of meeting physical activity guidelines. Readiness to invest in exercise as a medical treatment for a chronic condition is also affected by both literacy and health literacy. Thus, the health–exercise professional needs to be aware of the patient's educational level and health literacy, which are often correlated. The combination of low educational attainment and low health literacy can be particularly dire, because such individuals may have either an underlying cognitive impairment or adverse childhood experiences that put them at greatly increased risk for chronic conditions in adulthood. These situations can undermine an individual's ability to participate in an exercise or lifestyle intervention program.

Spirituality

Many people believe that spirituality plays an important role in motivation or satisfaction or adherence to an exercise plan. Although this remains more a belief than a well-founded fact grounded in observational data, spirituality (whether on a personal level or in an organized religious faith) is mostly overlooked by technologically biased health professionals. Some of this may be a bias against a spirituality-oriented mindset or may simply be due to the paucity of research in this area. Nonetheless, if a patient has a strong spiritual mindset, this can be an asset in helping the patient be motivated. Also be aware of cultural or religious aversions to exercise.

Integration Into a Medical Home Model

This chapter has discussed how extremely challenging it can be to use one's exercise knowledge and counseling skills to help a patient be more active. For patients who have been adversely affected by their chronic condition, with a substantial burden of disease or disability, the main goal should be to use exercise as a treatment to prevent or slow their future decline and to maintain their quality of life and independent living.

The artful application of knowledge about exercise, and—more important—an understanding of and facility with lifestyle counseling, will have a big impact on how well patients process counseling given their life experiences and socioecological milieu. Overreliance on any technological aspect of exercise, specific chronic condition(s), or theory on counseling or human development is not likely to succeed.

Taken in this perspective, exercise management for patients with a chronic disease or disability is far more art than science, and it's easy to see why it takes a professional a lifetime to master. One major reason that the Basic *CDD4* Recommendation aims at the low end of the Health and Human Services and ACSM Guidelines is that according to expert consensus, few severely affected patients will do more than this. Further, expert consensus is that many such patients will have significant socioecological barriers to overcome, and it is better to focus efforts on these barriers than it is to push a little harder on the amount of exercise.

One important challenge for health professionals in the medical home is developing the ability to let go of counseling approaches that are not succeeding. When such interventions succeed in only 10% to 20% of individuals, continuing to use these techniques consigns >80% of patients to failure. Carefully controlled innovations in counseling interventions may be a better approach, and to that end we encourage readers to extend their counseling knowledge and skills.

Suggested Readings

Counseling

Artinian NT, Fletcher GJ, Mozaffarian D, Kris-Etherton P, Van Horn L, Lichtenstein AH, Kumanyika S, Kraus WE, Fleg JL, Redeker NS, Meininger JC, Banks J, Stuart-Shor EM, Fletcher BJ, Miller TD, Hughes S, Braun LT, Kopin LA, Berra K, Hayman LL, Ewing LJ, Ades PA, Durstine L, Houston-Miller N, Burke LE; on behalf of the American Heart Association Prevention Committee of the Council on Cardiovascular Nursing. Interventions to promote physical activity and dietary lifestyle changes for cardiovascular risk factor reduction in adults: a scientific statement from the American Heart Association. *Circulation.* 2010;122:406-441. Available at: http://circ.ahajournals.org [Accessed July 17, 2010]. doi: 10.1161/CIR.0b013e3181e8edf1.

Dweck CS. *Self-Theories: Their Role in Motivation, Personality, and Development.* New York: Psychology Press, Taylor & Francis Group; 2000.

Fortier MS, Duda JL, Guerin E, Teixeira PJ. Promoting physical activity: development and testing of self-determination theory-based interventions. *Int J Behav Nutr Phys Act.* 2012 Mar 2;9:20. Available at: www.ncbi.nlm.nih.gov/pmc/articles/PMC3353256 [Accessed August 14, 2013]. doi: 10.1186/1479-5868-9-20.

Frølich A. Identifying organisational principles and management practices important to the quality of health care services for chronic conditions [PhD thesis]. Available at: www.danmedj.dk/portal/page/portal/danmedj.dk/dmj_forside/PAST_ISSUE/2012/DMJ_2012_02/B4387 [Accessed August 14, 2013].

Hauer KE, Carney PA, Chang A, Satterfield J. Behavior change counseling curricula for medical trainees: a systematic review. *Acad Med.* 2012 Jul;87(7):956-968. Available at: www.ncbi.nlm.nih.gov/pmc/articles/PMC3386427 [Accessed August 14, 2013]. doi: 10.1097/ACM.0b013e31825837be.

Knowler WC, Barrett-Connor E, Fowler SE, Hamman RF, Lachin JM, Walker EA, Nathan DM, for the Diabetes Prevention Program Research Group. Reduction in the incidence of type 2 diabetes with lifestyle intervention or metformin. *N Engl J Med.* 2002;346(6):393-403. PMCID:PMC1370926.

Lundberg M, Grimby-Ekman A, Verbunt J, Simmonds MJ. Pain-related fear: a critical review of the related measures. *Pain Res Treat.* 2011. Article ID 494196.

Orrow G, Kinmonth AL, Sanderson S, Sutton S. Effectiveness of physical activity promotion based in primary care: systematic review and meta-analysis of randomised controlled trials. *BMJ.* 2012 Mar 26;344:e1389.

Vlaeyen JWS, Linton SJ. Fear-avoidance and its consequences in chronic musculoskeletal pain: a state of the art. *Pain.* 2000;85:317-332.

Wagner EH, Austin BT, Davis C, Hindmarsh M, Schaefer J, Bonomi A. Improving chronic illness care: translating evidence into action. *Health Aff.* 2001;20(6):64-78. Available at: http://content.healthaffairs.org/content/20/6/64.full.html [Accessed August 8, 2013]. doi: 10.1377/hlthaff.20.6.64.

Disparities

Agardh E, Allebeck P, Hallqvist J, Moradi T, Sidorchuk A. Type 2 diabetes incidence and socio-economic position: a systematic review and meta-analysis. *Int J Epidemiol.* 2011 Jun;40(3):804-818. doi: 10.1093/ije/dyr029. Epub 2011 Feb 19.

Berkman ND, Sheridan SL, Donahue KE, Halpern DJ, Crotty K. Low health literacy and health outcomes: an updated systematic review. *Ann Intern Med.* 2011 Jul 19;155(2):97-107. doi: 10.7326/0003-4819-155-2-201107190-00005.

Birnie K, Cooper R, Martin RM, Kuh D, Sayer AA, Alvarado BE, Bayer A, Christensen K, Cho S-I, Cooper C, Corley J, Craig L, Deary IJ, Demakakos P, Ebrahim S, Gallacher J, Gow AJ, Gunnel D, Haas S, Hemmingsson T, Inskip H, Jang S, Noronha K, Osler M, Palloni A, Rasmussen F, Santos-Eggimann B, Spagnoli J, Starr J, Steptoe A, Syddall H, Tynelius P, Weir D, Whalley LJ, Zunzunegui MV, Ben-Shlomo Y, Hardy R, on behalf of the HALCyon study team. Childhood socioeconomic position and objectively measured physical capability levels in adulthood: a systematic review and meta-analysis. *PLoS One.* 2011;6(1):e15564. Available at: www.plosone.org [Accessed August 26, 2013].

Charreire H, Weber, C, Chaix B, Salze P, Casey R, Banos A, Badariotti D, Kesse-Guyot E, Hercberg S, Chantal Simon C, Oppert J-M. Identifying built environmental patterns using cluster analysis and GIS: relationships with walking, cycling and body mass index in French adults. *Int J Behav Nutr Phys Act.* 2012, 9:59. doi:10.1186/1479-5868-9-59 [Accessed August 26, 2013].

Clement S, Ibrahim S, Crichton N, Wolf M, Rowlands G. Complex interventions to improve the health of people with limited literacy: a systematic review. *Patient Educ Couns.* 2009 Jun;75(3):340-351. doi: 10.1016/j.pec.2009.01.008. Epub 2009 Mar 3.

Cooper R, Kuh D, Cooper C, Gale CR, Lawlor DA, Matthews F, Hardy R, the FALCON and HALCyon Study Teams. Objective measures of physical capability and subsequent health: a systematic review. *Age Ageing.* 2011;40:14-23. doi: 10.1093/ageing/afq117 [Accessed August 26, 2013].

DeWalt DA, Hink A. Health literacy and child health outcomes: a systematic review of the literature. *Pediatrics.* 2009 Nov;124 (Suppl 3):S265-S274. doi: 10.1542/peds.2009-1162B.

Dubbert PM, Robinson JC, Sung JH, Ainsworth BE, Wyatt SB, Carithers T, Newton R Jr, Rhudy JL, Barbour K, Sternfeld B, Taylor H Jr. Physical activity and obesity in African Americans: the Jackson Heart Study. *Ethn Dis.* 2010;20(4):383-389.

El-Sayed AM, Scarborough P, Galea S. Unevenly distributed: a systematic review of the health literature about socioeconomic inequalities in adult obesity in the United Kingdom. *BMC Public Health.* 2012 Jan 9;12:8. doi: 10.1186/1471-2458-12-18.

Felitti VJ, Anda RF, Nordenberg D, Williamson DF, Spitz AM, Edwards V, Koss MP, Marks JS. Relationship of childhood abuse and household dysfunction to many of the leading causes of death in adults. The Adverse Childhood Experiences (ACE) Study. *Am J Prev Med.* 1998 May;14(4):245-258.

Kania J, Kramer M. Collective impact. *Stanford Social Innovation Review.* Winter 2011:36-41.

Michael M, Allen J, Goldblatt P, Boyce T, McNeish D, Grady M, Geddes I. *Fair Society, Healthy Lives: The Marmot Review.* London: UCL; 2010.

Sanders LM, Federico S, Klass P, Abrams MA, Dreyer B. Literacy and child health: a systematic review. *Arch Pediatr Adolesc Med.* 2009 Feb;163(2):131-140. doi: 10.1001/archpediatrics.2008.539.

Strand BH, Cooper R, Hardy R, Kuh D, Guralnik J. Lifelong socioeconomic position and physical performance in midlife: results from the British 1946 birth cohort. *Eur J Epidemiol.* 2011 June;26(6):475-483. doi: 10.1007/s10654-011-9562-9 [Accessed August 26, 2013].

Tremblay MS, LeBlanc AG, Kho ME, Saunders TJ, Larouche R, Colley RC, Goldfield G, Connor Gorber S. Systematic review of sedentary behaviour and health indicators in school-aged children and youth. *Int J Behav Nutr Phys Act.* 2011 Sep 21;8:98. doi: 10.1186/1479-5868-8-98.

Additional Resource

Motivational Interviewing, www.motivationalinterviewing.org [Accessed August 14, 2013].

II

COMMON CHRONIC CONDITIONS AND COMORBIDITIES

For this fourth edition of this book, we have taken a new approach to the most common conditions and comorbidities (chapter 5). In large part this is because the vast majority of research focuses on one condition at a time, but that's not how patients present in the clinic. Many patients have several problems, and the most common ones occur in many combinations. Accordingly, both physicians and exercise specialists need to think of the exercise program in multiple contexts with a variable mixture of condition-specific risks and benefits. Sadly, there is not much research from this perspective, and this is the main reason we have presented the topic in this fashion. Exercise in persons with multiple chronic conditions is increasingly common, and readers need to know how to deal with these situations.

Another new aspect in this edition is that these common conditions are clustered by their most significant lifestyle attribute—inactivity (chapter 6) or tobacco use (mainly smoking) (chapter 7). The purpose is to emphasize the huge roles that physical inactivity and tobacco have in the pathogenesis of these chronic conditions. Discussions about exercise in a particular condition, without mention of the broader and exacerbating consequences, does not adequately address most people's life situation. Exercise is not an isolated aspect of life, and a well-implemented program accounts for other issues in patients' lives that affect their health.

Cancer and mental wellness are considered separately, though these too are associated with inactivity and smoking. Cancer has its own chapter (chapter 8) due to the complex range of manifestations, whether from the primary illness or secondary effects or as a result of medical and surgical treatment.

Depression is discussed in chapter 9 mainly as a comorbidity of the common chronic conditions (as compared to a primary mental health diagnosis). This is of note because a majority of chronic disease patients exhibit some element of affective disorder, but generally not depression, which is addressed as a primary problem in part VII.

Amputation and frailty are also included in chapter 9, as sequelae related to chronic disease. Previous versions of this textbook have categorized amputation as a neuromusculoskeletal disability, but ever since the second edition we have not been satisfied with that positioning. While having an amputation is having a disability, there really are two kinds of persons with amputation—those who had traumatic amputations and those with vascular insufficiency. People with a traumatic amputation tend to be young and otherwise healthy, while the vast majority of those with amputation for vascular reasons have diabetes. So most individuals with traumatic amputation do quite well with adaptive exercise, with the use of prosthetics or wheelchair activities. Such individuals are very different from someone with diabetes who has, in effect, end-stage target organ disease of the circulation. This is why the focus in the chapter is on lower-limb amputation.

Lower-limb amputations due to military activity deserve special note here. As a consequence of recent military conflicts around the world, there are many people—soldiers and civilians—who have traumatic amputations sustained in combat or by accidental triggering of a mine. If such an individual is fortunate enough not to have been harmed in another way, he usually fits into the otherwise healthy amputation grouping. If the individual was not that fortunate, then she probably sustained multiple life-threatening wounds involving many organ systems. The editors felt that such persons don't fit into either of the two amputation categories; readers will need to refer to the multiple relevant chapters that address the given person's conditions.

Finally, we included frailty in chapter 9 in this part of the book because many people with the conditions discussed end up becoming frail, and many people who have multiple comorbidities discussed in part II are elderly. One might include frailty in part V, as a chronic metabolic condition, but on balance the editors elected to include it in part II. Frailty is likely to become a more prevalent problem as societies age, and persons who have multiple chronic conditions often end up being frail, so we felt that this condition fits best here.

Approach to the Common Chronic Conditions

The purpose of this chapter is to summarize the implications of issues introduced in the chapters of part I. Much about exercise management in persons with chronic health conditions is in an indeterminate state of understanding—a true gray zone—and this is especially true for persons with multiple chronic conditions. For most such individuals, there really is no substantial evidence base. If one wanted to find exercise studies in which the inclusion or exclusion criteria exactly matched a particular patient or client with her own unique combination of conditions and treatments, there is often no evidence base at all. Accordingly, it is extremely important to acknowledge that such a program is improvisational and that this is the best we have to offer right now, and the chronic care team (especially the clinical exercise specialist) must have a systematic method for approaching and solving these situations. Each team will develop their own approaches and algorithms, and this chapter simply provides suggestions on how to approach the problem.

Nature of Multiple Conditions and Related Comorbidities

During the last 25 to 40 years, the increase in noncommunicable diseases (particularly cardiometabolic conditions) in both developed and developing societies is believed to be related to changes in diet and physical activity levels. An aging population underlies some of the increased prevalence of cardiovascular conditions, but—independent of age—obesity, cardiovascular disease, and type 2 diabetes have become highly prevalent worldwide. In the United States, data from the Centers for Disease Control and Prevention show that about two-thirds of the population is overweight and one-third is obese by body mass

index criteria; cardiovascular disease remains a leading cause of disability and death; and type 2 diabetes will soon affect more than 10% of American adults. Also, in part because of aging and in part because of changes in habitual physical activity levels, musculoskeletal pain in the joints and the lumbar spine is the most common reason for visits to a doctor's office.

These trends have resulted in the phenomenon that a growing group of people have various combinations of a few common chronic conditions. This presents a challenge in exercise management since very few studies are aimed at combinations of conditions, mainly because this increases the complexity and difficulty of clinical research. The presence of multiple chronic conditions typically excludes people from clinical trials and also makes it less likely that a subject will complete a longitudinal study. Intercurrent illness often complicates the ability to follow the study protocol, and it is common for such subjects to just drop out because of complexities in life (such as transportation) that are often brought about by their ill health. Researchers are reluctant to include subjects with multiple conditions, and as a consequence, good objective outcome evidence is lacking in this population. Nonetheless, many such persons come to the clinic or gym or rehabilitation facility in search of help, and appropriately prescribed exercise will be beneficial for almost all of them. The challenge is to effectively interpret their situation and successfully implement a program.

General Recommendations for Exercise

For persons with common and multiple chronic conditions, the overriding objective is to promote a physically active lifestyle and the most fitting exercise program that will successfully attain this goal. Chapters 1 and 2 clarify the basis for this goal.

Exercise professionals must be aware of the individual patient's or client's needs and limitations, in light of scientifically guided exercise principles, and use their experience to adapt and implement a safe program that optimizes health and fitness benefits. See chapters 2 through 4 for further information on exercise management for persons with a chronic condition. These are major aspects to keep in mind:

- Goals of the individual and the objective of increasing his activity level
- Careful attention to the contraindications and limitations
- Variability in exercise responses and adaptations to training
- Modification of exercise prescriptions in accordance with the observed exercise responses
- Flexible interpretation of guidelines (as described in chapters 2 and 3)

Recommendations for Exercise Assessment

Chapters 2 and 3 provide a deeper discussion of how exercise testing is used as a part of medical management. Diagnostic exercise testing (e.g., Bruce treadmill protocol with 12-lead electrocardiogram [ECG] for establishing a diagnosis of coronary artery disease) has a different purpose than functional testing (e.g., 30-s sit-to-stand or other performance test that involves an action necessary for activities of daily living). Health professionals need to understand how specific exercise tests can be used for diagnostic purposes and for exercise management, recognizing that essentially all exercise tests contribute important information that is useful in exercise management.

In most situations, there is little need to compare a patient or client to norms on standardized protocols. Most people with the common chronic conditions discussed in this chapter have mild severity with minimal impact on their functional ability to do exercise or physical work. There are, however, many persons who are severely affected by their chronic conditions and have low physical functioning and a deconditioned state. Diagnostic exercise testing is recommended for many of the conditions in this chapter, but for people who have

low physical functioning and who are advised to start their program *at their usual level of activities of daily living* (ADLs), the Basic *CDD4* Recommendation is to proceed with a functional exercise trial without referral for diagnostic exercise testing. This protocol is especially applicable to the individuals most severely affected by chronic conditions, for the following reasons:

- Most exercise tests in this population are terminated due to peripheral limitations at well below 85% of predicted maximal heart rate, reducing the diagnostic utility of a 12-lead ECG for establishing cardiac ischemia (see chapter 2).
- Many conditions include pathophysiology that limit the ability to interpret the 12-lead ECG (i.e., chronic left ventricular hypertrophy, electrolyte abnormalities).
- Most people in this situation will not be prescribed vigorous activity—the goal is typically to increase low-level physical activity, at least as the starting point.

Preprogram exercise testing presents a significant time and cost burden to the individual and often presents a barrier to participation. A good physical examination and functional assessments are usually adequate to get someone started and to provide objective baseline measures.

For many people, the *CDD4* standard recommended strength and physical performance functional assessment tests are appropriate as a starting point (see chapters 2 and 3). For those who can't complete these recommended exercises, they serve as a good goal as a minimum level of physical functioning. Each therapist has his own favorite set of assessments, but commonly used tests include the following:

- Gait speed, or timed get-up-and-go
- 30 s sit-to-stand test, or chair stand 10 test
- Stair climb or some functional test pertinent to the individual's life situation
- A functional assessment of flexibility (e.g., chair sit and reach)

A variety of other measures can be useful components of physical performance assessment, and an exercise specialist or therapist may prefer to use the assessments listed next (or others not listed), noting the presence or absence of fatigue or discomfort to help customize training programs:

- Mobility (walking tests such as gait speed, 6-min walk, shuttle walk, 400 m walk)
- Lower-extremity strength (chair stand, stair climb–descent)
- Flexibility–range of motion (sit and reach, trunk, back, and leg flexibility)
- Balance (functional reach, tandem walk)
- Combination of tests (get up-and-go, bend and lift tests, Short Physical Performance Battery)

These are relative contraindications to physical performance testing:

- Extreme pain
- Risk of joint pain exacerbation
- Patient or client reluctance
- Risk of falls (use a waist belt or other precautionary measures)

In persons who have chronic conditions, it is often difficult to interpret what amount of functional deficit represents deconditioning versus how much is due to their chronic health problem, and it is possible that low physical performance is the result of lack of motivation due to comorbid depression, apathy, or common comorbid medical conditions such as sleep apnea. Common causes of early fatigue and volitional exhaustion include the following:

- Morbid obesity and inactive lifestyle can cause a low power/weight ratio.
- Cardiovascular disease can limit aerobic function or endurance.
- Weakness due to inhibition or pain can lead to falsely low strength or endurance.
- Poor effort can come from depression.
- Neurovascular sequelae, especially of diabetes, can reduce performance.
- The individual can be just very out of shape from a sedentary lifestyle.

Recommendations for Exercise Programming

It is absolutely essential for persons with any chronic condition to be physically active, in part to preserve their health but especially to maintain physical functioning and independence. Mildly afflicted and adherent individuals do well in a home-based exercise program and in seeking to achieve higher levels of performance, but those who are severely afflicted usually benefit from being in a mentored or supervised program. The type of program that is possible will be influenced by insurance coverage and available local resources (places to go, programs to participate in, people who can help), as well as external constraints placed on people by their disease, treatment, or both (e.g., requirement for regular treatments, ability to obtain transportation to facilities). As noted in chapter 1, with many of these issues, having a clinical exercise physiologist as an integral member of the chronic care management team would be extremely helpful, and it is unfortunate that many health care systems do not have clinical exercise physiologists as a part of the medical home.

In many persons with a chronic condition or disability, the factors contributing to exercise intolerance are not very clear at the time of an initial assessment. In a strict academic or scientific sense, this might seem to be a reason for performing a series of formal exercise tests. Health insurance typically does not cover these tests as a benefit, however, and because our goal is less to make a new diagnosis than to characterize how much the existing diagnosis impairs function, it provides better value to constrain formal exercise testing to situations in which the diagnosis is uncertain. In these situations, exercise testing can make the diagnosis (e.g., cardiac ischemia on a graded exercise test with 12-lead ECG), and the predictive value of the test is known. This is not to say that exercise testing does not provide much additional data of value or interest, quite the contrary, but these are data that can be gradually obtained over time during exercise training sessions.

CDD4 Alternative Recommendation: The Functional Exercise Trial

As noted in chapter 2, a paucity of clinical research information focuses on people with multiple chronic conditions that cause functional limitations (e.g., an ex-smoker with obesity, type 2 diabetes, mild chronic obstructive pulmonary disease, angina pectoris, chronic back pain, and knee–hip osteoarthritis). In any case, this information gives little guidance on the level of activity that should be advised for such persons in terms of duration, intensity, and progression. An alternative and logical approach in such situations is to perform a functional exercise trial, described in chapter 2 as part of the Basic *CDD4* Recommendation. People are started slowly with a low-dose, low-intensity bout of exercise at an intensity similar to that of their usual ADLs, with the goal of observing the physical accommodation to this during the session and during recovery over days to weeks. Any adverse effects of exercise (e.g., exertional dyspnea, knee pain, delayed-onset muscle soreness) become apparent and provide clues to how much each of the person's chronic problems contributes to low functional capacity. This approach—a series of low-level exercise sessions—is a functional exercise trial.

A diagnostic exercise test is not required before starting a functional exercise trial because ADL levels of physical activity (e.g., walking) are not high-risk activities. Diagnostic exercise testing is useful in those who are able to perform at a high enough level to achieve diagnostic criteria and who will be participating in higher-intensity exercise, but diagnostic exercise testing is not helpful for diagnostic or safety purposes when the individual is limited to low-level physical activity.

General Solutions for Common Chronic Conditions

One good alternative for many individuals who have a chronic condition is referral to an existing program designed for such persons. Many of these programs are designed to charge the health insurance payer for outpatient services and then seek to transition the individual into a self-pay client of the business. It is common to call these services a transitional, medically integrated program or physician-referred program or some variant thereof. It is important to note that such programs are designed for people who have a chronic condition, not for apparently healthy individuals. Therefore, most of these programs require a physician referral, so it is important that physicians and medical home staff keep these services prominently in mind.

Most such programs receive only a small fraction of the referrals that they should receive based on known prevalence of the common chronic conditions. The consequence is that these programs usually have few clients, referred by only a handful of physicians, and struggle financially. They would likely become much more readily available if physicians would refer appropriate patients more often. Allied health care or medical home staff should not hesitate to remind the physician that such a referral would be good for the person involved, and it is important that physicians provide their blessing and support. Fitness facilities that offer such services benefit from developing a close relationship with both primary care and specialist physicians, because patients of these practices become the clients of their programs. These programs provide the foundation of exercise services for persons with a chronic disease or disability, so without close collaboration between physicians and gyms, people who need medically directed exercise end up with nothing.

Top Program Choices

In populations of individuals with a cardiometabolic condition or a neuromusculoskeletal disability, cardiovascular disease is the number one cause of death. Accordingly, anyone with any such chronic condition is a good candidate for a multifactorial lifestyle risk-reduction intervention. Cardiac rehabilitation programs are an example of such programs, although cardiac rehabilitation is commonly misunderstood by clinicians. In the United States, one problem is that it is a covered benefit for only a limited number of diagnoses, and because of operational restrictions these sessions are usually delivered with a suboptimal balance of content.

When cardiac rehabilitation is not available for an individual, then a medically referred (i.e., a provider makes a referral to the program) multifactorial risk-reduction program is an appropriate choice. Some cardiac rehabilitation programs have a transitional component that bridges between cardiac rehabilitation and a gym membership, sometimes called phase III or phase IV, and one of these might be an option. There are also many community-based programs or home programs that operate out of local YMCAs and other facilities such as nursing homes or senior centers. People will need guidance as to what type of program best meets their needs. Likewise, programs may need guidance on the goals for certain individuals.

Community-Based Musculoskeletal Programs

In persons who don't have these first options, and for those who have musculoskeletal pain (back pain, arthritis), often the best place to start is a referral to a good physical therapy program that has staff skilled in multifactorial chronic conditions.

Before starting, one key challenge is to distinguish if the individual's limitations are an acute problem that can rapidly be overcome, or if she has become deconditioned and faces a longer period of rehabilitation:

- Use maneuvers to see if the muscles are strong when the joint doesn't hurt.
- Look for atrophy and muscle wasting that suggests deconditioning.
- Determine whether modified exercise programs (e.g., aquatic therapy) that unload the skeleton can enable exercise with less discomfort.

Aquatic Programs

In persons who have joint pain or overweight or obesity, non–weight-bearing and aquatic exercise programs are a mainstay for the exercise–physical activity aspect of multifactorial risk reduction. These are the major long-term goals for people with overweight-obesity and joint pain:

- Maintaining robust functional capacity, quality of life, and independent living
- Preserving mental wellness and averting progression to long-term disabilities
- Delaying the need for surgery to implant prosthetic joints
- Preventing progression of obesity and physical inactivity into type 2 diabetes
- Preventing heart disease and stroke as causes of disability and death
- Providing a social support to maintain long-term exercise behaviors

Program Options for Frailty and for Seniors Who Are Aging

In both frail and nonfrail seniors, gait speed, balance, and lower-extremity functional strength are good predictors of future disability. It therefore makes sense for an exercise program for persons with multiple chronic conditions (even those who are young) to emphasize these aspects:

- Walking endurance
- Functional lower-extremity strength (e.g., sit to stand, stair climbing)
- Core stability and balance activities, particularly those that simulate ADL challenges

- Graduation to attempting short interval-style efforts to increase gait speed

Persons who are frail or prefrail may have difficulty doing specific exercises; they may need to

- start with just muscular endurance such as sit to stand to facilitate ADL mobility, and
- set a goal of slowing their rate of deterioration.

It is worth noting that many persons with multiple chronic conditions who don't have sarcopenia or weight loss will have three or four of the Fried criteria for frailty (see chapter 9) and should probably be managed as if they were frail.

Nonresponders to Exercise Training

In all people—those who are healthy and those with chronic health problems alike—there are genetic factors that influence the response to exercise training, with some people gaining more benefit from exercise than others. It can thus be expected that some persons with chronic conditions will not respond robustly to exercise training, not because of their underlying illness or deconditioning but because they are intrinsically not good responders to exercise training. This can be disappointing, but it is not sufficient reason to discontinue exercise training: Exercise has beneficial effects on psychosocial function and quality of life, which by themselves justify the program. It is good to remember that in people with chronic conditions, in whom the expected course is further deterioration in physical functioning, an exercise intervention that simply maintains current levels of physical functioning is a positive outcome.

Integration Into a Medical Home Model

The common chronic conditions present a significant burden to society in terms of direct health care expenses and many other economic costs. Much of this societal burden is due to trends in lifestyle, including the increase in being sedentary. Accordingly, as health care evolves to the medical home model (see chapter 1), it is essential that staff in the medical home understand how to make full use of assessments and referrals for exercise management. The medical home must incorporate physical function assessments and provide recommendations and encouragement for increasing physical activity as a part of routine care in order to affect the lifestyle-related burdens of obesity, cardiovascular disease, and type 2 diabetes, as well as other conditions related to physical inactivity.

Suggested Readings

Ainsworth BE, Haskell WL, Herrmann SD, Meckes N, Bassett Jr DR, Tudor-Locke C, Greer JL Vezina J, Whitt-Glover MC, Leon AS. 2011 Compendium of Physical Activities: a second update of codes and MET values. *Med Sci Sports Exerc.* 2011;43:1575-1581.

Artinian NT, Fletcher GF, Mozaffarian D, Kris-Etherton P, Van Horn L, Lichtenstein AH, Kumanyika S, Kraus WE, Fleg JL, Redeker NS, Meininger JC, Banks J, Stuart-Shor EM, Fletcher BJ, Miller TD, Hughes S, Braun LT, Kopin LA, Berra K, Hayman LL, Ewing LJ, Ades PA, Durstine JL, Houston-Miller N, Burke LE. Interventions to promote physical activity and dietary lifestyle changes for cardiovascular risk factor reduction in adults: a scientific statement from the American Heart Association. *Circulation.* 2010;122(4):406-441.

Balady GJ, Ades PA, Comoss P, Limacher M, Pina IL, Southard D, Williams MA, Bazzarre T. Core components of cardiac rehabilitation/secondary prevention programs: a statement for healthcare professionals from the American Heart Association and the American Association of Cardiovascular and Pulmonary Rehabilitation Writing Group. *Circulation.* 2000;102:1069-1073.

Blair SN, Kohl HW III, Paffenbarger RS Jr, Clark DJ, Cooper KH, Gibbons LW. Physical fitness and all-cause mortality. A prospective study of healthy men and women. *JAMA.* 1989;262:2395-2401.

Celli BR. Pulmonary rehabilitation in patients with COPD. *Am J Respir Crit Care Med.* 1995;152:861-864.

Cress ME, Buchner DM, Prohaska T, Rimmer J, Brown M, Macera C, Dipietro L, Chodzko-Zajko W. Best practices for physical activity programs and behavior counseling in older adult populations. *J Aging Phys Act.* 2005;13:61-74.

Franklin BF. Preventing exercise-related cardiovascular events: is a medical examination more urgent for physical activity or inactivity? *Circulation.* 2014;129:1081-1084. doi: 10.1161/CIRCULATIONAHA.114.007641.

Gordon NF, Gulanick M, Costa F, Fletcher G, Franklin BA, Roth EJ, Shephard T. Physical activity and exercise recommendations for stroke survivors: an American Heart Association scientific statement from the Council on Clinical Cardiology, Subcommittee on Exercise, Cardiac Rehabilitation, and Prevention; the Council on Cardiovascular Nursing; the Council on Nutrition, Physical Activity, and Metabolism; and the Stroke Council. *Circulation.* 2004;109:2031-2041.

Hand C, Law M, McColl MA. Occupational therapy interventions for chronic diseases: a scoping review. *Am J Occup Ther.* 2011;65:428-436.

Heran BS, Chen JM, Ebrahim S, Moxham T, Oldridge N, Rees K, Thompson DR, Taylor RS. Exercise-based cardiac rehabilitation for coronary heart disease. *Cochrane Database Syst Rev.* 2011;CD001800.

Hung WW, Ross JS, Boockvar KS, Siu AL. Recent trends in chronic disease, impairment and disability among older adults in the United States. *BMC Geriatr.* 2011;11:47.

Lacasse Y, Goldstein R, Lasserson TJ, Martin S. Pulmonary rehabilitation for chronic obstructive pulmonary disease. *Cochrane Database Syst Rev.* 2006;CD003793.

McDonald CM. Physical activity, health impairments, and disability in neuromuscular disease. *Am J Phys Med Rehabil.* 2002;81:S108-S120.

Norton K, Norton L, Sadgrove D. Position statement on physical activity and exercise intensity terminology. *J Sci Med Sport.* 2010;13:496-502.

Pescatello LS, ed. *ACSM's Guidelines for Exercise Testing and Prescription.* Philadelphia: Lippincott Williams & Wilkins; 2014:42-46.

Ries AL, Kaplan RM, Limberg TM, Prewitt LM. Effects of pulmonary rehabilitation on physiologic and psychosocial outcomes in patients with chronic obstructive pulmonary disease. *Ann Intern Med.* 1995;122:823-832.

Rijken PM, Dekker J. Clinical experience of rehabilitation therapists with chronic diseases: a quantitative approach. *Clin Rehabil.* 1998;12:143-150.

Sawatzky R, Liu-Ambrose T, Miller WC, Marra CA. Physical activity as a mediator of the impact of chronic conditions on quality of life in older adults. *Health Qual Life Outcomes.* 2007;5:68.

Whitfield GP, Gabriel KKP, Rahbar MH, Kohl HW III. Application of the American Heart Association/American College of Sports Medicine Adult Preparticipation Screening Checklist to a nationally representative sample of US adults aged ≥40 years from the National Health and Nutrition Examination Survey 2001 to 2004. *Circulation.* 2014;129:1113-1120. doi: 10.1161/CIRCULATIONAHA.113.004160 [Accessed October 28, 2014].

Chronic Conditions Strongly Associated With Physical Inactivity

t is important to consider how to approach exercise management in individuals with multiple chronic conditions, because there is a large group of people with various combinations of just a few common chronic conditions:

Cardiometabolic

Hypertension

Dyslipidemia

Atherosclerosis

Obesity

Type 2 diabetes

Musculoskeletal

Arthritis

Back pain

Osteoporosis

Amputations

Pulmonary

Chronic obstructive pulmonary disease

Cancer

>200 forms

Mental Health

Depression

Integrative

Frailty

The clustering of conditions by physiological system is the traditional medical manner for categorizing a patient's condition(s); these are largely grouped by medical subspecialty to facilitate the division of physician labor, assigning medical management roles by discipline. For people who have only one or two of these conditions,

that system works pretty well. The disadvantage of this approach is that when individuals start having three, four, or more of these conditions, the care tends to become fragmented and difficult to coordinate.

Historically, sorting conditions by medical specialties has also compartmentalized exercise management, mostly by cardiopulmonary, neuromusculoskeletal, and mental health and cognitive conditions. Diabetes and cancer, too, have typically been approached as constituting their own categories of "special cases" for exercise management. Increasingly, however, individuals with multiple chronic conditions are referred to an exercise specialist for rehabilitation, and the nature of the person's situation transcends these historical approaches.

For the purposes of *CDD4*, we have elected to move away from the traditional approach, and we categorize the common chronic conditions by the two main lifestyle circumstances that are strongly associated with these conditions: physical inactivity and tobacco (primarily cigarettes). In some cases, the association is causal (i.e., tobacco and physical inactivity are strong independent risk factors for almost all of these conditions), and in some cases physical inactivity is a common consequence that must be avoided (e.g., traumatic amputation).

Hypertension and Dyslipidemia

Hypertension and dyslipidemia are separate independent risk factors for coronary artery disease (CAD), and in the United States, 33% of Americans have dyslipidemia and 29% have hypertension. These diseases frequently occur together, however, and the combination speeds the process of atherosclerosis. When both conditions are present, the likelihood of having clinically significant CAD is drastically increased because atherosclerosis develops in areas of arteries that experience high blood pressure when high levels of low-density lipoprotein (LDL) are present. Because of this interrelationship and the role of exercise in mitigating these two conditions, hypertension and dyslipidemia are discussed here as linked CAD risk factors.

Basic Pathophysiology

The basic pathophysiology of these conditions is closely interrelated but is discussed separately to highlight key issues.

Hypertension

Hypertension (HTN) is a very common chronic condition, affecting about one in five people; it is defined as either

- systolic blood pressure (BP) over 139 mmHg or
- diastolic BP over 89 mmHg.

Table 6.1 has a more complete description of the classes of hypertension. High BP increases the risk for nonfatal and fatal cardiovascular disease,

particularly CAD, kidney disease, and stroke. Even BP that is increased but not meeting the criteria for hypertension causes cardiovascular disease.

While hypertension is a common condition, the cause is not very well understood. In more than 95% of hypertension cases, the etiology is unknown and the condition is referred to as *primary* or *essential* or *idiopathic* hypertension. The remaining 5% of cases are *secondary* hypertension, so named because there is an identifiable underlying cause such as sleep apnea, drug-induced or drug-related causes (e.g., chronic corticosteroid therapy), chronic kidney disease, renovascular disease, aldosteronism, Cushing's syndrome, pheochromocytoma, coarctation of the aorta, and hyperthyroidism.

Dyslipidemia

Dyslipidemia—a high level of blood lipids—is a complex set of conditions, but the main concern is a high level of cholesterol. Cholesterol is critically important to every cell in the body, as a constituent of cell membranes that helps regulate their stability. Cholesterol is also an intermediary compound in steroid hormones, but the amount of cholesterol needed for hormones is many orders of magnitude lower than that needed for cell membranes. Cholesterol is synthesized and metabolized in the liver and then transported between the liver and the rest of the body. Choles-

Table 6.1   Classification and Management of Blood Pressure for Adults*

BP classification	SBP mmHg*	DPB mmHg*	Lifestyle modification	INITIAL DRUG THERAPY* Without compelling indications	INITIAL DRUG THERAPY* With compelling indications†
Normal	<120	*and* <80	Encourage if not living a heart-healthy lifestyle	No drugs indicated	Drug(s) added for compelling indications‡
Prehypertension	120-139	*or* 80-89	Yes		
Stage 1 hypertension	140-159	*or* 90-99	Yes	Antihypertensive drug(s) indicated	Additional anti-hypertensive drugs, as needed‡
Stage 2 hypertension	≥160	*or* ≥100	Yes	Antihypertensive drug(s) indicated, and most patients need a two-drug combination†	

BP = blood pressure

DBP = diastolic blood pressure

SBP = systolic blood pressure

* Treatment determined by highest BP category.

† Initial combined therapy should be used cautiously in those at risk for orthostatic hypotension.

‡ Compelling indications include heart failure, status postmyocardial infarction, high coronary artery disease risk, diabetes, chronic kidney disease, and stroke survivor. Chronic kidney disease or diabetes treated to BP goal of <130/80 mmHg.

terol is not soluble in blood plasma and therefore has to be transported in the blood by *lipoproteins.* There are many different lipoproteins, each functioning to guide the cholesterol it carries to the proper metabolic pathway. This is the basic triad of lipoproteins:

- Very low-density lipoprotein (VLDL)
- Low-density lipoprotein (LDL)
- High-density lipoprotein (HDL)

The VLDL fraction carries about 80% triglycerides and 20% cholesterol, while the LDL and HDL mainly carry cholesterol. The total blood cholesterol is the sum of the cholesterol bound to VLDL, LDL, and HDL, and these cholesterol subcomponents are indicated with a -C suffix (e.g., LDL-C). Several subclassifications also exist, including two notable subclasses of LDL:

- Lipoprotein (a)
- Small dense LDL

Cholesterol metabolism is extremely complex, but the most important issue is that elevated total cholesterol and low-density lipoprotein cholesterol (LDL-C) are associated with an increased risk of CAD. Lipoprotein (a), or Lp(a), increases risk of CAD and of developing a thrombus, and small dense LDL also increases CAD risk. In contrast, high-density lipoprotein cholesterol (HDL-C)

Terms Often Used to Refer to Blood Lipids

- dyslipidemia—elevated triglycerides and cholesterol
- hypercholesterolemia—only cholesterol is elevated
- hypertriglyceridemia—only triglycerides are elevated
- exaggerated postprandial lipemia—prolonged elevation of triglycerides following consumption of dietary fat
- hyperlipoproteinemia or dyslipoproteinemia—high lipoprotein concentrations from genetic abnormalities or an underlying condition such as diabetes, renal disease, hypothyroidism, biliary obstruction, or dysproteinemia

decreases CAD risk, through a cardioprotective effect that is partially related to this lipoprotein's role in the reverse cholesterol transport. In reverse transport, the cholesterol in HDL-C is transported to the liver where it is catabolized and excreted as bile.

Management and Medications

Medical management in persons who have the common chronic conditions often involves one or more medications for each condition, such that people with cardiometabolic conditions are often on five to eight prescription medicines. Again, this discussion treats these independently, but readers should expect to see multiple medications in persons in this group.

Hypertension

The U.S. guidelines for medical management of hypertension are drafted by the Joint National Committee on Prevention, Detection, Evaluation, and Treatment of High Blood Pressure (JNC). The most recent JNC 8 report was released in autumn of 2013, and the BP goal remains at ≤140/90 mmHg, with the exception of persons ≥60 years of age who don't have diabetes or chronic kidney disease, for whom JNC 8 recommends ≤150/90 mmHg.

The European Society of Hypertension (ESH) and European Society of Cardiology (ESC) also published new guidelines in 2013, which are quite consistent with the recommendations of JNC 8. The ESH/ESC recommends that all individuals achieve <140 mmHg systolic BP, with exceptions for special populations such as those with diabetes and persons who are elderly. For individuals with diabetes, ESH/ESC recommends <85 mmHg diastolic BP; for older individuals up to 80 years, ESH/ESC supports a relaxation of the guideline to a systolic BP of 140 to 150 mmHg. For people older than 80 years, ESH/ESC notes that achieving a systolic BP ≤140 mmHg is recommended only for those who are fit and healthy.

Both JNC 8 and the ESH/ESC Guidelines state that initial attempts to reach BP goals should be through dietary modification, weight loss, and exercise:

- Losing weight if person is overweight
- Reducing salt intake as much as possible, ideally to 65 mmol/day (corresponding to 1.5 g/day of sodium)
- Eating a diet rich in fruits, vegetables, and low-fat dairy (e.g., the DASH diet, see next paragraph)

- Maintaining daily potassium intake at 120 mmol/day (4.7 g/day)
- Limiting alcohol intake to no more than two drinks a day in most men and one drink a day in women and lighter-weight individuals
- Performing aerobic physical activity (per the ACSM Guidelines—see chapter 2)
- If the individual uses tobacco, stopping use

Lifestyle intervention has a noteworthy role in the management of BP, as exercise and the DASH diet (Dietary Approaches to Stop Hypertension) independently can produce 5 to 10 mmHg improvements in BP. People who are 10 mmHg over goal thus have a very good chance to control their BP with lifestyle, and even those who are up to 20 mmHg above goal have a chance of getting their BP in control with just dietary changes, regular exercise, and modest weight loss. Many people who do require medications could reduce the number or doses (or both) that they have to take if only they persisted with a heart-healthy lifestyle. Unfortunately, many people are primarily taking medications to control BP and do not use lifestyle intervention to maximal benefit for their BP (and

cholesterol). See table 6.2 for a list of commonly used medications to treat hypertension.

For people who do go on antihypertensive medication, the old JNC 7 recommendations were to start with a thiazide diuretic, a β-adrenergic antagonist (β-blocker), or both. The JNC 8 and 2013 ESH/ESH recommended medications are broader:

- Nonblack persons can start with an angiotensin-converting enzyme (ACE) inhibitor, angiotensin receptor blocker (ARB), thiazide diuretic, or calcium channel blocker (CCB).
- Black patients or clients will likely do better with a thiazide diuretic or CCB.
- Persons with chronic kidney disease should take an ACE inhibitor or ARB.

The reason for these changes is that data now show that newer medications also improve outcomes. Among exercise specialists, this is likely to be welcomed because ACE inhibitors and ARBs have an ergomimetic (exercise mimicking), afterload-reducing effect that facilitates cardiac output, whereas β-blockers and diuretics work more by dampening cardiac output, which can

Table 6.2   Commonly Used Medications to Treat Hypertension

Generic name	Class of drug	Mechanism of action
Lisinopril Enalapril Captopril Benazepril Moexipril Perindopril Quinapril Ramipril Trandolapril	Angiotensin-converting enzyme (ACE) inhibitors	Suppress renin–angiotensin system by inhibiting the conversion of angiotensin I to angiotensin II (a potent vasoconstrictor)
Losartan Azilsartan Candesartan Eprosartan Irbesartan Olmesartan Telmisartan Valsartan	Angiotensin II receptor blockers (ARB)	Bind to angiotensin I receptors in vascular smooth muscle to prevent the vasoconstrictor effects of angiotensin II
Aldactone Eplerenone	Aldosterone blockers	Competitively bind to aldosterone receptors, blocking aldosterone Primarily for complex hypertension, such as heart failure, liver or renal disease, nephrotic syndrome, and K^+-wasting conditions

Generic name	Class of drug	Mechanism of action
Aliskiren	Renin inhibitor	Direct renin inhibitor; inhibits conversion of angiotensinogen to angiotensin I
Prazosin Doxazosin Terazosin	α-1-adrenergic receptor blockers	Block α-1 receptors in vascular smooth muscle, causing arteriolar vasodilation Useful in high sympathetic activity (chronic renal disease)
Guanfacine	α-2-adrenergic agonist	Decreases sympathetic drive from central vasomotor center to cardiovascular tissues (cardiac and smooth muscle)
Clonidine Methyldopa	α-Adrenergic antagonists (centrally acting)	Stimulate vasomotor center α-adrenergic receptors and decrease sympathetic outflow to cardiovascular tissues (cardiac and smooth muscle) Useful in high sympathetic activity (chronic renal disease)
Phenoxybenzamine	α-Adrenergic blocker (peripherally acting)	Reduces vascular tone in viscera and skin, reducing peripheral vascular resistance
Propranolol Timolol Nadolol	β-Adrenergic receptor blockers (nonselective)	Inhibit β_1- and β_2-adrenergic receptors, inhibiting inotropic, chronotropic, and vascular responses to β-adrenergic stimulation
Atenolol Metoprolol Acebutolol Betaxolol Bisoprolol Nebivolol	β-Adrenergic receptor blockers (selective)	Cardioselective β_1-adrenergic receptor blocking agents, decreasing inotropic and chronotropic responses to β-adrenergic stimulation
Nifedipine Clevidipine Felodipine Isradipine Nisoldipine	Calcium channel blockers (dihydropyridines)	Inhibit calcium ion influx with membrane depolarization, reducing contractility; also act in smooth muscle as an arterial vasodilator
Diltiazem Verapamil	Calcium channel blockers (nondihydropyridines)	Similar mechanisms to dihydropyridines; also act in atrio-ventricular node to slow conduction/increase refractoriness (for supraventricular tachycardia)
Hydrochlorothiazide Chlorothiazide Methyclothiazide	Thiazide diuretics	Block reabsorption of Na and Cl in proximal tubule
Indapamide	Indoline diuretic	Primarily for treatment of HTN in persons with CHF
Furosemide Bumetanide Torsemide	Loop diuretics	Inhibit renal reabsorption of Na and Cl in proximal and distal tubules and the loop of Henle Usually secondary agents in HTN; more useful for CHF
Chlorthalidone	Monosulfamyl diuretic	Acts on ascending limb of loop of Henle to reduce H_2O reabsorption and increase excretion of Na and Cl
Metolazone	Quinazoline diuretic	Inhibits Na reabsorption at cortical loop and in proximal tubule
Hydralazine	Peripheral vasodilator	Relaxes peripheral arteriolar smooth muscle, likely through inhibiting calcium channels

reduce exercise capacity. Also, for hypertensive persons who have hyperlipidemia, thiazides and β-blockers can increase very low-density cholesterol (VLDL-C) and LDL-C levels.

For people who are more than 20/10 mmHg over their goal, it is likely that they will need two antihypertensive medications in addition to lifestyle intervention, and one of the drugs will likely be a thiazide diuretic. Treatment of hypertension that occurs in combination with certain comorbid conditions (congestive heart failure, chronic renal disease, diabetes) is usually treated with antihypertensive medications that have been proven to improve outcomes in the comorbid conditions (table 6.3).

Notable adverse effects of antihypertension agents:

- β-Blockers (particularly nonselective) can reduce exercise capacity and blunt the chronotropic exercise response.
- β-Blockers or diuretics may impair thermoregulatory function.
- α-Blockers, calcium channel blockers, and arterial vasodilators may cause postexertional hypotension.

Dyslipidemia

Like the recommendations for hypertension, all current guidelines recommend lifestyle interven-tion as the number one line of therapy for persons with dyslipidemia. Lifestyle recommendations for dyslipidemia are very similar to those for hypertension, though they are more specific with regard to dietary fat:

- 25% to 35% of daily calories from fat
- <7% of daily calories from saturated fat
- ≤10% of daily calories from polyunsaturated fat
- <20% of daily calories from monounsaturated fat

Recommendations on when to institute drug therapy are evolving; initial treatment most often is a hydroxymethlyglutaryl coenzyme-A (HMG-CoA) reductase inhibitor (statin) with the primary goal of reducing the LDL-C. Today, most people who take a lipid-lowering agent are on a single statin or a statin with fish oil supplementation, which is added to help reduce triglycerides. Modern statins are quite potent, reducing LDL-C by 20% to 60% and triglycerides by as much as 40% while increasing HDL-C by 6% to 10%. Individuals who have very high risk, such as someone with familial hypercholesterolemia, might take multiple lipid-lowering agents. Because of the high risk for drug interactions and adverse side effects, these individuals are usually managed by a lipid specialist.

Evolving Dyslipidemia Guidelines

In 2013 (as this book was being drafted), the American College of Cardiology (ACC) and the American Heart Association (AHA) released a joint ACC/AHA Guideline on the Treatment of Blood Cholesterol to Reduce Atherosclerotic Cardiovascular Risk in Adults. The existing guideline, by the Adult Treatment Panel III (ATP III), had been issued by the U.S. National Heart, Lung and Blood Institute (NHLBI) in 2001 and updated in 2004. The NHLBI appointed a group to draft ATP IV, but stopped the project. The ACC/AHA went forward with some of the members of the NHLBI writing group. The approaches taken in ATP III and the ACC/AHA Guidelines are quite different, and are a source of controversy because there is not a consensus of opinion on the best way to medically manage cholesterol. Until this is resolved, physicians are likely to follow protocols that they judge to work best for their patients.

There is abundant clinical trial and epidemiological evidence to support the notion that the lower the LDL-C, the lower the risk of cardiovas-

Table 6.3   Recommended Medications for Hypertension With Common Cardiovascular Comorbidities

Comorbid condition	Recommended antihypertension agents
Heart failure	Diuretics, β-blockers, angiotensin converting enzyme inhibitors, aldosterone antagonists
Status postmyocardial infarction	β-Blockers, ACE inhibitors, angiotensin receptor blockers, aldosterone antagonists
High risk for coronary artery disease	Diuretics, β-blockers, ACE inhibitors, angiotensin receptor blockers, calcium channel blockers
Stroke survivor	Diuretics, ACE inhibitors, angiotensin receptor blockers
Chronic kidney disease	ACE inhibitors, angiotensin receptor blockers

cular mortality. Adult Treatment Panel III, based on the totality of the data, established somewhat arbitrary thresholds as goals of treatment. Further, the ATP III approach was—depending on the traditional Framingham-based categorical risk factors (family history, personal history of coronary heart disease or diabetes, sex, age, smoking, high BP, and cholesterol levels)—for people to use lifestyle and pharmacotherapy to achieve an LDL-C under the following thresholds:

<160 mg/dL (4.1 mmol/L)—<2 traditional risk factors

<130 mg/dL (3.4 mmol/L)—2 or more traditional risk factors, but no clinical disease

<100 mg/dL (2.6 mmol/L)—presence of atherosclerotic disease or equivalent

<70 mg/dL (1.8 mmol/L)—diabetes with heart disease (optional goal in 2004 update)

There are, however, no randomized controlled clinical trials to substantiate that medical management into one of these treatment groups reduces morbidity and mortality in comparison to the group(s) that are above it.

The ACC/AHA authors noted that this approach of establishing "categorical thresholds" of a continuous variable (LDL-C) has not been tested and that it has a variety of pragmatic problems in implementation, and they therefore developed an approach that would reduce the complexity of titrating medication doses to achieve a particular LDL-C blood level. The ACC/AHA group proposed a different way, establishing five groups of people based on their need for risk reduction:

1. Individuals who *do not need cholesterol lowering* or

2. Individuals aged 40 to 75 for whom *modest cholesterol lowering* is recommended:

   Presence of diabetes but no known coronary heart disease and also a 10-year risk of coronary heart disease (CHD) mortality <7.5% (as estimated by a new formula)

3. Individuals for whom *intensive cholesterol lowering* is recommended:

   • Known to have CHD, or

   • Have LDL-C >190 mg/dL (e.g., familial hypercholesterolemia), or

   • Aged 40 to 75 with a diagnosis of diabetes and also a 10-year risk of CHD mortality >7.5% (using the new formula)

In this paradigm, physicians should individualize treatment so that persons who are in the modest cholesterol-lowering group can be treated more intensively if it is deemed appropriate, and persons who are in the intensive cholesterol-lowering group can be treated less intensively if they have adverse effects from medications. The aim of high-intensity statin therapy is to achieve ≥50% reduction in LDL-C. The aim of moderate-intensity statin therapy is to achieve 30% to 49% reduction in LDL-C.

Cardiovascular Risk Assessment

The Framingham formula is known to many as the oldest and most robust risk formula, and the ATP III system mainly used the traditional Framingham risk factors to categorize individuals as having >2 risk factors, 1 or 2 risk factors, or 0 risk factors. In the new ACC/AHA Guidelines, rather than using the number of risk factors to create categories of risk, the calculations in the formula were updated to assess the risk for an initial cardiovascular event. The reason for this was that the population used for the Framingham formula is not representative of the United States. The ACC/AHA therefore created a new formula, using more racially and geographically diverse study populations:

• Framingham Heart Study (FHS)

• Atherosclerosis Risk in Communities (ARIC)

• Coronary Artery Risk Development in Young Adults (CARDIA)

• Cardiovascular Health Study (CHS)

The new equation is believed to better predict the risk of cardiovascular disease and stroke as a 10-year CHD risk for non-Hispanic Caucasian as well as African American women and men. Unfortunately, the new formula has been tested only against population databases that had only 4 years of follow-up at the time the formula was published. Strengths of this equation are that it uses age, sex, race, total cholesterol, HDL-C, systolic BP, whether individual is being treated for BP, diabetes, and smoking. The weakness of this score (also of the Framingham formula) is that this particular risk assessment model does not consider

• physical activity and exercise habits,

• diet,

• family history of premature CHD,

• elevated triglycerides (notably postprandial triglycerides),

- waist circumference,
- body mass index, and
- history of smoking.

In summary, the strength of the ATP III approach is that it took in a comprehensive set of databases, including epidemiological studies that gave strength of association but that could not show causality. It is noteworthy that while these databases are now old, there have not been studies refuting the validity of the ATP III regression relationship between LDL-C and cardiovascular mortality. Indeed, the new ACC/AHA recommendation uses ATP III thresholds (70 mg/dL and 190 mg/dL). These are the weaknesses of the ATP III model:

- The grouping thresholds (70, 100, and 130 mg/dL) are somewhat arbitrary.
- Costs and complexities are associated with titrating medications to a target LDL-C.
- No randomized trials validate the model.

The strength of the new ACC/AHA approach is that many lipid-lowering pharmaceutical trials with HMG coenzyme-A reductase inhibitors (statins) can support the concept of modest versus intense cholesterol lowering. These are the weaknesses of the ACC/AHA approach:

- It heavily favors randomized controlled trials.
- It discounts powerful epidemiological studies.
- It uses a new and unproven risk assessment formula.
- This formula does not include lifestyle issues (most notably physical activity) in risk calculations.

In addition, one unresolved issue regarding the risk assessment is that many groups have regarded a 20% 10-year risk of cardiovascular mortality as "high risk," whereas the ACC/AHA reduced that to 7.5%. The ACC/AHA definition of "high risk" has the strength of being statistically based, and the prior threshold of 20% risk really is quite high. It seems likely that the new definition of high risk will prevail, but this will likely take some time.

Thus, the choice physicians face in choosing a paradigm for medical management of cholesterol is whether they prefer to use the unproven threshold model of ATP III, the unproven risk assessment method of the ACC/AHA method, or some combination thereof. The definition of "high risk" seems to remain an unresolved matter of opinion that, at this point, physicians must individualize for each patient. The net effect is that until these dilemmas are resolved, there is likely to be much variability in the medical management of cholesterol.

Most antihypertensive and lipid-lowering medications (table 6.4) can be taken together with minimal side effects, but statins have some notable side effects (especially when combined with fibric acid derivatives):

- Fibric acid derivatives and statins can cause muscle weakness or myalgia (especially when used in combination).
- Niacin, cyclosporine, macrolide antibiotics, and azole derivatives can also exacerbate rhabdomyolysis when combined with a statin.
- Grapefruit and grapefruit juice interfere with hepatic metabolism of most statins and should be avoided.
- Statins exacerbate postexercise muscle soreness and exertional rhabdomyolysis.

Effects on the Exercise Response

The effects that the conditions in this chapter have on the exercise response can vary greatly, depending on the severity of the condition as well as the magnitude of damage on target organs. Some individuals show minimal effects and have a very normal exercise response, whereas others have a dramatically altered response (especially if they are on certain medications such as a β-adrenergic antagonist).

Hypertension

Blood pressure increases during exercise; the normal pressor response in young healthy persons is an increase in systolic BP of about 10 mmHg for every MET (metabolic equivalent) increase in exercise work rate (see chapter 2 for discussion of METs). In most persons who have hypertension, the pressor response will rise faster than normal, and the maximum systolic BP attained during exercise can be extremely high (often much greater than 250 mmHg in persons with chronic renal disease). The diastolic BP should remain constant or go slightly down during exercise, despite an increasing work rate. But many people with hypertension show a modest increase in diastolic BP.

Table 6.4   **Commonly Used Medications to Lower Blood Lipids**

Generic name	Class of drug	Mechanism of action
Cholestyramine Colestipol Colesevelam	Bile acid sequestrants	Remove bile from enterohepatic circulation, requiring increased turnover of cholesterol to make bile
Ezetimibe	Cholesterol absorption inhibitor	Inhibits absorption of cholesterol at small intestine brush border
Lovastatin Pravastatin Fluvastatin Simvastatin Atorvastatin Pitavastatin Rosuvastatin	HMG-CoA reductase inhibitors	Inhibit enzymatic activity on hydroxymethlyglutaryl coenzyme-A (HMG Co-A), which is reduced to form mevalonate in the key rate-limiting step of cholesterol synthesis
Fenofibrate Gemfibrozil	Fibric acid derivatives	Activate peroxisome proliferator-activated receptor α (PPAR-α) to activate lipoprotein lipase and increase lipolysis of triglyceride-rich very low density lipoproteins as well as increase synthesis of high-density lipoproteins
Niacin	Nicotinic acid	Mechanism unknown; raises HDL
Omega-3 fatty acids	Fish oils	Decrease triglycerides

Immediately following 30 to 45 min of moderate dynamic exercise, most people experience about a 10 mmHg reduction in systolic BP that lasts for several hours, with some people getting as much as a 20 mmHg reduction. It is worth noting that these acute postexercise effects of exercise do not markedly affect the pressor response during a subsequent bout of exercise. There is also essentially no long-term adaptation; this acute postexercise reduction in BP occurs every time after a bout of exercise.

Dyslipidemia

In contrast to hypertension, dyslipidemia generally does not alter the exercise response. Dyslipidemia can affect the exercise response in rare congenital cases, where blood lipids are so elevated as to alter the viscosity of blood and impair blood flow (e.g., familial hypertriglyceridemia). Triglycerides improve after a single exercise session of modest duration and intensity, but cholesterol is relatively unaffected. The amount of exercise needed to affect the LDL-C or HDL-C with a single acute bout of exercise is sufficiently large that most people get most of their cholesterol-lowering effects from medication.

Effects of Exercise Training

Exercise training benefits both hypertension and dyslipidemia. Endurance exercise training reduces the systolic pressor response (the rate at which BP increases in response to physical work), both in individuals with hypertension and for those at increased risk for developing hypertension. Aerobic exercise training causes an average reduction of roughly 5 to nearly 10 mmHg in resting BP (systolic and diastolic) in persons with stage I and stage II hypertension. Physiologically, the mechanisms that cause reductions in BP experienced by persons with hypertension are not fully understood, but potential mechanisms include the following:

- Alteration in renal function
- Amelioration of hyperinsulinemia
- Decrease in plasma norepinephrine
- Increase in circulating vasodilator substances

People with hyperlipidemia also experience modest but beneficial effects of aerobic exercise training on blood lipids, typically on the order of 5% to 10% decrease in LDL-C and increase in

HDL-C. Notable benefits of exercise on blood lipids include these:

- Decreased concentrations of small dense LDL particles
- Increased number of larger-sized LDL particles
- Usually a higher HDL-C concentration
- Lower triglyceride concentrations
- Reduced postprandial lipemia

Increases in lipoprotein enzyme activity are linked to these changes, including increases in lipoprotein lipase, lecithin-cholesterol acyltransferase, and cholesteryl ester transfer protein.

Recommendations for Exercise Programming

Exercise training recommendations for dyslipidemia and hypertension are similar. Both hypertension and dyslipidemia benefit from increasing total energy expenditure or weekly exercise volume, with greater improvements in BP and blood lipids after greater volume of exercise training. Benefits will likely be even greater if the person loses weight during the exercise training program. Thus, if the goal is to maximize benefits on BP and blood lipids, it is advisable to accumulate more than the Basic *CDD4* Recommendation of 150 min/week.

In persons who have only hypertension, dyslipidemia, or the combination, with no other chronic conditions,

- follow the ACSM Guidelines, and
- encourage 150 to 300 min/week of moderate to vigorous physical activity.

In persons who have hypertension or dyslipidemia along with other chronic conditions that have a severe disease burden or disability,

- use the modified *CDD4* Recommendation of
- 150 to 300 min/week of self-paced physical activity.

Individuals who have a comorbidity but generally normal physical functioning will require clinical judgment by the health care team. It is not unreasonable to start someone on a supervised light-intensity walking program to observe the chronotropic and pressor responses as well as any signs or symptoms of cardiovascular instability. Clients can gradually increase their exercise

Tips for Exercise Training With Hypertension or Dyslipidemia

- Interactions between hypertensive or hyperlipidemic medications and exercise training are not well studied.
- If xanthomas are present, biomechanical problems may arise, so proceed cautiously.
- Be aware of potentially undiagnosed metabolic syndrome, diabetes, or renal disease.
- Persons with uncontrolled severe hypertension (i.e., resting systolic BP ≥180 mmHg, diastolic BP ≥110 mmHg, or both) should do only light-intensity activities until their BP has been controlled with medication.
- Resting systolic BP ≥200 mmHg or diastolic BP ≥110 mmHg is a relative contraindication to exercise (especially vigorous exercise).
- Systolic BP ≥250 mmHg or diastolic BP ≥115 mmHg is an absolute contraindication, and exercise should be stopped.
- The Valsalva maneuver, which can dramatically increase BP, should be avoided.

intensity, but the physician should have a low threshold for obtaining a graded exercise test with 12-lead electrocardiogram (ECG).

Overweight, Obesity, Prediabetes, and Type 2 Diabetes Mellitus

Overweight, obesity, and type 2 diabetes mellitus are traditionally separated into different categories, but for the purposes of this text we consider them in tandem. Certainly one can be overweight or obese but not have diabetes, and some people with type 2 diabetes have a normal body weight. In the last two decades, however, the prevalence of persons who are obese and physically inactive has increased dramatically, and with this has come an increase in the prevalence of type 2 diabetes due to the strong relationship between inactivity, obesity, and diabetes. Additionally, compelling data support the notion that, in and of itself, being heavy causes only about a 10% increase in cardiovascular risk after adjusting for physical fitness. However, it is known that ~10% of individuals with prediabetes convert to type 2 diabetes mellitus every year and that exercise and modest weight loss are more effective than medicine to prevent this, especially in people who are older. Accordingly, for *CDD4* we have elected to consider obesity and type 2 diabetes in tandem, as linked metabolic conditions.

It is worth noting that type 2 diabetes occurs in normal-weight individuals. Before the obesity pandemic of the last three decades, about 1 in 4 persons with type 2 diabetes had a normal body mass index (BMI). Today, it's more like 1 in 20 persons, likely due to an increase in the prevalence of obesity and physical inactivity. There are likely differences in the pathogenesis of diabetes between persons who are obese and normal-weight individuals, but the nature of these differences remains poorly understood. For purposes of this text, the main differences between normal-weight individuals and persons with obesity who have diabetes are biomechanical–thermal considerations and the issue of weight loss.

Basic Pathophysiology

As with hypertension and blood lipid disorders, overweight, obesity, prediabetes, and type 2 diabetes are closely interrelated as to pathophysiology. They are considered independently here, but the reader should realize that there is great overlap in the biology of these conditions.

Obesity Alone

Obesity is often oversimplified, and many people overestimate its role as an *individual* cardiovascular risk factor. With regard to population health, obesity is a very serious problem because it has become highly prevalent and is often a precursor to the serious chronic conditions discussed in this chapter.

One can be both heavy and healthy, particularly if physical fitness is high. Simplifying obesity to a BMI understates the differences between morphologic body types, as it has been well shown that an android obese body habitus is associated with much higher cardiovascular risk than does a gynoid obese body habitus. Weight and body habitus have some direct consequences with regard to performing exercise, but in the absence of other chronic conditions, biomechanical and thermal aspects of exercise are the main obesity-related issues. Accordingly, the primary considerations about exercise in a person with obesity should mainly be their other chronic conditions.

Obesity With Prediabetes or Type 2 Diabetes

Obesity with type 2 diabetes is a far more complicated situation, in part since the two are interrelated and in part because the presence of diabetes dramatically increases cardiovascular risk and other long-term sequelae of diabetes such as neuropathy, nephropathy, retinopathy, and cerebrovascular disease with cognitive impairment. Combined with the increased risk of musculoskeletal complications of obesity, such as osteoarthritis of the knees, hips, and spine, someone with obesity and type 2 diabetes is at very high risk for developing multiple chronic conditions with disability and a downward spiral of deconditioning and worsening chronic conditions.

The pathophysiology of type 2 diabetes is complex, involving the following factors:

- Decreased insulin sensitivity in target tissues (especially muscle and fat cells)
- High fasting blood glucose or an elevated postprandial glucose or both
- Altered gastrointestinal signaling to decrease insulin production and affect satiety
- Activation of proinflammatory pathways that promote atherosclerosis
- Nonalcoholic fatty liver disease
- Ultimately, a reduced pancreatic ability to produce insulin
- Increased glycosylation of body tissues and proteins, including hemoglobin
- Accelerated microvascular disease in the brain and nerves
- Vascular dysfunction and stiffening

Many diabetologists now believe that an increased fasting blood glucose involves a reduced ability to secrete insulin in the fasting state and that a high postprandial glucose involves insulin resistance in target cells. Persons with prediabetes or diabetes can have either or both of these responses, meaning that there are three kinds of physiological patterns of blood glucose in prediabetes-diabetes:

- Increased fasting glucose
- Increased postprandial glucose
- Both increased fasting and postprandial glucose

These are the current American Diabetes Association criteria for diabetes (at least two separate measurements):

- Fasting glucose >125 mg/dL (>6.9 mmol/L), or
- Two-hour glucose tolerance test >200 (>11.1 mmol/L), or
- Glycosylated hemoglobin (Hb A1c) >6.5%

These are the current American Diabetes Association criteria for prediabetes:

- Fasting glucose of 100 to 125 mg/dL (5.6-6.9 mmol/L), or
- Two-hour glucose tolerance test of 140 to 200 (7.8-11.0 mmol/L), or
- Glycosylated hemoglobin (Hb A1c) of 5.7% to 6.5%

Management and Medications

Although genetics is a key factor in development of diabetes, the vast majority of cases of type 2 diabetes are strongly related to the modifiable risk factors of overweight-obesity and inadequate physical activity (sedentary lifestyle). Thus, the primary recommendation for preventing and delaying progression of diabetes is the combina-

tion of weight loss and increased physical activity. It is also known to be very important to have persons with diabetes participate in diabetes education programs, though in the United States less than 1 in 13 people with diabetes receive such training.

The medical management of type 2 diabetes is becoming increasingly complex as more is known about the condition and more varieties of medications become available to treat it (table 6.5). Because the trajectory of type 2 diabetes is primarily insulin resistance leading to hyperinsulinemia and eventual islet cell failure with subsequent requirement for exogenous insulin, the general paradigm for diabetes medical management is to start with drugs that increase insulin sensitivity, advancing to drugs that increase insulin levels and then to providing exogenous insulin.

It is easy for health care providers and patients alike to fall into a pathway that focuses on medications to control the diabetes. For both type 2 and type 1 diabetes, however, people need to spend quite a bit of emotional and cognitive energy on managing their diet. One tendency that many develop is thinking that diet management is about counting carbohydrate calories, commonly called counting carbs, but diet self-management in diabetes is quite a bit more complex than that and can be quite difficult to master. The activities in life that involve food are extensive, with many subconscious stimuli and processes based on the full socioecological realm for each individual (see chapter 4).

Table 6.5 Medications Used to Treat Diabetes

Generic drug	Class of drug	Mechanism of action
Alogliptin Linagliptin Saxagliptin Sitagliptin	Dipeptidyl peptidase-4 inhibitors	Slow degradation of incretin hormones
Acarbose	α-Glucosidase inhibitor	Inhibits amylase and glucosidase enzymes
Lantus Levemir	Very slow-release (basal) insulin	Provides "floor" of very long-acting insulin
Apidra Humalog Humulin Novolin Novolog	Insulins of various durations of activity	Fine-tuning of insulin levels via injection and half-life of particular formulation
Insulin pump	Insulin	Short-acting insulin; primarily used in type 1 diabetes

Participants in diabetes self-management education often discover that stress-coping skills are central to diabetes control, not only to avoid blood glucose spikes due to fight-or-flight responses, but also to help regulate counterproductive eating behaviors that can be deeply ingrained and provoked by an enormous number of things in the individual's life (including the marketing of food products and cultural influences). Failing to address these socioecological aspects of diabetes makes the medical management substantially more difficult, since it is likely that medications will be used in escalating doses in order to counteract lifestyle (including lack of physical activity). For this reason, persons with diabetes should be strongly encouraged to attend diabetes education and diabetes self-management training.

Surgical Treatment of Type 2 Diabetes

In persons who have diabetes with severe (BMI >35) or morbid obesity (BMI >40) and who have diabetes that is not well controlled and is resulting in progressive renal, cardiovascular, and retinal or neurological complications, bariatric surgery is often the treatment of choice. Bariatric surgery, especially a Roux-en-Y gastric bypass, typically causes a rapid resolution of diabetes within a week. This phenomenon is sustained for a prolonged period of time (usually years), but studies suggest that it is probably not permanent for many people who undergo bariatric surgery.

Effects on the Exercise Response

The most common issues in exercise with obesity are related to biomechanics and the extra work of carrying more weight:

- Symptoms—few to none
- Effect on exercise response—mechanical and thermal stress increased
- Effect on exercise capacity—decreased aerobic capacity in mL O_2/kg body weight per minute
- Key health sequelae—back and joint pain, hypertension, dyslipidemia, diabetes

Moreover, in most persons with stable and well-managed type 2 diabetes who have normal blood glucose, the exercise response is more affected by the sequelae of diabetes and any unrelated comorbidities than to the diabetes per se. Comorbid conditions related to diabetes that can influence the exercise response can be multiple and complex, so individualization of exercise for anyone with type 2 diabetes should include these considerations:

- Cardiac conditions including prior myocardial infarctions or revascularization
- Silent ischemia (ECG signs of ischemia in the absence of symptoms)
- Other atherosclerotic disease or prior strokes
- Exaggerated pressor response to exercise
- Neuropathy impairing proprioception, motor control
- Retinopathy impairing visual acuity
- Fall risk associated with neuropathy or retinopathy
- Skin concerns or foot trauma, especially in someone with neuropathy

Clinical consequences of these considerations are that persons with diabetes should ideally go through at least a mental (if not actual) checklist before every exercise session. The checklist should always include the following:

- Do they feel "well" that day (or at least well enough to exercise)?
- Have they inspected their feet for sores and blisters?
- Have they checked their blood glucose?

Some people with type 2 diabetes that is in good control through lifestyle (and maybe a modest dose of metformin) may not feel the need to check their glucose before every bout of exercise, but they should at least review whether or not they have varied from their usual medication pattern and consider checking their glucose before starting exercise. If they have any reason to believe that their glucose might not be in good control, they should check it.

Although the most common concern about exercising with diabetes is that exercise will cause either hypoglycemia or hyperglycemia, in practice, for people with type 2 diabetes this concern is usually more theoretical than real. The following considerations are relevant to individuals with medically stable diabetes and without acute illness sufficient to cause clinically meaningful glucose abnormalities:

- The sympathetic stimulation of exercise is typically too small to cause hyperglycemia.
- Few people with type 2 diabetes exercise long or hard enough to cause hypoglycemia.
- Exertional hypoglycemia is more related to insulin or insulin secretagogue medications.

Accordingly, persons with type 2 diabetes should confirm, before exercising, that their glucose is in a reasonable range (not <80 and not "high" on their glucometer) and that they have not taken insulin or an insulin secretagogue that is likely to cause hypoglycemia when combined with exercise.

According to one anecdotal observation, people often say they feel they're hypoglycemic, though checking their blood glucose reveals a value in the normal range (or even high). In such individuals, we recommend a shortened and easy day (but not a day off) and being alert for symptoms or unusual vital signs, because anecdotal experience has been that there is rarely a problem.

Effects of Exercise Training

Exercise training is perhaps the most important medical prescription for any person with type 2 diabetes. It should be combined with a weight loss intervention if the person is also obese.

Inasmuch as severe and morbid obesity with a sedentary lifestyle involves a risk of about 11% per year of developing diabetes, one can reasonably hold that a similar situation applies to the individual with obesity and a sedentary lifestyle who does not have diabetes. In such individuals it may be paradoxically less important that they combine exercise with a weight loss intervention, since weight loss is an essential element of treating diabetes but only a desired element in a person who is obese.

Recommendations for Exercise Programming

Exercise recommendations for obesity and obesity-related conditions are similar, but they vary slightly depending on the magnitude of the burden of comorbidities.

- Obesity with no disability or comorbidity: use ACSM Guidelines

- 150 to 250 min/week of moderate physical activity
- Optimum goal of ≥250 min/week of moderate physical activity
- Exercise specialist supervision to begin with if comorbidities, biomechanical difficulties, or other concerns are present
- Obesity with disability or high comorbid burden: Modified *CDD4* Recommendation
  - 150 to 250 min/week of moderate physical activity
  - Optimum goal of ≥250 min/week of moderate physical activity
  - Supervision of a skilled exercise specialist (e.g., Registered Clinical Exercise Physiologist, physical therapist) strongly advised
- Diabetes: Basic *CDD4* Recommendations
  - 150 to 250 min/week of moderate physical activity
  - Optimum goal of ≥250 min/week of moderate physical activity
  - May need to start at light intensity and gradually increase
  - May be possible for some persons who are successfully doing moderate physical activity to add vigorous activities, particularly if they have few comorbidities
  - Supervision of a Registered Clinical Exercise Physiologist strongly advised, particularly for people who have multiple comorbidities

Integration of Cardiometabolic Conditions Into a Medical Home Model

Lifestyle intervention for cardiometabolic conditions involves four main factors:

1. Exercise training, increasing physical activity
2. Dietary change and weight management
3. Tobacco cessation
4. Enhanced tools for coping with stress

This text focuses primarily on the exercise component, but the other components are very important, particularly for cardiometabolic conditions. Essentially everyone with a major chronic condition or disability needs a risk-reduction program that addresses physical activity, diet, tobacco use, and psychosocial well-being. Remember that cardiovascular events such as heart attacks and strokes are the number one cause of death for persons with cardiovascular disease or those with a disability.

If the community doesn't have a resource that can address all lifestyle risk domains with behavioral interventions (including exercise), then the medical home may need to hire staff or a consultant to create a set of integrated resources for these purposes. Another option is to contract with local professionals who can do this in collaboration with the medical home.

Arthritis and Back Pain

Arthritis and chronic joint pain are the second and third most prevalent conditions in seniors, according to the U.S. Centers for Disease Control and Prevention, and arthritis is the number one cause of disability. Back pain lasting for a period of at least 3 months affects about one in three adults during their life and is the chief complaint in about 2.5% of all office visits to a physician. It is thus very predictable that musculoskeletal pain either will be the primary reason for referral to

an exercise therapist or will be a comorbidity in someone with cardiometabolic disease.

Basic Pathophysiology

Pain afflicting joints and the spine can be thought of in three main categories: (1) pain from the joint itself, (2) pain from the surrounding tissues, and (3) referred pain (that feels like pain in the joint but is from other sources). This section addresses the first two types of pain.

Pain from the joint itself, or arthritis-arthralgia, is classically divided into two main categories: (1) osteoarthritis (OA) and (2) rheumatic forms of arthritis (RA). Either form can affect large and small joints or the spine.

Osteoarthritis is believed to be

- primarily a biomechanical failure of the articular cartilage, with
- secondary exacerbation by alteration in the biochemical milieu of the joint tissues.

Osteoarthritis is often asymmetrical and classically affects high-load–bearing joints in the upper and lower extremities and spine (commonly the hips, knees, hands and feet, cervical spine, and lumbar spine). Osteoarthritis is usually mild but can lead to complete joint destruction and need for prosthetic joint replacement.

Back Pain and Physical Exercise

The vast majority of people experience back pain in their lifetime, a high proportion have recurrent episodes. Most back pain research focuses on the low back, or lumbar region, but almost everyone occasionally awakens to an achy stiff neck. For this textbook, we limit the discussion to subchronic and recurrent low back pain.

The pathophysiology of low back pain is difficult to pinpoint, because so many possible sources of pain exist, ranging from discs to nerve roots to facet joints to bone to musculotendinous and connective tissues. Isolating the source(s) of pain is uncommon, so conservative use of analgesic and anti-inflammatory medications are mainstay treatments, as are referrals to treatments such as massage, yoga, acupuncture, physical therapy and regular physical activity.

Exercise has modest benefits, but there is no clearly superior system of exercises for back pain. Most spine-specific programs involve strengthening, range-of-motion, and integrative functional exercises intended to achieve better core stability and, hopefully, pain relief. However, regular physical activity alone has comparable efficacy to spine-specific exercises. Thus, it is not clear whether spine exercises or regular all-around physical activity is preferred. Perhaps back pain just waxes and wanes no matter what exercises are done. Most spine specialists believe that current back pain episodes are shortened and future episodes can be thwarted through the expert use of spine exercises. Unfortunately, this remains more expert opinion than scientific fact, am more research in this are is needed.

Rheumatic arthritis (of many forms) is understood to be primarily

- an autoimmune-mediated inflammatory degradation of joint tissues, with
- secondary exacerbation due to biomechanical alteration of the affected joints.

Rheumatic conditions are often symmetrical (affecting the same joints on both sides); rheumatoid arthritis typically affects hands and feet, but many other rheumatic conditions can affect large joints or the spine. Rheumatic conditions are highly variable; most are typically mild, but can be highly destructive of joints or other affected vital organs (commonly kidney) and can lead to very severe disability and even be fatal.

It should also be noted that many people have bursitis, tenosynovitis, tendonopathies, and enthesopathies (abnormalities at the site of tendon insertion onto bone) that are often present with arthritis or even are the primary cause of joint pain.

Bursitis, tenosynovitis, tendonopathies and enthesopathies are

- inflammation-alteration of soft tissues that are involved in joint function, with
- secondary exacerbation due to both inflammatory and mechanical factors.

These perijoint conditions are very common and are believed to arise from overload (often called overuse), alignment or weakness issues, or a combination of the two, leading to an overcompensation injury. It is more common to think of these kinds of issues as a sport- or work-related injury, but anecdotal experience suggests that they are highly prevalent in persons with chronic conditions or disability as well as people who are overweight or obese and inactive.

Abnormal joint alignment (either as a cause or as a result of joint deterioration) is thought to contribute to joint degradation through altered loading characteristics. Such alterations are a most likely cause for secondary bursitis and tendon problems that are often present alone or in combination with OA and rheumatoid arthritis (RA). These perijoint problems can be very painful but typically do not progress to a life-threatening situation. Anecdotally, it is observed that these problems lead to reduced physical functioning and a more sedentary lifestyle.

Hallmarks of OA, RA, and perijoint conditions:

- Pain, inflammation, effusions, or swelling (or more than one of these) in the affected areas
- Usually a waxing–waning course (occasionally rapid progressive deterioration)
- Fatigue (primarily in RA secondary to immune-mediated systemic effects)
- Weakness (due to both inhibition to avoid pain and deconditioning)
- Loss of functional capacity
- High prevalence of disability

Management and Medications

Medical management of joint pain conditions varies, and treatments are primarily aimed at alleviating the symptoms of pain and swelling and blocking the pathophysiological process.

Medications and devices used to reduce pain include analgesic or nonsteroidal anti-inflammatory drugs, anti-inflammatory steroids, immune-modulating rheumatologic-remitting drugs, pain medications, and implantable neural stimulators (see table 6.6).

Effects on the Exercise Response

Individuals with painful joints have a strong tendency to downregulate their activity level to avoid having problems or pain. One dilemma the exercise specialist must address is whether a person's pain is acute pain with weakness mainly due to inhibition of motor command or chronic pain and deconditioning with low functional capacity, atrophy, or loss of mobility.

In an acutely painful joint, the effect on the exercise response is that loading the affected joint causes pain and *neuromotor inhibition* to reduce muscular force and load on the painful part. Muscular contractions are weak primarily because of inhibition, not because of intrinsic muscle weakness. In the long term, however, it is common for fear avoidance and downregulation of regular physical activity to cause muscle atrophy, weakness, and low functional capacity or deconditioning that is not related to inhibition (see chapters 3 and 4). Thus the following are the clinical consequences of chronic joint pain on the exercise response:

Table 6.6   **Medications Used to Treat Arthritis**

Generic drug	Class of drug	Mechanism of action
Aspirin Ibuprofen Naproxen	Nonsteroidal anti-inflammatories (NSAIDs)	Block inflammatory pathway and analgesic
Acetominophen	Nonnarcotic analgesic	Unknown; possibly elevates pain threshold
Celecoxib	Cyclo-oxygenase-2 inhibitor	Blocks arachadonic acid pathway and analgesic
Prednisone Triamcinolone	Corticosteroids	Inflammatory suppressants
Gold	Gold salts	Unknown
Enbrel	Disease-modifying arthritis-remitting drug (DMARD)	Tumor necrosis factor-α receptor blocker
Anakinra	DMARD	Interleukin-1 receptor antagonist
Actemira	DMARD	Interleukin-6 receptor antagonist
Methotrexate	Dihydrofolate reductase inhibitor	DNA synthesis inhibitor (folate)
Leflunomide	Pyrimidine synthesis inhibitor	DNA synthesis inhibitor (folate)
Tofacitinib	Kinase inhibitor	Inhibits cell–cell signaling by blocking Janus kinase
Oxycodone Hydrocodone	Narcotic pain medications	Narcotic analgesics
Gabapentin	Nonnarcotic pain medication	Central nervous system pain pathways
Hyaluronic acid	Viscoelastic supplementation	Unknown
Chondrocytes Platelet-rich plasma	Autologous transplantation	Regrowth of chondrocytes in area of cartilage defect

- Muscle atrophy and loss of flexibility
- Slow gait speed, poor stair climb and descent, low 30-s sit to stand
- Low aerobic capacity and functional performance

Effects of Exercise Training

In persons who primarily have neuromuscular inhibition due to pain, therapeutic exercise often restores good function with astounding rapidity in a small number of sessions. People who have more profound functional losses typically face a much longer time to recovery. Even in people who have had long-standing deconditioning, therapeutic exercise can rapidly yield modest functional gains and stop the decline. In arthritis,

- neuromuscular inhibitions are often rapidly reversed,
- atrophy and deconditioning require adaptations to training that take much longer,
- resistance training can improve musculoskeletal and joint integrity, and
- exercise can increase strength-to-weight ratio, which may enhance symptom relief.

The interactions between exercise training and the course of chronic joint conditions are not fully known for all conditions. Examination of outcomes research for osteoarthritic, rheumatic, and soft tissue conditions suggests that there are differences between conditions and that they cannot be viewed interchangeably. This generalized discussion is primarily based on findings in the following clinical conditions; all other conditions have lesser levels of supporting evidence:

- Osteoarthritis in the knee
- Rheumatic arthritis in the ankles, knees, hips, and shoulders

In OA, the standard *CDD4* exercise recommendations, supervised or home based, can be expected to cause mild exacerbations of pain and swelling but not accelerate joint destruction. The following key points should be remembered:

- In OA and RA, people with good preservation of articular cartilage are more likely to benefit than those with advanced deterioration of articular cartilage.
- In OA, knees are more likely than hips to benefit from exercise training.
- Knee extensor strengthening with general aerobic activity seems to be the most effective approach for OA in the knee.
- In OA and RA, supervised and self-directed programs both work, but greater benefit seems to be associated with having a higher number of supervised sessions.

In RA, higher-intensity weight-bearing exercise accelerates subtalar (ankle) and shoulder joint destruction, and perhaps in knees and hips as well. Thus, weight-bearing exercise should probably be limited to moderate intensity walking in RA that affects large joints. Persons with such problems may be better off doing exercise training in non–weight-bearing and aquatic modes and limiting walking to activities of daily living. Aquatic and non–weight-bearing exercise are commonly used for persons with chronic joint conditions, as a means of maintaining function with minimal irritation of the joints.

There is a paucity of research on joint pain from soft tissue conditions (bursitis, tenosynovitis), and it is not known how prevalent this source of pain is in people with a chronic disease or disability. Anecdotal experience suggests that mild exercise, that is, physical therapy, is beneficial in most of these conditions. Such individuals should be referred to physical therapy as a first step toward increasing their physical activity.

Other factors to consider in exercise programming for individuals with arthritis:

- Blunting reflexive neuromotor inhibitory pathways

- Restoring musculoskeletal "balance," especially core stability
- Overcoming fear avoidance
- Exercise and manual therapies to reduce symptoms and restore mobility
- Using adaptive techniques and devices to assist function in severely affected individuals

Recommendations for Exercise Programming

The recommendations for exercise programming in arthritis are straightforward:

- Mild arthritis: follow the ACSM Guidelines
- Arthritis with disability or diminished physical functioning: Basic *CDD4* Recommendations
- Non–weight-bearing repetitive–endurance exercises involving the large muscles
- Knee extensor strengthening
- Core strengthening and attention to balance issues; balance often impaired
- Individualized exercises that meet each person's unique needs
- Moderate-intensity walking sufficient to maintain independence and quality of life
- Supervision by a physical therapist recommended

Special Considerations for Back Pain

The *CDD4* approach to subchronic and recurrent nonspecific back pain is the same as for osteoarthritis, both in regard to the medical management and exercise, with no specifics for the spine other than regular physical activity consistent with the Basic *CDD4* Recommendations. A prudent low-cost approach would be to advise physical activity first, and advance to physical therapy or an integrated spine health program if physical activity alone isn't satisfactory.

Osteoporosis

Bone health is extremely important in the maintenance of well-being. Not long ago, people in the industrialized world commonly suffered from malnutrition, which led to organic bone diseases such as rickets, because there was poor understanding of the mechanisms of bone metabolism and lack of awareness of vitamin D. In the developed world today, the most common noncongenital metabolic disease of bone is osteoporosis, which afflicts some 200 million people worldwide and one-third of postmenopausal women. About 20% to 25% of men will develop an osteoporotic fracture over their lifetime. The direct health care costs and the loss of independence and quality of life associated with osteoporotic fractures are substantial. With a growing population of citizens who are elderly, the current costs of osteoporosis are estimated at $10 to $20 billion annually in the United States.

Basic Pathophysiology

Bone health is a balance of bone deposition and resorption, and one can lose bone content because of decreased deposition, increased resorption, or both. Regulation of bone deposition and resorption is a complex function of parathyroid hormone, 1,25-dihydroxyvitamin D, calcitonin, hormones in the adrenal–gonadal axis, and mechanical bone loading. Bone density is adversely affected by tobacco, excessive alcohol consumption, and in some cases as an effect of nutritional status.

Osteoporosis is a complex condition in which bone matrix and mineral content are both decreased, which reduces the strength of bone and thus increases the risk of fractures. Osteoporosis is commonly viewed as a postmenopausal condition secondary to estrogen deficiency (type I osteoporosis) but is also related to aging (type II osteoporosis) and thus is not limited to women. There are also many other conditions of altered bone metabolism and bone health; examples are loss of bone mineral related to poor energy balance (often seen in anorexia and in athletes), as well as hyperparathyroidism, renal failure, and other causes of osteodystrophy and osteomalacia. Essentially, any condition that causes a prolonged alteration in the metabolism of calcium and phosphorus or genetically alters the structure of bone (e.g., osteogenesis imperfecta) will alter the strength and resilience of bone.

One of the most common causes of osteoporosis is iatrogenic—through prolonged use of high-dose corticosteroids in people who require aggressive immunosuppression because of autoimmune disorders, organ transplantation, or conditions involving extreme inflammation that must be sup-

pressed (e.g., radiation therapy to tumors in the brain, which cannot be allowed to swell because of the confined space in the skull). Modern immunosuppressant drugs and treatments are designed to avoid this kind of complication, but many people nonetheless have osteoporosis as a complication of treatment.

Management and Medications

Osteoporosis has a wide range of presentation:

- Asymptomatic
- Loss of height
- Back pain (from vertebral compression fractures)
- Stress fractures (most commonly in lower extremities)
- Hip fractures
- Extreme kyphosis (extensive vertebral compression)

One feared aspect of osteoporosis stems from the fact that osteoporotic hip fractures, usually from a fall, are associated with a high rate of subsequent mortality (up to 20-25% after 1 year following a hip fracture). Death may be a consequence of complications from the fall and injury (e.g., deep venous thrombosis and pulmonary embolism), or osteoporosis may simply be a marker of failing health and increasing frailty. Disability and loss of functional status are also a common complication after osteoporotic fractures.

Management of osteoporosis is primarily aimed at prevention, though individuals who have the more severe manifestations such as back pain or fractures are treated for those complications. Thus, the primary approach to osteoporosis is radiographic diagnosis and medical management. The diagnostic tool of choice is dual-energy x-ray absorptiometry (DEXA), which yields a three-dimensional image like a computed tomography (CT) scan and is typically graded by both absolute and relative bone density.

Absolute bone density has been categorized into three basic groups by a so-called T-score:

- Normal—within 1 standard deviation of the mean for normal adults
- Osteopenia—ranging from 1 to 2.5 standard deviations of the mean for normal adults
- Osteoporosis—greater than 2.5 standard deviations of the mean for normal adults

Relative bone density has also been categorized by a Z-score, which adjusts the same ranges for age and sex.

Once osteopenia or osteoporosis has been diagnosed, medical treatment is approached in a graded fashion to suit the circumstances.

- Vitamin D supplementation
- Estrogen replacement therapy (postmenopausal women)
- Androgen replacement therapy
- Bisphosphonates
- Calcitonin
- Anti-RANKL antibody (Prolia)
- Parathyroid hormone (Forteo)
- Pain medications

Vitamin D, calcium, and steroid hormone replacement improve the balance of bone resorption and deposition. Bisphosphonates and calcitonin decrease the resorption of bone. Pain management with medications is used when the individual has fracture-associated pain, but analgesics are not a mainstay of osteoporosis prevention.

Orthotics and prosthetics are commonly used for this condition:

- Orthotics—for reducing weight bearing in acute fractures
- Prosthetics—for skeletal disabilities from fractures and pain

Effects on the Exercise Response

Exercise research in osteoporosis has primarily been geared toward type I and type II osteoporosis, so the reader should be aware that recommendations based on research in exercise and osteoporosis may be addressing a biological phenomenon different from osteomalacia, osteodystrophy, iatrogenic corticosteroid-induced bone loss, or a bone disorder related to other causes. Nonetheless, it is reasonable for a person with one of these less common causes of osteoporosis to cautiously follow the exercise advice in this chapter, since there are few other sources of guidance.

There are many limitations in our knowledge on the interrelationships of exercise and bone health in the less common causes of osteoporosis, and caution should be used with such cases.

In persons without bone pain or fractures, osteoporosis has little effect on the exercise response. Bone pain in osteoporosis is usually due to fractures—most commonly stress fractures, but also radiographically visible fractures. Bone pain and macroscopic fractures reduce the ability to make a good effort.

Kyphosis may impair the ventilatory response to aerobic exercise. Bone pain and deconditioning or loss of muscle mass may decrease strength.

Severe osteoporosis may require volitional reduction of bone loading to avoid risking fracture, and thus can reduce physical functioning. It is widely advised that persons with bone pain, kyphosis, or a history of osteoporotic fractures avoid high-impact, high-resistance (as percentage of 1-repetition maximum [1RM]), and high-torque (twisting, bending) activities because of the risks of causing a fracture.

Effects of Exercise Training

Physical exercise is a mainstay of osteoporosis prevention and treatment. Exercise has multiple benefits to skeletal health and overall well-being. Wolff's law is a biological phenomenon that bone hypertrophies in response to bone loading, and thus the normal response to gradually progressive exercise is for bone strength to increase. In addition, it is important for everyone to maintain bone integrity and health in order to avoid loss of physical function that comes with pathologic fractures from osteoporosis.

Weight-bearing aerobic and resistance exercise, following the ACSM Guidelines (i.e., 150 min/week of moderate-intensity aerobic activity and 2 or 3 days/week of resistance exercises at 60% to 80% of 1RM), help maintain and even marginally increase bone density in mild to moderate osteoporosis. Data are insufficient to establish the ability of exercise to prevent fractures.

In persons with more severe osteoporosis as evidenced by high T- and Z-scores, a history of fractures, and kyphotic changes in the spine, the dose–response relationship is less clear. Such individuals may be limited by the inability to load the skeleton, but more research on this topic is needed. Persons in this situation today are in somewhat of a circular bind in that their bone might benefit from higher loading but is not strong enough to safely withstand such loading. To avoid this situation, exercise training should be instituted early in the clinical course of osteoporosis.

Recommendations for Exercise Programming

The Basic *CDD4* Recommendations for exercise testing and training, with a focus on maintenance of physical functioning, are very reasonable for almost anyone with the following general starting points:

Osteopenia: use the ACSM Guidelines

Mild to moderate osteoporosis (T- and Z-scores <3): use the ACSM Guidelines

For those individuals who have high T- and Z-scores (>3 standard deviations), bone pain, kyphosis, or fractures, especially if combined with loss of strength or muscle mass:

Severe osteoporosis: use the Basic *CDD4* Recommendations

Relative contraindications to exercise (in proportion to the severity of osteoporosis):

Twisting movements and high-impact activities
Extreme spine flexion, extension, twisting

Absolute contraindications to exercise:

Bone pain
Unhealed fractures

If weight-bearing activities cause bone pain, non–weight-bearing activities should be substituted.

Activities designed to improve and maintain overall strength and balance may help prevent falls.

Integration of Arthritis, Back Pain, and Osteoporosis Into a Medical Home Model

Astute medical management for any musculoskeletal condition will involve most if not all of the strategies discussed in the preceding sections. This is best done through a coordinated team approach, involving pharmacologic treatment to reduce symptoms and block the progression of the condition, as well as physical and manual therapies that seek to restore normal function—all leading to a maintenance physical activity and exercise program. The objective should be for individuals

to maintain themselves in as hardy and robust a state as possible, maximizing their physical functioning and preserving their independence.

Because of its role in the maintenance of bone integrity and health and in preservation of physical functioning, exercise is a hugely important aspect of chronic care management in persons with osteoporosis. Follow-up care visits should routinely include evaluation of

- daily physical activity and
- current functional capacity.

Suggested Readings

Amati F, Dubé JJ, Shay C, Goodpaster BH. Separate and combined effects of exercise training and weight loss on exercise efficiency and substrate oxidation. *J Appl Physiol.* 2008;105:825-831.

Artinian NT, Fletcher GF, Mozaffarian D, Kris-Etherton P, Van Horn L, Lichtenstein AH, Kumanyika S, Kraus WE, Fleg JL, Redeker NS, Meininger JC, Banks J, Stuart-Shor EM, Fletcher BJ, Miller TD, Hughes S, Braun LT, Kopin LA, Berra K, Hayman LL, Ewing LJ, Ades PA, Durstine JL, Houston-Miller N, Burke LE. Interventions to promote physical activity and dietary lifestyle changes for cardiovascular risk factor reduction in adults: a scientific statement from the American Heart Association. *Circulation.* 2010;122(4):406-441.

Burton AK, Balagué F, Cardon G, Eriksen HR, Henrotin Y, Lahad A, Leclerc A, Müller G, van der Beek AJ, on behalf of the COST B13 Working Group on Guidelines for Prevention in Low Back Pain. European Guidelines for prevention in low back pain. Eur Spine J. 2006;15(Suppl. 2):S136-S168. doi: 10.1007/s00586-006-1070-3.

Casale R, Alaa L, Mallick M, Ring H. Phantom limb related phenomena and their rehabilitation after lower limb amputation. *Eur J Phys Rehabil Med.* 2009;45:559-566.

Choi BKL, Verbeek JH, Tam WWS, Jiang JY. Exercises for prevention of recurrences of low-back pain. Cochrane Database of Systematic Reviews 2010, Issue 1. Art. No.: CD006555. doi: 10.1002/14651858.CD006555.pub2 [Accessed January 4, 2016].

Chou R, Qaseem A, Snow V, Casey D, Cross JT Jr, Shekelle P, Owens DK for the Clinical Efficacy Assessment Subcommittee of the

American College of Physicians and the American College of Physicians/American Pain Society Low Back Pain Guidelines Panel. Diagnosis and treatment of low back pain: a joint clinical practice guideline from the American College of Physicians and the American Pain Society. Annals Internal Med 2007;147:478-491.

Chou R, Huffman LA. Nonpharmacologic Therapies for Acute and Chronic Low Back Pain: A Review of the Evidence for an American Pain Society/American College of Physicians Clinical Practice Guideline. Ann Intern Med 2007;147:492-504.

Church TS, Blair SN, Cocreham S, Johannsen N, Johnson W, Kramer K, Mikus CCR, Myers V, Nauta M, Rodarte RQ, Sparks L, Thompson A, Earnest CP. Effects of aerobic and resistance training on hemoglobin A1c levels in patients with type 2 diabetes: a randomized controlled trial. *JAMA.* 2010;304(20):2253-2262.

Church TS, Earnest CP, Skinner JS, Blair SN. Effects of different doses of physical activity on cardiorespiratory fitness among sedentary, overweight or obese postmenopausal women with elevated blood pressure: a randomized controlled trial. *JAMA.* 2007;297:2081-2091.

Colberg SR, Albright AL, Blissmer BJ, Braun B, Chasan-Taber L, Fernhall B, Regensteiner JG, Rubin RR, Sigal RJ. Exercise and type 2 diabetes: American College of Sports Medicine and the American Diabetes Association: joint position statement. Exercise and type 2 diabetes. *Med Sci Sports Exerc.* 2010;42(12):2282-2303. doi: 10.1249/MSS.0b013e3181eeb61c.

Deshpande AD, Harris-Hayes M, Schootman M. Epidemiology of diabetes and diabetes-related complications. *Phys Ther.* 2008;88:1254-1264.

Fransen M. Osteoarthritis. In Saxon JM, ed. *Exercise and Chronic Disease: An Evidence-Based Approach.* New York: Routledge; 2011:156-174.

Freedman VA, Ayken H. Trends in medication use and functioning before retirement age: are they linked? *Health Aff.* 2003;22(4):154-162.

Fried LP, Tangen CM, Walston J, Newman AB, Hirsch C, Gottdiener J, Seeman T, Tracy R, Kop WJ, Burke G, McBurnie MA. Frailty in older adults: evidence for a phenotype. *J Gerontol A Biol Sci Med Sci.* 2001;56:M146-M156.

Galley R, Allen K, Castles J, Kucharik J, Roeder M. Review of secondary physical conditions associated with lower-limb amputation and long-term prosthetic use. *J Rehabil Res Dev.* 2008;45(1):15-30.

He XZ, Baker DW. Body mass index, physical activity, and the risk of decline in overall health and physical functioning in late middle age. *Am J Public Health.* 2004;94:1567-1573.

Herring MP, Puetz TW, O'Connor PJ, Dishman RK. Effect of exercise training on depressive symptoms among patients with a chronic illness. *JAMA.* 2012;172(2):101-111.

James PA, Oparil S, Carter BL, Cushman WC, Dennison-Himmelfarb C, Handler J, Lackland DT, LeFevre ML, MacKenzie TD, Ogedegbe O, Smith SC Jr, Svetkey LP, Tale SJ, Townsend RR, Wright JT Jr, Narva AS, Ortiz E. 2014 evidence-based guideline for the management of high blood pressure in adults: report from the panel members appointed to the Eighth Joint National Committee (JNC 8). *JAMA.* doi: 10.1001/jama.2013.284427.

Last AR, Hulbert K. Chronic low back pain: evaluation and management. Am Fam Physician 2009;79(12):1067-1074.

Karpansalo M, Lakka TA, Manninen P, Kauhanen J, Rauramaa R, Salonen JT. Cardiorespiratory fitness and risk of disability pension: a prospective population based study in Finnish men. *Occup Environ Med.* 2003;60:765-769.

Mancia G, Fagard R, Narkiewicz K, et al. 2013 ESH/ESC guidelines for the management of arterial hypertension. *Eur Heart J.* 2013. doi: 10.1093/eurheartj/eht.151.

Naci N, Ioannidis JPA. Comparative effectiveness of exercise and drug interventions on mortality outcomes: metaepidemiological study. *BMJ.* 2013;347:f5577. doi: 10.1136/bmj.f5577 [Accessed October 18, 2013].

Nathan DM, Buse JB, Davidson MB, Ferrannini E, Holman RR, Sherwin R, Zinman B. Medical management of hyperglycemia in type 2 diabetes: a consensus algorithm for the initiation and adjustment of therapy. *Diabetes Care.* 2009;32(1):193-203. doi: 10.2337/dc08-9025.

Parker BA, Thompson PD. Effect of statins on skeletal muscle: exercise, myopathy, and muscle outcomes. *Exerc Sports Sci Rev.* 2012;40(4):188-194.

Praet SFE, Rozenberg R, Van Loon LJC. Type 2 diabetes. In Saxon JM, ed. *Exercise and Chronic Disease: An Evidence-Based Approach.* New York: Routledge; 2011:265-296.

Sansam K, Neumann V, O'Connor R, Bhakta B. Predicting walking ability following lower limb amputation: a systematic review of the literature. *J Rehabil Med.* 2009;41:593-603.

Smith SC, Benjamin EJ, Bonow RO, Braun LT, Creager MA, Franklin BA, Gibbons RJ, Grundy SM, Hiratzka LF, Jones DW, Lloyd-Jones DM, Minissian M, Mosca L, Peterson ED, Sacco RL, Spertus J, Stein JH, Taubert KA. AHA/ACCF secondary prevention and risk reduction therapy for patients with coronary and other vascular disease. *J Am Coll Cardiol.* 2011;58(23):2432-2446.

Tsauo JY. Osteoporosis. In Saxon JM, ed. *Exercise and Chronic Disease: An Evidence-Based Approach.* New York: Routledge; 2011:175-192.

Viester L, Verhagen EALM, Hengel KMO, Koppes LLJ, van der Beek AJ, Bongers PM. The relation between body mass index and musculoskeletal symptoms in the working population. *BMC Musculoskel Dis.* 2013;14:238.

Wrobel JS, Najafi B. Diabetic foot biomechanics and gait dysfunction. *J Diabetes Sci Technol.* 2010;4(4):833-845.

Wu SC, Driver VR, Wrobel JS, Armstrong DG. Foot ulcers in the diabetic patient, prevention and treatment. *Vasc Health Risk Manag.* 2007;3(1):65-76.

Additional Resources

Related ACSM Position Stands

American College of Sports Medicine. American College of Sports Medicine position stand: exercise for patients with coronary artery disease. *Med Sci Sports Exerc.* 1994;26(3):i-v.

Donnelly JE, Blair SN, Jakicic JM, Manore MM, Rankin JW, Smith BK. American College of Sports Medicine position stand: appropriate physical activity intervention strategies for weight loss and prevention of weight regain for adults. *Med Sci Sports Exerc.* 2009;41(2):459-471.

Kohrt WM, Bloomfield SA, Little KD, Nelson ME, Yingling VR. American College of Sports Medicine position stand: physical activity and bone health. *Med Sci Sports Exerc.* 2004;36(11):1985-1996. doi: 10.1249/01.MSS.0000142662.21767.58 [Accessed November 12, 2013].

Pescatello LS, Franklin BA, Fagard R, Farquhar WB, Kelley GA, Ray CA; American College of Sports Medicine. American College of Sports Medicine position stand: exercise and hypertension. *Med Sci Sports Exerc.* 2004;36(3):533-553.

Web Resources

ACC/AHA Joint Guidelines

Treatment of Cholesterol. http://circ.ahajournals.org/content/early/2013/11/11/01.cir.0000437738.63853.7a [Accessed November 9, 2015].

ACSM Position Stands

Hypertension, CAD, Obesity, T2DM, Osteoporosis. www.acsm.org/access-public-information/position-stands [Accessed November 9, 2015].

American Diabetes Association

www.diabetes.org [Accessed November 9, 2015].

American Heart Association

www.heart.org [Accessed November 9, 2015].

www.heart.org/HEARTORG/Conditions/Conditions_UCM_001087_SubHomePage.jsp [Accessed November 9, 2015].

American College of Rheumatology

http://www.rheumatology.org/Practice-Quality/Clinical-Support/Clinical-Practice-Guidelines [Accessed January 1, 2016].

Assessment of Cardiovascular Risk

http://circ.ahajournals.org/content/early/2013/11/11/01.cir.0000437741.48606.98 [Accessed November 9, 2015].

ATP III

http://www.nhlbi.nih.gov/files/docs/guidelines/atglance.pdf [Accessed November 9, 2015].

JNC 8

http://jama.jamanetwork.com/article.aspx?articleid=1791497 [Accessed November 9, 2015].

National Center for Complementary and Integrative Health

https://nccih.nih.gov/health/providers/digest/chronic-low-back-pain [Accessed January 1, 2016].

http://circ.ahajournals.org/content/early/2013/11/11/01.cir.0000437740.48606.d1 [Accessed November 9, 2015].

National Center on Health, Physical Activity and Disability

www.nchpad.org [Accessed November 9, 2015].

Chronic Conditions
Very Strongly Associated With Tobacco

This chapter addresses the most common chronic conditions that are very strongly associated with tobacco (primarily cigarettes). These conditions, especially coronary artery disease, are also associated with physical inactivity; but chronic obstructive pulmonary disease (COPD) and, to a slightly lesser degree, peripheral arterial disease (PAD) are very strongly associated with smoking tobacco (particularly cigarettes). Tobacco cessation programs are beyond the scope of this book, but the linkage between tobacco and these conditions is sufficiently strong that anyone with one of these diagnoses should be closely questioned about tobacco use.

Chronic Obstructive Pulmonary Disease

Chronic obstructive pulmonary disease is the third leading cause of death in the United States, responsible for more than 3 million deaths in 2012 (about 6% of all deaths). Chronic obstructive pulmonary disease is also the second leading cause of disability-adjusted life-years lost. More important for the purposes of this book, the World Health Organization (WHO) estimates that about 65 million people now suffer from COPD, and that morbidity and mortality from COPD are expected to rise to the third leading cause of death by 2030. The vast majority of people who have COPD got it by smoking tobacco, through environmental exposure from indoor air pollution (mainly from burning biomass fuels for home cooking or heating), or occupational exposure, with small percentages for congenital reasons and from recurrent pulmonary infections. Mainly because of the continuing high prevalence of tobacco smoking, COPD is an important chronic condition that will confront any reader of this book.

Everyone who uses tobacco should have a tobacco cessation intervention coupled with the exercise program.

Basic Pathophysiology

Chronic obstructive pulmonary disease imposes multiple pathophysiological problems, not only through obvious ventilatory and gas exchange impairments but also through complex interactions with the cardiovascular and musculoskeletal systems. To appreciate the nature of these impairments, it is instructive to divide the pathophysiology into a number of categories.

Ventilatory Impairments

Increased airway resistance is seen in obstructive pulmonary diseases such as chronic bronchitis and asthma (see chapter 17). Airflow obstruction is compromised mainly during expiration because the airways tend to become smaller as lung volume decreases. While smoking-related chronic bronchitis is slowly progressive, asthma can be intermittent and of varying severity. In the special case of exercise-induced bronchoconstriction, pulmonary function can be normal between attacks.

The outflow obstruction creates a number of biomechanical factors that are essential elements of COPD, much of which is related to hyperinflation that occurs with expiratory airflow obstruction:

- Work of breathing increases (for all obstructive pulmonary diseases).
- Hyperinflation places inspiratory muscles at a mechanical disadvantage.
- Increased dead space contributes to an inappropriately high ratio between physiological dead space and tidal volume (V_D/

VT). Thus there is an increase in "wasted ventilation" in COPD.

The combination of these factors can lead to ventilatory muscle fatigue resulting from increased work of breathing, ventilatory muscle weakness, ventilatory inefficiency, or a combination of these factors. In the more severe stages of COPD, this fatigue can progress to ventilatory failure with inadequate alveolar ventilation, hypoxemia, and hypercapnia.

Abnormalities of Gas Exchange

In COPD, anatomical and physiological changes in the lung undermine gas exchange:

- Destruction of the alveolar–capillary membrane in emphysema
- Ventilation–perfusion inequality (particularly in COPD)

The loss of alveolar–capillary interface impairs diffusion and can cause hypoxemia during exercise or even at rest in advanced disease, while ventilation–perfusion inequalities also contribute to hypoxemia, which usually worsens during exercise.

Cardiovascular and Musculoskeletal Impairments

Because COPD typically progresses very slowly over time, people tend to curtail their physical activity and become deconditioned. In addition, systemic and pulmonary cardiovascular effects impair integrated physiological function during exercise. Poor nutritional intake is often a consequence of severe breathlessness and can lead to muscle atrophy. Together, this set of factors has very detrimental effects:

- Cardiovascular deconditioning
- Peripheral muscle deconditioning, wasting, and weakness
- Subsequently increased accumulation of lactic acid at low work rates, increasing ventilatory demand and sense of breathlessness
- Reduced pulmonary vascular conductance from chronic hypoxemia leading to secondary pulmonary vascular disease (pulmonary hypertension)
- Right ventricular systolic dysfunction

Breathlessness (dyspnea) is a frightening symptom with a complex etiology. Dyspnea results from a summation of neurological inputs from pulmonary receptors, chemoreceptors, and mechanoreceptors in the chest wall and limbs. Some individuals have dyspnea that is disproportional to their ventilatory limitations. The frightening nature of these subjective sensations that the person experiences can lead to chronic anxiety and depression. The psychological impacts can be profound, because people feel limited in their ability to pursue normal daily activities, so they end up experiencing social isolation and often have feelings of helplessness and despair.

Management and Medications

Exercise training is a crucial aspect of clinical management and rehabilitation, because individuals with COPD have varying degrees of physical deconditioning. The primacy of exercise training in COPD cannot be overstated, since an inability to breathe precludes all activities. Improving pulmonary function and the matching of respiratory and cardiovascular physiology are essential in restoring and preserving the person's ability to do activities of daily living. Pulmonary rehabilitation goals are directed toward enabling an individual to perform exercise at higher intensity, thus increasing the potential for obtaining a reconditioning effect from exercise training. The main intents are as follows:

- To correct physical deconditioning
- To reduce lactic acidosis during exercise
- To reduce ventilatory requirement during exercise

Another important element of medical management of COPD is to optimize pulmonary medications and use of oxygen to avert further decline and facilitate physical activity. Such elements of care include the following:

- Optimization of respiratory system mechanics with
  - breathing retraining to improve ventilatory efficiency and
  - careful attention to bronchodilator therapy, inhaler dosage, and technique
- Correction of hypoxemia whether it occurs at rest, during exercise, or during sleep
- Desensitization to dyspnea, addressing fear avoidance and other symptoms
- Energy conservation through improved coordination, balance, and mechanical efficiency

Lastly, it should go without saying that people with COPD should be very strongly encouraged to avoid tobacco smoke and indoor or outdoor air pollution (including secondhand tobacco smoke). Smoking cessation programs are an important element in persons with COPD who have not quit smoking.

Individuals with COPD are often taking several medications that could have implications for exercise testing and training. Furthermore, COPD often coexists with cardiovascular diseases such as coronary artery disease, hypertension, and PAD. The potential effects of medications for these conditions are described in tables 7.1, 7.2, and 7.3.

Effects on the Exercise Response

Some, but not all, people with COPD have true ventilatory limitation whereby the ventilatory requirement at maximum exercise matches the maximal voluntary ventilation (measured over 12 to 15 s, or calculated from spirometry). As noted in the overview, individuals with obstructive disease have impeded expiration requiring a longer time for adequate lung emptying. In these persons, increased breathing frequency during exercise tends to cause dynamic hyperinflation and reduced inspiratory capacity—circumstances that worsen breathing efficiency by increasing VD/VT.

In the absence of true ventilatory limitation, exercise can be limited by cardiovascular factors such as deconditioning, impaired left ventricular function due to hypoxemia, or reduced pulmonary blood flow secondary to chronic hypoxemia. Peripheral muscle deconditioning can lead to lactic acid accumulation at low work rates, increased CO_2 output from bicarbonate buffering, and consequently an increased ventilatory requirement (which compounds the situation).

Table 7.1   Common Drugs Used in Management of Chronic Obstructive Pulmonary Disease

Medications	Mechanism of action	Adverse effects
Albuterol Metaproterenol Salmeterol Formoterol Indacaterol	Selective β-2-adrenoceptor sympathomimetic agonists that produce bronchodilation	• Tachycardia, palpitations, and tremulousness • Nonselective sympathomimetic drugs (e.g., epinephrine and isoprenaline) should not be used
Theophylline Aminophylline	Methylxanthines have potent bronchodilator activity	• Tachycardia, cardiac dysrhythmias • Central nervous system stimulation • Increased respiratory drive • Risk of seizures
Hydrochlorothiazide Furosemide (see chapter 10)	Diuretics, primarily thiazide and loop diuretics, to control right heart failure in cor pulmonale	• Intravascular volume depletion leading to hypotension • Hypokalemia with risk of cardiac dysrhythmias and muscle weakness
Prednisone (see chapter 17)	Glucocorticoid for reducing inflammation Often prescribed for COPD exacerbations with the hope of improving pulmonary function and speeding recovery	• Skin atrophy and fragility • Osteoporosis • Muscle atrophy • Myopathy (including ventilatory muscles) • Obesity (increases work of breathing)
Bupropion Other antidepressants (see chapter 32)	Antidepressants can improve mood and affect Can improve success with smoking cessation efforts	• Tachycardia (especially tricyclics)
Nicotine gum or patch Varenicline	Nicotine replacement and nicotine receptor agonists improve success with smoking cessation efforts	• Possible increase in suicidal ideation
Pneumococcal and influenza vaccines	Prevent infection with strep pneumococcus and influenza virus	• Tenderness at injection site • Other side effects extremely rare

Impairment of gas exchange occurs in emphysema because of destruction of the alveolar–capillary membrane, which can cause hypoxemia during exercise, particularly when increases in pulmonary blood flow increase shunting through areas of incomplete gas exchange. Chronic hypoxemia also causes erythrocytosis, and the resulting increase in blood viscosity can further compromise the circulation during exercise. In smokers, increases in carboxyhemoglobin, which cannot bind with oxygen, impair blood oxygen transport. Moreover, in some individuals, symptoms (predominantly dyspnea) and psychological factors might limit exercise capacity independently.

Effects of Exercise Training

Regular participation in physical activity provides many benefits for persons with COPD. These changes include the following:

- Facilitation of activities of daily living
- Cardiovascular reconditioning
- Improved ventilatory efficiency
- Improved lactate and ventilatory thresholds
- Desensitization to dyspnea and the fear of exertion
- Increased muscle strength and endurance
- Improvements in flexibility and balance
- Improved body composition
- Enhanced body image

Regular participation in exercise is critically important, because the benefits of aerobic training on improving insulin sensitivity go away in just a few days.

The accomplishment of these changes requires careful attention to medications to obtain optimal respiratory mechanics and may require use of oxygen therapy to maintain adequate oxygenation during exercise.

Recommendations for Exercise Testing

Exercise testing is extremely valuable in persons with COPD to distinguish between several possible causes of limited exercise capacity as described in the preceding sections. It is also essential to identify coexistent exercise-induced hypoxemia, hypertension, cardiac dysrhythmias, or myocardial ischemia. Cardiopulmonary exercise testing, whether using a cycle ergometer or treadmill, can be administered in the following ways (details follow):

- Maximal or submaximal test
- Cycle ergometer with low-level ramp protocol
- Treadmill walking with a low-level ramp or steady-state protocol

With appropriate monitoring, maximal exercise testing is safe and provides the best definition of the limitations, including psychological problems and symptoms. The cycle ergometer offers the best means of controlling external work rate, measuring gas exchange, and blood sampling. A typical protocol might include 3 min of unloaded pedaling followed by a ramp increase in work rate of 5, 10, 15, or 20 W/min, depending on the degree of impairment in the person being tested. The aim should be to obtain between 8 and 10 min of exercise data. A constant work rate protocol on a cycle ergometer can be used repetitively to allow accurate comparison of physiological responses and to show improvements after therapeutic interventions including exercise training.

Treadmill walking is a more familiar mode of exercise to many people with COPD and is probably less likely to be limited by leg fatigue than cycle ergometry. Treadmill walking is therefore more likely to be limited by abnormalities of respiratory system mechanics. For this reason, constant work rate treadmill testing has been advocated to test the effects of inhaled bronchodilator therapies on exercise endurance. Treadmill exercise testing is also helpful for the oxygen-dependent person and can be repeated with different oxygen systems and flows to determine the best means of correcting hypoxemia during ambulation.

Standard contraindications to exercise testing in COPD include these:

- Exacerbations of COPD
- Refractory hypoxemia with desaturation <85% by pulse oximetry
- Unstable cardiovascular conditions (dysrhythmias, myocardial ischemia, or heart failure)

Recommendations for Exercise Programming

The Basic *CDD4* Recommendations are appropriate for most persons with COPD. Modification may be required for these reasons:

- Individual variations and situations
- Inability to do prolonged continuous activity
- Fluctuating clinical status

The exercise program should be individualized and flexible, and any significant change in the medical condition of the person requires reassessment of the goals and risks of the program.

Key goals unique to programs for persons with COPD include the following:

- Improved body mechanics that reduce oxygen requirement of specific activities
- Improved breathing efficiency (pursed lips and diaphragmatic breathing)
- Slowing of the respiratory rate

Because of these unique parameters, exercise rehabilitation for individuals with COPD is best started in a center-based environment. An intensive 6-week exercise program with group interaction is helpful to begin the process of physical reconditioning and should involve several different professionals:

- Respiratory therapists—evaluate, teach, and ensure effective use of bronchodilator medications and oxygen therapy
- Physical therapists or clinical exercise physiologists—evaluate exercise endurance, muscle strength, flexibility, and body composition
- Occupational therapists—evaluate activities of daily living and teach energy conservation

Oxygen should be administered during exercise to individuals who have had a fall in arterial oxygen tension to less than 55 mmHg, or in oxyhemoglobin desaturation to less than 88%. Preventing exercise-induced hypoxemia not only improves exercise capacity but also enhances the effects of exercise training. The goal of oxygen therapy during exercise is to maintain an oxyhemoglobin saturation above 90%.

As in the Basic *CDD4* Recommendation, the mode of training can be walking, cycling, swimming, or conditioning exercises based on energy centering and balance, such as tai chi. In addition to the specificity of training effect on performance, desensitization to dyspnea may also be task specific.

Modifications to the duration and frequency of exercise might be necessary. Commonly the person with chronic respiratory disease is unable to sustain 20 or 30 min of exercise. Interval-type exercise consisting of 5- or 10-min sessions might be necessary until adaptations have occurred that allow reduction of rest intervals and gradual increases in work intervals.

Integration Into a Medical Home Model

The importance of maintaining a high level of physical activity in persons with COPD cannot be overemphasized. Individuals with COPD are at particular risk of relapsing into a state of inactivity and physical deconditioning, and during follow-up visits the health care provider should confirm that they are not relapsing.

- Every individual with COPD should go through a pulmonary rehabilitation program.
- After completing this program, the individual should consider membership in a health and fitness facility that can cater to the needs of someone with COPD.
- Lifestyle modifications such as weight management and dietary modifications are important considerations for people with comorbidities like obesity, diabetes mellitus, and metabolic syndrome.

Here are some special considerations:

- Corticosteroids may contribute to psychological disturbances.
- Rating of perceived exertion (RPE) and dyspnea are the preferred methods of monitoring intensity since many people are unable to achieve a peak heart rate but still show physiological improvement.
- Coronary artery disease, PAD, and musculoskeletal problems (e.g., arthritis, osteoporosis) are common in persons with COPD.

- Myopathy (including respiratory muscles) may be present due to corticosteroids and disuse atrophy.
- People with COPD usually respond best to exercise in mid to late morning.
- Avoid extremes in temperature and humidity.
- Supplemental O_2 flow rate should be adjusted to SaO_2 >90%.
- Anxiety, depression, or fear or some combination of these is common due to dyspnea and physical disability.

Coronary Artery Disease and Atherosclerosis

Coronary artery disease (CAD) is the number one cause of death in the world. In the United States, 17.6 million persons have CAD, including 8.5 million who have had a myocardial infarction and 10.2 million with angina pectoris (chest pain from cardiac ischemia). The American Heart Association estimates that nearly 50% of all American men under the age of 55 and nearly one-third of all American women under the age of 65 will develop some symptom of CAD. In addition, silent ischemia, a cardiac ischemia episode that doesn't cause chest pain, is estimated to account for up to 75% of all ischemic episodes, leading to myocardial infarctions or sudden cardiac death with no warning signs. These facts explain why many people consider CAD the most important chronic condition to target with aggressive risk reduction.

Basic Pathophysiology

The atherosclerotic process starts early, during childhood. The focal injury progresses to a fatty streak during young adulthood and turns into a fully formed fibrous plaque in adulthood. The atherosclerotic process occurs in all arteries but is most prominent in arteries that incur the highest mechanical strain, mostly caused by the systolic pulse pressure. Accordingly, the most common sites of atherosclerosis are the arteries that supply the heart, the aorta, and the carotid and vertebrobasilar arteries that supply the brain. Atherosclerosis is thus the pathophysiological mechanism that causes most heart attacks and strokes.

Coronary artery disease is an accumulation of fibrous plaques on the walls of the arteries that supply oxygen to the heart. The disease starts with some type of focal injury to the endothelial lining of the arterial wall (e.g., hypertension stretching the artery wall or caustic toxins oxidizing the endothelium). Once this injury occurs, platelets adhere to the injury site and release growth factors that attract low-density lipoprotein-cholesterol (LDL-C), which oxidizes and then attracts immune cells that augment the injury site and eventually transform it into a fatty streak. In the fatty streak, fibroblasts and smooth muscle cells from the vessel's intima migrate to the media layer of the artery, creating a rigid fibrous cap on the atherosclerotic plaque. This cap is vulnerable to breaking off in response to shear stress (blood flow) and thereby triggering an acute coronary syndrome. Symptoms or complications of CAD that arise from the plaque include the following:

Angina pectoris

Myocardial ischemia

Myocardial infarction

Sudden cardiac death

Over the past 50 or more years, both nonmodifiable and modifiable risk factors for CAD have been determined. Nonmodifiable risk factors include age, sex, and family history of premature CAD. Modifiable risk factors of CAD amendable to lifestyle changes include:

- Hypertension
- Dyslipidemia
- Physical inactivity
- Smoking
- Obesity
- Diabetes mellitus

Management and Medications

As with hypertension and hyperlipidemia, the treatment of the CAD should always begin with lifestyle changes to decrease modifiable risk factors. Thus, all persons with CAD are advised to increase their physical activity, lose weight and get leaner, stop using tobacco, improve their blood glucose, eat a more heart-healthy diet, and manage hypertension or hyperlipidemia to recommended levels.

Pharmaceutical treatments and surgical operations for people with CAD are available when it's too late for lifestyle changes. The various pharmaceutical treatments, including β-blockers, angiotensin-converting enzyme (ACE) inhibitors, statins, and platelet antagonists such as clopidogrel and aspirin, are all effective at reducing cardiovascular-related morbidity and mortality. When lifestyle interventions are successfully combined with pharmaceutical treatments, even greater reductions in morbidity and mortality may be achieved. For a list of commonly used medications to treat atherosclerosis, please see tables 6.2 and 6.4.

Percutaneous and surgical procedures (e.g., coronary angioplasty or coronary artery bypass grafting) are used with great success, with new and better technologies every year. As a result, there are more survivors of acute coronary syndromes now than in years past; these situations and conditions are addressed in part III.

Effects on the Exercise Response

Coronary artery disease sometimes alters the cardiorespiratory and hemodynamic responses to submaximal and maximal exercise, usually when CAD is sufficiently advanced to significantly reduce blood flow in coronary arteries:

- Aerobic capacity reduced to 40% to 60% of age and sex predicted is primarily due to diminished cardiac output (stroke volume, heart rate, or both).

- A decreased contractile force of the left ventricle due to silent ischemia can cause a progressive loss in ejection fraction and stroke volume that is often manifest as a blunted or decreasing systolic blood pressure response (exertional hypotension).

- Potentially fatal exercise-induced ventricular dysrhythmias occur less commonly in individuals with CAD than in those who have had a myocardial infarction and have residual exercise-induced ischemic ST-segment depression, angina pectoris, or left ventricular dysfunction.

Effects of Exercise Training

There are many potential benefits of exercise training for persons with CAD, including (see also table 7.2):

- All the benefits listed previously for hypertension and hyperlipidemia
- Improvement in the ventilatory response to exercise and ventilatory threshold
- Increased maximal oxygen consumption ($\dot{V}O_2$max) (mean of ~20%)
- Improved psychosocial well-being and self-efficacy

Table 7.2   Effects of Common Cardiac Drug Classes on Exercise

Class of drug	Physiological effect	Exercise-related consequences
Diuretics	Decreased preload	Possible increased heart rate response with decreased aerobic capacity
β-adrenergic blockers	Decreased heart rate	Especially nonselective β-blockers may decrease peak heart rate and thus aerobic capacity and tolerance of angina pectoris
Angiotensin converting enzyme inhibitors; angiotensin receptor blockers; vasodilators	Decreased afterload	Ergomimetic properties facilitate cardiac output; no significant effect on exercise response
Calcium channel blockers (nifedipine, diltiazem, verapamil)	Decreased heart rate	May decrease chronotropic response and aerobic capacity, improving tolerance of angina pectoris
α-adrenergic blockers; clonidine	Peripheral vasodilation	May decrease vasomotor tone and result in exertional or (more likely) postexertional hypotension
Antiarrhythmics	Prevent cardiac conduction or rhythm instabilities	No effect on exercise response
Digoxin	Improves contractility in congestive heart failure	False-positive ST-segment depression

- Protection against the triggering of a myocardial infarction by physical exertion ≥6 METs (metabolic equivalents)
- Decreased coronary inflammatory markers (e.g., C-reactive protein, Lipoprotein-associated Phospholipase A$_2$)
- Increased numbers of cells that promote angiogenesis and vascular regeneration
- Decreased blood platelet adhesiveness, fibrinogen, and blood viscosity, and increased fibrinolysis
- Increased vagal tone and decreased adrenergic activity, resulting in increased heart rate variability

Recommendations for Exercise Programming

The optimal program for persons diagnosed with CAD is cardiac rehabilitation (whether they were treated medically, with angioplasty, or with coronary bypass surgery). In some areas (e.g., the United States), most payers require that the diagnosis has been made within the last 12 months.

If the costs of cardiac rehabilitation are not covered and individuals cannot afford it on their own, it is highly recommended that they initiate their exercise program under the most expert exercise specialist available, preferably an ACSM Registered Clinical Exercise Physiologist (RCEP).

Whether the person participates in cardiac rehabilitation or has a program individually guided by an exercise specialist, the nature of the program will be determined by the severity of the CAD and the burden of other comorbidities:

- Minimal severity (e.g., postangioplasty with no cardiac damage): use ACSM Guidelines
- High severity (e.g., congestive heart failure, valvular heart disease): see pertinent chapters and use Basic *CDD4* Recommendation

Individuals between these extremes are strongly advised to obtain medical supervision for purposes of optimizing their program.

Integration Into a Medical Home Model

Cardiac rehabilitation should be part of the default plan of care for persons with CAD. Most cardiac rehabilitation services require that the program start within 1 year from the date of diagnosis. For a myriad of reasons, some people will not fall within that time period, and cardiac rehabilitation will not be a covered service (though in most societies they could pay for it themselves with cash). Home-based walking programs are an alternative, especially in areas with limited access to a formal rehabilitation program. Home-based programs can be monitored with pedometers and physical activity logs (including electronic media and social networks) or with regular phone or electronic communication. Individuals who follow their home exercise recommendations can achieve benefits similar to those in people who participate in formal exercise or rehabilitation programs.

Take-Home Message

Keep these factors in mind when working with people with CAD:

- Cardiac rehabilitation is the default referral when someone receives the diagnosis.
- Underlying CAD risk factors should be intensively managed.
- Persons with CAD who smoke should be strongly encouraged to stop smoking.

Angina and Silent Ischemia

When the heart doesn't get enough arterial blood supply (cardiac ischemia), the individual experiences chest discomfort known as *angina pectoris* (commonly known as *angina*). Some people, mainly persons with diabetes or those who are elderly, can experience cardiac ischemia and myocardial hypoxia but without pain. Because the person is asymptomatic but still has cardiac ischemia, this is known as *silent ischemia.*

Basic Pathophysiology

The heart relies almost exclusively on aerobic energy metabolism, requiring a continuous and reliable supply of oxygen. The heart closely regulates coronary blood flow (CBF) in response to variations in myocardial oxygen demand, because the myocardium extracts nearly 80% of the available oxygen from the blood. Increasing blood flow is thus the primary means by which the heart meets its oxygen demand. Accordingly, when the oxygen supply is reduced by a constraint in blood flow, the myocardium becomes hypoxic.

Symptomatic and asymptomatic myocardial ischemia are usually the result of an obstruction in one or more coronary arteries, which may be caused by the following:

- Atherosclerosis of conductance-sized coronary arteries (majority of cases)
- Coronary artery spasms (Prinzmetal's angina)
- Microvascular disease (syndrome X)
- Vasculitis (uncommon)
- Tunneling of coronary arteries (very uncommon)

There are other uncommon situations that might cause angina, such as extreme anemia, thyroid storm (severe hyperthyroidism), pheochromocytoma, malignant hypertension, sepsis, and other conditions in which myocardial oxygen supply is inadequate to meet myocardial demand. These situations are medically unstable, need immediate medical care, and should not involve exercise and are thus not relevant to this chapter. In addition, angina that is present in someone at rest and known as unstable angina requires immediate care and is not relevant to this chapter.

Angina and silent ischemia in the stable chronic state are associated with a predictable level of physical exertion, emotional stress, or exposure to cold and are relieved promptly with rest or sublingual nitroglycerin. The etiology is most commonly a fixed stenosis, or narrowing, in a segment of a coronary artery. Stenoses sufficient to cause angina most often have reduced the lumen diameter by more than 70%. Although cardiac blood supply is impaired, blood flow is sufficient to serve the myocardial oxygen needs at rest. Myocardial oxygen requirements increase up to 10-fold during exercise. As CAD progresses, the frequency, intensity, and duration of angina episodes often increase as well.

Stable angina symptoms generally last 2 to 10 min and are classically characterized as

- heaviness, squeezing, or constricting that
  - originates behind the sternum and
  - sometimes radiates to the shoulders, arms, neck, or jaw.

Other people exhibit what is known as *atypical angina* and experience

- shortness of breath,
- nausea,

- diaphoresis, or
- headache, toothache, or some other sensation.

Chest pain is a nonspecific symptom that can have both cardiac and noncardiac causes. Noncardiac causes include gastritis (esophageal spasm or reflux), musculoskeletal pain (from the chest wall, back, shoulder, or arms), hiatal hernia, and panic disorder. In general, chest pain that is cardiac in origin is reproducible with exertion. The presence of angina during exercise is not sufficient to make a diagnosis of angina pectoris (myocardial ischemia and hypoxia). Along with chest pain, clear physiological signs of myocardial ischemia must be demonstrated, such as ST changes on a 12-lead electrocardiogram (ECG).

Management and Medications

The major goals for medical management of chronic, stable angina pectoris are best achieved through a combination of medications, exercise training, and lifestyle modifications. These are the major goals:

- Increasing myocardial oxygen supply
- Decreasing myocardial oxygen demand
- Reducing vasospastic sources of ischemia
- Reducing risk of acute thrombotic events such as myocardial infarction and stroke

Accepted strategies include aggressive control of CAD risk factors (e.g., tobacco cessation, hypertension control, weight loss and lipid-lowering dietary changes, stress management), as well as cardiac revascularization with balloon angioplasty (with or without stenting) and coronary artery bypass graft surgery (CABG). In addition, people with angina commonly take a lipid-lowering drug, blood pressure medications, aspirin or other antiplatelet agents, and possibly a tobacco cessation agent such as nicotine replacement, bupropion, or varenicline.

Persons with angina or silent ischemia are commonly taking medications to treat the angina itself; the following are the classes of agents:

- β-Blockers
- Calcium channel antagonists
- Nitrates (short- and longer-acting nitroglycerin)
- A recently approved sodium channel agent called ranolazine

In addition to these classes of drugs, lipid-lowering therapy using statins, ACE inhibitors, and angiotensin receptor blockers are of value in treating risk for CAD and have been shown to improve long-term outcomes.

Effects on the Exercise Response

Because of the relatively high oxygen extraction of the myocardium (at any heart rate), when myocardial demand increases during exercise, an increase in CBF is the primary way the myocardium receives additional oxygen. During vigorous exercise, CBF can increase four to six times above resting levels in an effort to keep pace with the high metabolic demand. Coronary blood flow is rarely a limiting factor for exercise in healthy individuals, but in the presence of diseased arteries, angina or silent ischemia occurs when myocardial oxygen supply cannot meet the demand.

The degree to which myocardial ischemia limits exercise tolerance is largely dependent on the severity of coronary artery obstruction and the extent to which collateral blood flow supports affected areas of the heart. Since the heart relies almost exclusively on aerobic energy metabolism, myocardial ischemia alters functional, metabolic, and morphological factors in the myocardium, which can lead to dysrhythmias, impaired contractility, and electrophysiological abnormalities. Consequences of a hypoxic myocardium also include reductions in stroke volume, left ventricular ejection fraction, and cardiac output, as well as a drop in blood pressure; these collectively limit skeletal muscle perfusion and overall exercise tolerance. Decreased stroke volume may also lead to a compensatory increase in heart rate, putting a further strain on the heart.

Effects of Exercise Training

The primary goal of exercise training for people with angina is to improve physical functioning by increasing the level of exertion that causes angina. This is called raising the ischemic threshold, or the point at which angina symptoms occur. Following exercise training, most people with angina are able to perform at a higher work rate without symptoms, reducing the frequency and duration of symptoms during activities of daily living. This is due in part to reductions in heart rate (HR) and systolic blood pressure (SBP) at any given level of submaximal work. The product of HR and SBP, termed the *rate–pressure product (RPP)* or *double product,* closely parallels myocardial oxygen demand. Also, exercise training exposes the lining of the coronary arteries to shear stress (from blood flowing through the arteries), and regular stimulation of the arteries in this way increases their ability to make nitric oxide and thus to vasodilate in response to exercise. This mechanism is very likely important in reducing coronary vasospasm as a cause of exertional angina.

The RPP may be used to document specific workloads at which signs and symptoms of angina occur, and therefore track improvements in exercise capacity and physical functioning relative to the onset of angina. The RPP is thus the best way to track improvements in the ischemic threshold.

The majority of individuals with stable angina achieve other benefits of exercise, including these:

- Reduced overall cardiac risk
- Slowing or preventing the atherosclerotic process
- Improved quality of life (particularly because of fewer angina episodes)

Similar responses to training have been demonstrated with aerobic, interval, and home-based exercise programs.

Recommendations for Exercise Testing

The exercise tolerance test (ETT) is a valuable tool for diagnosing angina and silent ischemia, evaluating and guiding medical and exercise management, and estimating prognosis. For diagnostic purposes, current guidelines rate ETT as a class I recommendation for the following patient populations with signs or symptoms of cardiac ischemia:

- Known or suspected CAD
- Low-risk stable angina, after 8 to 12 h without ischemic or heart failure symptoms
- Intermediate-risk stable angina, after 2 or 3 days without ischemic or heart failure symptoms
- Known angina or silent ischemia, medically stable and free from unstable angina or symptoms of decompensated heart failure for at least 1 week before ETT

For purposes of exercise management, a submaximal ETT can be used to quantify the submaximal cardiovascular response to exercise and to define the work rate and RPP at the ischemic threshold.

Such a submaximal ETT may be terminated once the ischemic threshold has been identified.

In addition to documenting standard clinical measures obtained during exercise testing (HR, ECG, BP, and signs and symptoms), particular attention should be paid to the onset and duration of angina symptoms, including pain rating. The 4-point angina scale and safety criteria for terminating an ETT that relate specifically to angina and silent ischemia are highlighted in "Angina Scale" and "Considering Terminating a Test."

The presence or absence of ST-segment changes alone does not confirm or rule out CAD. Other ETT responses including exercise capacity, HR and BP responses, signs and symptoms of ischemia, and presence of serious rhythm abnormalities may help determine the diagnosis of CAD. The sensitivity of the ETT increases when clinical testing guidelines (ACSM, American College of Cardiology [ACC], American Heart Association [AHA]) are adhered to and multivariate scoring is used to quantify diagnosis and prognosis.

Cardiopulmonary exercise testing (CPX), or standard ETT with the addition of expired gas analysis, can provide enhanced evaluation of cardiopulmonary function. Variables of pump function such as O_2 pulse ($\dot{V}O_2$/HR) can be helpful to detect the ischemic threshold, and indices of cardiorespiratory efficiency such as the $\dot{V}E/\dot{V}CO_2$ slope are important prognostic markers.

If ETT or CPX data are not available, the Basic *CDD4* Recommendations for assessing physical functioning should be considered and may even be more appropriate if the individual also has other chronic conditions and low physical functioning (see parts I and II).

If the individual has angina, has been medically stable for at least a week as explained earlier, and is not high risk, one can improvise using a submaximal ETT by choosing an appropriate ramp–incremental protocol and measuring HR and blood pressure at increasing work rates until the angina is at level 1 (4-point scale). This can be done with or without telemetry monitoring. If the exercise specialist or client is not comfortable with this informal submaximal ETT, a 6-min walk

Angina Scale

1. Light, barely noticeable
2. Moderate, bothersome
3. Severe, very uncomfortable
4. Most severe pain ever experienced

Terminating an Exercise Test

Absolute Indications to Terminate a Test

- Drop in SBP >10 mmHg from baseline, despite an increase in workload, accompanied by other evidence of ischemia
- Moderately severe angina (defined as 3 on standard 1-4 angina scale)
- ST elevation (+1.0 mm) in leads without diagnostic Q-waves (other than V1 or aVR)

Relative Indications to Terminate a Test

- Drop in SBP >10 mmHg from baseline, despite an increase in workload, in the absence of other evidence of ischemia
- ST or QRS changes such as excessive ST depression (>2.0 mm horizontal or downsloping ST-segment depression) or marked axis shift

test or shuttle walk test can be used in similar fashion. Close monitoring is required to ensure proper performance and safety during these tests.

Recommendations for Exercise Programming

The Basic *CDD4* Recommendations for exercise programming are applicable to persons with angina or silent ischemia, but the aerobic activity recommendation should be considered a minimum. Physical activity guidelines for secondary prevention of CAD have been published by the AHA/ACC and are similar to the Health and Human Services Guidelines for the apparently healthy:

- Following an acute event such as myocardial infarction, daily walking is encouraged immediately after discharge.
- All persons with CAD are encouraged to perform 30 to 60 min of moderate-intensity aerobic activity, such as brisk walking, on most, preferably all, days of the week.
- Regular aerobic activity should be supplemented with an increase in daily lifestyle activities (e.g., walking breaks at work, gardening, and household work) and a decrease in sedentary activity.
- Resistance training of major muscle groups 2 or 3 days per week is encouraged.

Note that the Basic *CDD4* Recommendation for strength training is more explicit than the AHA/ACC Guideline, though the Basic *CDD4* Recommendation is geared more toward a minimum for persons with low physical functioning, and many people with angina or silent ischemia may be substantially more robust.

One aspect of exercise training that is quite notable in people with angina is that a long and slow warm-up period often improves exercise performance that very day, known as a "warm-up phenomenon." A slow warm-up provides a more moderate increase in cardiac demand, giving time for coronary vessels to dilate and collateral circulation to develop, resulting in an improvement in oxygen supply. This is often compared to ischemic preconditioning, the phenomenon in which temporary low-pressure inflation of an angioplasty balloon, before the actual procedure, reduces the effects of ischemia during the procedure itself. Thus, a gradual warm-up is recommended for persons with angina.

In the absence of an ETT or supervision resources, a sensible approach would be to do a functional exercise trial based on subjective and objective evaluation during several sessions of low-level exercise. Inpatient exercise guidelines for intensity include the following:

- RPE <13 (6-20 scale)
- HR <120 beats/min or HRrest + 20 to 30 beats/min (arbitrary upper limit, but never above the ischemic threshold)

Integration Into a Medical Home Model

Medical home providers should encourage home-based physical activity with the goal of reducing anginal symptoms and the overall burden of CAD. People with stable angina should understand that they can safely perform exercise at home, especially if they have been to a supervised program of cardiac rehabilitation where they can learn how to self-monitor their HR as well as RPE and angina scales. This will aid them in knowing their safe limits of exercise. Additionally, individuals should be able to articulate the proper use of nitroglycerin and when to call 911 if they have further signs of myocardial ischemia or acute coronary syndrome. Caregivers also need to be educated about identifying early signs of ischemia and infarction.

Take-Home Message

Remember these important points when working with people with angina or silent ischemia:

- Persons with angina or silent ischemia should always be referred to cardiac rehabilitation when it is available.
- Before unsupervised exercise training, anyone with a diagnosis of angina must be able to define angina, describe his unique symptoms, identify provoking factors, describe the immediate treatment for angina, and understand his upper limits of exercise.
- The individual must be able to describe the effects of his medications on his exercise capacity and symptoms.
- An ETT to quantify the ischemic threshold is very important to establish parameters of exercise training. If a formal ETT is not available, the exercise professional should consider whether to perform a low-level submaximal symptom-limited ETT to obtain such data.
- A prolonged warm-up (≥10 min), consisting of active range of motion or low-intensity aerobic activities or both, has an antianginal effect.
- Early in the training period, especially for individuals who are deconditioned, exercise can consist of short-duration (5-10 min) bouts of exercise performed several times per day. As exercise tolerance improves, the bouts can be lengthened and the frequency reduced.
- Any change in frequency, type, or severity of anginal symptoms before, during, or after an exercise session should be evaluated, and the individual's cardiologist should be notified.

Peripheral Arterial Disease

Peripheral arterial disease is atherosclerosis of the peripheral arteries, distal to the aorta. Like atherosclerosis of the coronary and cerebral arteries, PAD is predominantly in areas where the arteries incur wall strain. In the coronary and cerebral arteries, the dominant source is the systolic pulse, but in peripheral arteries it is mainly the bending and kinking of arteries that are a consequence of movement, primarily flexion and extension of the extremities.

Basic Pathophysiology

Peripheral arterial disease is a disorder characterized by atherosclerotic lesions in the arteries of the lower extremities that restrict distal blood flow. Peripheral arterial disease can progress to critical limb ischemia and is associated with a two- to sixfold increased risk of CAD and a four- to fivefold increased risk of strokes. These are risk factors associated with PAD:

- Smoking
- Hypertension
- Diabetes
- Dyslipidemia
- Age

A symptom commonly associated with PAD is intermittent claudication, which is an aching, cramping pain in the lower extremity that can involve the calves, thighs, or buttocks or some combination of these and that is brought on by exercise (especially walking or running) but subsides with rest. Only a small percentage (≈10%) of individuals with PAD develop intermittent claudication, as about 40% won't develop any symptoms and the remaining 50% will experience atypical intermittent claudication. The severity of PAD may be classified into the following categories:

- Grade 0 = asymptomatic
- Grade 1 = intermittent claudication
- Grade 2 = ischemic rest pain
- Grade 3 = minor or major tissue loss from the foot

Management and Medications

Diagnosis of PAD includes a history and physical exam, vascular ultrasound, and angiography or magnetic resonance imaging angiography (MRA). A useful aid in a physical exam screening for PAD is the ankle–brachial index (ABI). The ABI uses a Doppler ultrasound device along with a standard sphygmomanometer to compare the systolic pulse pressure of the posterior tibial and dorsalis pedis arteries to systolic pulse pressure in the brachial artery.

The ABI is obtained after the individual has been in a supine position for 15 min. Blood pres-sure is measured in the brachial artery of both arms and in the posterior tibial and dorsalis pedis arteries of both legs (via Doppler). The arteries in the ankles and arms that yield the highest systolic pressure are used to calculate ABI as a ratio of ankle to brachial artery pressures. A normal ABI score is ≥1.0, whereas a score of <1.0 is abnormal. An ABI ≤0.9 is associated with increased cardiovascular risk even without PAD symptoms, as follows:

- <0.90: significant narrowing of one or more blood vessels
- <0.8: pain in the foot, leg, or buttock may occur during exercise (intermittent claudication)
- <0.3: symptoms may occur when at rest
- ≤0.25: severe limb-threatening circulatory compromise

The single most dominant risk factor for PAD is tobacco smoking, and management of PAD must include an attempt to persuade the individual to quit smoking, including the use of nicotine replacement, varenicline, or bupropion in concert with a comprehensive tobacco cessation program.

Common medications to treat intermittent claudication are listed in table 7.3. Cilostazol and pentoxifylline are approved in the United States for the treatment of intermittent claudication; in Europe, naftidrofuryl is also used. Cilostazol is more effective than any other medication for the treatment of intermittent claudication, increasing walking distance by 40% to 60% after 12 to 24 weeks of use. Cilostazol also improves ABI, walking ability, and quality of life. Naftidrofuryl, too, increases pain-free walking distance by 30% to 40%. Unfortunately, both cilostazol and naftidrofuryl are effective only in a minority of people with intermittent claudication. Pentoxifylline has marginal benefit for pain-free and maximal walking distances but has no effect on ABI. Statins are commonly used in persons with PAD for the purpose of preventing atherosclerosis. This is fortunate, because while statins are not Food and Drug Administration approved for treatment of intermittent claudication, they may improve walking tolerance, ABI, and claudication symptoms.

Individuals with PAD typically have diffuse and widespread atherosclerosis, including coronary artery and cerebrovascular disease. Thus, it is common for such persons to take multiple medications for cardiovascular disease and risk factors (e.g., hypertension, hyperlipidemia, diabetes), especially statins and antiplatelet agents for

Table 7.3   **Medications to Treat Intermittent Claudication**

Medication	Mechanism of action	Adverse effects
Cilostazol	Antiplatelet agent, likely has vasomotor effects	• Diarrhea, loose stools • Headache, dizziness • Contraindicated in persons with heart failure
Naftidrofuryl	Spasmolytic, may have serotonin antagonist activity	• Gastrointestinal upset • Calcium oxalate stones
Pentoxifylline	Phosphodiesterase inhibitor, increases deformability of red cells, improving flow	• Gastrointestinal upset • Headache, dizziness
Clopidogrel Aspirin	Antiplatelet agents	• Bleeding, bruising
Simvastatin	Hydroxymethylglutaryl Co-A reductase inhibitor, likely enhances vascular function	• Musculoskeletal soreness and pain

general atherosclerosis prevention and to lower risk of cardiovascular and cerebrovascular events and mortality. These medications are discussed in chapter 6. People with PAD are also likely to be taking β-blockers, ACE inhibitors, or angiotensin II receptor blockers as cardio- and renoprotective therapy. Antiplatelet agents, including aspirin and clopidogrel, are recommended.

Effects on the Exercise Response

The primary effect of PAD during a single exercise session is the development of claudication pain, due to insufficient blood flow to meet the metabolic demands of the leg musculature. The time or distance or both to onset and to maximal claudication pain during walking are used as criteria for assessing the functional severity of disease. Ankle systolic pressure and ABI are reduced following exercise because blood flow is shunted into the proximal leg musculature at the expense of the periphery and distal areas of the leg.

Medications used in individuals with PAD may affect exercise testing. For example, β-blockers may decrease the time to onset of claudication and decrease maximal HR response to exercise.

Effects of Exercise Training

Exercise training is critically important in the prevention and treatment of PAD. Most important, individuals with PAD who attain high levels of physical activity have one-third the mortality rate of their inactive counterparts. Following supervised exercise training, the majority of individuals with PAD report less severe claudication

at any given level of work. Proposed mechanisms for the increase in exercise tolerance include the following adaptations:

- Increase in leg blood flow because of increased capillary number and improved vasodilation
- More favorable redistribution of blood flow
- Improved hemorheological and fibrinolytic properties of blood (e.g., reduced viscosity)
- Higher concentration of oxidative enzymes in skeletal muscles
- Less reliance on nonoxidative (glycolytic) energy metabolism
- Improvement in the efficiency of walking economy and oxygen uptake kinetics
- Increased free-living daily energy expenditure
- Increased strength

The interaction of exercise training and medication therapy for treatment of intermittent claudication (notably cilostazol, pentoxifylline, and naftidrofuryl) is not well understood.

Recommendations for Exercise Testing

These are the primary objectives of graded exercise testing for individuals with PAD:

- To obtain claudication pain data (pain-free and maximum walking distances)
- To obtain reliable measures of ankle pressure following exercise

- To assess for underlying CAD
- To obtain information that can be used to develop an exercise prescription

Treadmill testing is the preferred mode for exercise testing in people with PAD because it mimics the symptom-provoking activity of daily living and more vigorously stresses the calves than cycle ergometry. The following procedure is used for treadmill testing individuals with PAD:

- Obtain an ABI (as described earlier).
- Conduct a constant-speed test—nominally 2 mph (3.2 km/h), but may need to be individualized.
- Begin at 0% grade and increase 2% every 2 min.
- Monitor and record HR by 12-lead ECG.
- Record brachial blood pressures during the last 30 s of each stage.
- Obtain pain scale ratings (table 7.4), noting time (converted to distance) for each pain score.
- Obtain repeat ABIs immediately postexercise and every 2 to 5 min for 15 min.
- Record the time elapsed from the start of recovery to the relief of claudication pain (a score of 0).

Holding on to the handrails is discouraged because it alters metabolic demands and causes variability in claudication distance. Measurement of expired gases is generally undesirable for this test, because easy and open communication about the pain is the central purpose.

If treadmill testing equipment is not available, the 6-min walk is an excellent alternative for measuring improvements in onset to claudication and maximal distance walked. The blood pressures are impossible to obtain in a 6-min walk, but continuous HR can be measured by telemetry and the distances to claudication ratings can be obtained. This test is less optimal than a treadmill for this purpose because of the self-selected walking speed and because fear avoidance affects the volitional nature of walking speed.

Special considerations include the following:

- Persons with PAD have a high risk for CAD.
- Diabetic neuropathy and spinal stenosis can mimic claudication.
- Feet should be inspected in individuals with resting leg pain (grade 2).
- Ankle–brachial index measures during exercise testing may help with cases that are difficult to diagnose.

Recommendations for Exercise Programming

The Basic *CDD4* Recommendation needs to be modified for people with PAD, with an aerobic exercise program specifically designed for

- improving claudication pain symptoms and
- reducing cardiovascular risk factors.

The Basic *CDD4* Recommendations for strength, flexibility, and balance training are appropriate; it is mainly the aerobic exercise that must be modified.

The type of aerobic exercise training recommended to improve symptoms of claudication is different than for any other chronic condition, for this reason:

Exertion to the point of experiencing leg pain is required for maximum benefit.

Most individuals with PAD should participate in a walking program according to the following parameters:

- Mode: interval walking, pole striding, or stair climbing
- Frequency: three times a week
- Intensity for claudicators:
  - Aim is to cause a pain score of 3 on a 4-point scale (see table 7.4)
  - Onset of claudication should occur in ~5 min.
  - Full recovery is allowed between exercise intervals.

Table 7.4 Claudication Pain Scale

Rating	Description of pain
0	No discomfort at all
1	Minimal discomfort
2	Moderate pain
3	Intense pain
4	Unbearable pain (person has to stop)

- Intensity for nonclaudicators:
  - Start at 40% to 50% of heart rate reserve.
  - Duration: Start with 15 to 30 min of exercise per session, as tolerated.

Most people with PAD, especially those with claudication, are very deconditioned, move slowly, and have very limited endurance to start with. Chronic obstructive pulmonary disease is a common comorbidity and may limit exercise capacity (see "Chronic Obstructive Pulmonary Disease" section). As claudication symptoms allow, exercise intensity can gradually progress to 40 min at 70% to 80% of heart rate reserve over a period of about 6 months. Non–weight-bearing tasks (e.g., cycling) may be used for warm-up and cool-down.

These are special considerations for exercise programming in persons with PAD:

- Improvement in functional capacity may unmask coronary ischemia.
- Changes in comorbidities should be monitored.
- Cold weather may worsen symptoms, necessitating a longer warm-up.
- Cardiovascular responses to exercise may be exaggerated.

Integration Into a Medical Home Model

Regular exercise, especially walking, is critically important for persons with PAD. These individuals, especially those who experience claudication, should initially be referred to a supervised exercise program to monitor for claudication symptoms and severity.

The main reason to refer people with PAD to a supervised exercise program is to provide encouragement and support for overcoming fear avoidance of claudication pain.

Home-based walking programs are an alternative, especially in areas with limited access to a formal rehabilitation program. Home-based programs can be monitored with pedometers and physical activity logs (including electronic media and social networks) or with regular phone or electronic communication. People who follow their home exercise recommendations can achieve benefits similar to those in persons participating in formal exercise or rehabilitation programs but have challenges of overcoming fear avoidance. Facilitating the individual's ability to overcome fear avoidance without face-to-face contact requires artful counseling skills.

Take-Home Message

When working with people who have PAD, remember these key concepts:

- Exercise programs for PAD involve walking regimens that take the individual up to the point of lower-extremity pain, which can improve the time to onset of pain.
- Walking intervals should involve rest periods that allow full recovery of leg pain.
- Treadmill testing with a constant speed (2 mph [3.2 km/g]) and grade increases of 2% every 2 min can be helpful in identifying claudication pain times and disease severity.
- Cardiovascular risk factors should be intensively managed in persons with PAD.
- Individuals with PAD who smoke should be strongly encouraged to stop smoking.

Suggested Readings

COPD

Casaburi R, Patessio A, Ioli F, Zanaboni S, Donner CF, Wasserman K. Reductions in exercise lactic acidosis and ventilation as a result of exercise training in patients with obstructive lung disease. *Am Rev Respir Dis.* 1991;43:9-18.

Cooper CB. Long-term oxygen therapy. In Casaburi R, Petty TL, eds. *Principles and Practice of Pulmonary Rehabilitation.* Philadelphia: Saunders; 1993:183-203.

Cooper CB. Determining the role of exercise in patients with chronic pulmonary disease. *Med Sci Sports Exerc.* 1995;27:147-157.

Cooper CB, Celli B, Wise R, Legg D, Guo J, Kesten S. Treadmill endurance during 2-years treatment with tiotropium in patients with COPD. *Chest.* 2013;144(2):490-497.

Cooper CB, Storer TW. *Exercise Testing and Interpretation: A Practical Approach.* Cambridge: Cambridge University Press; 2001.

Criner GJ, Celli BR. Effect of unsupported arm exercise on ventilatory muscle recruitment in patients with severe chronic airflow obstruction. *Am Rev Respir Dis.* 1988;138:856-861.

Davidson AC, Leach R, George RJD, Geddes DM. Supplemental oxygen and exercise ability in chronic obstructive airways disease. *Thorax.* 1988;43:965-971.

Lozano R, Naghavi M, Foreman K, Lim S, Shibuya K, Aboyans V, Abraham J, Adair T, Aggarwal R, et al. Global and regional mortality from 235 causes of death for 20 age groups in 1990 and 2010: a systematic analysis for the Global Burden of Disease Study 2010. *Lancet.* 2012;380:2095-2128.

Mahler DA. The measurement of dyspnea during exercise in patients with lung disease. *Chest.* 1992;101(Suppl 5):2425-2475.

Minino A, Xu J, Kochanek M. Deaths: preliminary data for 2008. *Natl Vital Stat Rep.* 2010;59:1-71.

Murray CJ, Lopez AD. Measuring the global burden of disease. *N Engl J Med.* 2013;369:448-457.

Ries AL. Position paper of the American Association of Cardiovascular and Pulmonary Rehabilitation: scientific basis of pulmonary rehabilitation. *J Cardiopulm Rehabil.* 1990;10:418-441.

CAD, Angina, Silent Ischemia

Alosa Foundation. Antiplatelet therapy. Available at: www.rxfacts.org [Accessed August 15, 2015].

American Association of Cardiovascular and Pulmonary Rehabilitation. *Guidelines for Cardiac Rehabilitation and Secondary Prevention Programs* (4th ed.). Champaign, IL: Human Kinetics; 2004.

Balady GJ, Williams MA, Ades PA, et al. Core components of cardiac rehabilitation/secondary prevention programs: 2007 update. *Circulation.* 2007;115:2675-2682.

Canty JM Jr. Coronary blood flow and myocardial ischemia. In Bonow R, Mann DL, Zipes DP, Libby P, eds. *Braunwald's Heart Disease: A Textbook of Cardiovascular Medicine* (9th ed.). Philadelphia: Saunders; 2012:1049-1075.

Chaitman BR. Exercise testing, In Zipes D, Libby P, Bonow R, Braunwald E, eds. *Heart Disease* (7th ed.). Philadelphia: Saunders; 2005:153-186.

Durstine JL, Moore GE, LaMonte, MJ, Franklin BA, eds. *Pollock's Textbook of Cardiovascular Disease and Rehabilitation.* Champaign, IL: Human Kinetics; 2008.

Fletcher GF, Mills WC, Taylor WC. Update on exercise stress testing. *Am Fam Physician.* 2006;74(10):749-754.

Franklin BA, Gordon NF. *Contemporary Diagnosis and Management in Cardiovascular Exercise.* Newton, PA: Handbooks in Healthcare Co.; 2009.

Froelicher V, Shetler K, Ashley E. Better decisions through science: exercise testing scores. *Prog Cardiovasc Dis.* 2002;44:395-414.

Hambrecht R, Walther C, Mobius-Winkler S, et al. Percutaneous coronary angioplasty compared with exercise training in patients with stable coronary artery disease: a randomized trial. *Circulation.* 2004;109:1371-1378.

King ML, Williams MA, Fletcher GF, et al. AHA/AACVPR scientific statement. Medical director responsibilities for outpatient cardiac rehabilitation/secondary prevention programs. *Circulation.* 2006;112:3354-3360.

Kones R. Recent advances in the management of chronic stable angina I: approach to the patient, diagnosis, pathophysiology, risk stratification and gender disparities. *Vasc Health Risk Manag.* 2010;6:635-656.

Lee TH. Approach to the patient with chest pain. In Zipes D, Libby P, Bonow R, Braunwald E, eds. *Heart Disease* (7th ed.). Philadelphia: Saunders; 2005:1129-1140.

Lee TH. Chronic coronary artery disease. In Zipes D, Libby P, Bonow R, Braunwald E, eds. *Heart Disease* (7th ed.). Philadelphia: Saunders; 2005:1281-1355.

Morrow DA, Scirica BM, Karwatowska-Prokopczuk E, et al. Effects of ranolazine on recurrent cardiovascular events in patients with non-ST elevation acute coronary syndromes: the MERLIN-TIMI 36 randomized trial. *JAMA.* 2007;297:1775-1783.

Parker BA, Thompson PD. Effect of statins on skeletal muscle: exercise, myopathy, and muscle outcomes. *Exerc Sports Sci Rev.* 2012;40(4):188-194.

Shaw LJ, Bugiardini R, Merz NB. Women and ischemic heart disease. *J Am Coll Cardiol.* 2009;54:1561-1575.

Smith SC, Allen J, Blair SN, et al. AHA/ACC guidelines for secondary prevention for patients with coronary and other atherosclerotic vascular disease: 2006 update. *Circulation.* 2006;113:2363-2372.

Smith SC, Benjamin EJ, Bonow RO, Braun LT, Creager MA, Franklin BA, Gibbons RJ, Grundy SM, Hiratzka LF, Jones DW, Lloyd-Jones DM, Minissian M, Mosca L, Peterson ED, Sacco RL, Spertus J, Stein JH, Taubert KA. AHA/ACCF secondary prevention and risk reduction therapy for patients with coronary and other vascular disease. *J Am Coll Cardiol.* 2011;58(23):2432-2446.

Zuchi C, Tritto I, Ambrosio G. Angina pectoris in women: focus on microvascular disease. *Int J Cardiol.* 2013 Feb 20;163(2):132-140.

PAD

Aboyans V, Criqui MH, Abraham P, et al. Measurement and interpretation of the ankle-brachial index: a scientific statement from the American Heart Association. *Circulation.* 2012;126:2890-2909.

Aboyans V, Criqui MH, Denenberg JO, Knoke JD, Ridker PM, Fronek A. Risk factors for progression of peripheral arterial disease in large and small vessels. *Circulation.* 2006;113(22):2623-2629.

Dawson DL, Cutler BS, Hiatt WR, et al. A comparison of cilostazol and pentoxifylline for treating intermittent claudication. *Am J Med.* 2000;109:523-530.

Gardner AW, Afaq A. Management of lower extremity peripheral artery disease. *J. Cardiopulm Rehabil Prev.* 2008;8:349-357.

Gardner AW, Katzel LI, Sorkin JD, Sorkin JD, Bradham DD, Hochberg MC, Flinn WR, Goldberg AP. Exercise rehabilitation improves functional outcomes and peripheral circulation in patients with intermittent claudication: a randomized controlled trial. *J Am Geriatr Soc. 2001*;49:755-762.

Gardner AW, Montgomery PS, Flinn WR, Katzel LI. The effect of exercise intensity on the response to exercise rehabilitation in patients with intermittent claudication. *J Vasc Surg.* 2005;42:702-709.

Izquierdo-Porrera AM, Gardner AW, Powell CC, Katzel LI. Effects of exercise rehabilitation on cardiovascular risk factors in older patients with peripheral arterial occlusive disease. *J Vasc Surg.* 2000;31:670-677.

Lau JF, Weinberg MD, Olin JW. Peripheral artery disease. Part 1: clinical evaluation and noninvasive diagnosis. *Nat Rev Cardiol.* 2011;8:405-418.

Makris GC, Lattimer CR, Lavida A, Geroulakos G. Availability of supervised exercise programs and the role of structured home-based exercise in peripheral arterial disease. *Eur J Vasc Endovasc Surg.* 2012;44:569-575.

Murabito JM, D'Agostino RB, Silbershatz H, Wilson PW. Intermittent claudication. A risk profile from The Framingham Heart Study. *Circulation.* 1997;96:44-49.

Norgren L, Hiatt WR, Dormandy, JA, Nehler MA, Harris KA. Inter-Society Consensus for the Management of Peripheral Arterial Disease (TASC II). *J Vasc Surg.* 2007;45(Suppl S):S5-S67.

Ritta-Dias RM, Meneses AL, Parker DE, Montgomery PS, Khurana A, Gardner AW. Cardiovascular responses to walking in patients with peripheral artery disease. *Med. Sci Sports Exerc.* 2011;43:2017-2023.

Rooke TW, Hirsch AT, Misra S, et al. ACCF/AHA 2011 focused update of the guideline for the management of patients with peripheral arterial disease (updating the 2005 guideline). *J Am Coll Cardiol.* 2011;47:2020-2045.

Treesak C, Kasemsup V, Treat-Jacobson D, Nyman JA, Hisrch AT. Cost-effectiveness of exercise training to improve claudication symptoms in patients with peripheral arterial disease. *Vasc Med.* 2004;9:279-285.

Womack CJ, Sieminski DJ, Katzel LI, Yataco A, Gardner AW. Improved walking economy in patients with peripheral arterial occlusive disease. *Med Sci Sports Exerc.* 1997;29:1286-1290.

Additional Resources

Related ACSM Position Stands

American College of Sports Medicine. American College of Sports Medicine position stand: exercise for patients with coronary artery disease. *Med Sci Sports Exerc.* 1994;26(3):i-v.

Evidence-Based Guidelines

A report of the American College of Cardiology/American Heart Association Task Force on Practice Guidelines Writing Group. 2007 chronic angina focused update of the ACC/AHA 2002 guidelines for the management of patients with chronic stable angina. *J Am Coll Cardiol.* 2007;50:2264-2274.

Suggested Websites

American Association of Cardiovascular and Pulmonary Rehabilitation (AACVPR), Cardiac Rehabilitation Patient Resources. www.aacvpr.org/CardiacRehabilitationPatients/tabid/503/Default.aspx [Accessed November 11, 2015].

American College of Cardiology, Cardiosource. www.cardiosource.org/acc [Accessed November 11, 2015].

American Heart Association, Heart Attack Patient Information. www.heart.org/HEARTORG/Conditions/HeartAttack/HeartAttack_UCM_001092_SubHomePage.jsp [Accessed November 11, 2015].

Diagnosis and Management of Stable Chronic Obstructive Pulmonary Disease: A Clinical Practice Guideline from the American College of Physicians, American College of Chest Physicians, American Thoracic Society, and European Respiratory Society (2011). www.thoracic.org/statements [Accessed November 11, 2015].

Global Initiative for Chronic Obstructive Lung Disease. Global Strategy for Diagnosis, Management, and Prevention of COPD. www.goldcopd.org/Guidelines/guidelines-resources.html [Accessed November 11, 2015].

National Library of Medicine (NLM) National Institutes of Health (NIH) Medline Plus, Patient Information, Angina. www.nlm.nih.gov/medlineplus/angina.html [Accessed November 11, 2015].

Preventive Cardiovascular Nurses Association (PCNA). http://pcna.net/clinical-tools/tools-for-healthcare-providers [Accessed November 11, 2015].

Cancer

Cancer is not a single condition based in a single organ system; on the contrary, there are over 200 different types of cancer. Many of the most common cancers (but not all forms of cancer) are very strongly associated with tobacco use and with physical inactivity. Worldwide, every year nearly 13 million people are diagnosed with cancer. In the United States, over 1.5 million people are diagnosed with cancer annually, with more than 500,000 Americans dying each year. But today in the United States, nearly two-thirds of all cancer patients live for at least 5 years, resulting in over 13 million Americans who are living with cancer as a major chronic condition. Some 40% of all Americans will be diagnosed with cancer during their lifetime.

The most common causes of cancer are strongly related to physical inactivity, tobacco, or both, with tobacco by far the single most important cause of cancer, estimated as the cause of one in five deaths from cancer:

- Four out of five cases of lung cancer are caused by tobacco.
- One out of five cases of stomach cancer are caused by tobacco.
- Obesity and physical inactivity increase risk of breast cancer in women by ~50%.
- Physical activity reduces risk of colorectal cancer by up to 50%.

Lung, stomach, colorectal, and breast cancer (in women) account for about 40% of all cancers, suggesting a very strong link between tobacco, physical inactivity, and cancer.

Basic Pathophysiology

Cancer is an unchecked accumulation of cells for which the normal life cycle of cellular reproduction and senescence has failed. This failure is gener-ally thought to be the result of multiple genetic changes (called the *multiple hit hypothesis*) that alter cell signaling and metabolism.

Cancer cells reproduce quickly and are metabol-ically active, and a tumor has a greater and greater effect on systemic metabolism as it progresses. For some cancers (e.g., lung and gastrointestinal cancers), diagnosis may stem from unexplained weight loss once the tumor grows large enough to require large amounts of energy to keep growing, have mass effects, or produce sufficient amounts of cytokines to affect appetite and satiety. Some tumors are relatively symptomless and silent until they are large enough to be palpable, at which point the tumor may have broken through the basement membrane of the tissue of origin and spread to local or even distant tissue.

Cancers can affect any body system and are often grouped by the type of tissue from which they arise:

- Carcinoma—epithelial tissues (e.g., mela-noma, colon, and breast)
- Sarcoma—bone or connective tissues
- Leukemia—blood cells
- Lymphomas—lymphatic system
- Central nervous system tumors

Symptoms of cancer vary widely according to organ site. Hallmark symptoms associated with a newly diagnosed cancer include the following:

- Unintentional weight loss
- Unexplained cough
- Blood in the stool or urine
- Pain or change in the function of organ system affected

That stated, the more common diagnoses of breast, colon, and prostate cancers often occur after detection by screening, before the onset of frank symptoms.

Over several decades, screening tests for the more common cancers (e.g., mammography for breast and fecal occult blood tests and colonoscopy for colon cancer) have allowed many cancers to be caught at an earlier stage of progression. Further, cancers are treated in an outpatient setting now, including same-day curative surgeries (e.g., breast lumpectomies). The combination of successful screening programs with improved treatments means that many cancer survivors live for years, even decades, after completing treatment. There are an estimated 14 million cancer survivors in the United States today and, given that one in three adults will have a cancer diagnosis and that most of them will live a long time after completing treatment, the number of cancer survivors is expected to swell to over 25 million by 2050.

Management and Medications

The task of curing cancer is often demanding for the medical team and can be all-consuming for patients and their caregivers. Curative cancer care generally includes surgery (for solid tumors), chemotherapy, radiotherapy, or more than one of these. Historically, cancer treatments were so toxic as to make it impossible for people to consider functioning in the rest of life while undergoing treatment. Newer treatments also include monoclonal antibodies and biological therapies, as well as immunologic and hormonal therapies. If a cancer is considered to be too advanced for curative care, some of these same treatments may be used for palliation. Many cancer survivors have persistent adverse effects of cancer treatment, which have well-established predisposing host factors, including age, sex, and other comorbid health conditions. The predisposing host factors and combinations of treatments may be synergistic with regard to incidence and severity of persistent adverse effects.

Chemotherapy

Chemotherapy is toxic to healthy cells as well as to cancer cells. The "collateral damage" to healthy cells forms the basis of the adverse treatment effects. The timing of cancer treatment is modified according to the individual response to treatment, balancing the curative benefit and adverse effects. If the size of a solid tumor decreases less than expected after the prescribed chemotherapy or radiotherapy, there may be a second course of the same treatment, or some additional treatment might be tried. Or, if the adverse effects are too profound to continue treatment (e.g., intolerable peripheral neuropathy or nausea), there may be a delay or change in planned therapy. Treatment alterations are common, occurring in up to one-third of patients, which makes it challenging to predict the exact course of treatment and associated adverse treatment effects.

Chemotherapy is generally delivered intravenously, in combinations of drugs that vary by mechanism of action, with a period of time for recovery after an intravenous injection. For example, someone may have four to eight clinic appointments for intravenous injections of combination chemotherapy, in tandem or sequentially, with 1 to 4 weeks of recovery between injections. The mechanisms of action of these drugs all relate to interrupting some aspect of cell signaling, reproduction, or cell death; and the drugs vary from antimetabolites to antitumor antibiotics and DNA topoisomerase inhibitors, as well as alkylating agents.

Acute symptoms during chemotherapy vary according to where the individual is within the treatment process and typically accumulate over additional doses. The most common symptoms are those related to the death of cells within tissues in which cells turn over rapidly, including endothelial tissues (hair, nails, skin, gut, mouth) and blood cells produced in bone marrow. Therefore, these are the most common acute symptoms:

- Fatigue
- Nausea and vomiting
- Mouth sores
- Changes in sensations of taste and smell
- Changes in skin, hair, and nail
- Anemia, neutropenia, and thrombocytopenia

Some neurotoxic chemotherapy agents can result in peripheral neuropathies and ototoxicity (producing tinnitus).

The persistent effects of chemotherapy vary according to the agent(s) and dosages used. The low blood cell counts that develop acutely during treatment can persist beyond the end of chemotherapy. Change in body composition also is common; however, the specific gains and losses (total weight, lean body mass, fat mass, and so on) may differ by chemotherapeutic agent.

Treatment with anthracyclines or traztusumab (Herceptin) is associated with cardiotoxicity. Several chemotherapeutic agents used to treat breast cancer are associated with pulmonary damage (e.g., cyclophosphamide, paclitaxel). In premenopausal women, there is potential for chemotherapy-related amenorrhea or premature menopause or both, particularly after treatment with alkylating agents (e.g., cyclophosphamide). Taxanes are associated with autonomic neuropathy (postural hypotension, constipation, bladder problems), as well as peripheral sensory and motor neuropathies. There are negative short- and long-term adverse effects on cognitive function (so-called chemo brain).

Radiation

Radiation therapy treats cancer by damaging the DNA of cancer cells with ionizing radiation. Ideally, radiation treatments would work only on the cancer cells that are the intended target, but—as with chemotherapy—there is collateral damage to healthy cells in the vicinity of the irradiated tissue. The science and technology behind radiotherapy have improved dramatically over the past 50 years, leading to the development of very advanced therapies such as proton therapy, which promises to more precisely target ionizing radiation to the cancerous cells, producing less collateral damage to surrounding healthy tissue. For the majority of persons with cancer, however, the most common type of radiotherapy is *external beam radiotherapy,* which does damage tissues surrounding the malignant tumor.

Acute adverse effects of radiation treatments vary somewhat according to the location of the treatments, but often include the following:

- Fatigue
- Nausea and vomiting
- Skin changes
- Anemias
- Swelling

Additional symptoms may include changes in sexual function, difficulty swallowing, diarrhea, bladder changes, mouth problems, and hair loss in the treatment area. Some of these effects persist beyond treatment, but this varies from one individual to another. The following treatment-related factors are associated with incidence of persistent radiation toxicities:

- Dose
- Volume of tissue irradiated
- Severity of acute effects from irradiation
- Previous surgeries
- Concomitant chemotherapy

Organ systems not targeted for treatment may experience negative effects of radiation because they happen to be in the same region of the body. Examples are the pulmonary scarring and cardiac changes that can appear as late effects after radiation treatment to the chest wall. Severe, long-term side effects include lymphedema and pulmonary and cardiac toxicities, as well as damage to gastrointestinal tissues and sustained immunosuppression. The risk of radiation pneumonitis is higher in people treated with combination or sequential paclitaxel chemotherapy and radiation. Radiation also may contribute to the arm and shoulder morbidities seen after breast cancer treatment through fibrotic damage to soft or contractile tissue or both.

Surgery

With any surgical procedure there is a possibility of having long-term lingering pain, changes in appearance, psychosocial effects, and impaired wound healing or tightness of skin at the site of the surgery. The rate of wound healing varies by host factors and extent of surgery. Some operations result in permanent changes (e.g., limb amputation or an ostomy). Body systems other than the organ or tissue where the surgery is performed may also be affected by the trauma of surgery. For example, breast cancer surgeries result in significant shoulder and arm morbidity due to the soft tissue effects of surgery (and radiotherapy). Surgery to remove cancerous tissue anywhere in the body can require severing muscles and cutting through multiple layers of soft tissue; this can result in scarring, altered range of motion, and altered strength and function of the affected site.

Hormone Therapies

Hormonal therapies include antiestrogens, aromatase inhibitors, antiandrogens, and luteinizing hormone-releasing hormone (LHRH) analogs and antagonists. The magnitude of the side effects and symptoms varies by host and treatment factors, including the mechanism of action for a specific drug. Common side effects from antiestrogens (e.g., tamoxifen) are similar to the symptoms of menopause and include the following:

- Fatigue
- Hot flashes
- Vaginal discharge
- Weight gain
- Mood swings

Premenopausal women who take tamoxifen may experience irregular menstrual periods. These are the side effects associated with taking aromatase inhibitors (prescribed for postmenopausal women with estrogen-linked cancers):

- Bone loss
- Joint stiffness
- Muscle pain

Medical or surgical ovarian ablation (risk-reducing salpingo-oophorectomy) may be used in premenopausal women who have hormone-sensitive reproductive cancers, because this reduces the risk of recurrence in these relatively young women. This treatment abruptly induces premature menopause in relatively young women, however, and the following changes are to be expected with a sudden and permanent decrease in ovarian steroid hormones:

- Increases in body weight
- Cardiovascular risk
- Vascular symptoms
- Bone loss

- Skin changes
- Vaginal dryness

The LHRH analogs and antagonists reduce the production of androgens (e.g., testosterone) and are associated with numerous symptoms and side effects, including loss of libido and potency, vasomotor symptoms, breast tenderness, osteoporosis, anemia, cognitive changes (acuity), loss of muscle mass, weight gain, fatigue, cardiovascular risk increases, and depression. When used in combination with LHRH analogs or agonists, antiandrogens also may cause diarrhea, nausea, liver problems, and fatigue.

Targeted Therapies

Comparatively recent additions to the arsenal of cancer treatment include agents that target cancer or more specifically the products of cancer. These agents can be delivered either alone or in combination with chemotherapy, orally or through an intravenous infusion. As these treatments are comparatively new, less is known about their long-term effects. Some of the more common targeted therapies have included Herceptin, lapatinib, and avastin (table 8.1).

Another important late effect from chemotherapy or radiotherapy includes the increased risk for secondary cancers that are caused by the treatment. Numerous types of chemotherapeutic agents, including alkylating agents, cisplatin, topoisomerase II inhibitors, and athracyclines, are linked to the development of leukemias. Specific links have been noted between radiation to the

Table 8.1 Targeted Therapy Medications for Cancer Treatment

Medication	Mechanism of action	Adverse effects
Herceptin	• Monoclonal antibody that blocks the receptor for human epidermal growth factor 2 (HER2) • Used in tumors that overexpress the HER2 receptor	• Cardiac dysfunction • Affected by medical factors and medications being used for comorbid conditions
Lapatinib	• Inhibits tyrosine kinase domain of both epidermal growth factor receptor (EGFR) and HER2	• Typically used orally after Herceptin • Diarrhea, flu-like symptoms, fatigue, and rash (foot–hand syndrome) • Liver damage can occur late after use • QT prolongation risk during active treatment
Avastin	• Binds to vascular epithelial growth factor (VEGF) ligand • Inhibits growth of new blood vessels	• Gastrointestinal perforation, wound complications, hemorrhage, and cardiotoxicities

breast and the later development of lung cancer and leukemia. Survivors of childhood cancers who received radiotherapy are at elevated risk for thyroid and breast cancer, as well as leukemias and sarcomas.

Effects on the Exercise Response

There are no typical effects of cancer on the exercise response. On the side of negligible impact, the effect of a particular cancer on exercise capacity is usually minimal for most early-stage cancers, with the possible exception of reduced exercise capacity due to cancer-related fatigue. But in people who have larger tumors, the location and magnitude of cellular overgrowth influence the effects on the body and the ability to exercise and can have a profound impact. Where the overgrowth happens and how big it gets before being detected will affect how the cancer influences physical function and ability to exercise. For example, a brain tumor, benign or malignant, will press on the surrounding neural tissue, altering specific physiological functions according to where the tumor is in the brain. A tumor in the gut will press on organs and may alter digestion or cause pain or both. Someone who receives an amputation after diagnosis of osteosarcoma may be cured but is forever disabled and at risk for becoming sedentary, secondarily obese, and diabetic.

The effects of diagnosis and treatment for cancer on exercise response will vary by

- tumor site,
- cancer stage,
- grade (e.g., severity),
- treatment types,
- timing since diagnosis, and
- interactions of all elements with general health, well-being, and fitness.

For people undergoing active treatment, whether curative or palliative, exercise tolerance will vary from day to day in ways that are not at all predictable. It is common for the period of treatment to yield deconditioning and loss of fitness; some of this is related to the treatment's causing loss of robustness (e.g., sarcopenia), and some is from physical inactivity and deconditioning. In individuals who cannot be cured, some loss of physical functioning can be related to progression of the malignancy. After completion of cancer treatment, it may be a period of months before energy levels recover (with the return of full function of the bone marrow) and exercise tolerance improves.

Effects of Exercise Training

Hundreds of intervention studies published in the peer-reviewed scientific literature show that exercise is beneficial in people with cancer, although the majority of these studies have been conducted with breast cancer survivors. Exercise training has benefits for those who are actively undergoing cancer treatment and those who have completed treatment. Established benefits of exercise training in persons with cancer during and after treatment include improvements in

- quality of life,
- fatigue,
- strength,
- cardiorespiratory fitness,
- flexibility,
- physical functioning,
- depression and anxiety,
- lymphedema,
- body composition,
- sleep, and
- energy level.

Most impressive of all, moderate to vigorous intensity progressive exercise training prevents recurrence and death from breast and colon cancers. The overriding message, with details still to be elucidated in most cancers, is that exercise training is beneficial.

In addition to the benefits on cancer-free survival and quality of life, with cancers that are in remission and believed to be cured, exercise training takes on the same role as in other people with a chronic condition. The majority of cancers occur in adults who are older and have comorbid chronic conditions, including the other common chronic diseases reviewed in this volume (e.g., diabetes, cardiovascular disease, hypertension, obesity); and in a cancer survivor these conditions end up being the dominant conditions. Prostate cancer survivors are more likely to die of cardiovascular disease than their cancer, and breast cancer survivors after 10 years of survival are more likely to die of cardiovascular causes than breast cancer. Thus the benefits of exercise in cancer survivors also include those relevant to most adults with common chronic illnesses, including the well-established benefits to cardiorespiratory, metabolic, musculoskeletal, immunologic, neurologic, and psychological well-being. Finally, it can be noted that even if exercise training does not prevent recurrence of a cancer, those who are more fit are better able to withstand strenuous medical interventions. It could be said that individuals with cancer have the motivation to be in the best shape of their lives for the fight of their lives.

Key gaps in knowledge to be aware of regarding exercise for those who have had cancer include a poor understanding of

- exercise safety,
- exercise tolerance,
- benefits among cancer survivors who are older, and
- benefits in those with metastatic, rare, or unstudied cancers.

The primary challenges regarding exercise in the cancer survivor relate to the effects of the treatments undertaken in the name of curative care more than to the cancer itself. There is a separate set of issues in meeting the exercise needs of those with incurable cancer. In an increasing number of cancers, metastatic no longer means "dying soon." Palliative care can be ongoing for a decade or more, as in the example of some subtypes of breast and ovarian cancers. Some people cycle in and out of palliative treatment for the remainder of their lives. Less is known about exercise training safety and benefits in this population, but later in the chapter we present some guidance for the practicing clinician faced with a person with metastatic cancer interested in exercise training. When one considers exercise in a cancer survivor, one must take toxicity of cancer-specific care into account. The adverse effects of cancer treatments may be acute, resolving over a period of days, weeks, or months, or they may be persistent, lasting years after treatment is completed. Further, there are some adverse cancer treatment effects that may not appear for years or even decades after the end of treatment. The reader is referred to the 2006 Institute of Medicine report on adult cancer survivorship titled "From Cancer Patient to Survivor: Lost in Transition" for a more in-depth review of this topic than is possible here.

> *Cancer is not one condition, but many. So the treatment of cancer is also quite varied. The effects of a specific cancer on the ability to be physically active are unique to that cancer and that individual's manifestation of cancer, and the effects of treatment are equally varied. The role of exercise programming is likely to vary during the course of treatment and recovery. Because of this complexity, persons with cancer should be referred to an exercise specialist trained in cancer rehabilitation.*

Recommendations for Exercise Testing

These are specific exercise testing recommendations:

- For persons in active cancer treatment: use the Basic *CDD4* Recommendation
- For high-functioning individuals or those in remission: use the ACSM Guidelines
- Exercise professionals working with people who have cancer should seek ACSM certification as a Certified Cancer Exercise Trainer.

One challenge for individuals with a chronic condition, noted many times in this book, is the need to balance the goal of increasing physical activity and fitness with the risks of exposing them to exercise-associated adverse effects and the risk of creating a barrier to exercise. An exercise test can help define risks, but it can also be a substantial barrier to becoming more active. Ideally, each individual affected by cancer would undergo a full physical examination of all body systems relevant to exercise, as well as evaluation of cardiorespiratory fitness, musculoskeletal strength and endurance, balance, flexibility, agility, and coordination. However, requiring even one of these assessments before recommending exercise would usually create an unnecessary barrier to accessing exercise programming in a population at high risk for sedentary behavior because of the historic "Rest, take it easy, don't push yourself" rhetoric from oncologists (rhetoric that is still common, particularly in community cancer treatment settings). Thus, if exercise testing poses a barrier to being more active, exercise oncology experts generally agree that testing should be skipped in favor of initiating low-intensity activity such as walking.

Cancer rehabilitation programs can be useful to restore function among those who have lost the ability to do even low-intensity activities or those requiring more supervision to learn how to recover from deconditioning associated with cancer and its treatments (e.g., adults who are older or who have multiple chronic conditions). If low-intensity activities such as walking are not feasible due to the level of deconditioning, the individual should be evaluated in a medical setting (rehabilitation medicine, physical or occupational therapy, or both), in order to regain basic levels of fitness required for independent living. If an individual is deconditioned and does not seem able to work independently to recover his baseline precancer level of function, have him evaluated for cancer rehabilitation services. After completing cancer rehabilitation, he should progress to a community-based program with the goal of advancing beyond basic functioning.

If the opportunity arises to evaluate those who've had a cancer diagnosis through exercise testing, the testing recommended should be individualized based on the person's goals. For example, if a breast cancer survivor wants to return to playing tennis after surgery and radiation in the upper body, it would be ideal to evaluate range of motion, strength, and endurance of the upper body, as well as cardiorespiratory fitness and balance-agility, before allowing her to return to the courts.

Another reason to consider exercise testing in the cancer survivorship population is to screen for the common persistent adverse effects of treatment that may be of particular pertinence with regard to the safety and efficacy of exercise training. One example would be to screen for lymphedema among those who have had lymph node removal or irradiation as part of their cancer therapy (including survivors of breast, bladder, melanoma, head and neck, and gynecologic cancers). Because this population is at risk for so many long-term sequelae after cancer treatment, it may be of value to do regular prospective surveillance for those common sequelae (see table 8.2). The frequency recommended for this kind of surveillance is not generally agreed on today and varies by adverse effect of treatment, so exercise clinicians will have to improvise and create their own protocols for now.

Recommendations for Exercise Programming

Exercise professionals working with people who have cancer should seek ACSM certification as a Certified Cancer Exercise Trainer.

The ACSM roundtable on exercise for those who have been diagnosed with cancer was convened in 2009 and the subsequent recommendations published in 2010. The first two words in the ACSM guidance for exercise among those affected by cancer are the same as the first two words in the 2008 U.S. DHHS Guidelines for physical activity:

Avoid inactivity.

Table 8.2 Recommendations for Persistent Adverse Effects of Cancer Treatment

Adverse effect of breast cancer treatment	Testing recommended
Lymphedema	Examination by a certified lymphatic therapist
Upper-body dysfunction	Examination by a physical therapist
Fatigue	Screening survey
Psychosocial distress, depression, anxiety	PHQ-9 (depression screening questionnaire)
Deconditioning	Cardiorespiratory fitness assessment, musculoskeletal strength and endurance testing
Peripheral neuropathy	Balance tests, nerve conduction study, neurology consultation
Bone pain	Bone densitometry
Cardiotoxicity	Electrocardiogram and echocardiogram
Weight gain, sarcopenia	Anthropometrics, body composition assessment

This is a complete reversal of the decades-long advice to "rest, take it easy, and don't push yourself" that is unfortunately still often used with cancer patients and survivors. The goal of starting with these two words was to ensure to communicate to clinicians caring for those affected by cancer that there is potential harm to being sedentary during and after cancer treatment, as great as or greater than the potential harm to being active during and after cancer treatment.

Beyond avoidance of inactivity, the exercise recommendations for those affected by cancer are generally the same as for all Americans. Therefore, this is the recommendation for exercise in individuals in treatment for cancer and for survivors:

- To start, use the Basic *CDD4* Recommendations.
- Build up as possible to the full Health and Human Services Guidelines.
- Be alert to the need to evaluate exercise symptoms that arise during the program.

One should consider a few factors when developing exercise programming for those affected by cancer:

- Evaluate the level of exercise independence for which the individual is ready.
- In people with severely impaired physical functioning, evaluate for rehabilitation services.
- In others, start with light to moderate walking.
- Progress to more vigorous walking to ensure all the benefits of exercise.
- Periodically screen for relevant persistent adverse effects of treatment.
- Find a fitness professional with adequate training (registry of ACSM Certified Cancer Exercise Trainers at http://members.acsm.org/source/custom/Online_locator/Online-Locator.cfm.

People who have undergone or are undergoing chemotherapy experience additional adverse effects that must be taken into account. In active treatment,

- never overtax the body enough to exacerbate symptoms of the adverse effects, and
- be prepared for extreme day-to-day variations in physical functioning.

Setbacks with symptoms prevent continuing treatment without interruption, and treatment delays or dose alterations are associated with worse survival outcomes. Also, acute symptoms vary across the time period of chemotherapy, altering exercise tolerance in a manner that is not always predictable.

Rehabilitative exercises for recovery of full function are not always prescribed, and the extent to which they are needed after specific types of surgeries is not well established. The potential for exercise to prevent second cancers among survivors treated with chemotherapy or radiation has not been a focus of the field of exercise and cancer thus far. However, exercise has been demonstrated to be useful for preventing or attenuating many of the persistent effects of cancer treatment and for rehabilitation.

Another special challenge for those who have been active before cancer diagnosis and who are eager to return to exercise is that most people don't understand the general principles of exercise training and deconditioning. Commonly, formerly active cancer survivors attempt to come back to their sport or activity too soon or too fast, resulting in injury and a setback. A general training principle often overlooked in this setting is that the loss of muscle mass and strength makes the musculoskeletal system weaker and less able to withstand physical activity. People who are motivated need to understand that their mind is willing but their body will take some time to get back to where it once was.

Conversely, progression is necessary, and many cancer survivors need to be pushed a little, if the exercise specialist feels that they can handle higher intensity. Doing only walking has benefits, but there is ample evidence in this population for progressing exercise beyond walking, and persons with cancer are well served to do so when they are able.

Integration Into a Medical Home Model

Anyone affected by cancer should be exercising regularly. The benefits of exercise are well documented in this population, but there is reason for caution in sending the average cancer survivor to the gym without any evaluation, screening, or supervision. Primary care providers should be telling their patients affected by cancer to be active in the following ways:

- Avoid inactivity: Remain as active as possible during treatment and recover activity as quickly as possible after completing treatment.
- Walk, if you are able. Increase walking to 150 min per week. If people can do more, they should do more.
- If someone is unable to do usual activities during or after cancer treatment due to an adverse treatment effect, it is not necessary to get used to this as the "new normal."
- Refer those who are debilitated to rehabilitation medicine or physical therapy or both for evaluation and treatment, followed by progressive exercise training.
- The preferred setting is a supervised program under the direction of a fitness professional with advanced training in working with cancer survivors (e.g., ACSM Certified Cancer Exercise Trainer).

Minimize the adverse effects of exercise by fully understanding the treatments that an individual has undergone. Review the person's treatment summary, including the summary of persistent adverse effects. Use this document to understand if the individual has risks for long-term cardiovascular, pulmonary, neurologic, immunologic, endocrine, metabolic, or musculoskeletal issues for which screening tests *and* exercise prescription are appropriate.

Suggested Readings

Brown JC, Troxel AB, Schmitz KH. Safety of weightlifting among women with or at risk for breast cancer-related lymphedema: musculoskeletal injuries and health care use in a weightlifting rehabilitation trial. *Oncologist.* 2012;17(8):1120-1128.

Hewitt M, Greenfield S, Stovall E, eds. *From Cancer Patient to Cancer Survivor: Lost in Transition.* Washington, DC: National Academies Press; 2006.

Irwin ML, ed. *ACSM's Guide to Exercise and Cancer Survivorship* (1st ed.). Champaign, IL: Human Kinetics; 2012.

Ligibel JA, Denlinger CS. New NCCN guidelines for survivorship care. *J Natl Compr Canc Netw.* 2013;11:640-644.

National Lymphedema Network Medical Advisory Committee. Position statement of the National Lymphedema Network. Topic: Exercise. 2008. Available at: www.lymphnet. org/pdfDocs/nlnexercise.pdf [Accessed November 11, 2015].

Rock CL, Doyle C, Demark-Wahnefried W, Meyerhardt J, Courneya KS, Schwartz AL, Bandera EV, Hamilton KK, Grant B, McCullough M, Byers T, Gansler T. Nutrition and physical activity guidelines for cancer survivors. *CA Cancer J Clin.* 2012;62(4):242-274.

Schmitz KH, Courneya KS, Matthews C, Demark-Wahnefried W, Galvão DA, Pinto BM, Irwin ML, Wolin KY, Segal RJ, Lucia A, Schneider CM, von Gruenigen VE, Schwartz AL; American College of Sports Medicine. American College of Sports Medicine roundtable on exercise guidelines for cancer survivors. *Med Sci Sports Exerc.* 2010;42(7):1409-1426.

Silver JK, Baima J. Cancer prehabilitation: an opportunity to decrease treatment-related morbidity, increase cancer treatment options, and improve physical and psychological health outcomes. *Am J Phys Med Rehabil.* 2013;92(8):715-727.

Silver JK, Baima J, Mayer RS. Impairment-driven cancer rehabilitation: an essential component of quality care and survivorship. *CA Cancer J Clin.* doi: 10.3322/caac.21186. Epub 2013 July 15.

Speck RM, Courneya KS, Mâsse LC, Duval S, Schmitz KH. An update of controlled physical activity trials in cancer survivors: a systematic review and meta-analysis. *J Cancer Surviv.* 2010;4(2):87-100.

Additional Resources

American Cancer Society, www.cancer.org [Accessed November 11, 2015].

Cancer Support Community, www.cancersupportcommunity.org [Accessed November 11, 2015].

National Cancer Institute, www.cancer.gov [Accessed November 11, 2015].

National Lymphedema Network, www.lymphnet.org [Accessed November 11, 2015].

U.S. Department of Health and Human Services, Physical Activity Guidelines for Americans, www.health.gov/paguidelines/guidelines [Accessed November 11, 2015].

Significant Sequelae Related to Common Chronic Conditions

The previous chapters have discussed how to approach exercise management in persons with very common chronic conditions that are strongly associated with physical inactivity and tobacco use. We have also noted the phenomenon of comorbid conditions and the fact that the conditions in this part of the book often appear in combination. In this chapter, we consider common chronic comorbidities that develop as a consequence of chronic disease and that are often seen in persons who have the other conditions discussed in this book. These related comorbidities are as follows:

- Depression as a comorbidity
- Lower-limb amputation
- Frailty

Depression and frailty are common in many people with chronic conditions and with aging. Lower-limb amputation is less commonly seen, but it is a related condition in two important ways. First, in persons with diabetes who have an amputation due to vascular insufficiency, neuropathy, or osteomyelitis, it remains a highly feared and very costly complication that substantially increases the energetic cost of locomotion and physical activity, leading to a more sedentary lifestyle. Second, in persons who suffer a traumatic amputation, due to the impact on physical inactivity there is a highly increased risk of developing a sedentary lifestyle and secondarily developing obesity and type 2 diabetes. Accordingly, we have included amputation as an important comorbidity of the common chronic conditions, but diabetes is the main concern.

In the following sections, we discuss each of these individual conditions.

Depression as a Comorbidity

This section primarily addresses persistent depressive disorder (previously called dysthymia) as a comorbidity. For more details specifically on major depression and anxiety disorders, see chapter 32.

Up to half of all persons with a chronic disease or disability experience depression as a comorbid condition. Critically, depression is a high-risk factor for noncompliance with prescribed treatments. People who are depressed are approximately three times more likely than others to not adhere to practitioner recommendations, which adversely affects outcomes. A sense of hopelessness, commonly experienced with depression, can reduce belief in the efficacy of treatment and can impede treatment progress because optimistic expectations are important for compliance. Additionally, depression is often accompanied by social isolation and withdrawal. Considerable research suggests that support from family and friends facilitates a person's attempts to comply with treatment.

Any person who has a chronic condition or disability should be screened for depressive symptoms, for these reasons:

- Depression can be a barrier to engaging in and sustaining an exercise program.
- Depression reduces adherence to medical treatment for other conditions.
- Exercise therapy can directly benefit depressive symptoms.

Basic Pathophysiology

While the biological causes of depression are likely multifaceted, an imbalance in the serotonergic or endocrine pathways (or both) in the brain is believed to be a central cause of depression. These pathways are involved in the neural integration of several physiological systems, so depression has many physical as well as mental and cognitive symptoms. New research shows that depression is also strongly associated with inflammation. Notably, both inflammation and the brain pathways linked to depression can be modified by exercise.

Hallmark symptoms of depression are many and varied, but they commonly include

- profound sadness,
- fatigue,
- malaise,
- insomnia,
- anhedonia (inability to experience pleasure),
- suicidal ideation,
- self-degrading feelings,
- irritability,
- confused or confounded thinking,
- loss of appetite, and
- involuntary weight changes.

These are examples of events that can trigger a depressive episode:

- Emotional trauma
- Hormonal changes
- Stress (from many causes, e.g., financial problems, family turmoil, death of a loved one)

These are factors that can increase the risk of developing depression:

- Having a family history of depression
- Being female, especially in a postpartum period
- Early childhood trauma
- Substance abuse
- Having a chronic condition or disability

Management and Medications

In treating someone who presents with likely depression as a comorbidity, one must exclude a variety of other psychiatric or medical conditions that can mimic some or all of the symptoms of depression. A misdiagnosis may result in ineffective and potentially harmful treatment that may worsen the underlying causal disorder. A missed diagnosis of depression may slow the treatment of both the depression and underlying disease. Once a diagnosis of depression has been made, traditional treatment options include antidepressant medications and counseling (individual or group).

Some people who are prescribed an antidepressant medication respond adequately, with no need for additional therapy. They typically remain on their medication for years. Others respond well with behavioral counseling or therapy or a combination of medication and therapy (table 9.1).

Because mood disorders are highly prevalent in these populations and can hinder medical treatment, it is worthwhile to use a depression screening and management tool such as the nine-item Patient Health Questionnaire (PHQ-9). The advantages of the PHQ-9 as a screening tool are that it is free to use without a license, is easy to score, is valid in multicultural groups, has a version in Spanish, and has been validated as a longitudinal monitoring tool. Additionally, the PHQ-9 has items representing somatic symptoms that have been associated with a benefit from exercise training; thus longitudinal changes in PHQ-9 scores may represent somatic benefits of exercise independent of the individual's depression status.

Exercise as an Adjunct to Optimal Medical Care of Chronic Conditions

Exercise programs can have an important role in the health management for many individuals with chronic physical health conditions, and should be used as a low-risk, adjuvant treatment for depressive symptoms that occur in conjunction with a chronic disease. Exercise can play a considerable role in improving quality of life for these persons. Research suggests that the largest exercise-related antidepressant effects may stem from improvement in physical functioning, but more research is needed to determine the optimal dose of exercise training in these individuals.

Effects on the Exercise Response

Depression is not believed to have direct cardiovascular or neuromotor effects, but people with depression may have a blunted exercise response because of

- early volitional exhaustion,
- possible increased perceived exertion relative to heart rate,
- peak exercise tolerance below maximal aerobic capacity related to an inability to motivate to push to maximal levels, and
- disorders in initiating or maintaining sleep and the occurrence of sleep apnea, which accelerate physical and cognitive fatigue and are associated with insulin resistance.

Effects of Exercise Training

Exercise has long been thought to be beneficial for persons with depression. Recent reviews and

Table 9.1 Antidepressant Medications

Generic name	Mechanism of action	Adverse effects
Citalopram Escitalopram Fluoxetine Paroxetine Sertraline	Selective serotonin reuptake inhibitors (SSRIs) block the reuptake of serotonin in the neural synapse, increasing serotonergic stimulation by reducing the recycling of serotonin.	• Stomach upset, weight change • Decreased sexual desire • Fatigue, dizziness • Insomnia, headaches
Desvenlafaxine Duloxetine Venlafaxine	Selective serotonin–norepinephrine reuptake inhibitors (SNRIs) increase serotonergic and norepinephrine stimulation by reducing the recycling of serotonin and norepinephrine.	As for SSRIs, plus: • Sweating • Dry mouth • Fast heart rate • Constipation
Bupropion	Norepinephrine–dopamine reuptake inhibitors (NDRIs) increase dopaminergic and norepinephrine stimulation by reducing the recycling of dopamine and norepinephrine.	Lower seizure threshold (high doses)
Amitriptyline Nortriptyline Protriptyline Amoxapine Desipramine Doxepin Imipramine Trimipramine	Tricyclic antidepressants (TCAs) act through nonselective serotonin and norepinephrine stimulation by reducing the recycling of serotonin and norepinephrine.	Generally more adverse effects than SSRI, SNRI, or NDRI medications Stomach upset, dizziness, dry mouth Changes in blood pressure Changes in blood sugar levels Constipation and nausea Confusion Life-threatening cardiac conduction abnormalities in an overdose
Mirtazapine Trazodone Vilazodone	Atypical antidepressants	Primary or adjunct medication
Tranylcypromine Phenelzine Selegiline Isocarboxazid	Monoamine oxidase inhibitors (MAOIs) inhibit the deamination of serotonin, melatonin, epinephrine, norepinephrine, or phenylethylamine.	Oldest class of antidepressants Adverse effects can be severe Extreme hypertension with some medications Side effects similar to those with SSRIs and SNRIs

Chronic diseases and disabilities are likely to be accompanied by persistent depressive disorder. This should hardly be surprising when the person's body is in pain, they physically can't do the things they want to do, and are financially vulnerable because of disability and high health care expenses. This situation does not foster positive views or a high quality of life. If such a person does not manifest suicidal ideation, referral to exercise therapy may be very beneficial. This is partly due to the direct effects of exercise, but may also be due to health care providers showing that they're not giving up on helping improve the patient's predicament.

meta-analysis show therapeutic benefit for mild to moderate depression from both aerobic and strength training regimens. Aerobic exercise may also indirectly alleviate depressive symptoms by improving sleep integrity and the circadian sleep cycle. New studies show that exercise can increase expression of several brain growth factors that favorably influence the chemical messengers in the brain responsible for modulating depression and influencing cognitive function.

Recommendations for Exercise Programming

The Basic *CDD4* Recommendations are a good starting point for an exercise program. In addition, readers should refer to chapter 32 for more detailed discussion of exercise in people with depression.

- Use the major chronic condition for primary guidance.
- Otherwise the Basic *CDD4* Recommendation is a reasonable default program.
- Participating in group programs or having an exercise partner can be helpful.

Integration of Depression Into a Medical Home Model

Depression, generally mild in form with somatic symptoms and sleep disturbance, is highly prevalent in people with these very common conditions. It is important for the medical home team to understand that depression can be a substantial barrier to successful incorporation of exercise and physical activity. In many people, increasing physical activity alone is sufficient to improve their affect and mood, but many others may also need treatment with antidepressant medications and counseling. Such counseling needs to be compatible with and complementary to lifestyle intervention efforts to improve diet, physical activity, stress coping, and tobacco or substance use.

Lower-Limb Amputation

At first glance, lower-limb amputation might seem out of place in part II of this book, which consists of predominantly cardiometabolic conditions. Lower-limb amputations are associated with a very high risk of disability, loss of mobility, and short life expectancy due to the development of cardiometabolic disease. Whether the amputation is related to trauma or to diabetes or vascular disease, a lower-limb amputation should serve as a "red flag" to all clinicians: This is someone who needs diligent efforts to help maintain mobility and independence, aggressive cardiovascular risk reduction, and good diabetes preventive care (whether the individual has diabetes or is just at risk for it).

In many reviews on research in this population, a very strong conclusion is that various measures of mobility and functional performance are dominant prognostic indicators—the vast majority of those who perform poorly will, in 2 years, be home-bound, will not be ambulating with their prosthesis (even at home), will be fully disabled, or will have died. Some experts even hold that maintaining mobility with a prosthesis is the central element in helping such persons avoid this fate. If so, then exercise is the most important medicine for someone with a lower-limb amputation.

Basic Pathophysiology

There are four main causes of lower-limb amputations:

- Diabetes (causing vascular disease, neuropathy, and osteomyelitis)
- Trauma

- Malignancy (usually of bone)
- Birth defects

Diabetes is the most common cause, because 5% to 10% of people with diabetes will end up having a lower-limb amputation. Diabetes is the underlying cause for about two-thirds to three-fourths of all lower-limb amputations. Traumatic amputations mostly occur in motor vehicle accidents (particularly motorcycles), at worksites, and in armed conflict. Such individuals tend to be younger, healthier, and more functional at the time of their amputation. Diabetes and traumatic amputation together account for some 90% to 95% of all lower-limb amputations.

There are six basic types of lower-limb amputations (which can be unilateral or bilateral):

- A so-called minor amputation (toe or toes)
- Transmetatarsal (so-called Symes), leaving variable amounts of rear- and midfoot
- BKA (below the knee)
- TKA (disarticulation at the knee, leaving the femur intact)
- AKA (above the knee)
- Hip disarticulation

In general, higher-level amputation and bilateral amputation are associated with greater disability, immobility, and cardiometabolic burden, particularly when an amputation is from diabetes or vascular causes. It is difficult to determine how much of this increased risk of a poor outcome is due to the underlying diabetes and how much is related to age, smoking, poor physical functioning, socioeconomic status, presence of disability, and other health factors; but all of these parameters are believed to be factors. One can see why prognosis is grim for many persons with diabetes who are elderly and who have a lower-limb amputation.

Younger, fitter individuals, with higher functioning, employment, and absence of preexisting disability, as well as a greater likelihood of having a traumatic cause of amputation, tend to have better outcomes and a much greater likelihood of reintegration into social and work life. These people do, however, have increased risk of becoming less physically active and of developing type 2 diabetes.

Management and Medications

The care and rehabilitation of an individual with an amputation is often very complex. In someone with diabetes and vascular insufficiency, there are often concerns about wound healing, the possibility of residual infection (in osteomyelitis), quality of diabetes care, fitting of a prosthesis and learning and adapting to ambulation with a prosthesis, depression and anxiety, and coping with the emotional trauma of undergoing an amputation. Another concern is pain management, because essentially all people in this group have phantom pain (perception of pain originating from the body part that has been amputated).

Nuances of medical management for people with an amputation include the following:

- Persons with an amputation need to work closely with orthotists and prosthetists, as well as their therapists.
- Stump health and skin integrity are paramount and must be watched closely.
- Correct fitting of orthotics must serve physical activity needs.
- Prosthetic or orthotic fitting may change over time.
- Drug therapy is not for the amputation but for the underlying conditions and pain.

Effects on the Exercise Response

The major impact of amputation on exercise response is that the body asymmetry makes activity much less efficient (especially walking). This increases the chronotropic and pressor responses in comparison to those in people without amputation, thus increasing perceived exertion. General cardiovascular demands are 20% to 50% higher.

This means that persons wearing a prosthesis will have difficulty keeping up with peers who do not have disabilities because

- aerobic economy is lower than in others, and
- perceived exertion and heart rate response are higher than in others at similar work rates.

Effects of Exercise Training

The health benefits of exercise training in persons with a lower-limb amputation are similar to those for other chronic conditions that have high cardiometabolic risk. Major expected benefits aimed at reducing cardiometabolic risk would thus include the following:

- All the benefits for hypertension and hyperlipidemia

- Improvement in the ventilatory response to exercise and ventilatory threshold
- Increased maximal oxygen consumption (mean of ~20%)
- Improved psychosocial well-being and self-efficacy
- Improved insulin sensitivity and reduced risk of prediabetes or diabetes
- Decreased blood platelet adhesiveness, fibrinogen, and blood viscosity and increased fibrinolysis
- Increased vagal tone and decreased adrenergic activity, resulting in increased heart rate variability

Persons who have had an amputation are at increased risk of developing and dying from cardiometabolic disease. Exercise is an important therapy, both to mitigate existing cardiometabolic conditions and to enhance robustness. Persons who have had an amputation and need help with getting started on exercise should be referred to a physical therapist, occupational therapist or exercise specialist who has training with adapted exercise.

Recommendations for Exercise Programming

Given that there are two basic types of clients or patients with lower-limb amputations, two basic approaches are reasonable depending on the type of patient. In all persons who have an amputation, these are key goals:

- To minimize loss of physical functioning and disability
- To maintain mental wellness

For traumatic amputations (generally, but not always, in people who are younger and healthier), additional objectives are as follows:

- To help to return to work and a productive life
- To prevent secondary development of cardiometabolic disease

For an amputation from cardiometabolic disease, these are key objectives:

- To reduce lifestyle-related cardiometabolic risk factors
- To minimize loss of independence and worsening of comorbid conditions

In persons with a traumatic amputation who have few or no comorbid conditions and are capable of higher-intensity activities, this approach is reasonable:

- Follow the ACSM Guidelines.
- Emphasize improving efficiency of locomotion with the prosthesis.

In individuals with a cardiometabolic amputation who have comorbid conditions and a more severely impaired level of physical functioning, the following approach is appropriate:

- Use the Basic *CDD4* Recommendation.
- Try to increase duration to 150 to 300 min/ week (see diabetes, chapter 6).
- It is advisable to start under the supervision of a physical therapist, occupational therapist, or clinical exercise physiologist who has experience with lower-limb amputation and multiple comorbid conditions.

Frailty

Frailty is a complex and poorly understood phenomenon that is a complex interplay between aging, physical and cognitive functioning, chronic disease, and in many cases habitual physical inactivity. Research suggests that it is probably easier to delay or slow the onset of frailty than to reverse frailty after it is fully manifest. Exercise training is a mainstay of frailty prevention, even though physical inactivity and deconditioning may not be causal factors in frailty.

Though studies have linked pathophysiological function to frailty, the condition remains best known as a syndrome characterized by Fried and colleagues (2001) as a phenotype based on five criteria:

- Weight loss
- Exhaustion
- Low grip strength
- Slow walking speed
- Low physical activity

In the Fried model (see Suggested Readings), persons are classified based on scoring of these five criteria:

- 0 of 5 traits = robust
- 1 or 2 of 5 traits = prefrail
- >2 of 5 traits = frail

Basic Pathophysiology

It is believed that a variety of physiological factors are involved in the pathogenesis of frailty, especially sarcopenia (or low muscle mass) due to skeletal muscle loss and declines in both testosterone and growth hormone levels. In addition, chronic proinflammatory states and immunologic deficiencies may have a role in precipitating or at least accelerating physiological decline. Deterioration of these physiological pathways in people who are physically inactive predisposes them to a downward spiral that includes these elements:

- Risk of falls
- Disability
- Cognitive decline
- Hospitalization
- Need for assistive living or nursing home admission
- Other conditions associated with physical inactivity

Management and Medications

Systematic review of the frailty literature suggests that slow walking speed and low physical activity are strong independent predictors of future disability in community-dwelling seniors. Additional traits, less strongly associated with future disability, include weight loss, poor balance, and poor lower-extremity functional tests (strength, sit to stand, and so on). Persons who score poorly in these areas are at high likelihood for subsequent chronic illness, disability, and need for aid.

Frailty as a fully expressed phenotype can be very difficult to reverse, and current management strategies focus on prevention and early detection more than treatment. Early detection is best achieved through assessment of the five Fried traits, as well as balance and lower-extremity function. People who are at risk should reasonably be screened as well as encouraged to participate in regular physical activity.

For assessing sedentary lifestyle, it may be useful to use a physical activity questionnaire, though it is not yet clear if there is a particular questionnaire best suited for specific sets of chronic conditions and age of the individual. It is reasonable to use a modified exercise vital sign

(EVS, see chapter 1), though none of these questions are specific to frailty. The EVS assesses how many days of the week someone performs moderate to vigorous physical activity and for how many minutes. To adapt the EVS to frailty, the question can be related to light or moderate physical activity. The modality of exercise or activity and the quantity can be readily used to estimate the physical and caloric workload, which are key elements of lifestyle modification programs. Medications are sometimes used to treat frailty, but the optimal therapeutic regimen is not yet known. Research supports use of these medications:

- Growth hormone
- Growth hormone-releasing hormone
- Testosterone or testosterone analogs

Supplements are sometimes used to treat frailty, but again the data remain inconclusive.

Effects on the Exercise Response

Frailty, by the Fried definition, involves an inability to do much physical exercise, even just at the intensity of activities of daily living. This is to a large extent mediated by sarcopenia, since people who have little muscle cannot generate much force. Effects of frailty on the exercise response are thus predictable in that muscle tests show weakness (i.e., low strength compared to norms) with low endurance and easy fatigability. Individuals who are frail generally cannot achieve a "true maximum" exercise test (i.e., a plateau in oxygen uptake despite increasing work rate) of their aerobic capacity. Thus, initial testing and tracking of performance-based measures of physical function such as gait speed, chair stand, and timed walk may be the most appropriate evaluations for those considered frail. The biggest diagnostic challenge to the exercise therapist is that not everything that moves slowly and has low endurance is frail.

The following are the exercise measures of greatest concern in frailty:

- Slow gait speed
- Loss of strength, power, and endurance
- Sedentary lifestyle
- Balance and fear of falling, which influences exercise, activities, and social life

Gait speed has been suggested to be a measure of great importance in people who are older, as it is highly and independently associated with poor outcomes. Slow gait speed represents failing lower-extremity strength and function and often concurrent poor dynamic balance during gait. It is not known which comes first—loss of lower-extremity strength and function or slow gait speed; however, either results in sedentary lifestyle and a downward spiral of deconditioning that fosters further loss of muscle mass, strength, and function.

Effects of Exercise Training

Most research on exercise training in frailty has focused on strength training, some on aerobic exercise, and some on the combination of aerobic and strength training. All three types of programs have proven useful in persons with frailty to improve gait speed, reduce risk of falling, and improve functional performance of various activities of daily living. Multimodal exercise programs that challenge multisegmental motor control used in activities of daily living have been shown to improve several indices of balance and to modestly reduce fall rates. Hence, inclusion of a combination of muscular endurance and functional balance training may prove useful for individuals with frailty, sarcopenia, and high fall risk.

Chronic diseases and disabilities are likely to evolve towards frailty, especially in elderly persons. It is important to intervene early, because frail individuals need lots of assistance with daily living, which is both expensive and—more important—nobody wants to a burden on family and friends. Because experience shows that frailty is hard to reverse, physical activity or an exercise program should be recommended before frailty becomes manifest.

The geriatrics literature also provides strong evidence that adequate daily protein intake, proper hydration during exercise, and nutrition after exercise (e.g., <20 min) can reduce muscle injury and fatigue and optimize healthy recovery. The optimal role of pharmacotherapy with androgenic or growth factors in combination with exercise training remains uncertain.

Recommendations for Exercise Programming

These are exercise programming recommendations for frailty:

- Follow the Basic *CDD4* Recommendation.
- Additionally emphasize strength and balance exercises.
- It is advisable for clients to start under the supervision of a physical therapist, occupational therapist, or clinical exercise physiologist who has experience with frailty and in geriatrics.

Integration of Frailty Into a Medical Home Model

Exercise training is less effective at delaying or retarding the rate of functional decline if the frailty syndrome already exists than it is if the individual has progressed only to the "prefrail" level. This suggests that exercise training may be better as a preventive therapy than as a rehabilitative therapy for frailty, and that early intervention is important. Nonetheless, assessment for frailty and recommendations and encouragement for exercise participation are essential for maintaining function and preventing deterioration.

Take-Home Message

Depression as a comorbidity is extremely common in persons with chronic disease or disability. In many cases, affect improves with just an exercise program and with being proactive about self-management of chronic conditions. Conversely, comorbid depression is also a common barrier to participation in exercise and self-management programs. And in someone whose health has deteriorated to the point of having an amputation or to the point of being frail, the prognosis is usually poor. So persons with these severe comorbid sequelae need to be followed closely and encouraged as much as possible to invest time and energy in maintaining physical functioning for independence. In most cases, health expenses and disability have left them with no money, so medical home staff will need to use ingenuity to keep them engaged with community and free resources to help keep them physically active.

Suggested Readings

Casale R, Alaa L, Mallick M, Ring H. Phantom limb related phenomena and their rehabilitation after lower limb amputation. *Eur J Phys Rehabil Med.* 2009;45:559-566.

Deshpande AD, Harris-Hayes M, Schootman M. Epidemiology of diabetes and diabetes-related complications. *Phys Ther.* 2008;88: 1254-1264.

Freedman VA, Ayken H. Trends in medication use and functioning before retirement age: are they linked? *Health Aff.* 2003;22(4):154-162.

Fried LP, Tangen CM, Walston J, Newman AB, Hirsch C, Gottdiener J, Seeman T, Tracy R, Kop WJ, Burke G, McBurnie MA. Frailty in older adults: evidence for a phenotype. *J Gerontol A Biol Sci Med Sci.* 2001;56: M146-M156.

Galley R, Allen K, Castles J, Kucharik J, Roeder M. Review of secondary physical conditions associated with lower-limb amputation and long-term prosthetic use. *J Rehabil Res Dev.* 2008;45(1):15-30.

He XZ, Baker DW. Body mass index, physical activity, and the risk of decline in overall health and physical functioning in late middle age. *Am J Public Health.* 2004;94:1567-1573.

Herring MP, Puetz TW, O'Connor PJ, Dishman RK. Effect of exercise training on depressive symptoms among patients with a chronic illness. *JAMA.* 2012;172(2):101-111.

Karpansalo M, Lakka TA, Manninen P, Kauhanen J, Rauramaa R, Salonen JT. Cardiorespiratory fitness and risk of disability pension: a prospective population based study in Finnish men. *Occup Environ Med.* 2003;60:765-769.

Naci N, Ioannidis JPA. Comparative effectiveness of exercise and drug interventions on mortality outcomes: metaepidemiological study. *BMJ.* 2013;347:f5577. doi: 10.1136/bmj.f5577 [Accessed October 18, 2013].

Sansam K, Neumann V, O'Connor R, Bhakta B. Predicting walking ability following lower limb amputation: a systematic review of the literature. *J Rehabil Med.* 2009;41:593-603.

Wrobel JS, Najafi B. Diabetic foot biomechanics and gait dysfunction. *J Diabetes Sci Technol.* 2010;4(4):833-845.

Wu SC, Driver VR, Wrobel JS, Armstrong DG. Foot ulcers in the diabetic patient, prevention and treatment. *Vasc Health Risk Manag.* 2007;3(1):65-76.

Additional Resource

National Center on Health, Physical Activity and Disability, www.nchpad.org [Accessed November 9, 2015].

CARDIOVASCULAR DISEASES

n contrast to the conditions in part II, the cardiovascular conditions in part III are generally not causally associated with physical inactivity. These comprise conditions of the heart as well as the vasculature, including diseases of the myocardium, the cardiac conduction system, and cardiac valves as well as arterial degeneration in the form of aneurysms. It is worth noting that congestive heart failure has many causes, including ischemic cardiomyopathy from severe coronary artery disease and cardiomyopathy from long-standing and inadequately controlled hypertension. While these two conditions are related to physical inactivity, many other causes of heart failure are not. Aside from these exceptions, the cardiovascular conditions discussed in part III are grouped together because they are generally not caused by a physically inactive lifestyle.

The conditions in this section often have, however, an immense impact on the patient's ability to do exercise. Many individuals with the conditions in part III have extremely diminished capacity to do aerobic activities, and they may experience cardiovascular instability if pressed too hard with intense exercise. Implantable pacemakers, cardioverter-defibrillators, valve repair or prosthetic valve implantation, and cardiac transplantation are commonly used to help improve exercise tolerance in persons with the conditions discussed in part III, and these procedures themselves alter the ability of the person to do aerobic activities.

Exercise management for someone with a condition discussed in part III is thus less a matter of trying to prevent cardiovascular events and delay mortality and more a matter of preserving physical functioning to maintain independence and quality of life. It is unfortunate that health insurance very rarely covers exercise programming services for these conditions, because many people with these conditions really need help with an assessment of how their condition affects their ability to do exercise. This translates into their having an understanding of what physical activities they can safely do and then figuring out how to maintain an active lifestyle and independent living.

Chronic Heart Failure

Chronic heart failure (CHF) is a serious, debilitating, and costly noncommunicable condition that presently affects an estimated 5.8 million Americans and many millions more around the world. As modern societies adopt a more sedentary lifestyle, coupled with less heart-healthy cuisines, CHF is expected to become a major public health problem across the globe. Currently, there are more than 670,000 new cases of CHF every year in the United States; the prevalence is projected to increase to more 8 million over the next 20 years. By 2030, it is expected that 1 in every 33 Americans will have CHF, with the direct and indirect costs of medical management projected to increase from $39 to $70 billion per year. The utilization and economic impacts of CHF on the world's health care systems are staggering now and threaten to become overwhelming.

Basic Pathophysiology

Chronic heart failure, broadly defined as the inability of the heart to adequately deliver oxygen to the body, is a not a disease but rather a clinical syndrome that affects virtually all organ systems. This syndrome occurs when cardiac output is reduced at rest, or especially during exertion, and attributable to left ventricular systolic or diastolic dysfunction or both.

In systolic dysfunction, left ventricular contraction is generally impaired because of

- myocardial ischemia (see chapter 7),
- myocardial infarction (MI) (see chapter 6), or
- idiopathic (often postviral) cardiomyopathy.

In systolic dysfunction, the left ventricular ejection fraction (LVEF) is referred to as reduced, sometimes denoted as HFrEF.

In diastolic dysfunction, resistance to filling is increased in one or both ventricles, causing

- elevated diastolic pressure in the ventricles, generally because of
- reduced ventricular compliance.

In diastolic dysfunction, the LVEF is referred to as preserved, sometimes denoted as HFpEF.

It is worth noting that there are many other causes of heart failure, mostly acute and subacute, but some can be chronic, including valvular conditions (see chapter 13), malnutrition (e.g., beriberi), endocrine disorders, anemia, cardiac toxins, and various other diseases that cause the heart failure syndrome. This chapter primarily addresses stable chronic heart failure due to systolic and diastolic dysfunction.

These systolic (contractile) and diastolic (relaxation or lusitropic) changes to heart function have both peripheral and central hemodynamic consequences. The central circulation effects include the following:

- Decreased cardiac output during exercise
- Compensatory ventricular volume overload
- Elevated pulmonary and central venous pressures
- In severe cases, elevated left ventricular filling pressures when the person is resting

In addition to central hemodynamic abnormalities, the reduced cardiac output in CHF results in peripheral circulatory effects as well:

- Major disturbances in skeletal muscle metabolism
- Impaired vasodilation
- Renal insufficiency (from poor perfusion), leading to
  - Abnormal neurohormonal responses, as well as
  - Pulmonary–ventilatory dysfunction

The consequences of this constellation of physiological changes result in the cardinal signs and symptoms of CHF:

- Fluid retention (resulting in pulmonary or peripheral edema)
- Orthopnea (due to pulmonary edema)
- Dyspnea on exertion (also from pulmonary edema)
- Fatigue (generalized or localized to leg skeletal muscles, or both)
- Reduced exercise tolerance (from all the above)

Management and Medications

Despite continued advances in medical therapy, the prognosis for CHF is poor, with 5-year survival rates of 25% for men and 38% for women. Chronic heart failure is the most common reason for emergency room visits as well as hospitalizations among persons who are elderly. To manage the growing CHF burden, a greater focus on adherence to evidence-based guidelines to prevent and treat CHF is necessary.

Prevention of Chronic Heart Failure

Since a majority of HFrEF (systolic dysfunction) is attributable to myocardial ischemia or infarction (or both), both primary and secondary preventive efforts must focus on reducing the modifiable risk factors for coronary artery disease (see chapter 6 on hypertension, hyperlipidemia, obesity, diabetes, cigarette smoking, and sedentary lifestyle). In contrast, HFpEF (diastolic dysfunction) is most commonly observed in women who are elderly and persons who have had long-standing and poorly controlled hypertension or diabetes.

Treatment of Chronic Heart Failure

Treatment for people with CHF has several key features:

- Identifying and treating the underlying causes and risk factors
- Regulating sodium and fluid balance

- Pharmacologic therapy to reduce symptoms
- Reducing adrenergic stimulation (decreasing cardiac stimulation and afterload)

As CHF becomes more and more severe, more invasive approaches to managing symptoms include the following:

- Oral or infused inotropic agents to improve contractility
- Ventricular pacing to improve cardiac performance
- Left ventricular assist device implantation
- Cardiac transplantation

Initial management of CHF involves identifying the underlying cause. For example, a stenotic valve may need to be repaired or replaced, hypertension or myocardial ischemia controlled, or alcohol or drug use discontinued. In some individuals, these measures alone may restore cardiac function to normal. The second goal is to reduce the overload pharmacologically; and since excessive salt and water retention is a pathophysiological component of CHF, most individuals will require the use of diuretics. Afterload reduction through the use of angiotensin-converting enzyme (ACE) inhibitors, angiotensin II receptor blockers (ARB), or other arterial vasodilators tends to reduce left ventricular end-diastolic pressures and improve stroke volume and thus cardiac output, improving symptoms of CHF. Angiotensin-converting enzyme inhibitors may lower mortality, while digoxin or other inotropic agents can increase myocardial contractility.

β-adrenergic blocking agents (particularly α-1 receptor blockers, such as carvedilol) are an effective treatment for CHF, improving symptoms and reducing mortality. β-blockers inhibit the excessive sympathetic activation that is part of CHF, preventing deleterious effects of chronically increased adrenergic stimulation on the failing heart.

The primary difference in pharmacotherapy between clients with HFrEF versus HFpEF is that persons with preserved LVEF (HFpEF) do not require positive inotropes. In addition, the use of calcium channel blockers has been shown to improve ventricular relaxation, increase end-diastolic volume, and increase functional capacity in individuals with HFpEF.

Biventricular pacemakers for cardiac resynchronization therapy (CRT) and implanted cardioverter-defibrillators (ICD) are other treatments common in the management of CHF in persons who have an LVEF of <35%. The synchronization of the right and left ventricles improves cardiac output and exercise performance in many persons with CHF, particularly those with a widened QRS (≥0.12 s) on the electrocardiogram (ECG). Implanted ICDs decrease mortality in individuals with CHF. See chapter 12 for more on pacemakers and ICDs.

For many individuals, heart failure is a chronic and stable condition that warrants minimal changes in therapy over time. However, for reasons not well understood, changes in left ventricular (LV) function can cause an abrupt worsening of signs and symptoms that requires hospitalization for diuresis and pharmacologic adjustments. It is now being recognized that acute decompensated heart failure (ADHF) results in severely reduced quality of life, markedly increased mortality and hospitalization, and $39 billion per year in medical costs in the United States alone. The goal of therapy for ADHF is reduction of symptoms (primarily to eliminate edema) and identifying the reason(s) for change in LV function.

Effects on the Exercise Response

The CHF syndrome results in a number of central, peripheral, and ventilatory abnormalities that adversely influence the exercise response, irrespective of the cause of CHF, including these:

- Reduction in cardiac output
- Mismatching of lung ventilation to perfusion, causing shortness of breath
- Leg fatigue during exercise due to inadequate blood flow to skeletal muscles

Abnormal neurohumoral mechanisms also reduce cardiac performance during exercise in individuals with CHF because of the following:

- Elevated catecholamine levels cause downregulation of sensitivity to β-adrenergic receptors to β-agonist stimulation.
- Altered inotropic regulation and reduced contractile function of the heart.

Altered baroreceptor reflexes have also been observed and may contribute to diminished chronotropic responses or to reduced systolic pressure and reduction in peripheral perfusion.

Peripheral abnormalities, including the following, also influence the response to exercise in individuals with CHF:

- Reductions in muscle blood flow
- Abnormal redistribution of blood flow
- Reduced vasodilatory capacity
- Endothelial dysfunction
- Abnormal aerobic metabolism biochemistry in skeletal muscle

Abnormalities in skeletal muscle metabolism include reduced mitochondrial enzyme activities, reduction of type I aerobic fibers and increase in type II glycolytic fibers, and greater contribution of glycolysis and reduced contribution of oxidative phosphorylation; the result is more metabolic acidosis.

As a consequence of these central and peripheral maladaptations, exercise tolerance and quality of life are dramatically reduced in people with CHF. They achieve only ~50% of the age-predicted $\dot{V}O_2$peak of healthy individuals, impairing their ability to participate in activities of daily living.

Effects of Exercise Training

Over the past three decades, numerous studies have documented the safety and benefits of endurance exercise training in individuals with heart failure. These studies have demonstrated improvements after endurance training because of peripheral adaptations:

- Improved skeletal muscle metabolism
- Improved endothelial function
- Increased vasodilatory capacity and redistribution of cardiac output

These peripheral changes overshadow central hemodynamic changes, which include cardiac chamber volumes, ejection fraction, and pulmonary pressures at rest and during exercise. Exercise therapy for individuals with CHF improves

- physical functioning,
- quality of life, and
- morbidity and mortality.

HFrEF (Systolic Dysfunction)

Most exercise research in CHF is in persons with HFrEF. Before the mid-1980s, individuals with

CHF were generally discouraged from participating in formal exercise training due to concerns over safety and questions about whether or not exercise training harmed an already weakened heart. There remains controversy over the possibility that exercise training in CHF could cause abnormal ventricular remodeling and infarct expansion, particularly in exercise early after a myocardial infarction. Studies using high-intensity training and assessment of the myocardium using Doppler two-dimensional echocardiography and magnetic resonance imaging (MRI) have dispelled these concerns. Some individuals with CHF after an MI deteriorate despite intensive intervention, but exercise training does not cause further myocardial damage and may actually prevent LV remodeling in people with CHF.

The HF-ACTION Trial of more than 2300 stable subjects with HFrEF proved that properly prescribed endurance exercise performed initially at a rehabilitation center followed by a home-based program is safe, improves both physiological and psychological outcomes, and yields modest reductions in hospitalization rates and mortality.

Resistance training, alone or in combination with aerobic exercise in stable individuals with HFpEF, improves physical functioning and strength without any adverse effects or worsening of LV function. Since muscle loss and weakness are major contributors to physical dysfunction in adults with CHF who are older, including strength training is vitally important in someone with CHF.

There is a great deal of interest in the potential role of high-intensity interval training (HIIT) as a strategy to stimulate skeletal muscle without additional stress on the failing heart. High-intensity interval training protocols can yield dramatic improvements in $\dot{V}O_2$peak and other physiological measures, but there are concerns over safety and long-term adherence to such regimens; these issues need to be evaluated before HIIT can be recommended for people with CHF.

HFrEF (Diastolic Dysfunction) and Acutely Decompensated Heart Failure

There is much less evidence to support endurance exercise training in the management of persons with HFpEF, but several studies show 20% to 25% improvements in functional capacity, similar to or better than what has been shown in people with HFrEF. The benefits of endurance exercise train-ing in HFpEF, like those in HFrEF, appear to mainly be the result of peripheral adaptations, with no favorable changes in LV function or volumes. At this time, neither resistance training nor HIIT has been evaluated in persons with HFpEF.

There is interest in the role of exercise therapy in acutely decompensated heart failure, given the substantial morbidity and mortality associated with hospitalization in decompensated heart failure. Very little is known about this type of therapy, though many believe that early mobilization after hospitalization for HF exacerbation may prevent further disability and provide a foundation for a formal exercise training program. This remains unproven and cannot be recommended at this time.

Advances in medical technology have dramatically reduced mortality from myocardial infarctions, but development of congestive heart failure is often the consequence. Exercise intolerance has a major impact on physical functioning and quality-of-life in persons with heart failure. Saving the lives of heart attack victims is not enough; the health care system must also help patients with heart failure preserve their quality of life through prescribing exercise and referral to exercise programs.

Recommendations for Exercise Testing

Exercise testing can be valuable in objectively characterizing the severity of CHF and evaluating the efficacy of therapeutic interventions. The standard exercise ECG usually offers little insight, and exercise intolerance can be more appropriately characterized via these methods:

- Respiratory gas exchange response to exercise
- Quantification of work capacity
- Identification of the pathophysiological abnormalities

Exercise capacity measured by respired gas exchange techniques accurately quantifies functional limitations, identifies ventilatory abnormalities, and helps to optimize risk stratification in persons with CHF.

Exercise can cause a drop in ejection fraction, stroke volume, or both, as well as exertional hypotension. Although exercise testing in these individuals has the potential for a higher rate of complications, limited data suggest that it is similar to that observed among persons with coronary artery disease.

The following considerations relate to exercise testing with this population (see also table 10.1):

- Symptoms are frequently observed under 5 metabolic equivalents (METs).

- Lower-level, moderately incremented, individualized protocols are recommended (Naughton or ramp).

- Unstable or acutely decompensated CHF is a contraindication.

- Respiratory gas exchange measurements (including $\dot{V}O_2$peak, VT, $\dot{V}_E/\dot{V}CO_2$ slope) increase precision, optimize risk stratifica-

tion, and permit assessment of breathing efficiency and patterns.

- Six-minute walk tests supplement the graded exercise test but are not a surrogate for measured respired gas analysis.

- Exertional hypotension, clinically significant dysrhythmias, and chronotropic incompetence may occur in CHF, so 12-lead ECG should be monitored during testing.

- Test endpoints should focus on symptoms, hemodynamic responses, and standard clinical indications for stopping (not a target heart rate).

Recommendations for Exercise Programming

Because of improvements in therapeutic and surgical techniques, individuals with CHF represent one of the fastest-growing segments in cardiac

Table 10.1   Exercise Testing Considerations for Chronic Heart Failure (CHF)

Factor	Considerations
Digoxin	Diffuse ST changes may result in false-positive interpretation.
Blood pressure	May be low or fail to rise due to LV dysfunction, afterload-reducing medications, or both.
Heart rate	20-30% of CHF patients have chronotropic incompetence that is associated with decreased myocardial β-receptor sensitivity. Peak heart rate should achieve or exceed ≥80% (≥62% if on β-blockers) of age-predicted heart rate reserve.
Oxygen consumption	$\dot{V}O_2$ peak is inversely related to mortality. $\dot{V}O_2$ less than 18 mL · kg^{-1} · min^{-1} or significant drop with serial testing is of particular concern.
$\dot{V}E/\dot{V}CO_2$ slope	Important measure obtained during collection of expired gases that has important prognostic value. $\dot{V}E/\dot{V}CO_2$ slope ≥34 indicates poor prognosis.
Ventricular arrhythmias	PVCs common in CHF. Increasing frequency and runs of VT warrant concern and test termination.
Atrial arrhythmias	Common in CHF. Atrial fibrillation may worsen hemodynamics and result in increased fatigue, hypotension, dyspnea.
Bundle branch block	Common in HF and may indicate LV dyssynchrony and patients who would benefit from biventricular pacemaker. Makes interpretation of ischemia difficult due to ST-segment changes.
Left ventricular hypertrophy	More likely to be seen in HFpEF patients. Makes interpretation of ischemia difficult due to ST-segment changes.
Exercise intolerance	Achieve only 50% of peak exercise level of age-matched healthy subjects. Limited by dyspnea or fatigue. Use low-level protocols with small increments.

LV = left ventricular

PVCs = premature ventricular contractions

VT = ventricular tachycardia

HFpEF = heart failure with preserved ejection fraction (diastolic dysfunction)

rehabilitation programs. Formal exercise training programs can result in a person's being able to live independently and continue to work instead of being disabled and can significantly enhance quality of life in individuals with CHF. For these reasons, most individuals with CHF should

- be referred to cardiac rehabilitation,
- follow the Basic *CDD4* Recommendations, and
- graduate to a home-based program aimed at maintaining physical functioning.

Potential complications and outcomes for individuals with CHF differ significantly compared to the standard cardiac rehabilitation population, including the following:

- Many individuals with CHF will deteriorate irrespective of interventions.
- Risk of sudden death is higher.
- Prevalence of psychosocial and vocational problems is higher.
- Some individuals experience prolonged fatigue after only a single exercise session.

Absolute contraindications to exercise in CHF include these:

- Left ventricular outflow obstruction
- Decompensated CHF
- Unstable dysrhythmias

Relative contraindications to exercise are the same for individuals with CHF as for those with normal LV function. Collectively, these considerations necessitate that programs be designed carefully and that the staff be trained to recognize specific needs of this population as well as specific precautions to be alert to these issues.

Special Considerations for CHF

- Status can change quickly, and clients should be reevaluated frequently for signs of acute decompensation, rapid changes in weight or blood pressure, worse than usual dyspnea or angina on exertion, or increases in dysrhythmias.
- Warm-up and cool-down sessions should be prolonged.
- Some clients may tolerate only limited work rates and may need lower-intensity and longer-duration exercise sessions.

- Duration of activity may need to be adjusted to allow clients more opportunity to rest and to progress at their own pace. Some clients may better tolerate discontinuous training involving shorter sessions of exercise interspersed with periods of rest.
- Perceived exertion and dyspnea scales should take precedence over heart rate and work rate targets.
- Isometric exercise should be avoided.
- Electrocardiogram monitoring is required for persons with a history of ventricular tachycardia, cardiac arrest, or exertional hypotension.
- Consider ancillary study data (e.g., exercise echocardiogram, radionuclide studies, hemodynamic studies, ventilatory gas analysis) when developing the exercise program. In general, do not exceed a work rate that produces wall motion abnormalities, a drop in ejection fraction, a pulmonary wedge pressure greater than 20 mmHg, or the ventilatory threshold.

Integration Into a Medical Home Model

People with CHF should be followed closely by home health providers or medical staff (or both) to monitor symptoms, changes in daily weight, and compliance with medication and dietary recommendations (including sodium and fluid restriction). Excessive fatigue, dyspnea, or pulmonary congestion, palpitations, and swelling of ankles may indicate a change in LV function and requires evaluation. Careful monitoring and adherence to appropriate lifestyle recommendations (diet, physical activity) improves outcomes in persons with CHF (fewer hospitalizations, reduced mortality).

Take-Home Message

Consider these points when working with people with CHF:

- Moderate-intensity endurance exercise programs are generally safe and beneficial for individuals with CHF who are stable.
- At this time, exercise for acute exacerbations of CHF should not be advised, as the data on this are insufficient.

- Exercise can be prescribed based on heart rate and $\dot{V}O_2$ techniques, but these are often superseded by symptoms of dyspnea and fatigue.
- Properly performed resistance exercise training appears safe and beneficial in persons with HFrEF who are stable but has not been studied sufficiently in individuals with HFpEF.
- High-intensity interval training appears to produce superior physiological adaptations, but further studies of safety and compliance are needed before this can be widely recommended.
- Exercise testing with respired gas analysis provides additional and more accurate information on training responses as well as important predictors of prognosis.
- During exercise testing and training session, observe for hypotension, ventricular arrhythmias, and dyspnea.

Suggested Readings

Arena R, Myers J, Guazzi M. The clinical and research applications of aerobic capacity and ventilatory efficiency in heart failure: an evidence-based review. *Heart Fail Rev.* 2008;13:245-269.

Braith RW, Beck DT. Resistance exercise: training adaptations and developing a safe exercise prescription. *Heart Fail Rev.* 2008;13:69-79.

Cahalin LP, Chase P, Arena R, Myers J, Bensimhon D, Peberdy MA, Ashley E, West E, Forman DE, Pinkstaff S, Lavie CJ, Guazzi M. A meta-analysis of the prognostic significance of cardiopulmonary exercise testing in patients with heart failure. *Heart Fail Rev.* 2013 Jan;18(1):79-94.

Conraads VM, Deaton C, Piotrowicz E, Santaularia N, Tierney S, Piepoli MF, Pieske B, Schmid JP, Dickstein K, Ponikowski PP, Jaarsma T. Adherence of heart failure patients to exercise: barriers and possible solutions: a position statement of the Study Group on Exercise Training in Heart Failure of the Heart Failure Association of the European Society of Cardiology. *Eur J Heart Fail.* 2012 May;14(5):451-458.

Feiereisen P, Delagardelle C, Vaillant M, Lasar Y, Beissel J. Is strength training the more efficient training modality in chronic heart failure? *Med Sci Sports Exerc.* 2007;39:1910-1917.

Haykowsky M, Brubaker P, Kitzman D. Role of physical training in heart failure with preserved ejection fraction. *Curr Heart Fail Rep.* 2012 Jun;9(2):101-106.

Haykowsky MJ, Timmons MP, Kruger C, McNeely M, Taylor DA, Clark AM. Meta-analysis of aerobic interval training on exercise capacity and systolic function in patients with heart failure and reduced ejection fractions. *Am J Cardiol.* 2013 May 15;111(10):1466-1469.

Keteyian SJ. Exercise training in congestive heart failure: risks and benefits. *Prog Cardiovasc Dis.* 2011 May-Jun;53(6):419-428.

Keteyian SJ, Leifer ES, Houston-Miller N, Kraus WE, Brawner CA, O'Connor CM, Whellan DJ, Cooper LS, Fleg JL, Kitzman DW, Cohen-Solal A, Blumenthal JA, Rendall DS, Piña IL; HF-ACTION Investigators. Relation between volume of exercise and clinical outcomes in patients with heart failure. *J Am Coll Cardiol.* 2012 Nov 6;60(19):1899-1905.

Kitzman DW, Brubaker PH, Morgan TM, Stewart KP, Little WC. Exercise training in older patients with heart failure and preserved ejection fraction: a randomized, controlled, single-blind trial. *Circ Heart Fail.* 2010 Nov;3(6):659-667.

McElvie RS. Exercise training in patients with heart failure: clinical outcomes, safety, and indications. *Heart Fail Rev.* 2007;13:3-11.

Myers J. Principles of exercise prescription for patients with chronic heart failure. *Heart Fail Rev.* 2008;13:61-68.

O'Connor CM, Whellan DJ, Lee KL, Keteyian SJ, Cooper LS, Ellis SJ, Leifer ES, Kraus WE, Kitzman DW, Blumenthal JA, Rendall DS, Miller NH, Fleg JL, Schulman KA, McKelvie RS, Zannad F, Piña IL; HF-ACTION Investigators. Efficacy and safety of exercise training in patients with chronic heart failure: HF-ACTION randomized controlled trial. *JAMA.* 2009 Apr 8;301(14):1439-1450.

Pina IL, Apstein CS, Balady GD, et al. Exercise and heart failure: a statement from the American Heart Association Committee on Exercise, Rehabilitation and Prevention. *Circulation.* 2003;107:1210-1225.

Additional Resources

American College of Cardiology, www.acc.org [Accessed November 11, 2015].

American Heart Association, www.american-heart.org [Accessed November 11, 2015].

Heart Failure Society of America, www.hfsa.org [Accessed November 11, 2015].

Atrial Fibrillation

n atrial fibrillation (AF), chaotic and rapid atrial contractions cause an irregular ventricular response, which can be very rapid when conduction through the atrioventricular (AV) node is not adequately suppressed. This situation impairs ventricular filling and cardiac pump function and thus can lead to a variety of symptoms:

- Fatigue
- Decreased exercise capacity
- Presyncope, or syncope, falls
- Stroke

Basic Pathophysiology

Atrial fibrillation is one of the most common cardiac arrhythmias, and it occurs more frequently with advancing age. The pathophysiology of AF is not completely understood, but it is believed to be caused by multiple reentrant circuits within the atria. Atrial fibrillation often occurs with other cardiovascular conditions, particularly chronic heart failure, cardiomyopathy, valvular disease, coronary artery disease, hypertension, and hyperthyroidism. Some of these disorders may be underlying causes of AF, and in some cases they may be manifestations of AF, but they are something to bear in mind in any individual with AF.

Management and Medications

There is currently some controversy regarding whether it is preferable to restore normal sinus rhythm or to accept the presence of AF and use drugs to manage the ventricular response rate. Thus, medical management of AF primarily involves two basic approaches:

- Converting the heart back to normal sinus rhythm
- Controlling the ventricular rate response

At the time when AF is initially diagnosed, an effort is often made to convert the individual back to sinus rhythm through electrical cardioversion, radiofrequency ablation, or a surgical method known as the maze procedure. While the initial success rate of electrical cardioversion is high, many individuals return to AF within 4 to 6 weeks. Conversion and maintenance of normal sinus rhythm is particularly difficult when the AF has been long-standing (which is why it is primarily attempted at the time of the initial diagnosis).

When pursuing a rate-control strategy, 24-hour ambulatory monitoring can be helpful for examining ventricular rate control during strenuous activities of daily living. Medicines used to control the ventricular rate in AF include the following:

- Digoxin
- β-blockers (e.g., propranolol, sotalol, metoprolol, atenolol)
- Nondihydropyridine calcium channel blockers (e.g., diltiazem, verapamil)

Other agents are sometimes used to convert AF to sinus rhythm or maintain sinus rhythm once cardioversion is successful, including these:

- Amiodarone, dronedarone, propafenone
- Dofetilide
- Disopyramide
- Procainamide

If the medical management strategy is to use the rate-control method, the individual is at high risk for stroke and must be anticoagulated. The irregular heart rhythm leads to small eddies of blood that don't flush through the chambers very well, which can lead to formation of a thrombus that can be ejected from the heart and cause a stroke. Antithrombotic therapy reduces the risk of stroke by up to ≈80%. The following are agents used to provide long-term anticoagulation:

- Warfarin (in the majority of people)
- Aspirin
- Aspirin, clopidogrel, warfarin in some combination
- Dabigatran
- Apixaban
- Rivaroxiban
- Edoxaban

Successful anticoagulation therapy is dependent on regular (e.g., at least monthly) measurements of the international normalized ratio (INR), with careful adjustment to maintain an INR between 2.0 and 3.0. Anticoagulation therapy causes easy bruising and can cause prolonged bleeding, which people find a nuisance. New agents have recently been approved in the United States that do not require monitoring of the INR, but most of them do not have an antidote (i.e., anticoagulation cannot be reversed if the person needs to undergo medical procedures that incur a bleeding risk). As antidotes are developed, these medications are likely to displace warfarin.

Effects on the Exercise Response

The most notable feature of the exercise response in persons with AF is a rapid, irregular ventricular rate. Heart rate is comparatively high at any level of exercise, in part to compensate for the diminished stroke volume. Maximal heart rate tends to be considerably higher in individuals with AF compared to subjects in normal sinus rhythm, although there is marked variability in the maximal heart rate response (standard deviations up to 30 beats/min). The heart rate response is also affected by comorbid conditions commonly associated with AF (e.g., coronary artery disease, chronic heart failure) and the use of AV nodal suppressant drugs such as β-blockers, calcium channel blockers, and digoxin.

Stroke volume is reduced in AF because there is no atrial kick to aid in ventricular filling during diastole. Because of the variability in the diastolic filling period, determination of systolic blood pres-sure can be difficult and is poorly reproducible. Korotkoff sounds may be heard more distinctly at rest, when long RR intervals lead to better ventricular filling and higher stroke volumes that increase flow turbulence.

Exercise tolerance is generally reduced in AF relative to that in age-matched normal subjects. The degree of this reduction is typically about 20%, but it is highly dependent on the presence and extent of underlying heart disease. Accordingly, conversion from AF to sinus rhythm improves exercise capacity on the order of 15% to 20%, although functional gains probably do not occur until at least 1 month after successful cardioversion. Individuals with *lone* AF (AF without any underlying heart disease) usually achieve peak oxygen uptake values typical of age-matched subjects in normal sinus rhythm.

Effects of Exercise Training

Insufficient scientific literature is available concerning the effects of exercise training specifically in people with AF. Because the prevalence of AF is comparatively high in men older than 60 years, individuals with AF have been included in many rehabilitation studies just due to the association of AF with coronary artery disease. This experience suggests that individuals with AF probably have a training response similar to that of their peers who are in normal sinus rhythm.

The major concerns during exercise training are symptoms of AF due to inadequate medical management (rate control). Other important concerns are the presence of any underlying heart disease, particularly valvular disease, chronic heart failure, and coronary artery disease. The presence of these underlying diseases is the most important consideration in exercise programming for individuals with AF.

Atrial fibrillation can be very difficult to endure. The trial of medical management, electrophysiology studies, and pacemaker adjustment sounds simple, but in practice it often takes months before there is successful and sustained rate control. Many patients with AF find this frustrating and can really benefit from an exercise specialist who can help them through this period.

Recommendations for Exercise Testing

Maximal exercise testing can be safely used to measure the functional capability of individuals with AF, and the *CDD4* Basic Recommendations for exercise testing are appropriate in this group. Exercise testing is also helpful in determining the effectiveness of rate-control therapy. The reduction in exercise capacity associated with AF is a direct function of the underlying heart disease, which is common. Accordingly, moderately incremented protocols are recommended:

- Naughton
- Ramp

In the absence of other clinical indications for stopping, persons with AF may be safely taken to fatigue or shortness of breath endpoints.

Contraindications to exercise testing in AF are mostly related to comorbidities and other underlying conditions, such as the following:

- Instability of chronic heart failure
- Valvular disease
- Complex ventricular arrhythmias

Interpretation of the exercise electrocardiogram (ECG) is made more difficult in AF because many individuals with AF take medications to control the rate response and have underlying heart disease. Digoxin helps to control the ventricular response during exercise, but it also has diffuse effects on the ST-segment response, including false-positive (ECG) changes. Other AV nodal suppressants, including calcium channel blockers and β-blockers, can mask ischemic changes, and β-blockers are likely to reduce exercise capacity (see appendix B).

Age-predicted maximal heart rate targets are particularly useless in AF because of the rapid and highly variable ventricular response. Because of the irregular ventricular response, heart rate is most accurately measured using calipers over at least a 6-s rhythm strip during exercise.

Considerations for Exercise Testing in Atrial Fibrillation

- Digoxin: may control ventricular response; diffuse ST effects
- Verapamil: may mask ischemia and decrease heart rate response to exercise
- Diltiazem, verapamil: help control ventricular response; may improve exercise capacity
- β-blockers: help control ventricular response; may reduce exercise capacity, particularly with nonselective medications; decrease submaximal and maximal heart rate and blood pressure
- Bundle branch block: common in people with AF; makes determination of ischemia difficult
- Left ventricular hypertrophy: common in persons with AF; makes determination of ischemia more difficult
- Age-predicted maximal heart rate targets: not valid
- Irregular ventricular response: may make blood pressure determination less precise or more difficult

Recommendations for Exercise Programming

As the population ages, the number of individuals with AF referred for exercise rehabilitation will increase. There are a few major factors to consider in exercise programming for individuals with AF:

- Daily variation in ventricular rate
- In some cases, the intermittent nature (presence or absence) of AF
- Inherent unreliability of the pulse rate for prescribing exercise intensity
- Adequacy of anticoagulation in those on anticoagulation therapy
- Concomitant or underlying heart disease

Since AF is frequently accompanied by underlying cardiac disease, exercise programming for these conditions is the major consideration in selecting goals and rate of progression. In addition, some individuals with AF referred to a rehabilitation program will have experienced a stroke, in which case the goals of the rehabilitation program change accordingly (see chapter 25).

Atrial fibrillation can have day-to-day variations in ventricular response that can lead to symptoms of low cardiac output (presyncope, syncope). Assessing this is thus an important precaution before beginning every exercise session. Because AF has a variably irregular ventricular response rate, exercise intensity cannot be based on heart or pulse rate but should be prescribed based on work rate and perceived exertion. Frequency, duration, intensity, and progression of exercise are similar to those for individuals in normal sinus rhythm and can follow standard ACSM Guidelines.

An important concern is that AF can be intermittent (i.e., the individual may be in AF one day and in normal sinus rhythm the next). This influences not only the person's heart rate response to exercise but also exercise tolerance and level of fatigue. The rhythm should therefore be determined every day. It is also important to note that AF has varied effects; some people experience fatigue while others do not. Finally, many people with AF are elderly, and one must consider comorbid conditions such as osteoporosis, coronary disease, diabetes, and obesity when developing the exercise program.

Several precautions should be considered during exercise programming:

- Longer sampling of the pulse may be needed to reliably measure heart rate.
- Atrial fibrillation has varied effects; some people experience fatigue, while others do not.
- Atrial fibrillation is frequently intermittent, so ascertain rhythm daily.
- Many with AF are elderly, so consider comorbid conditions.

Integration Into a Medical Home Model

Individuals with AF should be followed closely by staff in the medical home to monitor the INR and reduce risk of stroke. Physical activity should be encouraged during follow-up (sometimes by phone or e-mail). The medical home staff should be aware that excessive fatigue, shortness of breath, palpitations, swelling of the feet or ankles, and weight gain may indicate that the person's underlying conditions have worsened, particularly heart failure.

Individuals with AF can exercise safely, and case managers should encourage persons with AF to participate in daily physical activity commensurate with the Basic *CDD4* Recommendations. Clients who follow this guideline are likely

to improve their exercise tolerance and quality of life. Pedometers can be useful for documenting and encouraging daily activity, and activity recall instruments along with frequent phone contact (e.g., weekly) are useful to assess progress and compliance and to monitor any changes in clinical status.

Take-Home Message

Here are some considerations to remember when working with people with AF:

- Exercise programs for persons with chronic AF are generally safe.
- People with AF are likely to benefit from exercise similarly to individuals who are in normal sinus rhythm.
- Exercise prescription should be based on work rate and perceived exertion.
- Observe for exertional hypotension due to
  - rapid ventricular rates or
  - medication blunting of heart rate response.
- Atrial fibrillation can be intermittent, so it should be checked for before each exercise session.
- Atrial fibrillation is frequently accompanied by other cardiac conditions.
- Exercise programming for these other conditions should take precedence over AF.
- The exercise specialist should be cognizant of any underlying conditions and surveillance of INR and should be ready to alert the staff in the medical home.

Suggested Readings

Atwood JE, Myers J. Exercise hemodynamics of atrial fibrillation. In Falk RH, Podrid PJ, eds. *Atrial Fibrillation: Mechanisms and Management.* Philadelphia: Lippincott-Raven; 1997:219-239.

Caldeira D, David C, Sampaio C. Rate versus rhythm control in atrial fibrillation and clinical outcomes: updated systematic review and meta-analysis of randomized controlled trials. *Arch Cardiovasc Dis.* 2012;105:226-238.

Lip GY, Tse HF. Management of atrial fibrillation. *Lancet.* 2007;370:604-618.

Mertens DJ, Kavanagh T. Exercise training for patients with chronic atrial fibrillation. *J Cardiopulm Rehabil.* 1996;16:193-196.

Osbak PS, Mourier M, Kjaer A, Henriksen JH, Kofoed KF, Jensen GB. A randomized study of the effects of exercise training on patients with atrial fibrillation. *Am Heart J.* 2011;162:1080-1087.

Shea JB, Sears SF. Cardiology Patient Page. A patient's guide to living with atrial fibrillation. *Circulation.* 2008;117:e340-e343.

Ueshima K, Myers J, Graettinger WF, Atwood JE, et al. Exercise and morphologic comparison of chronic atrial fibrillation and normal sinus rhythm. *Am Heart J.* 1993;126:260-261.

Ueshima K, Myers J, Morris CK, Atwood JE, et al. The effect of cardioversion on exercise capacity in patients with atrial fibrillation. *Am Heart J.* 1993;126:1021-1024.

Vanhees LD, Schepers J, Defoor S, Brusselle S, et al. Exercise performance and training in cardiac patients with atrial fibrillation. *J Cardiopulm Rehabil.* 2000;20:346-352.

Wann LS, Curtis AB, January CT, Ellenbogen KA, Lowe JE, Estes M, Page RL, Ezekowitz MD, Slotwiner DJ, Jackman WM, Stevenson WG, Tracy CM. 2011 ACCF/AHA/HRS focused update on the management of patients with atrial fibrillation (updating the 2006 guideline): a report of the American College of Cardiology Foundation/American Heart Association Task Force on Practice Guidelines. *Circulation.* 2011;123:104-123.

Watson T, Shanstila E, Lip GY. Modern management of atrial fibrillation. *Clin Med.* 2007;7:28-34.

Additional Resources

American Heart Association

Living With Atrial Fibrillation. Available to order from www.americanheart.org [Accessed November 11, 2015].

Heart Conditions and Strokes. http://www.heart.org/HEARTORG/Conditions/Conditions_UCM_001087_SubHomePage.jsp [Accessed November 11, 2015].

Pacemakers and Implantable Cardioverter-Defibrillators

There are several classes of devices that assist in the electrical functions of the heart; some are temporary and almost always external for that reason. Though there are pacing devices with external leads placed on the chest, these are mainly used in acute life-threatening situations. Many external pacing devices use tempory leads placed transvenously. People with these often go on to a permanently implanted device, and many go straight to an implanted device at first diagnosis. This chapter discusses only permanently implanted devices, because exercise and physical activity are treatments for medically stable individuals and don't really pertain to unstable individuals in an emergent or urgent situation.

Permanent Pacemakers

A variety of factors contribute to optimal cardiac functioning, including atrioventricular (AV) synchronization and the chronotropic and inotropic responses to neurohormonal stimuli. Alterations in the normal sequence of atrial and ventricular filling and contraction can result in deterioration of hemodynamics and subsequent symptoms at rest, during exercise, or both. In persons who have light-headedness, syncopal spells, shortness of breath, and more rarely chest pain or other cardiovascular symptoms owing to these problems, a permanent pacemaker

- improves symptoms,
- enhances exercise performance, and
- improves quality of life.

According to guidelines developed jointly by the American College of Cardiology (ACC), the American Heart Association (AHA), and the Heart Rhythm Society (HRS), class I indications for a permanent pacemaker include the following:

- Sinus node dysfunction
- Third-degree block and advanced second-degree AV block
- Hypersensitive carotid sinus syndrome
- Symptomatic bradycardia
- Sustained pause-dependent ventricular tachycardia
- Left ventricular systolic dysfunction and New York Heart Association (NYHA) functional class III or ambulatory class IV

A typical pacemaker system consists of two basic components: a pulse generator and either one or two pacing wires. In a traditional pacemaker the pacing wires are insulated and are implanted transvenously into the right atrium, right ventricle, or both. With a biventricular pacemaker, a lead is also placed in the left ventricle. The leads are connected to the pulse generator, which is typically implanted subcutaneously just below the clavicle. The two main functions of the leads are sensing and pacing. Sensing involves detecting electrical signals (i.e., P-waves and R-waves) from the heart. When these signals are not sensed at the proper timing, the pacemaker generator fires an impulse that causes the atria or ventricles (or both) to contract. Optimally, the pacing system uses an atrial and ventricular lead to maintain AV synchrony, which serves to optimize cardiac output at rest and during exercise.

Pacemakers are described by a standardized code. The first letter represents the chamber paced; the second is the chamber sensed; and the third denotes the response to a sensed event. The fourth position is used to indicate that the pacemaker has rate-response capabilities. For example, VVIR is the abbreviation used when the ventricle (V) is the chamber being paced and sensed. When the pacemaker senses a normal ventricular contraction, the pacemaker is inhibited (I).

The "R" indicates that the pulse generator is rate-responsive during exercise. The response by the pacemaker is to either trigger or inhibit a pacing stimulus, depending on the absence or presence, respectively, of atrial or ventricular conduction, separately or in combination, relative to the range of heart rates that are programmed into the pacer.

A commonly used mode of pacing is the DDDR, which has dual-chamber (i.e., atrium and ventricle) pacing and inhibiting and has rate-response capability. The DDDR pacemaker is widely regarded as the optimal pacing mode in individuals who have normal sinoatrial (SA) node function, because it provides AV synchrony and uses the client's own sinus rhythm to guide ventricular stimulation. This results in a heart rate and cardiac output that, for the rest of the circulatory system, is very nearly normal.

Implantable Cardioverter-Defibrillators

An implantable cardioverter-defibrillator (ICD) is another electronic device that can be permanently implanted in individuals who either have a history of, or are at increased risk for, a life-threatening ventricular dysrhythmia. These devices impressively reduce mortality in persons with cardiomyopathy. Cardioverter-defibrillators can be just an ICD or can be a model that functions as both an ICD and a permanent pacemaker; like pacemakers, they are usually implanted subcutaneously just below the clavicle.

Implantable cardioverter-defibrillators electrically terminate life-threatening ventricular tachyarrhythmias. They consist of two basic parts: the lead system and the ICD itself. Implantable cardioverter-defibrillators have lead systems that are placed transvenously, typically by way of the subclavian vein. The ICD leads pick up the electrical rhythm of the heart and transmit this to the pulse generator, which senses the rhythm. Implantable cardioverter-defibrillators can detect atrial and ventricular arrhythmias, can provide antitachycardia pacing and defibrillation, and can be programmed with multiple protocols and the ability to record an electrocardiogram. Ventricular tachycardia (VT) and ventricular fibrillation (VF) are recognized by their rapid rates. If either VT or VF is sensed, the pulse generator delivers defibrillation or a synchronized cardioversion to terminate the rhythm. Other accelerated rhythms will initiate a pacing therapy intended to restore the rate within the preprogrammed limits.

According to guidelines developed jointly by the ACC, AHA, and HRS, class I indications for an ICD include the following:

- Survivor of cardiac arrest due to ventricular fibrillation (VF) or sustained ventricular tachycardia (VT)
- Structural heart disease with VT
- History of syncope of undetermined origin with clinically relevant VF or sustained VT induced during an electrophysiology study
- Left ventricular dysfunction due to myocardial infarction (post ≥40 days) with an ejection fraction (EF) ≤35% and NYHA class II or III or an EF ≤30% and NYHA class I
- Nonischemic dilated cardiomyopathy with an EF ≤35% and NYHA class II or III
- Nonsustained VT due to myocardial infarction with an EF ≤40% and VF or sustained VT induced during electrophysiology study

Combination Pacemaker–Defibrillator Devices

Some implantable devices are capable of providing both pacing and defibrillation. Recent evolution in terminology of pacemakers and defibrillators is toward calling them cardiac resynchronization therapy (CRT), with subclasses that provide pacing alone (CRT-P) and those that provide both pacing and defibrillation (CRT-D).

Management and Medications

The medical management of individuals with a permanent pacemaker or ICD is aimed at the control of accelerated ventricular rates and irregular rhythms and the management of comorbidities, such as coronary artery disease and cardiomyopathy. Common medications include the following:

- Angiotensin-converting enzyme (ACE) inhibitors
- β-adrenergic blockers
- Calcium channel blockers
- Vasodilators

The cardiologist or electrophysiologist who manages a pacemaker, ICD, or combination device uses a radio frequency machine that wirelessly interrogates the device on stored data that it has collected about the individual's heart rate and rhythm. This machine also programs the implanted device, with settings chosen by the physician to minimize symptoms. These are the key issues:

- Sensing the native heart rhythms (including changes during exercise)
- Responding to prolonged pauses or tachycardias
- Successfully stimulating the heart chambers to contract
- Successfully terminating life-threatening rhythms

Effects on the Exercise Response

Pacemakers and ICDs have very different functions, so the effects on the exercise response are considered separately.

Permanent Pacemakers

In the absence of normal SA node function, cardiac conduction, or neurohormonal regulation, pacemaker implantation generally improves exercise tolerance. Rate-adaptive pacemakers use a variety of physiological sensors to increase the pacing rate during periods of increased physical activity. These include accelerometers, vibration sensors, and ventilation sensors. Pacemaker manufacturers manage rate response based on proprietary algorithms that may be based on one or more of these sensing types. The responsiveness of these algorithms can vary depending on the mode of physical activity, resulting in over- and underestimates of the metabolic demand. Discrepancies have been noted during climbing versus descending stairs, walking at an incline, and exercising on a stationary cycle (leg ergometer). Individuals should be cautious when starting new activities and should be educated about symptoms that may represent an inappropriate heart rate response, such as shortness of breath and light-headedness.

Implantable Cardioverter-Defibrillators

Ideally, persons with an ICD would get shocked only when they had a life-threatening dysrhythmia; the rest of the time they would know they had an

ICD only because it is palpable under their skin. An ICD should have essentially no effect on the exercise response.

In practice, inappropriate shocks (i.e., a shock in the presence of a normal rhythm) occur in about 15% of individuals who have an implanted ICD. This is, of course, very disturbing to the individual and can cause trepidation about continuing with an exercise program. The risk of having an inappropriate shock has not been shown to be greater among individuals who participate in moderate-intensity exercise. Once someone has had an inappropriate shock, however, he is often less willing to push the intensity.

Effects of Exercise Training

The potential benefits of exercise training among individuals who receive a pacemaker or ICD are similar to those for other patient groups. These benefits include, but are not limited to, the following:

- Increased exercise capacity
- Reduced risk of mortality
- Improved cardiometabolic risk factors
- Reduced depression
- Improved quality of life

Pacemakers and ICDs are most commonly adjusted while the patient is in an exam room, and it is very helpful to have physical activity or exercise program data on heart rates and blood pressures, to help the physician provide optimal operation of the device. This helps the patient be the most free from symptoms of the pacing rate not being adequate, or the ICD from prematurely firing due to high exertional heart rates. Either of these is very disconcerting to patients, so the exercise specialist plays a key role in helping optimize operating parameters of these devices.

Recommendations for Exercise Testing

Exercise testing can be useful in the evaluation of a rate-responsive pacemaker and to guide the programming of the upper heart rate limit. There is considerable interinstitutional variability in the use of exercise testing for this purpose. Exercise testing can be useful to assess exercise capacity and the hemodynamic response to exercise in the presence of either device. Pacemakers, but not ICDs, limit the utility of an exercise test in the assessment of myocardial ischemia because they reduce the sensitivity to detect repolarization changes in the ECG. The exercise testing protocol for individuals with a pacemaker or ICD should be selected based on the purpose of the test and the anticipated exercise capacity.

These are special considerations for exercise testing in an individual with a pacemaker or ICD:

- Medications should be taken as prescribed and at least 3 h before testing.
- The device's programmed upper limits should be identified before testing.
- To avoid an inappropriate shock, the test should be stopped before the heart rate is within 10 beats of the antitachycardia pacing or defibrillation threshold of the ICD.

Recommendations for Exercise Programming

Recipients of a pacemaker or ICD should be encouraged to be physically active. Any underlying heart disease or comorbidities have more influence on the ability to exercise than the presence of a pacemaker or an ICD does. Many recipients of pacemakers are elderly and prefrail or frail. In this group, light- to moderate-intensity exercise (aerobic and strength training) consistent with the Basic *CDD4* Recommendations is safe. Since most people who have an ICD have heart failure, at this time high-intensity exercise is not recommended.

- Use the moderate-intensity Basic *CDD4* Recommendation as the default program.
- People who are more frail can use light to moderate activities to maintain physical functioning.

These are special considerations for exercise training among individuals with a pacemaker or ICD:

- Rigorous activities involving the upper extremities should be limited during the first 3 to 4 weeks after implantation to allow the incision to heal.

- Medications should be taken as prescribed and at least 3 h before exercise.
- Individuals should be aware of their pacemaker or ICD heart rate thresholds.
- Target heart rate can be determined from a maximal exercise test but may not be necessary for all individuals.
- Exercise heart rate should be kept 10 beats or more below the antitachycardia pacing or defibrillation threshold of the ICD.

Integration Into a Medical Home Model

It is important that individuals who receive an implanted pacemaker or ICD attend follow-up visits recommended by their electrophysiologist or cardiologist. This typically includes regular assessment of their implanted device through a "pacemaker clinic." Individuals who experience an inappropriate shock or symptoms of inadequate heart rate should immediately contact emergency medical services or their electrophysiologist or cardiologist (or more than one of these).

It is not uncommon for people who have a pacemaker or ICD to be sedentary. As a result, they are at increased risk of cardiometabolic disease and should be encouraged to join an exercise program. The following are good exercise resources:

- Cardiac rehabilitation
- Specialist clinical exercise physiologist who is knowledgeable about exercise considerations for individuals with pacemakers and ICDs

Suggested Readings

American College of Sports Medicine. *ACSM's Guidelines for Exercise Testing and Prescription* (9th ed.). Baltimore: Lippincott Williams & Wilkins; 2013:250-251.

Epstein AE, Dimarco JP, Ellenbogen KA, Estes NA 3rd, Freedman RA, Gettes LS, Gillinov AM, Gregoratos G, Hammill SC, Hayes DL, Hlatky MA, Newby LK, Page RL, Schoenfeld MH, Silka MJ, Stevenson LW, Sweeney MO; American College of Cardiology/American Heart Association Task Force on Practice; American Association for Thoracic Surgery; Society of Thoracic Surgeons. ACC/

AHA/HRS 2008 guidelines for device-based therapy of cardiac rhythm abnormalities. *J Am Coll Cardiol.* 2008;51(21):e1-e62. doi: 10.1016/j.jacc.2008.02.032.

Isaksen K, Morken IM, Munk PS, Larsen AI. Exercise training and cardiac rehabilitation in patients with implantable cardioverter defibrillators: a review of current literature focusing on safety, effects of exercise training, and the psychological impact of programme participation. *Eur J Prev Cardiol.* 2012;19(4):804-812.

Piccini JP, Hellkamp AS, Whellan DJ, Ellis SJ, Keteyian SJ, Kraus WE, Hernandez AF, Daubert JP, Piña IL, O'Connor CM. Exercise training and implantable cardioverter defibrillator shocks in patients with heart failure: results from HF-ACTION. *JACC Heart Fail.* 2013;1(2):142-148.

Rait MH. Inappropriate implantable defibrillator shocks. *J Am Coll Cardiol.* 2013;62(15):1351-1352.

Stewart KJ, Spragg DD. Cardiac electrical pathophysiology. In Ehrman JK, Gordon PM, Visich PS, Keteyian SJ, eds. *Clinical Exercise Physiology* (3rd ed.). Champaign, IL: Human Kinetics; 2013:297-314.

Tracy CM, Epstein AE, Darbar D, DiMarco JP, Dunbar SB, Estes NA 3rd, Ferguson TB Jr, Hammill SC, Karasik PE, Link MS, Marine JE, Schoenfeld MH, Shanker AJ, Silka MJ, Stevenson LW, Stevenson WG, Varosy PD, Ellenbogen KA, Freedman RA, Gettes LS, Gillinov AM, Gregoratos G, Hayes DL, Page RL, Stevenson LW, Sweeney MO; American College of Cardiology Foundation; American Heart Association Task Force on Practice Guidelines; Heart Rhythm Society. 2012 ACCF/AHA/HRS focused update of the 2008 guidelines for device-based therapy of cardiac rhythm abnormalities: a report of the American College of Cardiology Foundation/ American Heart Association Task Force on Practice Guidelines and the Heart Rhythm Society. *Circulation.* 2012;126(14):1784-1800. doi: 10.1161/CIR.0b013e3182618569.

Additional Resources

American College of Cardiology, www.acc.org [Accessed November 11, 2015].

American Heart Association, www.americanheart.org [Accessed November 11, 2015].

Boston Scientific, http://www.bostonscientific.com/en-US/medical-specialties/electrophysiology/resources.html [Accessed November 11, 2015].

Medtronic, www.medtronic.com/for-healthcare-professionals/products-therapies/cardiac-rhythm/index.htm [Accessed November 11, 2015].

St. Jude Medical, http://professional.sjm.com/ products/crm [Accessed November 11, 2015].

Valvular Heart Disease

The four valves (tricuspid, pulmonary, mitral, and aortic) of the human heart work in concert to ensure unidirectional flow of blood through the chambers of the heart and the pulmonary and systemic circuits. The valves themselves are avascular, thin fibrous structures that open and close completely and passively with changes in pressure during the cardiac cycle.

Basic Pathophysiology

Common causes of heart valve *disease* include

- congenital defects,
- connective tissue disorders,
- infective endocarditis,
- rheumatic heart disease, and
- calcific disease of aging.

Heart valve *dysfunction* may also result from cardiomyopathic processes that lead to dilation of the right or left ventricle (which tends to pull the leaflets too far apart). Valvular heart disease may be complicated by or associated with arrhythmias as well.

The tricuspid and mitral valves, which sit between the atria and the ventricles, open when ventricular pressure is lower than atrial pressure during diastole and allow ventricular filling. These close immediately with the onset of ventricular systole as ventricular pressure exceeds that in the atria; and the pulmonary and aortic valves, which regulate the outflow of the right and left ventricles, open once ventricular pressure exceeds arterial pressure. Following the ejection of blood from the ventricle, chamber pressure falls again, causing the pulmonary and aortic valves to close. This cycle repeats with every heartbeat.

Disease or damage to the valve(s) can cause

- stenosis—obstruction of forward flow, or
- regurgitation—inadequate closure, resulting in backward flow of blood.

The abnormal flow across the diseased valve can often be heard as a murmur, and when the abnormality is confirmed by echocardiography, the severity of the condition is also quantified and usually graded as

- mild—generally not clinically significant,
- moderate—may cause symptoms, especially in active individuals, or
- severe—typically associated with symptoms.

Symptoms of heart valve disease may include

- fatigue;
- decreased exercise capacity;
- dyspnea (especially with exertion);
- palpitations, angina pectoris, or both;
- presyncope, syncope;
- heart failure with nonproductive cough; and
- lower-extremity swelling (advanced cases).

Mitral Valve Disease

A variety of mitral valve disorders can cause incomplete closure during systole, which allows blood to flow back (regurgitate) into the left atrium. Symptoms from the ensuing pulmonary congestion include exertional dyspnea or, in advancing cases, dyspnea at rest. Eventually pulmonary hypertension with left heart dilation and failure develops. Physical exam will reveal a holosystolic murmur. In the case of mitral valve prolapse, a midsystolic click precedes the mitral regurgitation murmur. People with mitral valve prolapse appear to have an underlying connective tissue disorder.

Mitral stenosis refers to any narrowing of the mitral orifice. The increased resistance to ventricular filling causes an inability to augment cardiac output during exertion. In advanced cases, high left atrial pressure mimics the signs and symptoms of left ventricular failure, with dyspnea, pulmonary hypertension, marked fatigue, and lower-extremity edema.

Rheumatic heart disease deserves special mention, because it is the most common cause (60% of cases). Streptococcal infection (i.e., strep throat) can cause an inflammatory reaction in the heart, including the mitral valve, and result in thickening of the valve and mitral stenosis after a lag time of 10 years or more. Rheumatic heart disease is

- commonly associated with atrial fibrillation and
- has a high risk of thromboembolism from atrial fibrillation.

Aortic Valve Disease

Aortic regurgitation results when the aortic valve cusps fail to close securely during diastole, allowing blood in the aorta to flow back into the left ventricle. There are numerous causes. Aortic regurgitation classically has a long "latent" period, during which mild and moderate degrees of regurgitation are well tolerated without symptoms. When regurgitation progresses to a severe degree, decreased exercise tolerance, dyspnea, palpitations, and angina may result. A diastolic murmur and exaggerated arterial pulses are common.

Aortic stenosis is failure of the aortic valve cusps to freely open during systole, leading to high left ventricular pressures and limited cardiac output. High ventricular pressures are necessary to maintain arterial blood pressures downstream of the narrowing, and these high pressures cause left ventricular hypertrophy. The most common cause is gradual and progressive fibrosis and calcification. Mild stenosis is generally well tolerated, but higher degrees of stenosis cause dyspnea, angina, and syncope.

Because progression is insidious, individuals often downregulate their physical activities to avoid symptoms and are unaware that they have the condition.

Right-Sided Valvular Heart Disease

Valvular conditions on the right side of the heart, involving the tricuspid and pulmonic valves, have multiple origins but cause fewer clinical problems than aortic and mitral valve disease. Many right heart valve problems are sufficiently well tolerated to not require surgery, and are covered here mainly for completeness. There are few objective studies on exercise in right-sided valvular conditions, so a person with a pulmonic or tricuspid valvular condition should be referred to a cardiologist for guidance as to the safety and appropriate type and degree of exercise.

Tricuspid regurgitation is often a functional lesion caused by either pulmonary hypertension or dilation of the right ventricle, and in these situations it is rarely severe. Edema and fatigue are potential symptoms, although tricuspid regurgitation is often asymptomatic. The exam may reveal distended neck veins and a holosystolic murmur that increases with inspiration.

Tricuspid stenosis is a rare condition involving obstruction to flow from the right atrium to the right ventricle. It is almost always due to rheumatic heart disease and seen in association with mitral stenosis. Lower-extremity edema and ascites are common manifestations of the high sustained right atrial pressure.

Pulmonic regurgitation is also rare and can be due to pulmonary hypertension; it may involve a systolic flow murmur as well as the expected diastolic murmur from the abnormal regurgitant flow. Treatment should commence before the right ventricular enlargement becomes significant.

Pulmonic stenosis is a narrowing of the pulmonary valve causing fixed blood flow to the lungs and increased pressures in the right ventricle. The murmur of pulmonic stenosis is a harsh systolic crescendo–decrescendo murmur heard best at the left second interspace. Other heart conditions that cause such a murmur are more common, such as atrial and ventricular septal defects, so these conditions should be considered and echocardiography obtained to make the correct diagnosis.

Management and Medications

Treatment decisions for valvular heart disease are primarily based on symptoms. Ultimately, valvular heart disease is essentially a mechanical disorder, and definitive treatment requires surgery. Symptoms that can be attributed to valvular heart disease with severe valve dysfunction on an echocardiogram are generally a clear indication for surgery. Asymptomatic persons who have severe echocardiographic findings require more investigation to determine treatment; there may be no benefit in waiting, but the risks of surgery can be substantial.

For people who have moderate or severe disease but for whom the risks for surgery outweigh the benefits, β-blockers may help control associated dysrhythmias, and symptoms of heart failure can be controlled with

- diuretics,
- angiotensin-converting enzyme inhibitors, and, less frequently,
- digoxin.

Emerging percutaneous techniques are currently suitable for specific conditions, such as rheumatic mitral stenosis (discussed earlier); transcatheter aortic valve replacement is an option for people who are not candidates for surgery due to high operative risk. Rheumatic mitral stenosis may be amenable to percutaneous balloon valvuloplasty, which is less invasive than surgery and has comparable long-term outcomes. Individuals who are candidates for this therapy should be referred as soon as exertional symptoms and pulmonary hypertension develop, even if the degree of stenosis is only moderate.

In aortic regurgitation, echocardiography can quantify the regurgitation, but transesophageal echocardiography and sometimes computed tomography or magnetic resonance angiography may be needed to visualize the entire aortic root. Surgical treatment should be considered when symptoms are present or at the first sign of left ventricular dilation or failure.

- Aortic regurgitation is commonly associated with dilation of the aortic root.
- Calcium channel and angiotensin receptor blockers can reduce progression.
- See chapter 15 on aneurysms for guidance on aortic root disease.

Surgical repair of the native valve is possible in some instances, depending on the specific anatomic defect. Mitral valve prolapse and bicuspid aortic valve disease can be repaired by very highly trained and experienced heart surgeons. Valve replacement is the standard treatment for degenerative mitral and aortic disease, and surgery is required for aortic regurgitation combined with aortic root dilation. Annuloplasty for tricuspid regurgitation is often undertaken at the time of mitral valve surgery. Before most valve surgeries, coronary angiography should also be performed to exclude concomitant coronary disease.

The principal decision before surgery is the choice of a bioprosthetic or a mechanical (metallic) valve. Bioprosthetic valves require anticoagulation for only a few months after implantation, so they do not preclude participation in contact sports. However, they have limited durability in the body, especially in younger people, such that if the person's life span extends beyond the valve's usable life (usually 7-10 years), repeat surgery will be required. Metallic valves are durable enough to last a lifetime but require continuous, lifelong anticoagulation. This affects quality of life and precludes participation in high kinetic energy sports because of the risks of life-threatening hemorrhage in the event of an accident, a fall, or trauma.

Effects on the Exercise Response

Valvular heart disease affects the exercise response in proportion to the severity of the lesion. Mild lesions generally do not have a significant impact on the person's ability to exercise. The impact of intermediate degrees of disease (or multiple lesions) may be difficult to predict, especially in individuals who are asymptomatic during daily activities of living. Severe valvular disease is more readily characterized:

Severe Mitral Regurgitation

Hemodynamic response is altered.

Exercise capacity and peak $\dot{V}O_2$ are reduced.

High static loads (weightlifting, downhill skiing) may provoke symptoms by markedly increasing afterload and the regurgitant fraction of cardiac output.

Severe Mitral Stenosis

Cardiac output may be blunted or fixed.

Flow limitation may cause exercise-induced hypotension.

Exertional tachycardia can increase left atrial and pulmonary pressures, so high heart rate activities (cycling, running, swimming) are contraindicated.

Aortic Regurgitation

High static loads (weightlifting, downhill skiing) may provoke symptoms by markedly increasing afterload and the regurgitant fraction of cardiac output.

If left ventricle function is normal, high dynamic loads are often well tolerated.

Aortic Stenosis

Prevents augmentation of cardiac output.

Flow limitation increases risk of exertional hypotension and syncope.

> *Valvular heart diseases commonly cause early exertional fatigue, exertional dyspnea and a hypotensive pressor response to exercise, and symptoms of presyncope or syncope. Medical and surgical management improves these exertional symptoms to improve quality of life and preserve independent functioning. Exercise management must consider how to perform exercise without provoking symptoms and whether the specific valvular condition will worsen with a particular form of exercise. The recommendations are thus complex and highly dependent on the valve(s) involved, the degree of severity, or whether the valve has been surgically repaired or replaced.*

Effects of Exercise Training

There are no randomized studies of exercise training with regard to the benefit—or harm—of exercise for persons with valvular heart disease. Exercise is associated with a higher quality of life in individuals with valvular heart disease; and in those who undergo valve surgery, preoperative functional capacity is a strong predictor of postoperative functional capacity. Thus, active individuals are likely to achieve the best outcomes by remaining active despite their heart condition. Symptoms of valvular heart disease may well restrict activity, but development of such symptoms is usually a strong argument for surgery, not for reducing activity.

Recommendations for Exercise Testing

Exercise tolerance testing with close attention to the electrocardiographic and blood pressure response, as well as to symptoms such as dyspnea, chest discomfort, palpitations, or presyncope, is often very helpful in evaluating persons with valvular heart disease. Such testing quantifies functional capacity in addition to objectively demonstrating symptoms and their association with cardiovascular pathophysiology. Exercise testing is therefore valuable for both symptom diagnosis and risk assessment before unsupervised exercise.

Exercise echocardiography is particularly advantageous for people with mitral valve disease, because Doppler imaging can estimate pulmonary artery pressures immediately following exercise. These pressures often correlate with symptoms better than electrocardiographic or systemic blood pressure changes. Exercise echocardiography recommendations for mitral and aortic disease are as follows:

Mitral Regurgitation

Exercise echo to evaluate for exercise-induced pulmonary hypertension and define functional capacity

Mitral Stenosis

Exercise echo to evaluate the mitral pressure gradient and pulmonary pressure

Aortic Regurgitation

Exercise testing to assess exertional symptoms and the hemodynamic response to exercise, looking for systemic hypotension, pulmonary congestion, or just reduced peak $\dot{V}O_2$

Aortic Stenosis

Exercise testing to determine the clinical severity of aortic stenosis.

A drop in blood pressure during exertion is an ominous sign, and can reveal whether someone is truly asymptomatic or simply sedentary.

Note: Aortic stenosis reduces the sensitivity and specificity of exercise testing for coronary disease.

Recommendations for Exercise Programming

Exercise recommendations for persons with valvular heart disease should take into consideration the type and severity of valvular heart disease, the type of exercise (i.e., dynamic or static or a mix of dynamic and static), and the life circumstances and desires of the individual.

Dynamic exercise activities are associated with increases in oxygen consumption (high percentage of peak $\dot{V}O_2$) that increase heart rate, stroke volume, and systolic blood pressure but decrease peripheral resistance. Examples of high dynamic exercises include racquetball, running, swimming, and rowing.

Static exercises are associated with large muscle forces and an increase in total peripheral resistance with smaller increases in cardiac output. Examples of high static loads include weightlifting (to maximal repetitions), wrestling, downhill skiing, and gymnastics.

Most activities of daily living are low to moderate dynamic and static exercise, as are many common recreational activities such as walking, bowling, golf, and the exercises in the Basic *CDD4* Recommendations.

Valvular heart disease is commonly a disease of aging, but there are many younger and athletic individuals who were born with congenital heart conditions or who have acquired a valvular condition such as mitral valve prolapse or endocarditis.

At the extremes of mild or severe valvular disease, it is relatively easy to make a recommendation, but with moderate or complex conditions this can be difficult. Individuals who have moderate valvular stenosis or regurgitation should consult with a cardiologist before starting an exercise program, because guidance on the type and intensities of exercise remains a clinical judgment with an insufficient evidence base to make a general recommendation. The following are general recommendations for mild or severe valvular heart conditions and after valvular surgery.

Mild and Asymptomatic to Minimally Symptomatic, Any Valve

Any exercise the person wants to do is permissible.

Regular follow-up with the cardiologist is required.

People should avoid going hard enough to cause symptoms.

If increasing symptoms develop, they should see their cardiologist.

Moderate to Severe and Symptomatic, Mitral or Aortic Valve

Stenosis—avoid high dynamic exercises

Aortic stenosis—generally low-intensity activities only

Regurgitation—avoid high static exercises

Connective Tissue and Congenital Conditions

Mitral valve prolapse—avoid contact sports

Genetic disorders (Marfan, Ehlers-Danlos, Loeys-Dietz, bicuspid valve disease) of the aortic root—avoid contact sports and high arterial pressures

Dilated aortic root—see chapter 15 on aneurysms

Status Postvalve Repair or Replacement

Cardiac rehabilitation with supervised exercise is strongly advised.

If high-intensity exercise is the goal, exercise testing up to the proposed level of exertion should be considered.

Special Considerations

The exercise response in people with multiple valve lesions can be difficult to predict. Expert consultation and caution with exercise testing is advised.

Current guidelines (2007) recommend antibiotic prophylaxis against infective endocarditis before surgery or dental procedures for

- persons with prosthetic valves,
- persons with prior infective endocarditis,
- unrepaired congenital heart disease,
- repairs involving prosthetic material, or
- heart transplant recipients with valvular heart disease.

Individuals who have had rheumatic fever, especially if there was cardiac involvement (mitral stenosis), should receive ongoing secondary prophylaxis until age 40 or 10 years after infection, and longer if they have ongoing contact with children.

Integration Into a Medical Home Model

The medical home staff can coordinate interval follow-up and assessment, which is especially important for those individuals with moderate degrees of valvular heart disease. People should be periodically queried about new or progressive symptoms such as decreased exercise capacity, palpitations, angina, syncope, and dyspnea; phone or e-mail contact may be well suited for this purpose. Similar symptom monitoring may be valuable for truly asymptomatic individuals with severe disease who opt for a strategy of *watchful waiting.* Repeat echocardiography can then be safely reserved for changes in symptoms or every 2 to 3 years for moderate disease and every 1 to 2 years for severe disease. Palpitations in persons with mitral stenosis warrant investigation for atrial fibrillation.

At each encounter, whether by phone, e-mail, or in person, individuals should be encouraged to continue an exercise regimen that maintains overall functional status, strength, and well-being. People who are say they are cutting back on previously enjoyable activity may need reevaluation for progression of valvular heart disease.

Take-Home Message

These are important considerations to remember when working with people with valvular heart disease:

- Mild degrees of valvular heart disease are common and should not prevent people from starting or continuing an exercise program.
- β-blockers to slow the heart rate may reduce symptoms by increasing time for ventricular filling.
- People with moderate degrees of valvular heart disease, as graded by echocardiography, can safely perform most types of exercise, especially with exercise testing to determine a safe work rate before commencing an unsupervised exercise program, with the following caveats:
  - People with moderate regurgitant lesions should avoid high static load exercises such as weightlifting, downhill skiing, and cycling.
  - Persons with moderate stenotic lesions should avoid high dynamic load exercises such as cycling, running, and swimming.
- Those with severe, symptomatic valvular heart disease should receive definitive (surgical) treatment before participating in most sports and exercise regimens.
- Asymptomatic individuals with severe valvular heart disease require cautious exer-

cise testing to quantify the work rate their cardiovascular system can accommodate and should be restricted to low-intensity activities only.

- Aortic valve disease is often accompanied by disease of the aortic root, which can be more dangerous.
- Prosthetic heart valves generally have hemodynamics similar to mild valvular heart disease at rest because prosthetic valves have some stenosis when compared to a normal, native valve.
- Cardiac rehabilitation is recommended for return to exercise after heart valve surgery.

Suggested Readings

Armstrong W, Ryan T. *Feigenbaum's Echocardiography* (7th ed.). Philadelphia: Lippincott Williams & Wilkins; 2009.

Bonow RO, Mann DL, Zipes DP, Libby P. *Braunwald's Heart Disease: A Textbook of Cardiovascular Medicine* (9th ed.). Philadelphia: Saunders; 2011.

Lancellotti P, Magne J. Stress testing for evaluation of patients with mitral regurgitation. *Curr Opin Cardiol.* 2012;27:492-498.

Mitchell JH, Haskell W, Snell P, Van Camp SP. 36th Bethesda Conference Task Force 8: classification of sports. *J Am Coll Cardiol.* 2005;45:1364-1367.

Rahimtoola SH. Choice of prosthetic heart valve in adults: an update. *J Am Coll Cardiol.* 2010;55:2413-2426.

Siu SC, Silversides CK. Bicuspid aortic valve disease. *J Am Coll Cardiol.* 2010;55:2789-2800.

Additional Resources

Evidence-Based Guidelines

Bonow RO, Carabello BA, Chatterjee K, de Leon AC Jr, Faxon DP, Freed MD, Gaasch WH, Lytle BW, Nishimura RA, O'Gara PT, O'Rourke RA, Otto CM, Shah PM, Shanewise JS. 2008 focused update incorporated into the ACC/AHA 2006 guidelines for the management of patients with valvular heart disease: a report of the American College of Cardiology/American Heart Association Task Force on Practice Guidelines (Writing Committee to Revise the 1998 Guidelines for the Management of Patients With Valvular Heart Disease): endorsed by the Society of Cardiovascular Anesthesiologists, Society for Cardiovascular Angiography and Interventions, and Society of Thoracic Surgeons. *Circulation.* 2008;118:e523-e661.

Bonow RO, Cheitlin MD, Crawford MH, Douglas PS. 36th Bethesda Conference Task Force 3: valvular heart disease. *J Am Coll Cardiol.* 2005;45:1334-1340.

Heart Transplantation

Heart transplantation is the gold standard treatment for selected individuals with end-stage heart failure who continue to have symptoms despite optimal medical and device therapy. Heart transplantation significantly improves quality of life and dramatically increases survival (90% and 70% after 1 and 5 years, respectively, with a median survival of 10 years). Despite receiving a healthy heart in an orthotopic heart transplant (which involves the complete removal of the recipient's own heart), most people continue to experience some exercise intolerance due to peripheral muscle dysfunction and reduced aerobic capacity.

Effects on the Exercise Response

Because the donor heart's innervation is transected in order for the heart to be transplanted, immediately after transplantation there is no parasympathetic or sympathetic innervation to the heart. With time after transplantation, only some people demonstrate signs of partial cardiac reinnervation, so most heart transplant recipients must rely on circulating catecholamines to provide adrenergic stimulation to the heart. This results in heart transplant recipients having different heart rate (HR) and hemodynamics, both at rest and during exercise, when compared to healthy individuals with a normal heart. The classic differences in the chronotropic response are as follows:

- Resting HR is increased ~20 beats/min.
- Onset of increased HR is delayed for the first several minutes of exercise.
- This is followed by an increase in HR that is more gradual than normal.

- Peak HR is slightly lower than normal (~150 beats/min).
- Heart rate may remain near its peak value for several minutes during recovery.
- Return to resting levels is delayed.

The reason for this blunted response and low chronotropic reserve (the difference between maximal and resting HR) is that the HR response is driven by circulating catecholamines, which follow a more prolonged time course than does neural control of an innervated heart.

The systemic circulation is also affected by heart transplantation, again because the heart is denervated, which disrupts the baroreflex control mechanisms that maintain the balance between cardiac output and vascular resistance. These are the results:

- Blood pressure at rest is often mildly elevated.
- Systolic pressure is lower than normal at peak exercise.
- Cardiac output is lower than normal at peak exercise ($\approx$60-70% peak cardiac output compared to that in age-matched healthy controls).

Stroke volume is also lower than normal at maximal exercise, but this is due to persistent left ventricular dysfunction.

Effects of Exercise Training

Exercise training is an integral part of rehabilitation after heart transplantation, because improving exercise intolerance is just as important a goal as prolonging the individual's life. Participation in a formal exercise training program is very beneficial and provides physiological and clinical improvements in physical functioning. The clinical and physiological benefits of exercise training in heart transplantation recipients include the following:

- Resting HR and blood pressure are lower.
- Maximal HR is higher.
- Peak $\dot{V}O_2$ and $\dot{V}O_2$ at ventilatory threshold are higher.
- Submaximal exercise endurance is improved.
- Blood lactate concentrations are lower at the same work rate.
- Ventilatory efficiency improves.
- Exertional fatigue and dyspnea are diminished.

These improvements in physical performance are related to tissue markers of improved health, including improvements in aerobic characteristics of skeletal muscle, higher maximal muscle force production, and improved endothelial function. Exercise training also increases bone mineral density, which is important to counteract the deleterious effects of immunosuppressive therapy (see chapter 6 on osteoporosis).

Management and Medications

Medical management of organ transplantation requires immunosuppressive drugs unless the donor is an identical twin (which mainly applies to kidney or bone marrow transplantation). Immunosuppression carries numerous issues, including the need for prophylactic treatment of infections, treatment of drug side effects, and treatment of

associated pathologies such as metabolic disorders that are exacerbated by immunosuppressant drugs (table 14.1). In addition, many people have medication needs not related to the heart transplant (e.g., for arthritis).

Controlling immune system rejection of the donor heart and avoiding the adverse side effects of immunosuppressive therapy are thus primary concerns following heart transplantation. Acute graft rejection is common, especially within the first year. Treatment therefore includes aggressive medical management using immunosuppressive agents (e.g., prednisone, cyclosporine, azathioprine, mycophenolate, tacrolimus), and in rare cases, retransplantation. Adverse side effects of immunosuppressant drugs include infections, hyperlipidemia, hypertension, obesity, osteoporosis, renal dysfunction, and diabetes. Moreover, just 1 year after surgery, the probability of developing accelerated atherosclerosis increases.

With the magnitude of expense and life disruption that are associated with a heart transplant (really any organ transplant), it makes little sense for transplant recipients to not participate in an exercise program as part of his or her rehabilitation. Surprisingly, such programs are not always covered by health insurance and may require some innovative efforts for the transplant recipient to receive the measure of benefit from transplantation. Everyone on the health care team should be pushing for including exercise programming for everyone who undergoes organ transplantation.

Table 14.1 **Immunosuppressant Medication Side Effects**

Generic name	Class of drug	Major adverse side effects
Methylprednisolone Prednisone	Corticosteroids	Fluid retention, muscle atrophy, myopathy, osteoporosis
Cyclosporine Tacrolimus	Calcineurin inhibitors	Hyperglycemia, hypertension, nephrotoxicity, electrolyte abnormalities
Mycophenolate mofetil	Inosine monophosphate dehydrogenase inhibitor	Hyperglycemia, hypercholesterolemia, hypertension, peripheral edema
Sirolimus Everolimus	Mammalian target of rapamycin (mTOR) inhibitors	Hypertriglyceridemia, hypercholesterolemia, hypertension, peripheral edema

Recommendations for Exercise Testing

Exercise testing for heart transplant recipients can be done with either a treadmill or a stationary cycle ergometer and should follow a conservative exercise testing protocol that has relatively small increases in work rate per stage (e.g., 0.5-1.0 metabolic equivalents [METs]). Measuring respired gases allows accurate quantification of functional capacity and helps refine the exercise prescription, especially from identification of $\dot{V}O_2$ at the ventilatory threshold. Measurement of peak respiratory exchange ratio ($\dot{V}CO_2/\dot{V}O_2$) provides a highly objective measure of effort, while the $\dot{V}E/\dot{V}CO_2$ slope and peak $\dot{V}O_2$ are key variables for clinical and prognostic evaluation. Peak $\dot{V}O_2$ in untrained cardiac transplant recipients is generally ≤ 20 to 25 mL $O_2 \cdot kg^{-1} \cdot min^{-1}$.

Although isolated cases of chest pain associated with accelerated graft atherosclerosis have been observed, autonomic denervation reduces the likelihood of anginal symptoms, especially during the initial months or years following surgery when partial reinnervation is less likely. Exercise electrocardiography is also inadequate with respect to assessing ischemia, as evidenced by its low sensitivity (i.e., <25%) for detecting true disease in these individuals. Thus, radionuclide testing is more useful for diagnosis of ischemic heart disease. For these reasons, only professionals who are experienced with exercise testing in high-risk populations should perform an exercise test in a heart transplant recipient.

For muscle strength assessment, an adequate warm-up period is mandatory and can be accomplished using the 1-repetition maximum (1RM) method. Most individuals reach 1RM within three to five trials, and at least a 3-min recovery between trials is desirable. During upper-limb strength or flexibility evaluations, sternal precautions should be strictly observed in the initial weeks or months following heart transplantation.

Important considerations during exercise testing include these:

- Calf muscle cramps occur in ~15% of people.
- Leg strength deficits are common and decrease exercise time.

Persons who have received long-term, high-dose corticosteroids have a high probability of decreased bone mineral content and increased risk for fracture. As a result, certain exercise tests are relatively contraindicated (see chapter 6 on osteoporosis):

- High-resistance strength tests
- Forward flexion range of motion (compresses vertebral bodies)

Recommendations for Exercise Training

Exercise training is strongly recommended for this population, including both aerobic and resistance exercise. Aerobic exercise improves exercise capacity and quality of life and helps with modification of cardiovascular risk factors (most notably obesity, hypertension, and insulin resistance). Improvements in peak $\dot{V}O_2$ range between 15% and 40%. Resistance exercise is also strongly encouraged, with the goal of improving muscle strength and bone density and to prevent the adverse effects that antirejection medications have on skeletal muscle. Less established, however, is whether modification of cardiovascular risk factors through exercise alters the progression of accelerated graft atherosclerosis, which is the major factor limiting long-term survival in these individuals.

For aerobic exercise, the minimal threshold intensity (i.e., 40%, 50%, or 60% of peak $\dot{V}O_2$) needed to significantly improve peak $\dot{V}O_2$ is not known. However, general principles of exercise training (i.e., overload, specificity, and reversibility), as well as factors to consider for an individualized exercise prescription (frequency, intensity, time, and type), also apply to the heart transplant population.

- Heart transplant recipients should follow the Basic *CDD4* Recommendations.
- They should subsequently advance to higher exercise intensities as tolerated.

The use of HR alone to guide exercise intensity is not appropriate in this population. It is not uncommon for heart transplant recipients to achieve an exercise HR that exceeds 85% of measured peak HR. Combined assessment of systolic blood pressure and rating of perceived effort (RPE) (recommended range 11-14 out of 20) is a better gauge of exercise training intensity.

For resistance exercise, one or two sets of 10 to 15 repetitions at low to moderate intensity, for all major muscle groups, two times per week is generally recommended.

These are important considerations during exercise programming:

- Progress conservatively, as severe deconditioning is common, especially if prolonged bed rest was required before surgery.

- Intermittent exercise throughout the day may be needed until longer, continuous exercise can be tolerated.

- Range of motion and stretching exercises are important for the upper body to restore normal thoracic biomechanics after sternotomy. However, these exercises should be limited for up to 6 to 8 weeks after surgery.

- The RPE is the preferred method of monitoring exercise intensity, particularly as the individual progresses to an independent exercise program.

- Longer periods of warm-up and cool-down are indicated because the physiological responses to exercise are delayed.

Integration Into a Medical Home Model

Supervised exercise training is an important non-pharmacologic intervention for heart transplant recipients, leading to improved physical function, aerobic capacity, muscle strength, and quality of life. Heart transplantation recipients have the potential to return to work or other usual activities, although exercise capacity remains below age- and sex-predicted values for most individuals.

Take-Home Message

Remember these key points when working with people who have had cardiac transplants.

- Most transplant recipients receive the majority of their care through their transplant team, but all members of the health care team should be aware of the special needs of heart transplant recipients.

- All heart transplant recipients should be strongly encouraged to participate in a cardiac rehabilitation program.

- Heart transplant recipients should be encouraged to maintain their exercise program for the rest of their life.

- A decrease in physical functioning can be a sensitive indicator of early rejection, accelerated atherosclerosis, or adverse effects of immunosuppression on skeletal muscle and is a reason for referral to the transplant team.

- Exercise training is an important antidote to the adverse cardiometabolic effects of long-term immunosuppression.

Suggested Readings

Arena R, Myers J, Williams MA, et al. Assessment of functional capacity in clinical and research settings: a scientific statement from the American Heart Association Committee on Exercise, Rehabilitation, and Prevention of the Council on Clinical Cardiology and the Council on Cardiovascular Nursing. *Circulation.* 2007;116:329-343.

Braith RW, Edwards DG. Exercise following heart transplantation. *Sports Med.* 2000;30:171-192.

Carvalho VO, Bocchi EA, Guimaraes GV. Aerobic exercise prescription in adult heart transplant recipients: a review. *Cardiovasc Ther.* 2011;29:322-326.

Costanzo MR, Dipchand A, Starling R, et al. The International Society of Heart and Lung Transplantation Guidelines for the care of heart transplant recipients. *J Heart Lung Transplant.* 2010;29:914-956.

Guazzi M, Adams V, Conraads V, et al. EACPR/AHA scientific statement. Clinical recommendations for cardiopulmonary exercise testing data assessment in specific patient populations. *Circulation.* 2012;126:2261-2274.

Hermann TS, Dall CH, Christensen SB, Goetze JP, Prescott E, Gustafsson F. Effect of high intensity exercise on peak oxygen uptake and endothelial function in long-term heart transplant recipients. *Am J Transplant.* 2011;11:536-541.

Hsieh PL, Wu YT, Chao WJ. Effects of exercise training in heart transplant recipients: a meta-analysis. *Cardiology.* 2011;120:27-35.

Hunt SA, Abraham WT, Chin MH, et al. 2009 focused update incorporated into the ACC/AHA 2005 guidelines for the diagnosis and management of heart failure in adults: a report of the American College of Cardiology Foundation/American Heart Association Task Force on Practice Guidelines: developed in collaboration with the International Society for Heart and Lung Transplantation. *Circulation.* 2009;119:e391-e479.

Kato TS, Collado E, Khawaja T, et al. Value of peak exercise oxygen consumption combined with B-type natriuretic peptide levels for optimal timing of cardiac transplantation. *Circ Heart Fail.* 2013;6:6-14.

Nytroen K, Myers J, Chan KN, Geiran OR, Gullestad L. Chronotropic responses to exercise in heart transplant recipients: 1-yr follow-up. *Am J Phys Med Rehabil.* 2011;90:579-588.

Nytroen K, Rustad LA, Aukrust P, et al. High-intensity interval training improves peak oxygen uptake and muscular exercise capacity in heart transplant recipients. *Am J Transplant.* 2012;12:3134-3142.

Pierce GL, Magyari PM, Aranda JM Jr, et al. Effect of heart transplantation on skeletal muscle metabolic enzyme reserve and fiber type in end-stage heart failure patients. *Clin Transplant.* 2007;21:94-100.

Radford MJ, Arnold JM, Bennett SJ, et al. ACC/AHA key data elements and definitions for measuring the clinical management and outcomes of patients with chronic heart failure: a report of the American College of Cardiology/American Heart Association Task Force on Clinical Data Standards (Writing Committee to Develop Heart Failure Clinical Data Standards): developed in collaboration with the American College of Chest Physicians and the International Society for Heart and Lung Transplantation: endorsed by the Heart Failure Society of America. *Circulation.* 2005;112:1888-1916.

Squires RW. Exercise therapy for cardiac transplant recipients. *Prog Cardiovasc Dis.* 2011;53:429-436.

Additional Resources

Evidence-Based Guidelines

Mehra MR, Jessup M, Gronda E, Costanzo MR. Rationale and process: International Society for Heart and Lung Transplantation guidelines for the care of cardiac transplant candidates—2006. *J Heart Lung Transplant.* 2006;25(9):1001-1002.

Suggested Websites

American Heart Association, www.american-heart.org [Accessed November 11, 2015].

International Society for Heart and Lung Transplantation, www.ishlt.org [Accessed November 11, 2015].

Aneurysms

An aneurysm is a localized dilation or bulge of a blood vessel caused by disease or weakening of an arterial vessel wall. Aneurysms most commonly occur in arteries at the base of the brain and in the aorta (abdominal and thoracic) (table 15.1). Aneurysms can also stem distally off of the aorta or occur in other arteries in the body (popliteal, mesenteric, and splenic areas), but these are far less common.

They can be caused by congenital or acquired diseases, although most aneurysms are diseases of aging. They are usually asymptomatic and are often not discovered until they rupture or cause symptoms due to localized pressure on an adjacent tissue. Aneurysms can be caused by inflammation, infection, and injury or may also be acquired from iatrogenic causes (mainly as a complication of surgical attempts to dilate a vessel

Table 15.1 **Common Types of Aneurysms**

Type	Prevalence (% of population)	Symptoms	Risk factors	Associated survival rate after rupture (%)
Cerebral	~2-5	• Headache, facial pain • A dilated pupil • Change in vision or double vision • Numbness, weakness, or paralysis on one side of the face • A drooping eyelid	• Increasing age • Smoking • High blood pressure • Drug abuse (especially cocaine) • Heavy alcohol use • Head injury • Certain blood infections • Lower estrogen levels after menopause	33-67
Abdominal aortic	~4-10 >60 years (accounting for ~75% of aortic aneurysms)	• Sudden, severe, constant low back, flank, abdominal, or groin pain • Syncope • Pulsing in abdomen • Rarely, pain, discoloration, or sores on toes or feet from aneurysm material shed	• Smoking • High blood pressure • Immediate family history of aneurysms • Age (>60) • Sex (male) • Specific diseases (Marfan syndrome, tuberculosis, syphilis) • Blunt trauma	<50 (survival rate drops by 1% per minute after dissection)
Thoracic aortic	~3-4 >65 years (accounting for ~25% of aortic aneurysms)	• Pain in jaw, neck, upper back • Chest or back pain • Coughing, hoarseness, or difficulty breathing	• Smoking • High blood pressure • Family history of aneurysms • Specific diseases (Marfan syndrome, tuberculosis, syphilis) • Blunt trauma	20-30

with atherosclerosis). The major risk factors for aneurysms include the following:

- Smoking
- Age
- Male sex
- Hypertension
- Obesity
- Markers of inflammation

The major risk for a person with an aneurysm is progressive enlargement of the aneurysm, tearing of the arterial wall (dissection), and sudden rupture of the artery. These risks apply to any aneurysm, regardless of location in the body.

Management and Medications

Because aneurysms often go undiagnosed until a serious event such as a rupture occurs, many efforts have been made to assess mortality benefits and cost-effectiveness of screening for abdominal aortic aneurysms. A recent U.S. Preventive Services Task Force recommended that all men between the ages of 65 to 75 with a history of smoking be screened for the disease with ultrasonography. Because of the lower prevalence of the disease in women, disagreement exists regarding the appropriateness of screening, although some data suggest that women >65 years with a history of smoking or heart disease should be screened.

For people diagnosed with an aneurysm, the key step in medical management is to follow the progress of the enlarged area. Screening technologies include these:

- Echocardiography
- Magnetic resonance imaging
- Computed tomography (CT) or CT angiography (less commonly)

The next most important step in managing these individuals is to aggressively treat the risk factors that underlie the disease. This includes smoking cessation, lowering blood pressure, and management of lipids and obesity. Goals for resting blood pressure should be within the normal range per Joint National Committee 7 Guidelines (<120/<80). Management of blood pressure is particularly important, since LaPlace's law states that as a vessel radius becomes larger, a greater wall tension is required to withstand a given fluid pressure.

Thus, the area of the vessel around the aneurysm is exposed to particularly high wall tension and pressure, which may worsen the aneurysm. Medications for individuals with aneurysms primarily involve those needed to control cardiovascular risk, especially the following:

- β-blockers (e.g., atenolol, propranolol, metoprolol)—ideally suited to blood pressure management because they also reduce the strength of arterial pulses (even those with a low-normal baseline blood pressure should be treated with a β-blocker if there is no contraindication to this medicine).
- Statins—commonly prescribed because of their moderating influence on lipids and inflammatory markers.
- Angiotensin-converting enzyme (ACE) inhibitors—for blood pressure management; some evidence exists that ACE inhibitors may reduce aortic aneurysm rupture rate.

Aneurysm diameter is progressive in many people, and the decision regarding surgery for aneurysm repair should be made by a vascular surgeon when appropriate. It should be noted that aneurysms most commonly occur among individuals who are elderly, who are likely to be using other medications for various comorbidities (e.g., other cardiovascular conditions such as coronary artery disease or peripheral vascular disease, hypertension, diabetes).

Effects on the Exercise Response

Early in the course of aneurysmal disease when the dilation is minimal, there is usually no effect on the exercise response. Even in persons with existing aneurysms, little change in the exercise response is likely other than a delay or blunting in pulse pressure. Heart rate responses and exercise capacity in persons with aortic aneurysms have been shown to be similar to those of age-matched controls, while a slightly higher incidence of claudication and hyper- and hypotensive responses may be seen in the aneurysm population (specifically abdominal aortic aneurysm).

Exercise has, in rare circumstances, been associated with aneurysm dissection or rupture. This complication has been reported for cerebral, aortic, renal, and coronary arteries, and any artery is theoretically at risk (including in persons with Marfan syndrome). In most case reports of such incidents, it was not known whether the aneurysm was present before the onset of symptoms. One study of 262 patients with abdominal aortic aneurysms undergoing maximal exercise testing reported one rupture 12 h after the test (an event rate of 0.4%). Because this was a retrospective survey of a selected population, however, the true risk is not known. A recent study among 306 patients with presurgical abdominal aortic aneurysms (3-5 cm) who underwent maximal cardiopulmonary exercise testing reported no complications and suggested that maximal exercise testing in this population is safe.

Effects of Exercise Training

Inflammation is an important risk factor for aneurysms, and regular exercise is known to favorably modify the systemic inflammatory state. Thus, it is thought that exercise may reduce aneurysm incidence and, when an aneurysm is present, potentially limit its growth rate. In addition, intermittent periods of increased wall stress through regular exercise have been shown to improve vascular function, and it has been suggested that training could potentially diminish aneurysm expansion rates. In abdominal aortic aneurysms, studies have shown that acute bouts of submaximal exercise counteract these abnormal hemodynamic conditions by attenuating retrograde flow and increasing wall shear stress in the abdominal aorta. These studies have consistently observed that unfavorable hemodynamic conditions at rest are improved with even modest levels of acute exercise. Regular exercise also has favorable effects on the major risk factors for aneurysm disease (blood pressure, obesity, lipids). In addition, studies have shown that patients in chronic sedentary states (spinal cord injury, amputation) are at higher risk for abdominal aortic aneurysm disease. Thus, there is justification for recommending moderate exercise in these individuals.

A recent large clinical trial demonstrated no paradoxical increase in abdominal aortic aneurysm growth rate or adverse clinical events occurring as a consequence of exercise training. The results of this trial showed that exercise was well tolerated in people with small abdominal aortic aneurysms (3-5.5 cm). While the association between regular exercise and abdominal aortic aneurysm progression has yet to be fully explored, these persons appear to achieve benefits from training that are

Given that the true prevalence of aneurysms has only recently been appreciated, that aneurysms are common in persons who are elderly, and that most are asymptomatic (and undiagnosed), these people have undoubtedly participated in rehabilitation programs safely for many years without knowledge of the condition. But in light of the paucity of data on the effects of training on aneurysm disease, a conservative approach is appropriate at this time, with frequent follow-up by the individual's physician to monitor aneurysm progression.

similar to those in individuals after myocardial infarction.

Recommendations for Exercise Testing

Current data suggest that the risk of complication is low for diagnostic exercise testing before surgery. Blood pressure must be closely monitored during exercise testing, and hypertensive responses should elicit cessation of testing. Although maximal exercise testing for individuals with an abdominal aortic aneurysm <5.0 cm has been shown to be safe, those with known intracranial aneurysms of any type should not undergo maximal exercise testing. For these individuals, it is important to avoid an excessive rise of the heart rate–blood pressure product (HR × systolic BP), which is a measure of stress on the arteries. Submaximal testing can be used to optimize therapeutic control of heart rate and blood pressure in these individuals. For presurgical abdominal aortic aneurysms (3-5 cm), a reasonable upper limit for systolic blood pressure is 200 mmHg.

Alternate testing methods include the following:

- Submaximal exercise tests (optimize therapeutic control of HR and BP)
- Pharmacologic stress tests (e.g., adenosine, dipyridamole, dobutamine)
- Other functional tests, such as 6-min walk or similar timed walking tests

There are instances when pharmacologic stress testing may be best suited to a particular individual (for evaluation of coronary artery disease); however, the size and location rather than simply the presence of an aneurysm may be a more significant factor when one is determining the appropriateness of pharmacologic stress testing.

Unfortunately, there are few to no data to define criteria for "excessive" in the context of aneurysms, so this remains an individual judgment by the professionals administering exercise testing. No data are available regarding the likelihood that an increase in rate–pressure product will cause an aneurysm to rupture, but it is thought that the larger the aneurysm, the more vulnerable it may be to such an occurrence. Thus, testing endpoints should be conservative. Maximal strength testing is contraindicated.

Recommendations for Exercise Programming

The American College of Cardiology/American Heart Association (ACC/AHA) practice guidelines for the management of people with peripheral vascular disease recommend modest activity in those with abdominal aortic aneurysm, in part to counteract the reductions in fitness associated with poor outcomes in people who eventually require surgery. In general, aerobic exercise is recommended for individuals with aneurysms. Contact sports and competitive activities should be avoided. The larger the diameter of the aneurysm relative to the normal diameter of the vessel, the more exercise should be restricted, and any aerobic activity should be performed at a moderate intensity.

Special Considerations

- Medications
  - People taking β-blockers should not use heart rate for intensity, but rather perceived exertion.
  - Persons with hypertension or hypertensive responses to exercise must have this risk factor controlled before continuing with an exercise program.
- Intensity
  - Moderate resistance training (low resistance, high repetitions) is appropriate for persons with abdominal aortic aneurysm but contraindicated in those with cerebral aneurysms.

- While individuals with abdominal aortic aneurysm can safely exercise at 50% to 70% of heart rate reserve, those with cerebral aneurysms should reduce exercise intensity.
- Marfan syndrome
  - People in this group should avoid flexibility training (e.g., some yoga positions) due to risk of joint dislocation (these persons are more likely than others to have orthopedic concerns such as joint contractures, hypermobility, scoliosis, hyperlordosis, and kyphosis).
  - They need to limit, regulate, or stop competitive sport altogether (sport psychologists can often be helpful during the difficult transition from competition activity to moderate physical activity).

Integration Into a Medical Home Model

Persons with aneurysms should be followed closely by medical staff to monitor blood pressure, medications, and symptoms (if present) in order to reduce the risk of dissection or other adverse events. The presence of increased pain or swelling or areas of bulging likely indicate that conditions have worsened, and the individual should be evaluated by a vascular specialist.

Exercise programs and medication management should be overseen (in accordance with Basic *CDD4* Recommendations) by home medical staff. Pedometers, written logs, and phone calls are useful tools for motivating people and tracking activity. Integration of a smoking cessation program and optimizing nutrition are also essential components of risk reduction to limit aneurysm progression. Clinicians (generally nurses and exercise physiologists) should take a comprehensive approach to cardiac rehabilitation (including telehealth programs) to assess progress, compliance, and changes in clinical status and optimize quality of life.

Aneurysm rupture is a life-threatening situation that requires immediate medical attention. Severe pain, low blood pressure, rapid heart rate, lightheadedness, nausea, and confusion may occur. The risk of death after a rupture is high, and emergency medical services should be contacted immediately.

Take-Home Message

Remember these key points when working with people with aneurysms:

- Moderate aerobic exercise for persons with aneurysms is generally safe as long as blood pressure is well controlled.
- Symptom-limited exercise testing has been shown safe in individuals with small (pre-surgical) abdominal aortic aneurysms.
- Submaximal exercise testing should be used in persons with cerebral aneurysms.
- Maximal strength testing is contraindicated in persons with aneurysm disease.
- Aneurysm disease is closely associated with cardiovascular disease; thus related risk factors should be closely monitored and aggressively treated. Regular ultrasound imaging evaluations to monitor aneurysm progression should be performed as appropriate.
- The exercise specialist should be prudent in prescribing exercise and comprehensively work to improve overall risk profile and quality of life.
- Be aware of signs and symptoms of an aneurysm rupture:
  - Severe pain
  - Nausea
  - Low blood pressure
  - Rapid heart rate
  - Light-headedness
  - Confusion

Suggested Readings

Best PA, Tajik AJ, Gibbons RJ, Pellikka PA. The safety of treadmill exercise stress testing in patients with abdominal aortic aneurysms. *Ann Intern Med.* 1998;139(8):628-631.

Braverman AC. Exercise and the Marfan syndrome. *Med Sci Sports Exerc.* 1998;30(10 Suppl):S387-S395.

Cheitlin MD, Douglas PS, Parmley WM. 26th Bethesda Conference: recommendations for determining eligibility for competition in athletes with cardiovascular abnormalities. Task force 2: acquired valvular heart disease. *J Am Coll Cardiol.* 1994;24(4):874-880.

Dalman RL, Tedesco MM, Myers J, Taylor CA. Abdominal aortic aneurysm disease: mechanism, stratification, and treatment. *Ann NY Acad Sci.* 2006;1085:92-109.

DeRubertis BG, Trocciola S, Ryer EJ, Pieracci FM, McKinsey JF, Faries PL, Kent KG. Abdominal aortic aneurysm in women: prevalence, risk factors, and implications for screening. *J Vasc Surg.* 2007;46:630-635.

Egelhoff CJ, Budwig RS, Elger DF, Khraishi TA, Johansen KH. Model studies of the flow in abdominal aortic aneurysms during resting and exercise conditions. *J Biomech.* 1999;32:1319-1329.

Ellis CJ, Haywood GA, Monro JL. Spontaneous coronary artery dissection in a young woman resulting from an intense gymnasium "work-out." *Int J Cardiol.* 1994;47:193-194.

Fleming C, Whitlock EP, Beil TL, Lederle FA. Screening for abdominal aortic aneurysm: a best-evidence systematic review for the US Preventive Services Task Force. *Ann Intern Med.* 2005;142:203-211.

Graham TP Jr, Bricker JT, James FW, Strong WB. Task force 1: Congenital heart disease. 26th Bethesda Conference: recommendations for determining eligibility for competition in athletes with cardiovascular abnormalities. *J Am Coll Cardiol.* 1994;24(4):867-873.

Hirsch AT, Haskal ZJ, Hertzer NR, Bakal CW, Creager MA, Halperin JL, Hiratzka LF, Murphy WRC, Olin JW, Puschett JB, Rosenfield KA, Sacks D, Stanley JC, Taylor LM Jr, White CJ, John White J, White RA; American Association for Vascular Surgery, Society for Vascular Surgery; Society for Cardiovascular Angiography and Interventions; Society for Vascular Medicine and Biology; Society of Interventional Radiology; ACC/AHA Task Force on Practice Guidelines. ACC/AHA guidelines for the management of patients with peripheral arterial disease. *JVIR.* 2006;17:1383-1398.

Myers J, Dalman R, Hill B. Exercise, vascular health, and abdominal aortic aneurysms. *J Clin Exerc Physiol.* 2012;1;1-8.

Myers J, Powell A, Smith K, Fonda H, Dalman R. Cardiopulmonary exercise testing in small abdominal aortic aneurysm: profile, safety, and mortality estimates. *Eur J Cardiovasc Prev Rehabil.* 2011;18:459-466.

Myers J, White J, Narasimhan B, Dalman R. Effects of exercise training in patients with abdominal aortic aneurysm. *J Cardiopulm Rehabil Prev.* 2010;30:1-10.

Provenzale JM, Barboriak DP, Taveras JM. Exercise-related dissection of cranio-cervical arteries: CT, MR, and angiographic findings. *J Comput Assist Tomogr.* 1995;19(2):268-276.

Pyeritz RE, Francke U. Conference report: The Second International Symposium on the Marfan Syndrome. *Am J Med Genet.* 1993;47:127-135.

Sherrid MV, Mieres J, Mogtader A, Menezes N, Steinberg G. Onset during exercise of spontaneous coronary artery dissection and sudden death. *Chest.* 1995;108(1):284-287.

Taylor CA, Hughes TJ, Zarins CK. Effects of exercise on hemodynamic conditions in the abdominal aorta. *J Vasc Surg.* 1999;29:1077-1089.

Thomas MC, Walker RJ, Packer S. Running repairs: renal artery dissection following extreme exertion. *Nephrol Dial Transplant.* 1999;14:1258-1259.

Additional Resources

Medscape, www.emedicine.medscape.com [Accessed November 11, 2015].

National Heart Lung and Blood Institute, What Is an Aneurysm? www.nhlbi.nih.gov/health/health-topics/topics/arm [Accessed November 11, 2015].

VascularWeb, www.vascularweb.org [Accessed November 11, 2015].

PULMONARY DISEASES

In the same way that the cardiovascular disorders discussed in part III are generally not caused by a physically inactive lifestyle, there are many pulmonary disorders that are not primarily a consequence of tobacco use. These include disorders of the lung parenchyma, the airways, and pulmonary vasculature. These conditions are more a matter of environmental exposure, genetics, and gene–environment interactions. Of course, some people have these conditions in addition to tobacco use, and those folks tend to have very poor pulmonary function.

In a normal healthy human being, pulmonary function rarely limits the ability to do exercise, and even then this is typically in very fit athletes or during exercise at high altitude. But in persons with lung disease, pulmonary function is very often the limiting factor in their ability to be active. The limitation is often a consequence of inadequate gas transfer due to arteriovenous shunting, ventilation–perfusion inequalities, or a diffusion limitation; but it can also be related to an increased work of breathing because it's harder to move air. Ventilation during high-intensity exercise requires that a substantial amount of cardiac output go to the respiratory muscles. In someone who requires even more muscular effort to breathe, integrated cardiopulmonary function can impose a major limitation in the ability to do exercise.

As in the case of people with the cardiovascular conditions discussed in part III, persons who have the chronic pulmonary conditions discussed in part IV primarily need help with an assessment of how their condition affects their ability to do exercise and then help with how to achieve and maintain a physical activity program that preserves their physical functioning and ability to live independently. This is immensely important in maintaining their quality of life; and, again, it is unfortunate that few health insurance policies in the United States cover such a program. Exercise specialists and clinical staff in the medical home need to be creative in order to provide needed exercise management services in persons with chronic pulmonary disease.

Chronic Restrictive Pulmonary Disease

hronic restrictive pulmonary disease (CRPD) encompasses a range of heterogeneous disorders that are characterized by a low lung volume and reduced thoracic or lung compliance or both. These disorders result from diverse pathological processes involving the chest wall, respiratory muscles, and nerves serving the thorax, as well as the pleura and lung parenchyma. They are termed *restrictive* because they restrict the bellows mechanisms of the lung, which is in contrast to obstructive lung disease that prevents airflow by obstructing the airways. The major causes of CRPD are summarized next.

Basic Pathophysiology

The diverse pathophysiological mechanisms that cause CRPD can be categorized in the following manner.

Extrapulmonary disorders:

Neuromuscular

- Muscle—muscular dystrophy, myositis
- Nerves—phrenic nerve paralysis, neuritis
- Neuromuscular junction—myasthenia gravis, botulism, Eaton-Lambert syndrome
- Spinal cord disorders—amyotrophic lateral sclerosis, Guillain-Barré syndrome

Trauma and Mechanical

- Chest wall—kyphoscoliosis, ankylosing spondylitis, thoracoplasty, morbid obesity, and metabolic syndrome
- Pleural—fibrosis, effusion

Intrinsic pulmonary disorders:

Interstitial inflammation and fibrosis—idiopathic pulmonary fibrosis (IPF), connective tissue diseases, granulomatous inflammation such as sarcoidosis

Malignancy

Pulmonary vascular disease—hemorrhage, congestion or edema, vasculitis

Major lung resection—lobectomy, pneumonectomy

Radiation exposure

Infectious agents

Inhaled particles

- Inorganic—silica, asbestos, beryllium
- Organic—hypersensitivity pneumonitis (e.g., exposure to cotton, grains, or birds)
- Drugs—chemotherapy agents, amiodarone, phenytoin
- Miscellaneous other diseases—histiocytosis X, alveolar proteinosis, eosinophilic pneumonia

In the parenchymal conditions, alveolar–capillary units are destroyed or replaced by fibrous tissue. In conditions involving the pleura, chest wall, or neuromuscular system, the lung is intrinsically normal but respiratory efforts are unable to expand the lung parenchyma, leading secondarily to alveolar collapse, inflammation, and eventually fibrosis. Regardless of specific etiology, these are the common end results:

- Loss of alveolar–capillary surface area and volume
- Impaired oxygen transfer across the pulmonary blood–gas barrier

Development of clinical symptoms may be acute (days), subacute (weeks), or chronic (months to years). The most common symptom is shortness of breath upon exertion, and a dry cough and pleuritic chest pain may also be present. On examination, people hyperventilate in mild to moderate stages and have rapid shallow breaths in advanced stages.

Lung function testing shows reduced lung volumes, with airflow rates reduced in proportion to the reduction in lung volume. An obstructive component may coexist if the disease involves both the parenchyma and small airways (e.g., sarcoidosis, hypersensitivity pneumonitis), or if the restrictive disease develops on a background of chronic obstructive pulmonary disease. In the intrinsic lung disorders the diffusing capacity for carbon monoxide (DL_{CO}), a measure of the rate of gas transfer from alveolar air to alveolar capillary red blood cells, is the most sensitive parameter and is reduced even when lung volumes are normal.

One of the most important physiological consequences of CRPD is the effect it has on the work of breathing. Compliance of the lung or thorax is reduced, requiring a more negative airway pressure to inflate the lung. Hence, respiratory muscles work harder (i.e., increased work of breathing). Tidal volume is reduced while dead space is elevated, so ventilation becomes less efficient, further increasing the work of breathing needed to maintain a given oxygen uptake by the lungs. A larger fraction of total body metabolic energy must be diverted to respiratory muscles to sustain a given level of ventilation, leaving a smaller fraction available for working limb muscles during exercise. This leads to greater lactic acid production from exercising limb muscles, further stimulating ventilation and increasing the work of breathing. In the neuromuscular and skeletal disorders, the strength of respiratory muscles, measured from the maximal inspiratory and expiratory pressures generated at the mouth, is reduced and may not be able to generate the power demanded by the elevated work of breathing in CRPD.

Restriction of pulmonary vascular bed increases pulmonary vascular resistance, leading to secondary pulmonary hypertension, right ventricular strain, and a lower stroke volume during exercise. Cardiovascular deconditioning inevitably develops with progressive pulmonary disease and further exacerbates the disability. Long-term sequelae of restrictive diseases include respiratory failure, disordered breathing during sleep, right heart failure (cor pulmonale), and susceptibility to infections due to poor secretion clearance, immunosuppression, or both.

Management and Medications

Medical treatment of CRPD must be tailored to the specific pathology. The general approach is to first identify and treat the underlying cause:

- Orthopedically correct spinal deformity in kyphoscoliosis.
- Treat bacterial, viral, or fungal infections.
- Suppress inflammatory and immunologic activity in the intrinsic lung conditions (usually with inhaled or systemic corticosteroids, immunosuppressants, cytotoxic or antifibrotic agents).

Secondarily, management is aimed at minimizing pulmonary complications related to the abnormal lung physiology:

- Institute chest physiotherapy to clear secretions (mainly neuromuscular diseases).
- Prevent infections, for example, prophylactic antibiotics, regular vaccination, clearing secretions.
- Support oxygenation, for example, supplemental oxygen therapy.

Third, it is important to preserve normal physical and physiological functions as much as possible, to both maintain physical functioning and avert cardiometabolic conditions secondary to being sedentary:

- Optimize cardiopulmonary and muscular fitness with a rehabilitation program.
- Then maintain physical functioning with a regular exercise regimen.

In individuals with CRPD, management involves treating the underlying cause of restrictive pathology and using pulmonary rehabilitation to preserve respiratory function.

In many people it becomes necessary to manage complications of restrictive lung disease that become a severe burden to the individual:

- Pulmonary arterial hypertension
- Cor pulmonale
- Extrapulmonary involvement in sarcoidosis or connective tissue diseases
- Glycemic control in diabetes mellitus
- Weight loss measures for people who are obese
- Polysomnography if sleep apnea is suspected

- Evaluation for osteoporosis
- Psychiatric consultation for depression

Ultimately, some people develop respiratory failure or severe abnormalities in gas exchange requiring noninvasive positive pressure ventilation. They may end up being evaluated for lung transplantation.

Drugs commonly used in management of interstitial lung disease include the following:

- Anti-inflammatory and immunosuppressant drugs—prednisone, methotrexate, hydroxychloroquine, azathioprine, mycophenolate, infliximab
- Prophylactic antibiotics—trimethoprim/sulfamethoxazole, dapsone
- Supplemental oxygen—2 to 4 L/min by nasal cannula 24 h a day for people in advanced stages with chronic respiratory failure

Effects on the Exercise Response

Reduced exercise tolerance and dyspnea upon exertion are common manifestations of CRPD. The typical breathing pattern at rest consists of rapid shallow breaths, and with exercise the minute ventilation increases atypically because of increased forces required to expand the chest in persons who have CRPD.

Because of the atypical increase in minute ventilation with exercise in CRPD, exercise primarily increases respiratory rate rather than change in tidal volume.

Ventilatory factors contributing to dyspnea and exercise limitation include these:

- Inefficient ventilation with a high dead space

- Mechanoreceptor stimulation
- Heightened central respiratory drive

Secondary cardiovascular factors such as pulmonary arterial hypertension and right heart strain further exacerbate exercise limitation. In early restrictive disease, the alveolar–arterial oxygen tension gradient is typically elevated while arterial oxyhemoglobin saturation remains above 90%. Arterial carbon dioxide tension is reduced due to hyperventilation. Impairment in exercise capacity is associated with declines in

- exertional arterial oxygen tension and
- oxyhemoglobin saturation.

In moderate to severe restrictive disease, arterial oxygen saturation breathing room air may be above 90% at rest but decline to below 90% upon exercise. The extent of desaturation depends on the ability to recruit alveolar microvascular surfaces and blood volume for gas exchange. Normally, DL_{CO} increases from rest to exercise in a linear relationship with increasing cardiac output, but diffuse parenchymal diseases such as IPF are characterized by an inability to augment DL_{CO} during exercise, leading to a marked decline in arterial oxygen saturation that curtails exercise. In contrast, in parenchymal disease where the remaining lung units are capable of microvascular recruitment (e.g., scattered granulomatous inflammation in sarcoidosis or following partial lung resection), DL_{CO} continues to increase during exercise, and arterial oxygen saturation and exercise capacity are better maintained than in IPF (at comparable lung volumes).

Finally, in end-stage lung disease, the resting arterial oxygen saturation is reduced below 90% and arterial carbon dioxide tension begins to rise above normal, indicating respiratory failure. Other considerations are that medications for CRPD or for other conditions can alter the exercise response (see table 16.1).

Table 16.1 Potential Alterations to the Exercise Response in Chronic Restrictive Pulmonary Disease

Class of medication or factor	Potential effect on exercise response
Bronchodilators	May improve ventilatory response, ventilation–perfusion matching, and exercise capacity
Antihypertensive medication	β-blockers may blunt heart rate response during exercise
Systemic corticosteroid treatment	May increase blood pressure and induce muscle weakness
Severe pulmonary arterial hypertension	Increases risk of hypotension and arrhythmias upon exercise

Effects of Exercise Training

Allied health care services such as pulmonary rehabilitation and occupational therapy are extremely helpful for improving physical functioning and quality of life in persons who have CRPD. Exercise training has multiple benefits in restrictive lung disease:

- Improving submaximal exercise endurance
- Improving maximal oxygen uptake, depending on initial fitness level
- Improving ventilatory endurance, efficiency, and ventilation–perfusion matching
- Improving cardiovascular conditioning
- Increasing peak exercise DL_{CO} as peak cardiac output increases
- Improving oxygen extraction, skeletal muscle endurance, and efficiency
- Reducing oxygen and blood flow requirement of respiratory muscles, increasing the amount available to working limb muscles
- Reducing lactic acidosis to reduce ventilatory stimulation from exercise
- Desensitizing the perception of dyspnea and the fear of exertion

Physical training in individuals with chronic restrictive lung disease significantly enhances the sense of well-being even if large increases of peak oxygen uptake do not occur. Training ensures that the subject is not disproportionately debilitated, more than expected from the pulmonary dysfunction, and that all steps of oxygen transport are matched within the constraints imposed by the primary disorder. Optimizing the efficiency of oxygen transport is a critical issue because these individuals can ill afford any metabolic energy wastage. Careful adherence to physical conditioning allows people to realize the maximum benefit that can be derived from concurrent specific therapy aimed at correcting the underlying disorder.

Recommendations for Exercise Testing

Exercise testing is an important tool for assessing the impact of CRPD on physical functioning. Goals of exercise testing include these:

- Defining disability due to pulmonary dysfunction

- Detecting coexistent factors that aggravate disability
- Monitoring progression of impairment and response to therapy

The most common clinical exercise assessment is a 6-min walk test with concurrent transcutaneous measurement of pulse rate and oxygen saturation. The test is easy to conduct in any medical clinic or office, and the results are highly reproducible in persons with fibrotic lung disease. If desaturation occurs during the test, the test can be repeated while administering supplemental oxygen by nasal cannula in order to titrate the level of home oxygen therapy. Arterial blood gases may also be measured before and after exercise to determine the nature of gas exchange abnormalities.

For cardiopulmonary exercise testing, with or without respired gas analysis, either cycle ergometer or treadmill protocols may be used, though the prognostic value in this heterogeneous patient population has not been established. Following a resistance-free warm-up period, work rate should be increased in small increments (e.g., ≤25 W/stage and a Naughton treadmill protocol). Electrocardiogram, systemic blood pressure, and transcutaneous oxygen saturation should be monitored. Minute ventilation, respiratory rate, tidal volume, and basic ventilatory gas exchange analysis (oxygen uptake, CO_2 output) are commonly measured during exercise. Supplemental oxygen may be added as needed. Meticulous attention to breathing techniques and verbal reinforcement during testing will greatly reduce subject anxiety and enhance the quality of the measurements.

The usual clinical indications for terminating exercise apply—chest pain, faintness, dysrhythmias, drop in blood pressure below resting level or a hypertensive response, persistent inability to maintain oxygen saturation above 90%, reaching the age-predicted maximal heart rate, or volitional termination of exercise due to fatigue or shortness of breath.

Chronic restrictive pulmonary disease–related contraindications to exercise testing have to do with severe symptoms and secondary conditions such as these:

- Severe pulmonary arterial hypertension
- Unstable heart failure, ischemic or valvular cardiac diseases
- Life-threatening dysrhythmias
- Presyncope, syncope

Recommendations for Exercise Programming

The Basic *CDD4* Recommendations are generally advised for persons with CRPD. Emphasis on maintaining physical functioning and independent living is appropriate for most individuals with CRPD (especially in severely restrictive pathophysiology). The ACSM Guidelines may be more appropriate for those who are very early in the condition or who have fixed but mild restrictive pathophysiology (e.g., mild kyphoscoliosis). Evaluation and recommendations from the nutrition therapist, physical therapist, and occupational therapist can be very helpful for understanding the subject's home environment and work and lifestyle constraints before an individualized exercise program is formulated. This is especially true with respect to

- learning efficient breathing techniques and
- improving ergonomics during activities of daily living.

An initial period of intense training (20-30 min/day, 5 days a week for 6-8 weeks) is helpful in establishing a baseline level of fitness, and sustained training thereafter can be continued three times a week. For subjects with poor exercise endurance, a training session may be divided into several shorter segments with rest periods in between. Regular follow-up assessment of exercise and pulmonary function will quantify any improvement or decline in function, which helps to sustain the subject's motivation for continued training.

Special Considerations

- Worsening hypoxia during exertion may induce chest pain, arrhythmias, or both.
- People should take medications as usual to obtain the best exercise performance.
- Supplemental O_2 flow rate should be adjusted to maintain O_2 saturation >90%.
- Anxiety, depression, and fear avoidance are common barriers.

Integration Into a Medical Home Model

People with restrictive lung disease should be followed regularly by a pulmonologist for clinical and radiological evaluation as well as pulmonary function testing. They should be vaccinated against influenza yearly and against pneumococcus.

- Refer all individuals with CRPD to pulmonary rehabilitation.
- Encourage people to establish a regular program to maintain physical functioning.
- Smoking cessation and avoidance of secondhand tobacco exposure are very important.

Take-Home Message

Diet modification, weight, and blood glucose control are important considerations for persons taking systemic corticosteroids or those with comorbidities such as obesity and diabetes mellitus, which can develop because people with CRPD have a tendency to become sedentary.

- Use of metered-dose inhalers should be reviewed at intervals to ensure that proper techniques and frequency of dosing are followed.
- Avoid environmental extremes in temperature and humidity.
- Avoid exposure to chemicals, fumes, and other respiratory irritants.
- Medical devices, such as continuous and bilevel positive airway pressure (CPAP and BiPAP, respectively) machines and oxygen concentrators, should be checked regularly.

Suggested Readings

American Thoracic Society/American College of Chest Physicians (ATS/ACCP). Statement on cardiopulmonary exercise testing. *Am J Respir Crit Care Med.* 2003;167:211-277.

Arena R. Exercise testing and training in chronic lung disease and pulmonary arterial hypertension. *Prog Cardiovasc Dis.* 2011;53:454-463.

Eaton T, Young P, Milne D, Wells AU. Six-minute walk, maximal exercise tests: reproducibility in fibrotic interstitial pneumonia. *Am J Respir Crit Care Med.* 2005;171:1150-1157.

Fell CD, Liu LX, Motika C, Kazerooni EA, Gross BH, Travis WD, Colby TV, Murray S, Toews

GB, Martinez FJ, Flaherty KR. The prognostic value of cardiopulmonary exercise testing in idiopathic pulmonary fibrosis. *Am J Respir Crit Care Med.* 2009;179:402-407.

Hsia CC. Cardiopulmonary limitations to exercise in restrictive lung disease. *Med Sci Sports Exerc.* 1999;31:S28-S32.

Hsia CC. Coordinated adaptation of oxygen transport in cardiopulmonary disease. *Circulation.* 2001;104:963-969.

Mascolo MC, Truwit JD. Role of exercise evaluation in restrictive lung disease: new insights between March 2001 and February 2003. *Curr Opin Pulm Med.* 2003;9:408-410.

Palange P, Ward SA, Carlsen KH, Casaburi R, Gallagher CG, Gosselink R, O'Donnell DE, Puente-Maestu L, Schols AM, Singh S, Whipp BJ. Recommendations on the use of exercise testing in clinical practice. *Eur Respir J.* 2007;29:185-209.

Phansalkar AR, Hanson CM, Shakir AR, Johnson RL Jr, Hsia CC. Nitric oxide diffusing capacity and alveolar microvascular recruitment in sarcoidosis. *Am J Respir Crit Care Med.* 2004;169:1034-1040.

West JB. *Pulmonary Pathophysiology: The Essentials.* Baltimore: Lippincott Williams & Wilkins; 2007.

Asthma

Asthma is a chronic respiratory disease characterized by airway inflammation and hyperresponsive bronchoconstriction. It occurs in approximately 300 million people worldwide and 25 million people (or approximately 8-10%) in the United States. Approximately 7 million U.S. children currently have asthma, with non-Hispanic African American children having the highest prevalence of about 1 in 6 (17%). More than 3000 deaths annually are attributed to asthma in the United States. Although there is no cure for asthma, it can often be controlled with daily medication and by the avoidance of triggers. Sixty percent of those with asthma in the United States have allergic asthma, and 5% to 10% of persons who have asthma do not respond well to conventional treatment. These individuals are at greater risk for severe exacerbations and death.

Symptoms of asthma include the following:

- Wheezing from peripheral airways
- Cough
- Chest tightness
- Shortness of breath
- Excess sputum production (sometimes)

Symptoms are often worse at night and may be related to a specific allergen, air pollutant, or cold or dry air and as such may be seasonal. Symptoms are often triggered by exercise, in the condition known as exercise-induced bronchospasm (EIB), which typically occurs in exercise lasting longer than approximately 6 min.

Basic Pathophysiology

There are three basic phenotypes of asthma, which are characterized by their primary pathophysiological traits, but phenotypes often overlap (see table 17.1). Exercise-induced bronchospasm occurs in people with eosinophilic and neutrophilic types of asthma, but also in people who have normal airways and don't appear to have asthma. This group is commonly known as *EIB without apparent asthma.*

Exercise is thus itself a cause of bronchoconstriction in some people, and it also causes bronchoconstriction in persons who have asthma. The mechanisms by which exercise induces bronchospasm remain incompletely understood but are likely initiated by water loss from the airway

Table 17.1 Phenotypes of Asthma

Phenotype	Prevalence	Primary pathophysiological mechanisms
Eosinophilic	~50%	• Type 2 Helper cell cytokines (interleukins 4, 5, and 13) induce airway eosinophilia • Thickening of basement membrane • Responsive to inhaled corticosteroids
Neutrophilic	20-40%	• Associated with environmental irritants; often severe asthma • No thickening of basement membrane • Resistant to inhaled corticosteroids
Exercise-induced bronchospasm without apparent asthma	4-20%, but up to 50% in elite athletes	• Mast cell release of prostaglandins, leukotrienes, and histamine • Exacerbated by presence of environmental irritants • Triggered by thermal and osmotic stress of prolonged hyperventilation (as also occurs in EIB due to eosinophilic–neutrophilic phenotypes)

surface liquid, causing a change of osmolarity in airway cells. This is a consequence of warming and humidifying inspired air to body temperature and 99% relative humidity.

People with uncontrolled asthma can risk severe exacerbation, potential hospital admission, and death from exercise. However, when asthma is appropriately controlled, people can exercise normally, achieve the typical benefits of exercise, and compete at the elite level.

Management and Medications

Key steps in the management of asthma include the proper diagnosis of symptoms, particularly exercise-associated symptoms, as there are many possible causes of dyspnea during exercise. Differential diagnosis for conditions that mimic asthma should always be considered; these include vocal cord dysfunction, other glottal abnormalities, exertional dyspnea, paroxysmal atrial tachycardia, exercise-induced hypoxemia, pulmonary arteriovenous malformations, pulmonary vascular

disease, and gastroesophageal reflux. Confirming a diagnosis of EIB without apparent asthma can be particularly challenging.

Once someone has been properly diagnosed with asthma, the mainstay of preventing asthma attacks is to avoid any environmental irritants that precipitate an attack. Allergy testing can be quite helpful in identifying environmental causes. Very common allergens are cat dander, human dander, dust mites, chemical irritants (often perfumes and fragrances), and seasonal allergens. People with mild EIB can benefit from eating a low-sodium diet, supplementing with fish oil or ascorbic acid, and in some cases avoiding specific foods. Use of a high-intensity–interval-type preexercise warm-up and the use of a face mask during cold weather exercise can also help.

Most individuals end up needing medication to control their asthma, and there are several classes of agents (see table 17.2). In someone with apparent asthma (whether eosinophilic or neutrophilic), symptoms of bronchial hyperresponsiveness (BHR) suggest that control of the underlying inflammation is inadequate. Thus, the first-line therapy for a confirmed diagnosis of

Table 17.2 Medications Used for Asthma

Medication	Class	Mechanism of action
Beclomethasone Budesonide Flunisolide Fluticasone Mometasone	Inhaled corticosteroids	• Decrease inflammatory mediators and inflammatory cells • Reduce bronchial hyperresponsiveness to exercise
Albuterol Metaproterenol Terbutaline	Short-acting β_2-adrenergic receptor agonists	• Stimulate smooth muscle β_2-receptors • Inhibit release of mediators from mast cells • Rapid tachyphyllaxis; not for daily use
Formoterol Salmeterol	Long-acting β_2-adrenergic receptor agonists	• Stimulate smooth muscle β_2-receptors • Inhibit release of mediators from mast cells • Rapid tachyphyllaxis; not for daily use
Montelukast Zafirlukast	Leukotriene receptor antagonists	• Block lipoxygenase pathway • Useful in conditions of high oxidative stress (e.g., air pollution) • Have no bronchodilatory effects
Cromolyn sodium Nedocromil	Mast cell stabilizers	• Decrease mediator release from mast cells, eosinophils, neutrophils, monocytes, alveolar macrophages, and lymphocytes • Inhibit chloride ion flux, which regulates intracellular free calcium and prevents the calcium-mediated response • Class B agent; safe for use by childbearing females
Tiotropium Ipratropium	Anticholinergics	• Block muscarinic acetylcholine receptors to parasympathetic neural command

asthma is inhaled corticosteroids, which are used either alone or in combination with other medications. Inhaled corticosteroids decrease airway inflammation and attenuate exercise-related BHR, providing bronchoprotection during exercise in up to 60% of persons with asthma.

Additional daily medications such as short- or long-acting β_2-adrenergic receptor agonists (SABAs and LABAs, respectively), perhaps in combination with a leukotriene receptor antagonist, may be needed as add-on therapy to inhaled corticosteroids. For exercise-related symptoms, a SABA taken 5 to 20 min before exercise will provide optimal bronchoprotection for 2 to 4 h.

In an individual who has EIB without apparent asthma and has mild or intermittent symptoms, prophylaxis with SABAs alone can be effective. Daily use of SABAs is not recommended because tachyphylaxis rapidly reduces the efficacy. The use of SABAs as prophylaxis should be intermittent and not as a daily preexercise treatment. If a SABA is required every day, the individual probably has an inflammatory type of asthma and needs an inhaled corticosteroid.

Effects on the Exercise Response

Approximately 90% of people with asthma are symptomatic during or after exercise, and EIB is most severe in those with uncontrolled airway inflammation. The subgroup of individuals who react to exercise but do not have inflammatory asthma (EIB without apparent asthma) can be symptomatic depending on the exercise and environmental air quality. Classic EIB occurs after the cessation of exercise, but EIB also occurs during exercise and thereby limits exercise capacity. Several factors determine whether EIB is severe enough to limit exercise:

- The degree of chronic inflammation
- Airway responsiveness
- Mode of exercise (e.g., constant or interval)
- Intensity and duration of exercise
- Environmental conditions

During exercise, EIB causes "air trapping" and can cause restricted exercise ventilation, diminished decrease in $PaCO_2$, and reduced PaO_2, thereby limiting exercise capacity and oxygen uptake.

The classic EIB response observed during a 6- to 8-min exercise bout is bronchodilation during exercise and bronchoconstriction after exercise. The initial bronchodilation during exercise is believed to be due to vagal tone withdrawal, followed by a mechanical influence from exercising ventilation in which increased tidal volume mechanically couples lung tissue to small airways, causing airways to be stretched open from the larger tidal volumes. When tidal volume decreases after exercise, bronchoconstrictive influences dominate airway function.

During prolonged steady-state exercise, the bronchodilation within the first 6 min of exercise is followed by a steady decline in airway function. During interval-type exercise, of alternating moderate and light intensity, expiratory flows gradually decrease over time but show improvement in airway function during the moderate-intensity period with deterioration during the light-intensity period.

High-intensity exercise may provoke EIB, particularly when performed in an unfavorable environment (cold air, dry air, airborne irritant particulates). Increased airway responsiveness, airway mucosal membrane damage, increased airway inflammation, and increased thickening of the basement membrane in elite Nordic skiers without asthma have been attributed to long-duration, high-intensity exercise in cold, dry air. Likewise, the high prevalence of asthma in competitive swimmers has been attributed to the inhalation of chlorine by-products at swimming pools, and in ice rink athletes it is attributed to inhalation of ultrafine particles from ice resurfacing machines.

Fatal asthma exacerbations occur in both competitive and recreational athletes. Of 263 deaths reported over a 7-year period that were linked to sport, 61 were asthma related. Those who had fatal asthma attacks during exercise were typically white males between 10 and 20 years of age. Few had a history of persistent asthma, and only one had used inhaled corticosteroids. It is important to identify athletes with asthma and ensure that proper treatment is prescribed.

Effects of Exercise Training

Regular exercise can reduce the severity of EIB and increase the exercise level at which EIB occurs. This is probably because of an increase in ventilatory threshold as a consequence of regular training. Consistent exercise improves the quality of life, reduces EIB severity, and may reduce the need for medications to control asthma. Exercise alone does not improve airway inflammation, BHR to methacholine, or vital capacity. Exercise training does have a modest beneficial effect on asthma itself, comprising the following:

- Decrease in medication usage
- Decrease in intensity of exacerbations
- Decrease in postexercise falls in lung function
- Increase in the number of symptom-free days

The prevalence of EIB and asthma is higher in cold weather athletes, ice rink athletes, and swimmers, suggesting that long-term exposure to pollutants or cold, dry air at high ventilation rates damages airways. The EIB response may vary with seasonal climate and allergen loads. Approximately half of individuals suffering from EIB are refractory to a second episode of EIB 2 to 3 h after the initial episode.

Aerobic exercise training reduces migration of antigen-specific Helper T cells to the lung. Exercise training improves asthma-related quality of life (days without asthma symptoms).

Recommendations for Exercise Testing

Maximal exercise testing can be performed to measure physiological responses and monitor the status of individuals with asthma or EIB, but should be performed only if the asthma is well controlled or a SABA is used as prophylaxis. It is strongly recommended that individuals with suspected asthma or exercise-related symptoms of EIB undergo a bronchial challenge test before beginning an exercise program or participating in competitive sport. The severity of BHR should be documented with a reliable challenge test, and these data should be used to determine the appropriate medication(s).

Unfortunately, current challenge test methodology lacks standardization and varies from laboratory to laboratory and from clinic to clinic. The American Thoracic Society has published guidelines for the diagnosis of asthma, recommending both methacholine and an exercise challenge (2000), and more recently guidelines for the diagnosis of EIB (2013). Methacholine is a direct challenge that reflects the response of smooth muscle to a single agonist, independent of inflammatory cells, and thus is not recommended as a test for EIB. The indirect challenges such as exercise, eucapnic voluntary hyperpnoea (EVH), and mannitol inhalation powder are more specific to the mechanism of EIB and active asthma and are more effective in aiding the diagnosis than the direct challenge.

In an attempt to provide better diagnostics for their patients, some physicians have incorporated a treadmill or "field run" in combination with forced expiratory volume (FEV_1) measurements. Such tests are done during breathing of ambient air, not dry compressed medical-grade air, making such procedures quite unreliable, yielding low sensitivity (>50% of true positives will test negative at 50% relative humidity). An appropriate challenge requires high-intensity exercise for 6 to 8 min during breathing of dry compressed air.

- Dry air exercise and EVH positive tests: >10% fall in FEV_1 (15% for children)
- Inhaled mannitol powder: >15% fall in FEV_1

Contraindications to exercise testing in asthma include these:

- Baseline FEV_1 <70% of predicted, or <1.5 L
- Significant bronchodilation (>12% FEV_1) in response to a SABA
- History of cardiovascular problems
- History of severe hypertension
- Known aneurysm
- Conditions that affect the ability to perform a challenge test (e.g., costochondritis, chest injury, recent sternotomy)

Interpretation of challenge tests is sometimes difficult for many reasons, mostly due to problems in executing a proper test protocol:

- Exercise intensity <90% of maximal age-predicted heart rate
- Ventilation rates <21 times FEV_1 (last 6 min of an 8-min bout)
- Medical-grade, dry compressed air not used

For other factors confounding bronchial challenge test interpretation, see table 17.3.

Table 17.3 Factors That Confound Exercise Testing in Asthma

Factor	Considerations
Existing asthma medication	Use medication holiday before test.
β-blockers	These may block adrenergic impulses that cause bronchodilation and result in increased bronchoconstriction.
Caffeine	It decreases response to indirect challenge tests.
Activity before challenge test	Exercise before test may attenuate response; even a brisk walk through testing center parking lot may affect test results. Long warm shower may open airways and lead to larger than typical resting lung function.
Upper respiratory infection (URI)	A bronchial challenge test should not be performed on a patient with a URI. A period of 3-6 weeks should be observed after a URI before testing.
Comorbid respiratory disorders	Exercise-induced bronchospasm can occur with or without the above-mentioned mimics of exercise-induced bronchospasms and thus present a challenge to diagnostic interpretation.

Recommendations for Exercise Programming

Persons with asthma or EIB without apparent asthma should have their condition controlled, then undergo a challenge test and be subsequently referred to an exercise program. Many will already be active in sports and won't need an exercise prescription. Having asthma or EIB does not preclude one from regular exercise or participation in sports; in fact, as stated earlier, a high prevalence has been shown among elite and Olympic athletes. When asthma is well controlled, exercise programming should be dictated by other chronic conditions.

If asthma is the main chronic condition, follow the ACSM Guidelines.

Major Factors and Tips for Managing Exercise-Induced Asthma

- People should undergo a bronchial challenge test before beginning an exercise program.
- Severity of disease and level of control achieved should be known, as well as the response to exercise and effect of preexercise treatment.
- A warm-up consisting of interval-type exercise may attenuate the EIB response.

The exercise environment and ventilatory requirements of a particular activity are critical considerations in developing a program. Swimming in a pool, skating in an ice arena, and Nordic ski racing all are likely to exacerbate the EIB response, so other activities might be better choices.

Integration Into a Medical Home Model

Exercise should be included as part of asthma therapy. Individuals with moderate to severe asthma should be closely monitored to ensure the efficacy of treatment and medication compliance. Those with intermittent–mild EIB who use a SABA as prophylaxis before exercise should be aware of the potential to develop tachyphylaxis if the SABA is used daily. If a daily SABA is required, then a controller medication such as inhaled corticosteroids should be strongly considered. People can self-monitor their condition on a daily basis and pre- and postexercise using a peak flow meter.

Take-Home Message

Regular aerobic exercise will improve exercise tolerance and exacerbation severity for people with asthma and may allow medication to be titrated down.

- Exercise for someone with well-controlled asthma is generally considered safe, although in about half of this population a SABA may be necessary before exercise.

- In EIB without apparent asthma, a SABA before exercise is generally sufficient.
- Many persons with asthma train and compete at the elite level.
- Exacerbations of EIB may be related to seasonal climate change, specific allergen seasons, or air pollution.
- Individuals with asthma or EIB without apparent asthma should be aware of their specific triggers.

Suggested Readings

Anderson SD, Brannan JD. Bronchial provocation testing: the future. *Curr Opin Allergy Clin Immunol.* 2011;11(1):46-52.

Crapo RO, Casaburi R, Coates AL, Enright PL, Hankinson JL, Irvin CG, MacIntyre NR, McKay RT, Wanger JS, Anderson SD, Cockcroft DW, Fish JE, Sterk PJ. Guidelines for methacholine and exercise challenge testing-1999. This official statement of the American Thoracic Society was adopted by the ATS Board of Directors, July 1999. *Am J Respir Crit Care Med.* 2000;161(1):309-329.

Evans TM, Rundell KW, Beck KC, Levine AM, Baumann JM. Cold air inhalation does not affect the severity of EIB after exercise or eucapnic voluntary hyperventilation. *Med Sci Sports Exerc.* 2005;37(4):544-549.

Fahy JV. Eosinophilic and neutrophilic inflammation in asthma. *Proc Am Thorac Soc.* 2009;6(3):256-259.

Kippelen P, Fitch KD, Anderson SD, Bougault V, Boulet LP, Rundell KW, Sue-Chu M, McKenzie DC. Respiratory health of elite athletes - preventing airway injury: a critical review. *Br J Sports Med.* 2012 Jun;46(7):471-476. doi: 10.1136/bjsports-2012-091056.

Parsons JP, Hallstrand TS, Mastronarde JG, Kaminsky DA, Rundell KW, Hull JH, Storms WW, Weiler JM, Cheek FM, Wilson KC, Anderson SD; American Thoracic Society Subcommittee on Exercise-induced Bronchoconstric-

tion. An official American Thoracic Society clinical practice guideline: exercise-induced bronchoconstriction. *Am J Respir Crit Care Med.* 2013;187(9):1016-1027.

Rundell KW, Im J, Mayers LB, Wilber RL, Szmedra L, Schmitz HR. Self-reported symptoms and exercise-induced asthma in the elite athlete. *Med Sci Sports Exerc.* 2001; 33(2):208-213.

Rundell KW, Slee JB. Exercise and other indirect challenges to demonstrate asthma or exercise-induced bronchoconstriction in athletes. *J Allergy Clin Immunol.* 2008;122(2):238-246.

Weiler JM, Anderson SD, Randolph C, Bonini S, Craig TJ, Pearlman DS, Rundell KW, Silvers WS, Storms WW, Bernstein DI, Blessing-Moore J, Cox L, Khan DA, Lang DM, Nicklas RA, Oppenheimer J, Portnoy JM, Schuller DE, Spector SL, Tilles SA, Wallace D, Henderson W, Schwartz L, Kaufman D, Nsouli T, Shieken L, Rosario N; American Academy of Allergy, Asthma and Immunology; American College of Allergy, Asthma and Immunology; Joint Council of Allergy, Asthma and Immunology. Pathogenesis, prevalence, diagnosis, and management of exercise-induced bronchoconstriction: a practice parameter. *Ann Allergy Asthma Immunol.* 2010;105(6 Suppl):S1-S47.

Weiss P, Rundell KW. Imitators of exercise-induced bronchoconstriction. *Allergy Asthma Clin Immunol.* 2009; 5(1):7.

Additional Resources

American Academy of Allergy Asthma & Immunology. Asthma and exercise. www.aaaai.org/conditions-and-treatments/library/asthma-library/asthma-and-exercise.aspx [Accessed November 4, 2015].

An official ATS clinical practice guideline: exercise-induced bronchoconstriction. Online supplement. www.atsjournals.org/doi/suppl/10.1164/rccm.201303-0437ST/suppl_file/ATS_Document_Bronchoconstriction_Data_Supplement.pdf. [Accessed November 3, 2015].

Cystic Fibrosis

Cystic fibrosis (CF) is an extremely complicated disorder of the cystic fibrosis transmembrane conductance regulator (CFTR) gene. Over 1000 unique mutations have been described and linked to CFTR, which affects the lungs and digestive tracts. Impaired function of CFTR affects the secretion of chloride and bicarbonate, which results in the secretion of thick viscous mucus. The thickened mucus subsequently causes tissue and organ damage to the lungs and digestive tract. In the lungs, this thick mucus increases the risk of chronic inflammation and infection of the airways. An inflammatory response often results in tissue destruction of the lungs and deteriorating lung function. Because of this, CF is classified as a chronic lung disease, even though it affects other organs such as the pancreas.

Pancreatic involvement in CF mainly affects digestive enzymes, but also vulnerable to CF are the islets of Langerhans, and damage to islet cells leads to insulin deficiency and development of diabetes. Although the majority of people with CF develop exocrine pancreas insufficiency, only some in this group develop CF-related diabetes.

Management and Medications

Because of the complexity of the disease, CF is a difficult disorder to treat. The following principles are the pillars of CF treatment:

- Improving poor nutritional status (in particular in relation to fat-free body mass) with the use of oral pancreatic enzyme supplements to enhance digestion of dietary fats and proteins
- Improving general physical fitness
- Airway clearance techniques to facilitate removal of mucus

- Antibiotic therapy to fight pulmonary infection, colonization, and inflammation
- Influenza vaccination before every flu season

To treat and improve symptoms of CF, many people require multiple medications to achieve the goals of these pillars (table 18.1). Pancreatic enzyme supplements are used to aid nutritional status because CF impairs secretion of digestive enzymes and thus can lead to protein–calorie malnutrition. In individuals who lose pancreatic islet cell function, insulin is necessary to manage the resulting diabetes. The pulmonary aspects of CF are primarily a form of chronic bronchitis, requiring anti-inflammatory treatment (inhaled and oral corticosteroids), inhaled and oral bronchodilators, and mucolytic medications to help mobilize and expel airway secretions. This is commonly monitored at home with inexpensive peak expiratory flow meters, and in the clinic by measurement of the forced expiratory volume in 1 second (FEV_1). Supplemental oxygen is necessary for some people during exercise (sometimes at rest as well).

Effects on the Exercise Response

Effects of CF on exercise are highly dependent on the severity of the disease. Pulmonary, cardiac, and peripheral skeletal muscle function, along with physical inactivity, all contribute to exercise limitation. Logically, CF has detrimental effects on the exercise response under these circumstances:

- If ventilation is impaired
- If central circulation is secondarily impaired
- If nutritional deficiency and inactivity lead to low muscle mass

Table 18.1 Medications Used in Cystic Fibrosis

Medications	Mechanism of action	Adverse effects
Prednisone (see chapter 17 on asthma)	Glucocorticoid for reducing inflammation Often prescribed for airway obstruction exacerbations with the hope of improving pulmonary function and speeding recovery	• Skin atrophy and fragility • Osteoporosis • Muscle atrophy • Myopathy (including ventilatory muscles) • Obesity (increases work of breathing)
Albuterol Metaproterenol Salmeterol Formoterol Indacaterol	Selective β_2-adrenoceptor sympathomimetic agonists that produce bronchodilation	• Tachycardia, palpitations, and tremulousness • Nonselective sympathomimetic drugs (e.g., epinephrine and isoprenaline) should not be used
Theophylline Aminophylline	Methylxanthines have potent bronchodilator activity	• Tachycardia, cardiac dysrhythmias • Central nervous system stimulation • Increased respiratory drive • Risk of seizures
Cromolyn sodium Nedocromil	Mast cell stabilizers decrease mediator release from mast cells, eosinophils, neutrophils, monocytes, alveolar macrophages, and lymphocytes	• Stinging transiently on use
Guaifenesin Potassium iodide	Mucolytic therapy reduces viscosity of mucus to promote clearance of secretions	• Negligible side effects
Recombinant human deoxyribonuclease	Hydrolyzes deoxyribonucleic acid content in sputum, reducing mucus viscosity	• Pharyngitis, laryngitis • Upper respiratory symptoms, congestion
Ivacaftor	Facilitates chloride transport in patients who have the G551-CFTR genetic variant	• Headache • Abdominal pain, diarrhea • Upper respiratory symptoms, congestion

See also chapter 7 on chronic obstructive pulmonary disease and chapter 17 on asthma.

Individuals who have few pulmonary limitations and are able to obtain adequate protein and calories while remaining active are much less affected than someone who has severe pulmonary disease and malnourishment and is frequently hospitalized with respiratory infections. Exercise capacity thus has important prognostic value in CF and is a major determinant of quality of life in CF.

In healthier individuals, symptoms are muted and the exercise response is essentially normal. But in many cases, particularly noted in adolescents, CF is often associated with a low exercise capacity from multifactorial causes. The aerobic capacity (as indicated by peak oxygen uptake [$\dot{V}O_2$peak]) of adolescents with CF at the age of 12 years is comparable to that of age-matched healthy controls; however, it declines 20% during

adolescence. This longitudinal decline in exercise capacity is independent of age, pulmonary function, and body mass.

An early clinical feature of lung disease is the development of hyperinflation, which increases with further lung injury. Progressive hyperinflation changes the shape of the thorax, putting the inspiratory muscles (and particularly the diaphragm) at a mechanical disadvantage. Additionally, decreased chest wall compliance increases the energy and oxygen costs of breathing. Progressive airflow obstruction (caused by chronic mucus hypersecretion) involves an increase in airflow resistance, intrapulmonary gas trapping, and ventilation–perfusion mismatching. These changes may compromise respiratory muscle function, and individuals with CF are thus more susceptible to respiratory failure or ventilation limitation during exercise, resulting in oxyhemoglobin desaturation and carbon dioxide retention during exercise.

Effects of Exercise Training

The research evidence for exercise training in people with CF is limited. However, the research that currently exists suggests that aerobic exercise training can improve or enhance the following:

- Exercise capacity
- Muscular endurance and strength
- Mucus clearance
- Skeletal development and bone mineral density
- Posture
- Mobility of the chest wall
- Self-esteem

Exercise training in CF can positively affect pulmonary function, aerobic fitness, and muscle strength, but conclusions about the efficacy of exercise programs in CF are limited by small sample sizes, short duration, and incomplete reporting. Research is needed to comprehensively assess the benefits of exercise programs in persons with CF, particularly in children using a combination of modalities and respiratory muscle training.

Recommendations for Exercise Testing

The Basic *CDD4* Recommendations are not appropriate for most people with CF, who need a more specialized set of measurements. In part, this is because of the unique effects of CF on physiology and in part because the Basic *CDD4* Recommendations are geared more toward adults with chronic disease, not children, adolescents, and young adults.

The standard progressive exercise test used to assess fitness in adults is not sufficient for children and adolescents with CF. Children with CF, who typically have very different physical activity profiles than normal adults, typically require more extensive testing to enable full understanding of their limitations in physical functioning. Guidelines to perform exercise testing in CF are being developed by an international study group and

will likely specify measures of muscle function, cardiopulmonary function, and habitual physical activity. This will provide better identification of the impact of limiting factors and help tailor advice to patients or clients.

Pending these guidelines, an incremental cardiopulmonary exercise test to measure $\dot{V}O_2$peak using either a treadmill or cycle ergometer is recommended. Respired gas analysis accurately measures $\dot{V}O_2$peak and assesses the integrated physiological response of the pulmonary, cardiovascular, and metabolic systems. Only through use of this analysis can one see how the respiratory and nutritional functioning affect the physical functioning of the individual. In addition to the standard measurements of heart rate, blood pressure, and perceived exertion, the following measures should also be monitored for all people with CF:

- Pulse oximetry (if the system has breath-by-breath mode, also end-tidal gases)
- Breathing strategy (minute ventilation, tidal volume, and respiratory rate)
- Electrocardiogram preferred, but heart rate monitoring acceptable (low risk of cardiac ischemia)
- Perceived dyspnea ratings

The strength and flexibility measures in the Basic *CDD4* Recommendations are not ideally suited for persons with CF but could be used in those who are most severely affected. A high-quality accelerometer might be a good tool to measure habitual physical activity in children, since many of their activities are likely to be activities other than walking (an inexpensive pedometer might provide misleading data).

Recommendations for Exercise Programming

The Basic *CDD4* Recommendations for exercise programming can be used, but in general they are not adequate for children, adolescents, and young adults with CF.

No "one size fits all" principle should apply to persons with CF, and exercise programs, habitual physical activity, or both must be individualized as an integral part of the particular treatment regimen. Maintaining physical functioning, vitality, and as normal a childhood as possible are essential elements of optimizing the prognosis. Exercise specialists should thus consider factors that help or hinder long-term adherence to the program, including these:

- Social support
- Perceptions of competency and self-esteem
- Enjoyment of the activities (play)
- Availability of a variety of activities

All persons with CF have their own unique combination of physiological factors that limit their exercise capacity. The data from exercise testing should be used to focus on various goals and indications that best suit the individual. These indications depend not only on the individual, but also on the person's own disease progression and exacerbations. These inter- and intraindividual characteristics may require repeated exercise testing in order to give persons with CF safe training recommendations. In general, programs should be configured as listed next.

Mild to Moderate Disease (FEV₁ 65% to 50% Predicted)

Nonpulmonary factors usually limit exercise capacity, so exercise training should

- prevent lung function deterioration,
- optimize chest mobility and airway clearance techniques, and
- generally focus on peripheral muscle function (strength and power).

Severe Disease (FEV₁ <50% Predicted)

Pulmonary factors usually limit exercise capacity, so exercise training should

- continue the efforts just listed for those with mild to moderate disease,
- include inspiratory muscle training to improve ventilatory efficiency and work of breathing, and

> *In individuals with CF, building and maintaining muscle mass is important, because the work of breathing is hard and intercurrent illnesses are likely to cause atrophy. Having a reserve of strength and muscle mass defends against these vulnerabilities.*

- include high-intensity interval training to train muscles with a low ventilatory burden.

The following are contradictions to exercise for children with CF:

- Fever >38 °C
- Exercise-induced desaturation (SpO_2 < 90%)
- Cardiac dysfunction
- Scuba diving
- Avoidance of contact or collision sports in persons with an enlarged spleen or liver disease

Integration Into a Medical Home Model

Persons with CF will have multiple specialists, likely a pulmonologist, a gastroenterologist, and an infectious-disease doctor. One of these physicians is likely to be the main care provider, and office staff need to coordinate with the other providers and make sure that physical functioning remains a key goal for all. It is important to monitor physical activity level, muscle atrophy, and obesity and osteoporosis side effects of medications. For a person who has long-standing CF leading to multiple complications and diminished respiratory function, during an intercurrent illness it may be necessary to do low-intensity activities or even eliminate exercise entirely until her health status improves.

Special Considerations

- Bronchodilator premedication may benefit persons who respond to bronchodilators.
- SpO_2 should be monitored at the beginning to determine the level of O_2 supplement.
- Supplemental O_2 may help people obtain a training effect.
- Prolonged oral corticosteroid therapy may cause myopathy, obesity, elevated blood glucose, and high blood pressure.
- Severe disease with hypertrophic pulmonary osteoarthropathy can cause bone pain.
- End-stage lung disease may severely limit training intensity due to the development of cor pulmonale.

Take-Home Message

Individuals with CF greatly benefit from exercise training, and most people with CF do not require supervision or monitoring while performing exercise. Comprehensive exercise testing, including graded exercise with respired gas analysis, is extremely helpful for individualizing an exercise program to suit the unique physiological needs of each patient or client.

- Building skeletal muscle function (strength, mass, and power) is a key defensive strategy.
- Pulmonary hygiene (including nebulizer) is central to avoiding respiratory complications.
- For children, especially, exercise must be fun—play—and not a health chore or workout.
- Oxyhemoglobin saturation should be monitored in persons who exhibit hypoxemia.
- Bronchodilators or cromolyn will help people with exercise-induced bronchoconstriction.
- Long exercise sessions increase need for fluid and salt intake.
- People with severe lung disease may experience bone or joint pain in the legs.

Suggested Readings

Almajed A, Lands LC. The evolution of exercise capacity and its limiting factors in cystic fibrosis. *Paediatr Respir Rev.* 2012;13(4):195-199.

American Thoracic Society/American College of Chest Physicians (ATS/ACCP). ATS/ACCP statement on cardiopulmonary exercise testing. *Am J Respir Crit Care Med.* 2003;167(2):211-277.

Bradley J, Moran F. Physical training for cystic fibrosis. *Cochrane Database Syst Rev.* 2008; CD002768. doi: 10.1002/14651858.CD002768.pub2.

Dwyer TJ, Elkins MR, Bye PT. The role of exercise in maintaining health in cystic fibrosis. *Curr Opin Pulm Med.* 2011;17(6):455-460.

Ferrazza AM, Martolini D, Valli G, Palange P. Cardiopulmonary exercise testing in the functional and prognostic evaluation of patients with pulmonary diseases. *Respiration.* 2009;77(1):3-17.

Houston BW, Mills N, Solis-Moya A. Inspiratory muscle training for cystic fibrosis. *Cochrane Database Syst Rev.* 2008;CD006112. doi:10.1002/14651858.CD006112.pub2.

Nixon PA, Orenstein DM, Kelsey SF, Doershuk CF. The prognostic value of exercise testing in patients with cystic fibrosis. *N Engl J Med.* 1992;327(25):1785-1788.

Orenstein DM, Higgins LW. Update on the role of exercise in cystic fibrosis. *Curr Opin Pulm Med.* 2005;11(6):519-523.

Radtke T, Stevens D, Benden C, Williams CA. Clinical exercise testing in children and adolescents with cystic fibrosis. *Pediatr Phys Ther.* 2009;21(3):275-281.

Rand S, Prasad SA. Exercise as part of a cystic fibrosis therapeutic routine. *Expert Rev Respir Med.* 2012;6(3):341-351.

Ruf K, Winkler B, Hebestreit A, Gruber W, Hebestreit H. Risks associated with exercise testing and sports participation in cystic fibrosis. *J Cyst Fibros.* 2010;9(5):339-345.

Wheatley CM, Wilkins BW, Snyder EM. Exercise is medicine in cystic fibrosis. *Exerc Sport Sci Rev.* 2011;39(3):155-160.

Additional Resources

Cystic Fibrosis Foundation, www.cff.org [Accessed November 4, 2015].

Cystic Fibrosis Worldwide, www.cfww.org [Accessed November 4, 2015].

European Cystic Fibrosis Society, www.ecfs.eu [Accessed November 4, 2015].

Pulmonary Hypertension

ulmonary hypertension (PH) is identified in almost 2% of all hospitalized patients, usually as a result of other medical conditions such as left heart disease (heart failure, others), chronic lung disease, or nonresolved pulmonary emboli. Other causes of PH include an idiopathic form, as well as PH related to connective tissue disease and congenital heart disease, all three of which are grouped together in the World Health Organization (WHO) classification scheme as WHO Group I. Correctly identifying the type of PH is important, because therapy varies significantly.

Prognosis depends on the PH severity, including particularly whether the high pulmonary arterial pressures have led to right heart dysfunction. Signs and symptoms include the following:

- Dyspnea with exertion
- Decreased exercise capacity
- Presyncope and syncope, particularly during exertion
- Lower-extremity edema and jugular venous distension (when PH is severe enough to cause right heart failure)

Management and Medications

Individuals with PH related to one of the WHO Group I conditions (i.e., PH related to connective tissue disease and so on) are treated with advanced therapies. In contrast, those with PH related to left heart disease or to advanced lung disease are usually treated by optimizing the underlying condition, with particular emphasis on maintaining euvolemia (i.e., avoiding volume overload) and correcting both daytime and nighttime hypoxia, if present. In addition, some people whose PH is felt to be "out of proportion" to their underlying disease also receive advanced therapies, although this is a controversial area.

Currently approved medications include the following:

- Endothelin antagonists (ambrisentan, bosentan; oral)
- Phosphodiesterase inhibitors (sildenafil, tadalafil; oral)
- Prostacyclins (epoprostenol, treprostinil, iloprost; inhaled, intravenous, and subcutaneous)

These medications are used alone or in combination, with the intravenous and subcutaneous prostacyclins generally reserved for persons with the most advanced disease. Lifelong anticoagulation, generally with warfarin, is also indicated in all individuals with chronic thromboembolic PH, and most physicians recommend anticoagulation therapy for patients with idiopathic pulmonary arterial hypertension as well. When PH progresses despite therapy, referral for consideration of lung transplantation is recommended.

Effects on the Exercise Response

Individuals with PH have significant hemodynamic abnormalities, including a reduced stroke volume, an elevated pulmonary vascular resistance, and a failure of the pulmonary vasculature to undergo normal exercise-related pulmonary vasodilation. This results in a reduced ability to increase cardiac output with exercise and reduced overall exercise capacity, including specifically,

- reduced $\dot{V}O_2$peak,
- reduced ventilatory threshold,
- increased $\dot{V}E/\dot{V}CO_2$ slope (increased dead space and inefficient ventilation), and
- reduced peripheral oxygen uptake.

    Note: $\dot{V}E$ = volume of ventilation; $\dot{V}CO_2$ = volume of CO_2 produced

Desaturation during exercise is also common. Mechanisms may include the following:

- Shunting via a congenital heart defect. This is related to the increase in pulmonary arterial and right side heart pressures during exercise.
- Shunting via a patent foramen ovale (PFO), present in approximately 20% of the normal population. When PH develops in an individual with a PFO, intracardiac shunting may occur, particularly during exercise. Of note, this is *not* associated with worse outcome in PH, probably because it also tends to unload the right ventricle and allow for a higher cardiac output.
- Mismatch between lung perfusion and ventilation. This can occur in the presence of an underlying lung disease or because of disease- or medication-related abnormalities in blood flow.

Historically, some persons with PH have been counseled to avoid exercise altogether. However, current guidelines suggest that it is permissible for deconditioned individuals to undertake supervised exercise rehabilitation. These patients should be advised to

- avoid exercising to the point of distressing symptoms, including particular caution if light-headedness or dizziness has occurred, and
- in most cases, avoid Valsalva-like maneuvers that raise intrathoracic pressures and could lead to syncope, such as breath-holds during weightlifting.

Effects of Exercise Training

The scientific literature on exercise training in PH is limited. Small studies suggest that exercise capacity and quality of life do improve, including increases in distance walked on a 6-min walk test and increases in peak $\dot{V}O_2$ and ventilatory threshold on formal cardiopulmonary exercise testing. Animal studies have generally been supportive as well, at least in the absence of advanced right ventricular failure. Large studies are lacking, and the durability of the improvements in exercise capacity is also unknown.

Recommendations for Exercise Testing

Because exercise capacity is a strong predictor of prognosis, some form of exercise testing should be used routinely in tracking patients or clients with PH. In most cases this can be a 6-min hall walk (6MWD), with a better prognosis seen in persons with longer distances (>380-400 m). Formal cardiopulmonary exercise testing (CPX) may also be used for tracking exercise capacity in PH, but notably, CPX in less experienced centers has not been performed well in some studies. Cardiopulmonary exercise testing may also be considered in people with 6MWD over 450 m because 6MWD plateaus above this distance due to ceiling effects.

Cardiopulmonary exercise testing is also used in the diagnostic workup in some cases of PH or unexplained dyspnea (or both), including occasionally the use of combined right heart catheterization and CPX. This technique allows the measurement of exercise capacity and pulmonary arterial pressures simultaneously, and it may be useful in identifying people with marked "exercise-associated" elevations in either pulmonary pressures or left heart filling pressures. However, currently no consensus exists on how and whether to define and treat this condition.

Individuals with PH typically have significant cardiac limitations to exercise, so a moderately incremental protocol is recommended. Test termination is indicated for light-headedness, chest pain, severe dyspnea, pale or ashen appearance, and diaphoresis. Contraindications to exercise testing include the following:

- Unstable or acute right heart failure
- Recent or recurrent syncope with exertion

Problems with interpreting test data due to the nature of the disease or condition and issues to consider as potential complications are listed in table 19.1.

Recommendations for Exercise Programming

When PH accompanies other conditions (chronic obstructive lung disease, heart failure), exercise recommendations specific to those conditions should generally be followed with the caveat that

Table 19.1   Potential Complications and Issues to Consider During Exercise in Pulmonary Hypertension (PH)

Potential complication	Issues to consider
Hypoxia	Hypoxia during exercise is common in PH and in the absence of other concerning symptoms is not a contraindication to testing or an indication to terminate the test.
Supraventricular tachycardia	Nonsinus rhythm can significantly worsen exercise tolerance in PH and in some cases is seen predominantly during exercise.
Hypotension	Mild resting hypotension in PH is not uncommon, particularly among patients on PH therapies, and exercise-related increases in blood pressure may also be blunted.
Right bundle branch block	This is common in PH; electrocardiogram signs of ischemia are usually still interpretable.
Presyncope	Presyncopal symptoms are a highly concerning symptom in PH and in rare cases have progressed to cardiac arrest. The test must be stopped, and if dizziness or other symptoms persist on stopping the test, the patient should be placed in a supine position immediately.

the presence of PH is often an indicator of more advanced disease. For persons in whom PH is the primary issue, no established rehabilitation guidelines exist.

Several disease-specific factors should be considered:

- The underlying cause of PH (lung, heart, or other disease) often has a significant impact on response to exercise training.
- Day-to-day exercise capacity may vary, depending on volume status and other factors.
- Optimizing oxygen prescription may be of benefit.
- Exertional hypoxia is not usually a contraindication to exercise, and specific recommendations should be individualized based on the underlying condition(s).

Integration Into a Medical Home Model

Individuals with PH are followed frequently for complications related to their disease, medications, or both. Clinical staff routinely monitor for any change in physical activity level, dyspnea, or syncope and for any signs of right heart failure, such as weight change or swelling of the feet or ankles. Physical activity is encouraged, but limitations may be imposed depending on the individual's underlying disease severity. In some cases, limitations may be placed on the intensity of exercise, and some people may be advised to avoid exercising alone or to begin an exercise program only in a supervised setting.

Take-Home Message

Keep in mind these key points when working with people with PH.

- Clinical studies on exercise programs for PH are still limited, but existing studies are encouraging from both a safety and an efficacy standpoint.
- Worsening exercise capacity or syncope or presyncope during exertion can be a sign of disease progression, and the exercise specialist should be prepared to alert the staff in the medical home.
- Individuals with PH have significant limitations in aerobic exercise capacity, and exercise prescriptions and expectations should be adjusted accordingly.
- The exercise prescription should also take into account other cardiopulmonary medical conditions.

Suggested Readings

Burger C, Zeiger T. What can be learned in 6 minutes? 6-minute walk test primer and role in pulmonary arterial hypertension. *Adv Pulm Hypertens.* 2010;9. www.phaonlineuniv.org/Journal/Article.cfm?ItemNumber=812 [Accessed January 14, 2013].

Fox BD, Kassirer M, Weiss I, et al. Ambulatory rehabilitation improves exercise capacity in patients with pulmonary hypertension. *J Card Fail.* 2011;17:196-200.

Galie N, Hoeper MM, Humbert M, Torbicki A, Vachiery JL, Barbera JA, Beghetti M, Corris P, Gaine S, Gibbs JS, Gomez-Sanchez MA, Jondeau G, Klepetko W, Opitz C, Peacock A, Rubin L, Zellweger M, Simonneau G. Guidelines for the diagnosis and treatment of pulmonary hypertension. *Eur Respir J.* 2009;34:1219-1263.

Lewis GD. Pulmonary vascular response patterns to exercise: is there a role for pulmonary arterial pressure assessment during exercise in the post-Dana Point era? *Adv Pulm Hypertens.* 2010; 9. www.phaonlineuniv.org/Journal/Article.cfm?ItemNumber=810 [Accessed January 14, 2013].

Mereles D, Ehlken N, Kreuscher S, et al. Exercise and respiratory training improve exercise capacity and quality of life in patients with severe chronic pulmonary hypertension. *Circulation.* 2006;114:1482-1489.

Simonneau G, Robbins IM, Beghetti M, Channick RN, Delcroix M, Denton CP, et al. Updated clinical classification of pulmonary hypertension. *J Am Coll Cardiol.* 2009;54:S43-S54.

Triantafyllidi H, Kontsas K, Trivilou P, et al. The importance of cardiopulmonary exercise testing in the diagnosis, prognosis and monitoring of patients with pulmonary arterial hypertension. *Hellenic J Cardiol.* 2010;51:245-249.

Additional Resource

The PHA Scientific Leadership Council's Consensus Statement on Recommendations for Exercise in Patients with PAH, phassociation.org, Pulmonary Hypertension Association, www.phassociation.org/page.aspx?pid=1267 [Accessed November 4, 2015].

IMMUNOLOGICAL, HEMATOLOGICAL, AND ORGAN FAILURE

One of the great unintended inequities of modern medicine has been the division of clinical exercise services into cardiopulmonary and neuromusculoskeletal groups: coronary artery disease and chronic obstructive pulmonary disease in one group, and strokes, neurological conditions, and arthritis in the other group. Historically, these groupings have sent patients in the direction of cardiopulmonary rehabilitation or to physical therapy. The unintended inequity is that people who have a chronic condition that is not in one of these two categories often have nowhere to go for help. In the best of circumstances, the help they can get is fragmented. This situation applies to the conditions grouped together in part V, which might be considered metabolic disorders, though that name could create confusion with regard to diabetes. The important issue is that the conditions in part V have substantial impact on physical functioning, quality of life, and ability to live independently.

The conditions discussed in part V can affect exercise tolerance in a range from mild to extremely severe. In the case of organ failure, we have illustrated this in the sample case on kidney failure (see part VIII), which progresses from mild to end-stage to status/post-transplantation. In some conditions, such as a deep venous thrombosis, the individual may have a full recovery and in the long term be "normal" again. In well-managed HIV infection, the patient may have a normal exercise capacity but may have long-term side effects from medication.

As in parts III and IV, the main goal of exercise management in patients with these conditions is achieving and maintaining physical functioning and quality of life. As in many of the conditions discussed in this book, exercise programming services are not a covered benefit under most health insurance plans in the United States, so the clinical staff and exercise professionals must be creative in providing such services to these patients.

Chronic Kidney and Liver Disease

Kidney and liver disease don't receive as much media attention as heart disease, diabetes, obesity, emphysema, arthritis, and cancer, but people who have chronic insufficiency of the kidney or liver face an extremely challenging life. Kidney and liver are responsible for most of the metabolic processes that provide nourishment to all cells, and they rid the body of waste bioproducts as well as toxins to which we are exposed or that we unwittingly ingest. So once kidney or liver function is significantly impaired, all cells, tissues, and metabolic pathways are affected. Persons with these conditions often do not feel well and have impaired energy metabolism, so motivating them to exercise and be more active can be a challenge. Lack of motivation to exercise is made more challenging because our health system fosters an environment that facilitates inactivity and a sedentary lifestyle. As a result of the physiological changes and the inadequacies of our health system, exercise management for persons with chronic kidney and liver disease is among the most challenging problems an exercise professional can face.

Renal Disease

Chronic kidney disease (CKD) is a progressive deterioration of the filtration functions of the kidney. The most common etiologies for renal disease include the following:

- Uncontrolled hypertension
- Diabetes
- Genetic risk factors
- Congenital conditions
- Autoimmune diseases

In CKD, deterioration progresses until toxins that are normally removed from the bloodstream build up to levels that affect the functioning of multiple systems. Because normal kidneys have a great excess of filtration capacity, people generally do not sense that they are deteriorating until more than 80% of kidney function is lost. Symptoms of kidney disease include the following:

- Fatigue
- Nausea and anorexia
- Insomnia

Two major consequences of CKD are that fluid retention results in circulatory congestion, and both soluble and insoluble compounds build up in body fluids. This increase in compounds causes abnormally high levels of electrolytes, toxic by-products of metabolism, and altered neuroendocrine signaling mechanisms. This fluid and toxic compound buildup as well as altered neuroendocrine mechanisms culminates in extreme hypertension, fluid overload (often causing heart failure), endocrine abnormalities (particularly calcium metabolism), anemia, and increased levels of chronic inflammation and oxidative stress.

Because of this vast extent of metabolic derangement, persons with CKD are in the highest risk category for atherosclerotic cardiovascular disease, and because of their extreme hypertension they typically have left ventricular hypertrophy and increased risk of cardiomyopathy. Disturbances in bone and mineral metabolism are common because the kidney plays a critical role in vitamin D metabolism. This condition contributes to calcification of soft tissues and vessel walls, in turn contributing to atherosclerosis. As CKD progresses to end-stage renal disease (ESRD), individuals develop characteristic signs of renal failure:

- Edema in lower extremities
- Muscle wasting and cramping
- Progressive shortness of breath (congestive heart failure)
- Disorientation

The net effect of kidney failure is thus a tremendous alteration in essentially all physiological systems, and ESRD would be lethal except for modern medical technology known as renal replacement therapy (dialysis) or kidney transplantation.

Liver Disease

Hepatitis, or liver inflammation, occurs in both acute and chronic forms, and as with kidney disease, acute hepatitis can cause permanent and lethal liver failure. More commonly, however, chronic liver disease (CLD) develops more slowly and most often results from the following:

- Chronic liver inflammation (e.g., chronic viral hepatitis)
- Bile obstruction (cholestasis)
- Blood flow changes
- Toxic injury from
  - alcohol,
  - medications and drugs, or
  - poisons (including naturally occurring compounds, e.g., mushrooms).

Accumulation of fat (steatosis), hereditary hepatic conditions, and cancers can also result in CLD.

As in the kidney, some 75% of liver tissue must be affected before a decrease in overall function becomes symptomatic. Liver dysfunction affects digestion of food, glucose regulation, clotting, and conversion of waste products into urea and metabolism of medications. Damaged liver cells are replaced by scar tissue, resulting in cirrhosis. Symptoms associated with liver disease are specific to the etiology but often include these:

- Nausea and vomiting
- Abdominal pain
- Jaundice and ascites
- Bruising and itching
- Muscle wasting and weakness

For most causes of CLD, there currently is no cure. Management of CLD is aimed at avoiding the causes (e.g., alcohol), managing symptoms, and delaying progress to liver failure, or end-stage liver disease (ESLD). Unlike the situation with renal failure, there is no artificial replacement therapy, so liver transplantation is the only life-saving treatment.

Management and Medications for Kidney Disease

Medical treatment for the early stages of CKD focuses on management of blood pressure and anemia, with additional emphasis on improving glucose control in people who have diabetes. A special diet is recommended for persons with CKD to retard the rate of kidney deterioration. Individuals with CKD require antihypertensive agents (even after they have ESRD), with most requiring multidrug regimens, with a target of <130/80. The following are antihypertensive medications commonly used in renal disease:

- Diuretics (while some renal function remains)
- Angiotensin-converting enzyme inhibitors
- Angiotensin receptor blockers
- α-adrenergic blockers

The choice of antihypertensive agents is based on the effectiveness of slowing progression of kidney disease as well as reducing the risk of cardiovascular disease. In addition to hypertensive medication, mineral and bone disturbances are treated with dietary and pharmacologic interventions, depending on the stage of disease and the level of disturbance (Kidney Disease Outcomes Quality Initiative lipid management guidelines). Lastly, anemia is often treated with erythropoietin (EPO).

As CKD progresses to ESRD, patients, families, and their physicians decide whether the individual is a candidate for hemodialysis, peritoneal dialysis, or a renal transplant. Here are the three basic methods for renal replacement therapy:

- Hemodialysis: This technique passes the person's blood over a semipermeable membrane that filters and dialyzes water, electrolytes, and small molecules out of the blood. The vascular access is usually a surgically created arterial–venous fistula (in arm or leg), though subclavian catheters are often used until a permanent fistula is placed. Chronic hemodialysis typically is performed in outpatient centers, three times per week for 3 to 4 h per treatment, though technologies are beginning to be used to provide better hemodialysis at home.
- Peritoneal dialysis: This technique uses the abdominal cavity as a reservoir for dialysis fluid, and the dialysis occurs across the peritoneal mem-

branes. Individuals can choose one of two forms of peritoneal dialysis: continuous ambulatory peritoneal dialysis (CAPD), in which 1.5 to 2 L fluid is present in the abdomen and exchanged four times throughout the day; or continuous cycling peritoneal dialysis (CCPD), in which the dialysis fluid is cycled through the peritoneal cavity overnight. These exchanges occur through a catheter permanently placed in the abdomen to which the dialysis fluid bags are attached for introduction into the abdominal cavity. The exchanges are performed independently using sterile technique.

- Transplant: The source of a kidney transplant can be a living related or nonrelated source or a cadaver source. Kidney transplants require immunosuppressant medications for the life of the transplant unless the donor is an identical twin. These medications include broad-spectrum agents such as prednisone, in combination with medications that target specific steps in the rejection pathway.

Management and Medications for Liver Disease

Options in the management of liver disease are substantially more limited than those for kidney disease, partly due to the anatomy of these organs but mainly due to the extraordinarily complex metabolic functions of the liver. As with kidneys, even a partly functioning native liver is the best thing, so the initial goal is to preserve liver function via these means:

- Avoidance of inciting causes (e.g., alcohol)
- Surgical relief of obstructions of bile or portal circulation
- Minimizing the need to metabolize nitrogen-containing foods (i.e., proteins)

In most cases, except for gallstone disease and some viral infections such as hepatitis A or mononucleosis, liver diseases are medically managed and not cured. Antiviral treatments are used for chronic hepatitis, and medications to reduce protein absorption may be prescribed. Dietary sodium restriction or diuretics or both may be used to reduce fluid accumulation. As liver failure progresses to end-stage (ESLD), vitamin K may be provided to maintain production of coagulation factors, and liver production of protein may drop too low to maintain serum albumin. When this occurs, individuals develop severe edema and accumulation of fluid in the abdominal cavity (ascites). Ascites can be temporized by draining the fluid off in a procedure called paracentesis, but it will return rapidly if the individual cannot make albumin. People with liver disease that has progressed this far are very ill, and liver transplantation is the only life-sustaining treatment.

Effects on the Exercise Response

The courses of illness and treatments that people with chronic kidney or liver disease experience are different, but with regard to exercise, individuals with CKD and CLD are quite similar. Both show a significant reduction of exercise capacity and muscle strength as their disease progresses. It is worth reiterating that for both conditions, symptoms include the following:

- Severe skeletal muscle weakness
- Cramping
- Exertional muscle fatigue
- Shortness of breath (from secondary congestive heart failure and severe deconditioning)

Individuals with CKD are at high risk for cardiovascular disease, and up to 70% of those with progressive kidney dysfunction die before reaching end-stage disease—primarily due to cardiovascular events.

Due to the progressive nature and associated malaise of both kidney and liver disease, there is an almost universal reduction in physical activity. The goal of an exercise intervention in the earlier stages is to prevent people from becoming sedentary and prevent physical deconditioning, specifically to maintain muscle mass, strength, and endurance. As each disease progresses, the goal is to prevent further deterioration in physical function that often results in disability and bed rest.

The exercise response in CKD characteristically shows

- progressively lower peak exercise capacity as CKD gets worse,
- early muscle fatigue limiting exercise capacity,
- low maximal heart rates (especially in people on dialysis), and
- hypertensive pressor response.

It is worth noting that the chronotropic blunting is believed not to be due to skeletal muscle weakness and low peripheral demand for blood flow, but rather to some unexplained factor (in large part because this phenomenon reverses very rapidly after kidney transplantation).

Although there are limited published studies in persons with liver disease, exercise capacity is also markedly reduced in these individuals, primarily because of

- muscle weakness,
- early onset of fatigue, and
- low maximal heart rates.

Those with hepatopulmonary syndrome or fluid retention may show a cardiopulmonary limitation. Portal hypertension is exacerbated by even low levels of exercise (30% max) and may increase the risk of variceal bleeding in people with this condition.

Exercise capacity improves following both liver and kidney transplantation, although without appropriate rehabilitation interventions it remains impaired compared to age-predicted normal values.

Effects of Exercise Training

Three decades of experience prove that people with CKD treated with dialysis can safely participate in progressive cardiovascular and resistance exercise training. The adaptations vary, depending on comorbidities and adherence (which is often affected by frequent medical complications or hospitalizations in people on dialysis). In response to exercise training, most patients or clients subjectively report

- feeling more energetic,
- less limitation in physical functioning, and
- improved quality of life.

Cardiorespiratory fitness often improves in people on dialysis who have a lower disease burden, but even those who are more typical show improvements in

- physical performance tasks (6-min walk, gait speed, chair stand),
- muscle strength (after resistance training), and
- physical domains of quality of life.

Several small studies have reported lower blood pressure and reduced antihypertensive medication requirements and improved vascular function, as well as better lipid profiles and glucose management.

Individuals with earlier stages of CKD are less well studied with respect to exercise; however, preliminary data suggest that exercise training is safe and that the benefits include improved exercise capacity and physical function and possible reduction in cardiovascular risk factors. Attenuation of the decline in renal function (glomerular filtration) with exercise training is only speculative at this point.

Exercise training following kidney transplantation results in significant improvements in $\dot{V}O_2$peak, muscle strength, and physical domains of quality of life. Vigorous training for competition is well tolerated, and $\dot{V}O_2$peak that is similar to sedentary healthy levels and above has been reported.

There are few data on exercise training in persons with liver disease; however, in two studies with very small sample sizes, 12 weeks of exercise training improved estimated $\dot{V}O_2$peak by up to 29%. Exercise capacity improves following liver transplant, and if exercise training is continued following the transplant, further improvements are realized. Vigorous exercise training is well tolerated following liver transplant, and the individuals who train at high levels achieve $\dot{V}O_2$peak levels similar to or higher than normal healthy levels. There are no data supporting any "ideal" exercise training prescription for this population. However, given the low levels of physical function and low physical activity, there is no reason to assume that

Severe Debilitation in Persons With End-Stage Kidney or Liver Disease

People in these groups are typically sedentary, severely deconditioned, and most patients or clients suffer from fatigue and significant muscle weakness. Fatigue makes motivation to participate in physical activity more difficult. Depression, multiple comorbidities, and severe deconditioning are the norm. As a result of these circumstances, the main goal of exercise management is to prevent functional deterioration and preserve quality of life, though some individuals will improve health status.

guidelines for increasing physical activity would be any different than for people who are elderly or those who are severely deconditioned or have other comorbidities. In addition, the levels of exercise capacity achievable for people with kidney and liver disease are unknown. However, there are anecdotal reports of individuals with CKD before and after initiating dialysis who participated in high levels of exercise with no apparent adverse effects; and both liver and kidney transplant recipients are able to tolerate high levels of exercise training and adapt similarly to, albeit perhaps more slowly than, normal healthy individuals.

Recommendations for Exercise Testing

Because of the profound differences in metabolic status for persons with chronic kidney or liver insufficiency and those who have undergone organ transplantation and become much closer to being metabolically normal, we consider testing for those who have undergone transplantation separately.

Chronic Kidney and Liver Disease

Exercise testing for diagnostic purposes may be of limited value in persons with CKD treated with dialysis for the following reasons:

- Most stop exercising before achieving 85% of maximal heart rate.
- People often have left ventricular hypertrophy causing electrocardiogram (ECG) misinterpretation.
- They often have electrolyte abnormalities causing ECG changes.
- They exhibit significant muscle limitations.

Because the vast majority of persons with CKD and CLD have an exercise capacity well below average due to muscle weakness and fatigue, it is very unlikely they will "overexert" themselves; the goal of exercise programming is to increase light to moderate physical activity. Thus, exercise testing for diagnostic purposes presents a significant barrier to the goal of increasing physical activity. Additionally, the time burden of dialysis treatment and the logical consideration that exercise testing will be best tolerated on a midweek nondialysis day present a substantial barrier to many. For these reasons, for people on dialysis, the Basic

CDD4 Recommendation to use a functional exercise trial is advised.

For persons with CKD who do not yet require dialysis, exercise testing may be helpful due to the high prevalence of cardiovascular disease, and individuals who wish to pursue a moderate to vigorous exercise program should follow the ACSM Guidelines. Experience suggests that this will be a small number of people, as the majority of individuals with CKD are older with a low exercise capacity; for them the goal would be to increase physical activity, not participation in vigorous exercise training. For people in this group, a functional exercise trial with gradual progression and frequent follow-up has proven to be safe. Physical performance testing of persons with CKD with progressive renal disease and those who are on dialysis treatment can be used to monitor progress and adaptations. Although there are no data, similar recommendations should be appropriate for people with liver disease.

Status/Post-Transplantation

Individuals after transplantation typically have higher exercise capacity, the energy levels, and the desire to participate in more vigorous physical activity. Thus, in people who participate regularly in physical activity and are interested in increasing the intensity of exercise, exercise testing may be indicated and the ACSM Guidelines are advised. However, many people remain severely deconditioned after transplant and wish only to increase their physical functioning at a light to moderate level; such individuals can follow the Basic *CDD4* Recommendation.

The following are suggestions for exercise testing in persons with CKD and CLD:

- Use a treadmill or cycle ergometer.
- Start at very low work rates (10-25 W) on the cycle ergometer and progress in 25-W increments.
- Branching treadmill protocols that establish a comfortable walking speed with increases in intensity equal to 1 metabolic equivalent (MET) work well.
- Many individuals stop due to leg fatigue.
- Many CKD tests are terminated due to systolic blood pressure >250 mmHg.
- Measuring respiratory gases is usually not helpful.

Contraindications unique to these populations include the following:

- Liver transplant recipients with portal hypertension should avoid maximal exertion that may precipitate variceal bleeding.
- Persons with ESRD and a long-standing history of poor phosphate and calcium balance with secondary hyperparathyroidism are at risk for spontaneous tendon avulsions and pathologic fractures, and should avoid maximal strength testing and ballistic exercises.

Recommendations for Exercise Programming

The Basic *CDD4* Recommendations are appropriate for most persons with CKD, for those with ESRD treated with dialysis, and for people immediately after successful renal or liver transplantation.

Persons with progressive kidney or liver disease usually experience gradual reductions in physical activity and substantial deconditioning. All patients or clients should receive information on and encouragement to do progressive functional muscle strengthening exercises and progressive cardiovascular exercise, as well as Health and Human Services and American Heart Association advice on physical activity for adults who are older (including those with chronic conditions).

Recommendations following transplant should also be initially low level, with very gradual progression to recommendations for normal healthy adults. The adaptation may be delayed, so slower progression in intensity is advised, and the maximal levels achieved may be limited, depending on comorbid conditions and overall medical history and course of the disease. Renal patients who had long-standing renal osteodystrophy before successful transplantation remain at risk for stress fractures and tendon avulsion fractures; they should avoid high-resistance strength training and ballistic activities (running, jumping, and so on).

Summary of major factors to consider in exercise programming:

- Reduce as many barriers to physical activity as possible.
- Start with low-intensity activities, with gradual progression as tolerance improves.

- The goal may be to prevent deconditioning rather than improve exercise capacity.
- For persons on dialysis, consider the time demands needed for dialysis treatment.
- Individuals with severe muscle weakness may benefit from physical therapy before starting a cardiovascular program.
- More vigorous activity is well tolerated in persons who have developed a base of fitness (cardiovascular endurance and muscle strength) through regular moderate-level physical activity.

Integration Into a Medical Home Model

Individuals with kidney and liver disease are followed routinely by specialists who often function as their primary care physician. In such cases, this practice should function as the person's "medical home," and a plan for maintaining physical activity should be incorporated into the overall disease management plan. Transplant recipients and persons treated with home dialysis have physician visits every month, and people on in-center hemodialysis are seen by a physician three times a week. Routine assessment of participation in physical activity should be included as part of these periodic reevaluations.

Assessment of physical functioning (e.g., gait speed, 6-min walk) should be done early in the diagnosis, and follow-up reassessments should be done no less often than every 6 months. Persons who have low physical functioning or show evidence of a decline in physical functioning should receive counseling or counseling combined with referral to physical therapy or an exercise specialist. Encouragement and reinforcement of participation in physical activity can be assessed with regular phone follow-ups and at regular clinic visits. Because of multiple medical issues, frequent hospitalizations, and changes in medical status, the plan for exercise needs to be reassessed and modified as dictated by clinical condition, tolerance, and ability.

Take-Home Message

Several key points are important to keep in mind when working with people with kidney or liver disease:

- People in this group are typically sedentary and severely deconditioned.
- Depression, multiple comorbidities, and severe deconditioning are the norm.
- The main goal is to prevent functional deterioration, though some individuals will improve.
- Consider using gait speed as a vital sign, as is commonly done in geriatrics.
- Most patients or clients suffer from fatigue and significant muscle weakness.
- Fatigue often makes it difficult to become motivated to participate in physical activity.
- Consider the time commitment that these individuals must devote to other health care services.
- Exercise testing is not needed for those who start at low intensity and progress slowly.
- Watch for extreme exertional hypertension in persons with CKD.
- Functional strengthening exercises for muscle weakness must not be overlooked.
- People with organ transplant feel more energetic and are more apt to do vigorous exercise.
- Education of dialysis and transplant staff on encouraging patients or clients may be helpful.
- Consider use of stationary cycles during the dialysis treatment to optimize participation.

Suggested Readings

Cheema B, Singh MAF. Exercise training in patients receiving maintenance hemodialysis: a systematic review of clinical trials. *Am J Nephrol.* 2005;25:352-364.

Cheema BSB, Smith BCF, Singh MAF. A rationale for intradialytic exercise training as standard clinical practice in ESRD. *Am J Kidney Dis.* 2005;45:912-916.

Painter P. Exercise following organ transplantation: a critical part of the routine post transplant care. *Ann Transplant.* 2005;10:28-30.

Painter P. Physical functioning in end-stage renal disease patients: update 2005. *Hemodial Int.* 2005;9:218-235.

Painter P. Implementing exercise: what do we know? Where do we go? *Adv Chronic Kidney Dis.* 2009;16:536-544.

Painter PL, Carlson L, Carey S, Paul SM, Myll J. Physical functioning and health related quality of life changes with exercise training in hemodialysis patients. *Am J Kidney Dis.* 2000;35(3):482-492.

Painter P, Marcus RL. Assessing physical function and physical activity in CKD patients: in-depth review. *Clin J Am Soc Nephrol.* 2012;8(5):861-872.

Painter P, Roshanravan B. Association of physical activity and physical function with clinical outcomes in adults with chronic kidney disease. *Curr Opin Nephrol Hypertens.* 2013;22:615-623.

Additional Resources

The following booklets are available on http://lifeoptions.org/catalog [Accessed November 4, 2015]: Exercise: A Guide for People on Dialysis; Exercise for the Dialysis Patient: A Guide for the Nephrologist; and Exercise for the Dialysis Patient: A Prescribing Guide

KDOQI Guidelines for CKD Care, www.kidney.org/professionals/guidelines/guidelines_commentaries [Accessed August 19, 2015].

Acquired Immune Deficiency Syndrome

Acquired immune deficiency syndrome or acquired immunodeficiency syndrome (AIDS) is a group of symptoms and infections caused by the human immunodeficiency virus (HIV). The first definition of AIDS appeared in the September 24, 1982 issue of *Morbidity and Mortality Weekly Report,* published by the Centers for Disease Control and Prevention (CDC). Since 1982, the definition of AIDS has evolved as scientists and public health experts understand more about its etiology and epidemiology. To meet the surveillance case definition for HIV infection among adults and adolescents, laboratory-confirmed evidence of HIV infection is required. The most current and widely accepted definition of AIDS is the CDC AIDS case definition:

> HIV infection, Stage 3 (AIDS): CD4+ T-cell count below 200 cells/μL (or a CD4+ T-cell percentage of total lymphocytes less than 14%), or documentation of an AIDS-defining illness.

The CDC AIDS case definition is used to monitor trends in the number and distribution of AIDS cases in the United States and severe morbidity due to infection with HIV. The first two stages of HIV infection are classified by the CDC as follows:

> HIV infection, Stage 1: No AIDS-defining condition and either CD4+ T-lymphocyte count of ≥500 cells/μL or CD4+ T-lymphocyte percentage of total lymphocytes ≥29.

> HIV infection, Stage 2: No AIDS-defining condition and either CD4+ T-lymphocyte count of 200 to 499 cells/μL or CD4+ T-lymphocyte percentage of total lymphocytes 14 to 28.

Basic Pathophysiology

Human immunodeficiency virus infection weakens and reduces the function of the immune system, specifically the adaptive immune system. The specific, or adaptive, immune system is the most complex component of immune function and is activated by antigen-specific mechanisms that provide long-term immunologic "memory" of the foreign antigen. The adaptive immune system is dependent on the function of small lymphocytes, the most common being T- and B-lymphocytes (T-cells and B-cells). T-cells are involved in cell-mediated immunity that enlists a cellular immune response, with cells that physically attack and destroy infected cells, while B-cells act to neutralize microbes directly through the production of antibodies that attack the microbes. Most pertinent to HIV infection, the CD4+ T-cell (identified by the CD4 coreceptor on the T-cell surface) serves in a regulatory manner, releasing cytokines and growth factors that promote the growth of effector cell responses that are antigen specific and ultimately destroy infected cells and cancer cells carrying that particular antigen. CD4+ T-cells are thus critically important in regulating the immune response to specific infections.

Since CD4+ T-cells are a primary target of HIV, the most common immunologic abnormalities observed in HIV are seen in cells that are involved in cell-mediated immunity:

- Marked decrease in total T-cells
- Marked decrease in regulatory CD4+ T-cells
- Reduced response to infectious agents
- A normal or slight increase in effector CD8+ T cells
- A decrease in natural killer cell activity

The progression from initial HIV infection to AIDS moves through delineated stages, the boundaries of which have been distorted to some degree by antiretroviral drug therapy (ART). The progressive stages are as follows:

- Seroconversion: immediately postinfection, with a large production of HIV virus and initial immune response, often manifest as flu-like symptoms

- Asymptomatic (World Health Organization [WHO] Stage 1): can last for many years, and is associated with significant HIV activity in the lymph nodes and glands
- Symptomatic (WHO Stage 2): characterized by multisymptom disease, emergence of opportunistic infections and cancers
- AIDS (WHO Stage 3): severe weight loss and muscle atrophy, recurring or continuous infection of multiple organ systems, especially the respiratory, digestive, and nervous systems and the skin

Management and Medications

Many drugs have been approved by the Food and Drug Administration (FDA) for treatment of HIV infection; they are categorized by mode of function (see table 21.1). Typical treatment regimens include drugs from three or more of these categories. Adding to the difficulty of HIV medical management is the treatment of specific opportunistic infections, HIV symptoms, and sometimes HIV-related cancers, not to mention

Table 21.1   Medications for Human Immunodeficiency Virus

Drug	Mechanism of action	Adverse effects
NRTIs: Zidovudine Lamivudine Emtricitabine Abacavir Didanosine Stavudine Tenofovir Zalcitabine	Nucleoside reverse transcriptase inhibitors compete with substrate, terminating replication of viral DNA.	• Hepatotoxicity, pancreatitis, lactic acidosis • Neutropenia, anemia • Headache, fatigue, chills, insomnia • Nausea, diarrhea, vomiting, anorexia • Neuropathy, musculoskeletal pain • Depression
Non-NRTIs: Rilpivirine Etravirine Delavirdine Efavirenz Nevirapine	Non-nucleoside reverse transcriptase inhibitors also inhibit the conversion of HIV RNA to DNA.	• Nausea, diarrhea, vomiting, anorexia • Headache, rash, insomnia, fever • Hepatotoxicity, dyslipidemia • Neutropenia, anemia • Depression • Neuropathy
Protease inhibitors: Atazanavir Indinavir Tipranavir Saquinavir Lopinavir Ritonavir Fosamprenavir Darunavir Nelfinavir	PIs stop the processing of HIV genetic material within the infected cell.	• Abdominal pain, diarrhea, nausea, vomiting • Headache, rash, anemia, fatigue, weight loss, asthenia, insomnia • Hepatotoxicity • Dyslipidemia • Kidney stones • Lipodystrophy • Neutropenia, leukopenia
Fusion inhibitors: Enfuvirtide	FIs inhibit HIV virus from attaching to targeted cells, inhibiting initial infection.	• Diarrhea, nausea, abdominal pain • Anorexia, fatigue • Pancreatitis
Integrase inhibitors: Raltegravir Dolutegravir	Integrase inhibitors prevent HIV from integrating its genetic material into the infected cell's DNA.	• Headache, nausea, insomnia • Hepatotoxicity, pancreatitis, hyperglycemia • Neutropenia, thrombocytopenia
Entry inhibitor: Maraviroc	Entry inhibitors block CCR5 receptors that CCR5-tropic HIV-1 virions use to enter CD4 cells.	• Rash, fever, cough, upper respiratory infections • Hepatotoxicity • Dizziness

the prevalence of medication-related physical and psychological side effects. These include the metabolic syndrome (e.g., hyperglycemia, diabetes, hyperlipidemia) and an unusual condition called lipodystrophy. A number of generally accepted nutritional strategies exist for HIV infection and related comorbidities. Persons with HIV can be relatively normal or very ill, and medical management can be relatively simple or highly complex, with multisystem conditions and treatments.

Due to the number and type of medications, there are often significant issues with side effects and drug interactions, all of which must be considered and monitored as part of an exercise program. The most common drug-related side effects are gastrointestinal–abdominal cramps, nausea, vomiting, and diarrhea. Neurological side effects are also common; the most common of these is peripheral neuropathy. Other conditions such as lethargy, malaise, fatigue, anemia, mitochondrial toxicity, and myopathies are also common.

Role of Exercise in Preventing Cardiometabolic Disease in People With HIV

People who have HIV are at increased risk for cardiometabolic disease, in part because they are likely to have lipodystrophy, experience skeletal muscle atrophy, and become increasingly inactive as the disease progresses. Also, they are often prescribed medications that independently increase the risk of lipid disorders. These circumstances all increase the likelihood of developing cardiometabolic disease, so exercise is important in helping maintain skeletal muscle contribution to cardiometabolic status in persons with HIV.

Effects on the Exercise Response

The combination of HIV-related illness and medication side effects can lead people to an extremely sedentary lifestyle that often exacerbates the viral and medication-related symptomatology. As a result, the response to graded exercise testing in both trained and untrained HIV-infected individuals is typically an abbreviated test, usually terminated due to fatigue or volitional exhaustion with poor aerobic capacity and physical function-

ing. It is unknown currently how much the typical HIV-associated loss in physical functioning is due to the infection, the associated disease symptoms, and deconditioning. The following list illustrates how stages of HIV infection commonly affect exercise performance in comparison to age-adjusted norms:

Asymptomatic

From normal graded exercise test with physiological parameters within normal limits to potentially reduced exercise capacity related to sedentary lifestyle and deconditioning

Symptomatic

Significantly reduced exercise time, $\dot{V}O_2$peak, and ventilatory threshold

Increased heart rates at given absolute submaximal work rates

AIDS

Dramatically reduced exercise time and $\dot{V}O_2$peak

Possible failure to reach ventilatory threshold

Abnormal neuroendocrine responses at moderate- and high-intensity stages

Effects of Exercise Training

Exercise training is safe for individuals with HIV and AIDS who are medically stable. Aerobic exercise at low to moderate intensities does not increase the prevalence of additional infections, does not increase viral load, and actually increases CD4+ T-cell count. Several studies have reported exercise-enhanced immune function, particularly in asymptomatic participants. These activity-related improvements in immune function are controversial, however, and are contradicted by other studies.

As with most other clinical populations, aerobic and resistance training are beneficial for HIV-infected individuals. Almost without exception, studies involving the effects of aerobic exercise training in HIV-infected individuals demonstrate enhanced physical functioning, cardiovascular and muscular endurance, health-related quality of life, and general well-being. Although there is less evidence to support strength training in HIV-infected individuals, progressive resistance exercise programs increase lean tissue mass and improve strength in HIV-infected individuals both with and without muscle wasting.

While exercise and diet have been shown to be beneficial for many of the conditions associated with HIV infection and drug side effects, there is virtually no information on the interactions between HIV medications and exercise training.

Recommendations for Exercise Testing

Here are some exercise testing recommendations for working with people with HIV:

- For asymptomatic HIV: follow the ACSM Guidelines.
- For symptomatic to AIDS: use the Basic *CDD4* Recommendations if the patient or client has low physical functioning due to HIV, frailty, or some other condition.
- Observe strict adherence to blood-borne pathogen precautions.

While a majority of HIV+ individuals have lower functional capacities compared to uninfected individuals, most standard physical fitness tests are applicable to the HIV+ population.

Cardiorespiratory Fitness Testing

Given the common finding of lower than normal peak $\dot{V}O_2$ in this population, consider using a lower-intensity or submaximal rather than maximal protocol. For example, a Modified Bruce protocol or submaximal cycle ergometry test, such as the YMCA protocol, would be appropriate.

Muscular Fitness Testing

Assessing muscular strength and endurance and flexibility in HIV+ individuals should not differ significantly from testing in apparently healthy individuals. Because this population is generally untrained and likely unfamiliar with resistance training, using a 3RM, 6RM, or 10RM protocol versus 1RM (1-repetition maximum) may be more appropriate when assessing muscular strength. In lower-functioning individuals, use the Basic *CDD4* Recommendations.

Recommendations for Exercise Programming

These are the exercise programming recommendations for people with HIV:

- For asymptomatic HIV: follow the ACSM Guidelines.
- For symptomatic to AIDS: use the Basic *CDD4* Recommendations if the patient or client has low physical functioning due to HIV, frailty, or some other condition.

Advances in ART have contributed to increasing numbers of people middle-aged and older living with HIV. As a result, many conditions, including ART-related side effects such as the metabolic syndrome or frailty, as well as age-related chronic diseases, are calling for emphasis on the importance of exercise training and maintaining physical fitness for HIV-infected individuals.

Regardless of stage of the condition, HIV-infected individuals should obtain medical consultation before initiating an exercise program. Dyslipidemia, insulin resistance, obesity, cardiovascular disease, and bone disorders can complicate the exercise programming in this population. Individuals who have higher physical functioning can be evaluated in the same manner as apparently normal persons, including assessments of functional capacity, muscular strength, body composition, and flexibility. Lower-functioning individuals can be approached with the Basic *CDD4* Recommendations. Neuromuscular function should be assessed, because peripheral neuropathy is a common symptom of HIV infection and can influence the appropriate modes of training.

Listed next are principles to follow when prescribing exercise programs to HIV-infected individuals.

- The ultimate goal for individuals with HIV is to eventually meet the ACSM recommendations for aerobic and resistance exercise.
- Because many of the medications prescribed to HIV+ individuals have side effects that influence physical functioning, the goals may need to be modified or adjusted.
- Aerobic exercise programs should begin at lower volumes, and progression may occur at rates slower than seen in uninfected, apparently healthy populations.
- Participants are advised to choose a form of resistance training that is comfortable and provides a pain-free range of motion with appropriate progression.
- Age-appropriate, evidence-based exercise recommendations for HIV-infected adults who are older (age >50 years) are evolv-

ing, but a combination of endurance and resistance exercises three times per week is recommended initially, with progress reassessed after 6 weeks.

No adverse side effects of relatively high-intensity exercise interventions have been reported, but the absence of such data should be interpreted cautiously. Exercise programs should probably err on the side of undertraining, since overtraining syndromes may be associated with immunosuppression phenomena that should be avoided in an HIV-infected individual.

Integration Into a Medical Home Model

People infected with HIV benefit from regular physical activity by remaining robust in an effort to avoid the deconditioning-related sequelae of chronic disease. The metabolic effects of a sedentary lifestyle and deconditioning can be amplified by metabolic comorbidities of HIV (insulin resistance, lipodystrophy), as well as some of the medications these individuals have to use.

It is helpful for medical homes to have an HIV resource that supports physical activity and exercise training, as well as providing an HIV-specific social support group.

Take-Home Message

When working with people with HIV, keep in mind these special considerations:

- Know the stage of each individual with HIV, because this affects the exercise-related needs.

- Persons with HIV typically take a large number of medications that can have dramatic effects on exercise response and performance.

- These individuals are at heightened risk for opportunistic infections, especially if they overtrain at an intensity or volume level that is too high for their fitness status.

- Individuals on ARTs are at an increased risk for dyslipidemia, insulin insensitivity, lipodystrophy, increased central adiposity, and other cardiometabolic risk factors.

- Because of this increased cardiometabolic risk, cardiovascular exercise is extremely beneficial for people with HIV.

Suggested Readings

Bopp CM, Phillips KD, Fulk LJ, Hand GA. Clinical implications of therapeutic exercise in HIV/AIDS. *J Assoc Nurses AIDS Care.* 2003;14:73-78.

Bopp CM, Phillips KD, Sowell RL, Dudgeon, WD, Hand GA. Physical activity and immunity in HIV-infected individuals. *AIDS Care.* 2004;16:387-393.

Botros D, Somarriba G, Neri D, Miller TL. Interventions to address chronic disease and HIV: strategies to promote exercise and nutrition among HIV-infected individuals. *Curr HIV/AIDS Rep.* 2012;9:351-363.

Dudgeon WD, Phillips KD, Bopp CM, Hand GA. Physiological and psychological effects of exercise interventions in HIV disease. *AIDS Patient Care STDS.* 2004;18:1-16.

Dudgeon, WD, Phillips KD, Carson JA, Brewer RB, Durstine JL, Hand GA. Counteracting muscle wasting in HIV-infected individuals. *HIV Med.* 2006;7:299-310.

Fulk LJ, Kane BP, Phillips KD, Bopp CM, Hand GA. Depression in HIV-infected patients: allopathic, complementary, and alternative treatments. *J Psychosom Res.* 2004;57:339-351.

Gomes Neto M, Ogalha C, Andrade AM, Brites C. A systematic review of effects of concurrent strength and endurance training on the health-related quality of life and cardiopulmonary status in patients with HIV/AIDS. *Biomed Res Int.* 2013;2013:319534. doi: 10.1155/2013/319524.

Hand GA, Phillips KD, Dudgeon WD, Lyerly GW, Durstine JL, Burgess SE. Moderate intensity exercise training reverses functional aerobic impairment in HIV-infected individuals. *AIDS Care.* 2008;20:1066-1074.

Additional Resources

Evidence-Based Guidelines

Hammer SM, Saag MS, et al. Treatment for adult HIV infection 2006 recommendations of the International AIDS Society-USA Panel. *JAMA.* 2006;296 (7):827-843.

Lyles CM, Kay LS, et al. Best-evidence interventions: findings from a systematic review of HIV behavioral interventions for US populations at high risk, 2000–2004. *Am J Public Health.* 2007;97(1).

Yahiaoui A, McGough EL, Voss JG. Development of evidence-based exercise recommendations for older HIV-infected patients. *J Assoc Nurses AIDS Care.* 2012;23:204-219.

Suggested Websites

Centers for Disease Control and Prevention, HIV/AIDS, www.cdc.gov/hiv [Accessed November 4, 2015].

World Health Organization, HIV/AIDS, www.who.int/hiv/en [Accessed November 4, 2015].

Chronic Fatigue Syndrome

Chronic fatigue syndrome (CFS) is a commonly used name for a debilitating medical condition that lasts for a minimum of 6 months and is characterized by persistent fatigue and other specific symptoms. Many clinicians use other names, including myalgic encephalomyelitis (ME), postviral fatigue syndrome (PVFS), and chronic fatigue immune dysfunction syndrome (CFIDS). Because these conditions are syndromes and not a pathophysiological diagnosis, it is not clear whether or not these are all the same or are different entities. There is no laboratory test or biomarker that can be used to diagnose any of these maladies. Except where noted, the discussion here is based on research on CFS.

The overall prevalence of CFS is estimated at 235 per 100,000 persons, or about 1 million in the United States. Chronic fatigue syndrome is four times more common among women (373 per 100,000) than among men (83 per 100,000) and most common among white women 50 to 59 years of age (863 per 100,000). Among those with CFS, only 16% have received a diagnosis and medical treatment for their illness. The 1-year CFS incidence is 180 per 100,000 persons.

Basic Pathophysiology

In CFS, the cause of the fatigue is unclear, but the fatigue is not from overexertion, not significantly relieved by rest, and not caused by other medical conditions. Biological, genetic, infectious, and psychological causes have been proposed, and etiology may be related to multiple factors including oxidative stress, genetic predisposition, infection by viruses and bacteria, hypothalamic–pituitary–adrenal axis abnormalities, immune dysfunction, and psychological and psychosocial issues.

Myriad symptoms have been attributed to CFS, and several attempts to clarify the condition have been made by successive groups of international CFS experts, convened by the U.S. Centers for Disease Control and Prevention (CDC) in 1988, 1994, and 2003. By current CDC Guidelines, a CFS diagnosis requires three criteria:

- Severe chronic fatigue for 6 or more consecutive months
  - not due to ongoing exertion or other medical conditions associated with fatigue, and
  - for which other medical conditions have been ruled out by diagnostic testing
- Fatigue significantly interfering with daily activities and work
- At least four of the following eight symptoms:
  - Postexertion malaise lasting >24 h
  - Unrefreshing sleep
  - Significant impairment of short-term memory or concentration
  - Muscle pain (myalgia)
  - Joint pain (arthralgia) without swelling or redness
  - Headaches of a new type, pattern, or severity
  - Tender cervical or axillary lymph nodes
  - A sore throat that is frequent or recurring

The symptoms should have persisted or recurred during 6 or more consecutive months and cannot have first appeared before the onset of the fatigue.

Because of the very nonspecific nature of these symptoms, persons with symptoms resembling those of CFS need to have several treatable illnesses ruled out, including Lyme disease, sleep disorders, depression, alcohol or substance abuse, diabetes, hypothyroidism, mononucleosis (mono), lupus, multiple sclerosis (MS), and chronic hepatitis and various malignancies. Also, a variety of medications can cause side effects that mimic CFS symptoms.

Management and Medications

Managing CFS is as complex as the illness itself. Treatment is mostly directed at reducing symptoms rather than reversal of the underlying condition, mainly because the etiology and pathophysiology of CFS remain unknown. Many people with CFS have other medical problems, and comorbid fibromyalgia is common. Medication plays a minor role in management, is primarily aimed at reducing symptoms, and thus can include many kinds of drugs:

- Analgesics
- Antidepressants
- Gastrointestinal agents
- Immunosuppressive agents
- Sleep aids
- Stimulants
- Muscle relaxant agents

Persons with CFS often do not tolerate standard doses of many medications, so physicians often start with low doses and gradually increase the dose over time.

Cognitive behavioral therapy (CBT) and graded exercise therapy (GET) are moderately effective for many people and have been suggested by several analyses to be central to the treatment of CFS. Several authors on ME, however, have strongly opposed use of exercise for the management of ME. It's not really clear if there is disagreement here, mainly because of uncertainties about whether or not CFS and ME are the same condition, similar conditions, or perhaps subgroups of a broader definition.

People with CFS are usually advised to closely monitor their health and let their doctor know of any changes. A team approach is generally recommended, with physicians managing symptom-specific medications and rehabilitation specialists, mental health professionals, and physical or exercise therapists providing most of the treatments. Because traditional allopathic care often does not effect a cure, many individuals with CFS seek alternative remedies such as dietary manipulation, vitamin and mineral supplementation, herbal preparations, massage therapy, chiropractic, acupuncture, and aroma therapy. Despite the plethora of approaches to treatment, many people never fully recover from CFS.

Effects on the Exercise Response

The functional capacity of individuals with CFS varies greatly, with some leading relatively normal lives and others bedridden and unable to care for themselves. Employment rates vary, with over half unable to work and nearly two-thirds limited in their work because of CFS symptoms. Accordingly, individuals with CFS have lower than average cardiorespiratory and muscular fitness, but this varies widely due to multiple factors. Self-reports of fatigue do not closely correlate with $\dot{V}O_2$max, which is why experts have concluded that the underlying cause is something other than aerobic capacity. Although not all experts agree, the consensus is that cardiac, pulmonary, muscular, metabolic, immune, and endocrine responses to acute exercise are similar to those in normal individuals who have profound deconditioning. The nature of the fatigue is therefore viewed as central (i.e., either a currently unknown neurological cause, or an integrative perception stemming from cumulative afferent stimuli).

Effects of Exercise Training

Randomized trials typically show that graded exercise therapy (GET), especially when combined with CBT, can be safely added to standard medical care with moderate improvement of fatigue measures in individuals with CFS. Health-related quality of life and psychosocial functioning in individuals with CFS typically improve after several months of CBT and GET, often without much change in peak exercise capacity. Similarly, individuals with CFS who are educated on the benefits and encouraged to participate in regular physical activity at home tend to show significant improvement in functional capacity and quality of life.

Postexertion malaise can last more than 24 h in individuals with CFS. Thus a common complaint among individuals with CFS is that their fatigue and other symptoms are noticeably worse on the days following any amount of physical exertion. The basis for this phenomenon is currently not well understood. Despite the initial aggravation of symptoms caused by exercise, some overall improvement in symptoms has been reported for clients with CFS.

Recommendations for Exercise Testing

General recommendations for exercise testing for people with CFS are as follows:

- For testing to exclude other medical conditions: follow the ACSM Guidelines
- For programming purposes: follow the Basic *CDD4* Recommendations

While exercise testing does not establish a CFS diagnosis, an exercise test may be requested as part of an evaluation for excluding other conditions, such as cardiovascular diseases. Incremental exercise testing with monitoring of standard cardiovascular and ventilatory responses (electrocardiogram, blood pressure, heart rate, respiratory gas exchange, and ventilation) may be useful in these situations. Individuals with long-standing symptoms are very likely to have had a low level of exercise activity and have undergone significant deconditioning. Work rate increments used in testing are therefore usually low, so a low-level ramp of 10 to 25 W/increment or a Naughton type of protocol would be an appropriate choice.

Exercise testing may also be used for designing an individualized exercise program for clients who have a diagnosis of CFS, but this is generally not

While the nature of chronic fatigue remains somewhat of an enigma, clinical experience is that people who have chronic fatigue are quite vulnerable to "over-doing" a particular bout of exercise and can be quite exhausted for an extended time afterwards. This often results in non-adherence and dropping out of the exercise program. So the most important thing to know is to be certain to progress very slowly and err on the side of doing "too little" on any one day. Better to have the person exercising again tomorrow than for them to drop out of the program for weeks.

necessary because the focus of training programs is generally going to be light to moderate physical activity. Also, exercise testing may be requested by some patients or clients with an athletic background or those who have a deep interest in fitness assessment. In these situations, the nature of exercise assessments needs to be individualized across a wide range of physical functioning and fitness levels, but most persons with CFS have low physical functioning, so the Basic *CDD4* Recommendations are appropriate for most situations.

Recommendations for Exercise Programming

Most persons with CFS should follow the Basic *CDD4* Recommendations. Modifications are noted in "Take-Home Message."

The emphasis of exercise programming in CFS should be on improvement in quality of life and attenuation of fatigue through properly timed low-intensity exercise, with little focus on physiological improvements in aerobic capacity. The basic approach involves a gradual, planned, and individualized increase in physical activity, with the goal of improvement in physical functioning. Graded exercise therapy involves a careful assessment of the individual's current physical functioning and negotiation of individualized activity goals. Once a starting point is agreed upon, the physical activity should commence at a low intensity. The duration of the physical activity or exercise can then be increased slowly and carefully until patients or clients are doing 30 min in each session. The exercise intensity then can gradually be increased.

Experienced therapists typically recommend that the activity plan, not patient or client symptomatology, determine the day-to-day activity routine. Because persons with CFS tend to show a waxing and waning course, the plan should be mutually reviewed regularly and adjusted depending on health status and symptoms. The exercise therapist for patients or clients with CFS should have good counseling skills and knowledge of fear avoidance, because this will likely be a barrier to participation.

Pacing or energy management is considered useful by the majority of CFS participants in exercise programs. There are two basic forms of energy management:

- Symptom-contingent pacing
- Time-contingent pacing

In symptom-contingent pacing the decision to stop, rest, or change an activity is determined by symptom awareness. In time-contingent pacing these decisions are determined by a set schedule of activities that a patient or client estimates can be completed without triggering postexertional malaise. Thus, the principle behind either pacing strategy is to avoid overexertion and an exacerbation of symptoms.

The end goal is to help individuals with CFS reach at least 150 min/week of moderate-intensity physical activity (and if the most they can tolerate is for some or even all of these weekly minutes to be light intensity, that is better than being sedentary). People who become inactive because of CFS are highly likely to develop cardiometabolic conditions, type 2 diabetes, or both. Avoiding this outcome is important, as is the ability to participate in activities of daily living and to maintain an occupation.

The following general guidelines offer a conservative approach to exercise programming that takes into account some of the unique difficulties characteristic of the CFS population:

- The goal of exercise programming is to prevent further deconditioning that could worsen CFS symptoms and fatigue. Individuals and trainers alike should resist the temptation to adopt a traditional method of training aimed at optimization of aerobic capacity and should focus instead on modest goals of preventing progressive deconditioning.

- Individuals should be warned that they might feel increased fatigue in the first few weeks of an exercise program.

- Exercise should generally be initiated at very low levels, based on the client's current activity tolerance.

- Aerobic exercise should use a familiar activity, such as walking, that can be started at a low level.

- Flexibility exercises may be prescribed to preserve normal range of motion.

- Strength training should focus on preservation of levels of strength commensurate with daily living activities and should attempt to avoid activities and intensities

that induce delayed-onset muscle soreness (DOMS).

- The progression of exercise activity should focus primarily on increasing the duration of moderate-intensity activities in preference to increasing exercise intensity. Identification of the appropriate magnitude of progression from one exercise session to the next is the most challenging aspect of exercise programming for individuals with CFS. They should be "coached" not to overexert themselves on days when they are feeling well and to reduce their exercise intensity when their symptoms are increased.

Integration Into a Medical Home Model

Working with persons affected by CFS can be quite challenging for all the health care providers in the patient-centered medical home. People with CFS have many symptoms that are quite varied, that wax and wane, and that often never completely resolve, and it is easy to be frustrated by so many needs. All persons with CFS should be referred to a GET–CBT program and should be exercising regularly to help improve symptoms and quality of life. Individuals with CFS should be careful to not overload themselves and cause post-exercise myalgia, which can lead to prolonged nonparticipation and backsliding. People with CFS thus find themselves in a delicate balance:

- When in doubt, they should err on the side of not overdoing.

- They should avoid missing exercise sessions for prolonged periods.

Take-Home Message

When working with persons who have CFS, be sure to consider these points:

- Chronic fatigue syndrome is often accompanied by depression, although depression does not cause CFS.

- Because misunderstanding abounds concerning CFS, frustration and disillusionment are a common problem for both the patient or client and the care team.

- A supportive and understanding environment is therefore very important.
- People with CFS really need to plan their activities so as to budget their energy.

Suggested Readings

DeBecker PJ, Roeykens M, Reynders N, et al. Exercise capacity in chronic fatigue syndrome. *Arch Int Med.* 2000;160 (21):3270-3277.

Edmonds M, McGuire H, Price J. Exercise therapy for chronic fatigue syndrome. *Cochrane Database Syst Rev.* 2004;CD003200.

Fulcher KY, White PD. Strength and physiological response to exercise in patients with chronic fatigue syndrome. *J Neurol Neurosurg Psychiatry.* 2000;69(3):289.

Institute of Medicine. Beyond myalgic encephalomyelitis/chronic fatigue syndrome: Redefining an illness. Washington, DC: The National Academies Press, 2015. doi:10.17226/19012.

McCrone P, Sharpe M, Chalder T, et al. Adaptive pacing, cognitive behavior therapy, graded exercise, and specialist medical care for chronic fatigue syndrome: a cost-effectiveness analysis. *PLoS One.* 2012;7(8):e40808. doi: 10.1371/journal.pone.0040808.

Nijs J, Aelbrecht S, Meeus M, et al. Tired of being inactive: a systematic literature review of physical activity, physiological exercise capacity and muscle strength in patients with chronic fatigue syndrome. *Disabil Rehabil.* 2011;33(17-18):1493-1500.

Powell P, Bentall RP, Nye FJ, et al. Randomized controlled trial of patient education to encourage graded exercise in chronic fatigue syndrome. *BMJ.* 2001;322:387-390.

Reid S, Chalder T, Cleare A, et al. Chronic fatigue syndrome. *Clin Evid.* 2005;14:1366-1378.

Wallman KE, Morton AR, Goodman C, et al. Exercise prescription for individuals with chronic fatigue syndrome. *Med J Aust.* 2005;183(3):142-143.

White PD, Goldsmith KA, Johnson AL, et al. Comparison of adaptive pacing therapy, cognitive behavior therapy, graded exercise therapy, and specialist medical care for chronic fatigue syndrome (PACE): a randomized trial. *Lancet.* 2011;377(9768):823-836.

Additional Resources

Evidence-Based Guidelines

Chronic fatigue syndrome/myalgic encephalomyelitis (or encephalopathy): diagnosis and management. NICE guidelines [CG53]. Published date: August 2007, reviewed February 2014. Available at: http://www.nice.org.uk/guidance/CG53 [Accessed November 4, 2015].

Van Cauwenbergh D, De Kooning M, Ickmans K, et al. Interventions for the treatment and management of chronic fatigue syndrome: a systematic review. *JAMA.* 2001;286:1360-1368.

Working Group of the Royal Australasian College of Physicians. Chronic fatigue syndrome. Clinical practice guidelines–2002. *Med J Aust.* 2002 May 6;176(Suppl):S23-S56.

Suggested Websites

Centers for Disease Control and Prevention, Chronic Fatigue Syndrome (CFS), www.cdc.gov/cfs [Accessed November 4, 2015].

CFIDS Association of America, www.cfids.org [Accessed November 4, 2015].

IACFS/ME International Association for CFs/ME, www.iacfsme.org [Accessed November 4, 2015].

The PACE Trial, http://www.wolfson.qmul.ac.uk/current-projects/pace-trial [Accessed November 4, 2015].

Fibromyalgia

Fibromyalgia (FM) or fibromyalgia syndrome is a chronic disorder characterized by widespread musculoskeletal pain in addition to a variety of other multisystemic symptoms including fatigue, sleep disturbances, cognitive dysfunction, stiffness, and depressive episodes. Fibromyalgia is a complex multidimensional condition that commonly coexists with other conditions. An estimated 2% to 4% of the population are diagnosed with FM, about 5 to 10 million people in the United States, making it the third most prevalent rheumatologic disorder in the United States. Although the disease affects both men and women, 80% of those afflicted are women between 20 and 55 years of age.

Basic Pathophysiology

Fibromyalgia is a clinical syndrome (rather than a pathophysiological diagnosis) that historically was identified on the basis of the person's having certain specific tender points, but recently the diagnostic criteria have shifted toward chronic widespread pain and symptoms. The American College of Rheumatology (ACR) 2010 criteria for diagnosis are based on two subjective scales: the Widespread Pain Index (WPI) and the Symptom Severity scale (SS). The diagnosis is satisfied if the individual meets the following criteria:

- WPI ≥7 *and* SS ≥5, or
- WPI of 3 to 6 *and* SS ≥9

Symptoms must have been present at a similar level for at least 3 months, and the individual can have no disorder that would otherwise explain the pain.

The WPI is a score from 0 to 19, based on number of areas in the body with pain during the last week.

Areas counted in the score include the chest, abdomen, neck, upper back, lower back, right or left shoulder girdle, hip (buttock, trochanter), jaw, upper arm, lower arm, upper leg, and lower leg.

The SS score considers the severity of three symptoms over the past week:

- Fatigue
- Waking unrefreshed
- Cognitive symptoms

Severity of symptoms is rated using a symptom severity scale, whose parameters are as follows:

0 = no problem

1 = slight or mild problems, generally mild or intermittent

2 = moderate, considerable problems, often present or at a moderate level or both

3 = severe, pervasive, continuous, life-disturbing problems

The SS scale score is the sum of the severity of the three symptoms (fatigue, waking unrefreshed, cognitive symptoms) plus the extent (severity) of somatic symptoms in general, which are graded on the same scale. The final score is between 0 and 12.

The etiology of FM remains elusive, though theories on the causal mechanisms of FM include muscle abnormalities (e.g., muscle microtrauma, local muscular ischemia), neuroendocrine and autonomic system regulation disorders (e.g., stage IV sleep disturbance, hypothalamic–pituitary–adrenal [HPA] axis disturbance, diminished local muscular glucose metabolism), central augmentation of pain processing (serotonin metabolism abnormality, decreased activity of the descending antinociceptive pathways), and genetic predisposition. It is thought that certain events precipitate FM, such as

- physical trauma,
- emotional trauma,
- brain chemistry disturbances, and
- altered sleep patterns.

These or some other inciting events are believed to trigger the onset of FM by activating some underlying physiological abnormality to which the individual with FM is genetically susceptible. Like chronic fatigue syndrome (see chapter 22), FM is thought to be due to a disorder in central afferent processing (i.e., an altered integrative perception of afferent stimuli).

Chronic pain researchers are increasingly interested in the role of descending antinociceptive pathways. A modest body of research suggests that exercise induces a mild analgesia, or dampening of pain, which acts through the central nervous system's pain-inhibitory mechanisms. Dampening of pain is likely an adaptive response that allows an individual to not be deterred by the mild musculoskeletal trauma associated with exercise. A variety of studies suggest that these neural pathways may be dysfunctional in people who have widespread chronic pain syndromes (and likely chronic fatigue). Dysfunctional responses to stress as well as psychological and emotional problems may be involved in mediating this antinociceptive system. There are no good studies that connect this phenomenon with exercise (response or training), but this theory has the strength of unifying the psychiatric and musculoskeletal symptomatology of FM, and it would explain the interlocking roles of antidepressant medications, aerobic and resistance exercise, and cognitive behavior therapy with some emphasis on stress management and catastrophizing or fear avoidance.

Management and Medications

Because FM is a syndrome, conditions that have similar symptoms need to be excluded before the diagnosis can be properly made. There is no known cure, and a growing consensus is that the best approach to treatment involves the combination of pharmacologic and nonpharmacologic interventions. Recommended are multidisciplinary team efforts that use

- patient or client education (including stress management and relaxation),
- aerobic and strength exercise training,
- cognitive behavioral therapy, and
- pharmacologic treatment.

Pharmacologic interventions, including the following, are generally chosen to reduce symptoms and improve normal functioning:

- Antidepressants
  - Tricyclic antidepressants (e.g., amitriptiline)
  - Serotonin–norepinephrine reuptake inhibitors (e.g., duloxetine)
- Chronic pain medications (classified as anticonvulsants)
  - Pregabalin, gabapentin

Hypnotics, or medications to help promote sleep, are also commonly used for persons with FM.

Effects on the Exercise Response

The symptoms associated with FM directly and indirectly affect the acute response to exercise. Pain associated with basic activities of daily living, general fatigue, and altered perception of exertion contribute to the tendency for people with FM to become sedentary and deconditioned. Musculoskeletal symptoms interfere with the timing and type of activities that individuals with FM will do:

- Morning stiffness
- Exaggerated delayed-onset muscle soreness (DOMS)
- Poor recovery from exercise
- Difficulty with use of the arms in elevated positions

Eccentric muscle contractions, sustained overhead activities, and vigorous or high-impact activities are poorly tolerated because they evoke pain.

Effects of Exercise Training

Exercise training is safe and effective at producing the same general benefits for individuals with FM as it does for apparently healthy individuals (i.e., improved cardiorespiratory function, reduced risk for coronary artery disease, decreased cardiovascular mortality and morbidity, and improved psychosocial functioning). That many individuals with FM are sedentary and deconditioned is evidenced by the fact that many subjects who achieved some of these health-related benefits were in aerobic exercise studies involving only 20 min a day or just 2 days a week. Strength training studies typically involved two or three sessions a week using 8 to 12 repetitions per exercise. Strength and aerobic exercise have been combined, often in a 2 days + 2 days format. Strength training has consistently shown improvement in strength in the FM population comparable to that in healthy sedentary individuals. Less clear, however, is the specific exercise dose needed to improve FM symptoms. Exercise training does bring specific FM-related benefits, including these:

- Reduced number of tender points
- Decreased pain at tender points (reduced myalgic score)
- Decreased general pain
- Improved sleep and less fatigue
- Fewer feelings of helplessness and hopelessness
- More frequent and meaningful social interactions
- Lessened impact of the disease on daily activities

Fibromyalgia is distinctly different than chronic fatigue (see chapter 22), but individuals with either of these conditions are similarly vulnerable to "overdoing" exercise, which can result in non-adherence and dropping out of the exercise program. So in fibromyalgia it is better to progress very slowly and err on the side of doing "too little" on any one day. Better to have the person exercising again tomorrow, than to see them drop out of the program.

Recommendations for Exercise Testing

For most individuals with FM, follow the Basic *CDD4* Recommendations.

Individuals with FM tend to be sedentary and deconditioned, with poor ability to sustain exercise for extended periods. Therefore, tests of physical functioning and muscular endurance may be useful when one is developing an exercise prescription for individuals with FM.

Aerobic exercise is usually symptom limited rather than metabolically limited in the FM population, and thus individuals with FM tend to reach peak rather than maximal aerobic efforts. This issue and a lack of motivation tend to favor submaximal testing and graded exercise protocols with smaller increments (i.e., Naughton,

Balke-Ware) or low-level ramp protocols (10-25 W/increment). Walking or cycle ergometry is recommended, although individuals with gluteal tender points may have pain sitting on a standard ergometer and may do better with a recumbent cycle ergometer. Exercise testing should take place on a day when no other activities are scheduled to minimize its impact on later function, because most people with FM will be worn out afterward.

The sit and reach and other simple flexibility tests can identify specific areas that would benefit from routine stretching, and the Basic *CDD4* Recommendations for exercise assessments are appropriate for people with FM.

Consider the following factors before doing an exercise test in a person with FM:

- Subjects may be extremely deconditioned.
- Perception of increased exertion may alter rating of perceived exertion.
- Avoid early morning exercise because of morning stiffness.
- Test on a day when other activities are not scheduled.
- Postexertional muscle pain may be severe 24 to 48 h after testing.

Recommendations for Exercise Programming

For people with FM, follow the Basic *CDD4* Recommendations, and advance as tolerated to the ACSM Guidelines for individuals with FM.

The ACSM Guidelines for individuals with FM are consistent with the Basic *CDD4* Recommendations, as both recommend starting with light-intensity activities fewer than 3 days per week and progressing very slowly, as tolerated by each individual, to higher frequency and durations with a long-term goal of 150 min/week. The main difference between the Basic *CDD4* Recommendations and the ACSM Guidelines for people with FM is that *CDD4* is more oriented toward physical functioning than standard FITT (frequency, intensity, time, and type) programming.

Methodological inconsistencies and high attrition rates among participants in published studies on exercise and FM make it difficult to provide definitive recommendations for exercise programming in this population. The main goal of exercise training in this population is generally not to enhance cardiopulmonary fitness and lower cardiometabolic risk but rather to restore and maintain physical functioning. Because individuals with FM tend to be physically inactive and fearful of the pain associated with exertion, they may be unwilling or unable to participate at the same level as healthy individuals. Accordingly, they need to begin at lower levels of intensity, duration, and frequency and progress in a more gradual fashion than typically prescribed. The ultimate goal for individuals with FM is 150 min/week or more of moderate physical activity, though people who are successful in achieving that often take many months to get there.

Exercise programming for individuals with FM should therefore start with low- to moderate-intensity aerobic activities. Obtaining physical functioning assessments in the Basic *CDD4* Recommendations helps individualize a program. Programs should

- consist of no-impact or low-impact activities and
- minimize eccentric contractions.

Individuals with FM often have shoulder impingement and rotator cuff weakness and therefore do not tolerate sustained overhead activities. Thus, exercise programs should emphasize the lower body (e.g., water exercise, walking, cycling). Warm-water aquatic exercise programs may yield greater benefits on mood while conferring lower risk of adverse musculoskeletal effects than land-based programs; thus such programs should be used whenever possible.

Tai chi and yoga have also been shown to be beneficial and should be considered, particularly if instructors are available who are trained in programs for people with chronic pain conditions. These instructors tend to be very supportive and nurturing, which is helpful for fear-avoidance issues in individuals with FM. Integrative functional activities, like tai chi, yoga, or the exercises in the Basic *CDD4* Recommendations, might be favored over more unidimensional programs such as stretching or strength training, because limited long-term benefits regarding pain, tenderness, and global function have been reported for stretching or strength training alone.

Adherence rates among individuals with FM is often poor, with a high dropout rate. Supervised exercise or group exercise may be helpful at improving adherence, particularly if the supervisor can ensure that patients or clients do not overexert themselves. The group atmosphere should focus on maintaining warm, supportive attitudes and enjoyment. Offering other symptom management techniques, such as support groups and educational programs, can be helpful for maintaining engagement. Cognitive behavioral therapy with an emphasis on increasing self-efficacy may be especially helpful, since self-efficacy has been found to be the most powerful construct in predicting exercise initiation and maintenance in the FM population.

Integration Into a Medical Home Model

Often characterized as a "difficult" or "needy" population, individuals with FM may require extra time and attention, but this is important in order to obtain good adherence to the program. Clients with FM are usually hesitant to exercise or increase the intensity of activity due to fear avoidance of postexercise malaise and symptoms. Offering choices about the mode and timing of exercise, offering simple methods of assessing progress such as the use of a pedometer, and yielding to weather-related complaints may help to maximize participation in and adherence to the exercise program. Because symptoms may temporarily get worse after starting a program or increasing the intensity, it is important for participants to budget exercise in conjunction with other activities and obligations. Concomitantly, the importance of self-efficacy to exercise adoption and adherence reinforces the need to begin with exercise activities that are easily mastered and that progress slowly toward those of greater intensity or frequency. A supportive and understanding environment will also foster adherence to programs of exercise and ultimately the use of a comprehensive treatment program.

Take-Home Message

Keep these special considerations in mind when working with people who have FM:

- Gluteal trigger points may limit usefulness of cycling with a standard cycle.
- Avoid morning exercise because of morning stiffness.
- Supervision or group exercise may increase adherence.
- Clients may experience an increase in symptoms.

- Avoid sustained overhead activities.
- Minimize eccentric movements.
- Always remember that persons with FM often experience pain with stimuli that cause others to feel only mild pressure, so be gentle.

Suggested Readings

Arnold LM, Clauw DJ, Dunegan LJ, Turk DC. A framework for fibromyalgia management for primary care providers. *Mayo Clin Proc.* 2012;87:488-496.

Bidonde J, Busch AJ, Webber SC, Schachter CL, Danyliw A, Overend TJ, Richards RS, Rader T. Aquatic exercise training for fibromyalgia. *Cochrane Database Syst Rev.* 2014;10:CD011336. doi: 10.1002/14651858. CD011336.

Cazzola M, Atzeni F, Salaffi F, Stisi S, Cassisi G, Sarzi-Puttini P. Which kind of exercise is best in fibromyalgia therapeutic programs? A practical review. *Clin Exp Rheumatol.* 2010;28(6 Suppl 63):S117-S124.

Gowans SE, deHueck A. Pool exercise for individuals with fibromyalgia. *Curr Opin Rheumatol.* 2007;19:168-173.

Häkkinen A, Häkkinen K, Hannonen P, Alen M. Strength training induced adaptations in neuromuscular function of premenopausal women with fibromyalgia: comparison with healthy women. *Ann Rheum Dis.* 2001;60:21-26.

Jones KD, Burckhardt CS, Clark SR, Bennett RM, Potempa KM. A randomized controlled trial of muscle strengthening versus flexibility training in fibromyalgia. *J Rheumatol.* 2002;29:1041-1048.

Kingsley JD, Panton LB, Toole T, et al. The effects of a 12-week strength-training program on strength and functionality in women with fibromyalgia. *Arch Phys Med Rehabil.* 2005;86:1713-1721.

McCain GA, Bell DA, Mai FM, Halliday PD. A controlled study of the effects of a supervised cardiovascular fitness training program on the manifestations of primary fibromyalgia. *Arthritis Rheum.* 1988;31:1135-1141.

Nielens H, Booisset V, Masquelier E. Fitness and perceived exertion in patients with fibromyalgia syndrome. *Clin J Pain.* 2000;16:209-213.

Nijs J, Kosek E, Van Oosterwijck J, Meeus M. Dysfunctional endogenous analgesia during exercise in patients with chronic pain: to exercise or not to exercise? *Pain Physician.* 2012;15:ES205-ES213.

Smith HS, Barkin RL. Fibromyalgia syndrome: a discussion of the syndrome and pharmacotherapy. *Am J Ther.* 2010;17:418-439.

Smith HS, Harris R, Clauw D. Fibromyalgia: an afferent processing disorder leading to a complex pain generalized syndrome. *Pain Physician.* 2011;14:E217-E245.

Wolfe F, Clauw DJ, Fitzcharles MA, Goldenberg DL, Katz RS, Mease P, Russell AS, Russell IJ, Winfield JB, Yunus MB. The American College of Rheumatology preliminary diagnostic criteria for fibromyalgia and measurement of symptom severity. *Arthritis Care Res (Hoboken).* 2010;62:600-610.

Additional Resources

Evidence-Based Guidelines

Brosseau L, Wells GA, Tugwell P, Egan M, et al. Ottawa Panel evidence-based clinical practice guidelines for aerobic fitness exercises in the management of fibromyalgia: part 1. *Phys Ther.* 2008;88:857-871.

Brosseau L, Wells GA, Tugwell P, Egan M, et al. Ottawa Panel evidence-based clinical practice guidelines for strengthening exercises in the management of fibromyalgia: part 2. *Phys Ther.* 2008;88:873-886.

Carville SF, Arendt-Nielsen S, Bliddal H, et al. EULAR evidence based recommendations for the management of fibromyalgia syndrome. *Ann Rheum Dis.* 2008;67:536-541.

Wood PB. Symptoms, diagnosis, and treatment of fibromyalgia. *Virtual Mentor.* 2008;10(1):35-40.

Suggested Websites

American College of Rheumatology, www.rheumatology.org [Accessed November 8, 2015].

Centers for Disease Control and Prevention, http://www.cdc.gov/arthritis/basics/fibromyalgia.htm [Accessed November 8, 2015].

Mayo Clinic, http://www.mayoclinic.org [Accessed November 8, 2015].

National Fibromyalgia Association, www.fmaware.org [Accessed November 8, 2015].

Hemostasis Disorders

Hemostasis disorders, often colloquially referred to as bleeding and clotting disorders, are a group of conditions that result when the blood does not clot properly. Hemostasis involves a complex balance of platelet aggregation, coagulation, and clot dissolution known as fibrinolysis. These three processes involve many proteins and enzymes, continually working to maintain the integrity of the vascular system. In addition, the inflammatory response is an essential component of the initiation and resolution of hemostasis.

When a vessel is torn or cut, the open ends spasm, reducing blood flow and allowing blood platelets to aggregate at the site of injury. This aggregate of platelets forms a plug in the damaged area of the vessel. Simultaneously, proinflammatory substances released from the damaged tissue initiate a series of reactions known as the extrinsic pathway of the coagulation cascade. The coagulation cascade consists of proteolytic enzymes called coagulation factors that are suspended in blood. Upon damage to the vessel, these activated factors precipitate to form a web-like tangle of a protein called fibrin. The combination of platelet aggregation and the activated coagulation cascade produces a durable clot at the site of injury. As injured tissue heals, the fibrin in the clot is gradually disintegrated by the enzyme plasmin in a process known as fibrinolysis.

Diseases of hemostasis present clinically either as inadequate clotting (hemorrhage) or excessive clotting (thrombosis). Abnormalities occur in each of the three phases: platelet plug formation, fibrin formation and development of a clot, and fibrinolysis and resolution of the clot.

Physical activity causes trauma to the circulatory system, and thus bleeding and clotting are normal consequences of exercise. The blood hemostatic system serves to counteract this trauma, minimizing vascular damage and preventing blood loss. While the literature does not currently indicate effects of hemostasis disorders on the exercise response, proper preparation and precautions may be necessary to avoid exacerbating the pathological symptoms of hemostasis disorders while gaining the myriad benefits of exercise training.

Basic Pathophysiology of Hemorrhagic Disorders

Coagulation disorders generally cause hemorrhage, mainly due to deficiencies in clotting factors, which can be congenital or acquired. Hemorrhagic disorders are almost always inherited, although in rare cases they can develop later in life if the body forms antibodies that fight against the blood's natural clotting factors. A full discussion of the coagulation cascade is beyond the scope of this book, and we note only the most common conditions. Hemophilia (factor VIII deficiency) is perhaps the bleeding disorder most people know about, although it is relatively rare. The Centers for Disease Control and Prevention (CDC) estimates that 20,000 individuals, mostly males, in the United States have hemophilia. Many more people (1% of the general population) are affected by von Willebrand disease, the most common bleeding disorder in America. Another common congenital bleeding disorder is factor IX deficiency (Christmas disease). Symptoms of disorders in the coagulation cascade may include the following:

- Easy bruising
- Bleeding gums
- Heavy bleeding from small cuts or dental work
- Unexplained nosebleeds
- Heavy menstrual bleeding

Clotting factor deficiencies can be *acquired,* as opposed to congenital, by autoimmune illnesses and some cancers such as lymphoma. In addition, many people have medically induced factor deficiencies from taking anticoagulant drugs. Because of medical technologies such as stents and artificial heart valves, a huge number of people are on anticoagulants today. The purpose of medical anticoagulation is to prevent blood clots in persons who are at high risk for forming them. Fortunately, medical anticoagulation rarely causes spontaneous bleeding, although there is increased risk of bleeding after an injury, and such patients cannot undergo surgical or medical procedures without having the anticoagulation reversed.

Platelet disorders are another cause of inappropriate bleeding; the most common platelet bleeding disorder is *thrombocytopenia,* which is an insufficient number of platelets. Normally, platelet counts range from 120,000 to 600,000/mm³, and bleeding from platelet deficiency typically occurs when platelet counts fall below 50,000/mm³. The hemorrhages of thrombocytopenia are called *petechiae;* these are pinpoint-sized red spots in the skin that are too small to feel. Petechiae resemble a rash but are in fact small pools of blood that have leaked out of broken capillaries. Platelet disorders primarily cause bleeding, though they can also cause inappropriate clotting by aggregating too much. The most common disorders associated with platelet aggregation are myocardial infarction and stroke, but these conditions involve underlying vascular pathology. Nonetheless, normal arteries very rarely become occluded simply because of hyperactive platelet aggregation.

Basic Pathophysiology of Thrombotic Disorders

Although coagulation disorders generally cause excessive bleeding, some disorders in the coagulation cascade involve anticoagulation factors designed to keep coagulation in check. Such disorders cause excessive clotting because coagulation is insufficiently countered by anticoagulant activity. Congenital deficiencies of the anticoagulant system include deficiency of protein S or protein C, but some patients or clients acquire a thrombotic disorder when their immune systems starts making antibodies to their body's own coagulation proteins.

Abnormalities in fibrinolysis result in inadequate dissolution of the fibrin in clots, resulting in excessive clotting. Common abnormalities include deficiency in tissue plasminogen activator (tPA) or overexpression of plasminogen activator inhibitor

(PAI-1). The interaction of the platelet plug with the coagulation cascade potentiates the binding of fibrinogen and receptors known as GP IIb/IIIa receptors, as well as increasing P-selectin activity. Normal activation of GP IIb/IIIa activity requires binding with von Willebrand factor.

In summary, these are the general clinical manifestations of bleeding disorders:

- Thrombocytopenia: petechial bleeding
- von Willebrand disease: prolonged bleeding after minor trauma, gastrointestinal bleeding, heavy menses
- Hemophilia: hemarthrosis (bleeding into joints) or into muscle, retroperitoneal bleeding
- Medical anticoagulation: bruising easily, gastrointestinal bleeding

General clinical manifestations of clotting disorders are as follows:

- Thrombocytosis: deep venous thrombosis (DVT)
- Anticoagulant deficiencies: arterial occlusion and tissue infarction, DVT

Management and Medications

Bleeding disorders such as hemophilia are complex conditions. Good medical care can help prevent some serious problems and is best accomplished by a team that specializes in the care of people with bleeding disorders. The use of medications for bleeding disorders is determined by whether the problem predisposes to bleeding or to clotting, and also by the cause of the disease. Use of anticoagulants has become very common, almost ubiquitous. The reason is that the coagulation cascade is a well-defined target for modern pharmaceuticals, with broad clinical applicability. Recombinant biologic factors are being used more commonly as well, although less so because these medications are expensive. Here we consider persons with platelet disorders, then those with coagulation factor disorders, and finally those with prosthetic heart valves.

Thrombocytopenia

In persons with low platelets caused by an overly rapid destruction of platelets (most often by the immune system), the goal of treatment is to decrease the rate of platelet destruction with immunosuppressive medicines, such as prednisone, and sometimes through splenectomy. Platelet transfusion works only for a very short time and is therefore limited to hospitalized patients.

Thrombocytosis

In people with high platelets resulting from overproduction of platelets, the treatment is to decrease the rate of platelet formation, usually through chemotherapy.

Coagulation Factor Deficiencies

Purified factor transfusions are available for some factor deficiencies. Unfortunately, as with platelet disorders, such infusions are temporary because the factor gets consumed and individuals can't make enough on their own. Mild von Willebrand disease can be treated to increase the level of von Willebrand factor. In hemophilia, regular home infusions of factor VIII, up to three times a week, are now commonly used in young persons with hemophilia to prevent hemarthrosis and intramuscular bleeds. This approach is expensive but is less costly than standard care.

Medical Anticoagulation

Persons at risk for inappropriate clot formation because of mechanical heart valves, dilated hearts, or DVT require some form of anticoagulation. Inappropriate clots are highly dangerous because they can either grow in place and block blood flow or break off and lodge downstream in blood vessels critical to vital organs such as the brain or lungs. For this reason, such individuals are given anticoagulants (table 24.1 lists common medications). Historically, warfarin and heparin derivatives have been the most common drugs, but recently there have been several new drugs aimed at preventing iatrogenic thrombosis. These medications are commonly called blood thinners, though they do not alter the viscosity of blood.

There are some recommendations on the use of medications in athletes who have had a clotting disorder, such as DVT caused by protein C or protein S deficiencies or both. Such deficiencies in persons who are genetically heterozygous are usually not detected unless their family history is known or until they present with a clot. Because such cases are relatively rare even though these deficiencies are common in the general population, prophylaxis with anticoagulants has not been recommended because this would probably cause more harm than benefit.

Treatment as prophylaxis against future episodes is commonly recommended for individuals who have had a clot. Persons with homozygous factor deficiencies are treated with the appropriate factor replacement or anticoagulant (e.g., warfarin or heparin).

A number of case reports exist regarding DVT in athletes. Deep venous thromboses are potentially life threatening because of their potential to break loose and create a pulmonary embolus. For this reason, a break in training is indicated in an athlete who has recently developed a DVT. The duration of convalescence before resumption of sporting activities has not been objectively studied, but most sources advise an extended layoff of 6 months or more, until the thrombus has clearly resolved on vascular imaging studies. If a female athlete who has had a DVT is taking medications that increase her risk of DVT (i.e., oral contraceptives), she should be provided an alternative form of contraception.

Effects on the Exercise Response

Disorders of bleeding and clotting have little effect on the exercise response, but exercise markedly alters function of the thrombotic and fibrinolytic pathways. Moreover, the preponderance of evidence suggests that exercise has dose-dependent effects on platelet aggregation, thrombosis, and fibrinolysis. Therefore, it is less important to know how these disorders affect the exercise response than it is to know how exercise can trigger bleeding or clotting in persons with these disorders.

Exercise causes a complex mechanical and biochemical stimulation of platelet aggregation and fibrin formation that promotes the formation of a thrombus. Sympathetic stimulation (i.e., release of epinephrine) activates platelet aggregation and binding of fibrinogen receptors on platelets. In addition, shear stress from increased blood flow stimulates both the endothelium-dependent and platelet-dependent activation of the coagulation cascade.

Exercise also causes a transient increase in fibrinolysis, as reflected by an increase in fibrinolytic enzymes and serum fibrin degradation products. The risk of a clinically meaningful bleed or clot is

Table 24.1 Commonly Used Anticoagulant Medications*

Generic name	Mechanism of action	Adverse events
Aspirin	Irreversibly inhibits thromboxane A2, inhibiting platelet aggregation	• Bleeding, bruising, anemia • Nausea, vomiting, fever • Tinnitus
Coumadin	Blocks liver synthesis of vitamin K-dependent coagulation factors	• Bleeding, bruising, anemia • Nausea, vomiting, diarrhea • Hepatotoxicity • Hypersensitivity reactions
Heparin	Glycosaminoglycan anticoagulants inhibit multiple steps in fibrin formation and coagulation cascade	• Thrombocytopenia • Bleeding • Thrombotic thrombocytopenic purpura • Hypersensitivity reactions
Enoxaparin Dalteparin Tinzaparin	Low molecular weight heparins mainly act through enhancing binding of antithrombin to factor Xa and thrombin	• Bleeding, bruising, anemia • Nausea, vomiting, fever • Hepatotoxicity
Aminocaproic acid	Monoaminocarboxylic acids inhibit fibrinolysis by inhibition of plasminogen activators	• Headache • Abdominal pain • Nausea, vomiting, diarrhea • Hypersensitivity reactions • Rhabdomyolysis, weakness
Clopidogrel Ticagrelor Prasugrel	Class of platelet aggregation inhibitors that reversibly block $P2Y_{12}$ in activation of GP IIb/IIIa complexes	• Combined adverse effects (individual agents vary) • Bleeding • Nausea, vomiting, diarrhea, rash • Pain (headache, abdominal, back, joints) • Agranulocytosis, aplastic anemia • Thrombotic thrombocytopenic purpura
Anagrelide	Platelet-reducing agent for severe thrombocytosis; mechanism of action not known, believed to reduce platelet formation	• Headache • Abdominal pain • Nausea, vomiting, diarrhea, gas • Rash, urticaria
Rivaroxaban Apixaban Fondaparinux	Factor Xa inhibitors directly block coagulation cascade at factor Xa, inhibiting thrombin formation	• Bleeding, hematomas • Back pain • Hypersensitivity reactions
Abciximab Tirofiban Eptifibatide	Glycoprotein IIb/IIIa inhibitors inhibit platelet aggregation by blocking binding of fibrinogen and von Willebrand factor to glycoprotein IIb/IIIa on platelets	• Bleeding • Hypotension • Pain (headache, abdominal, pelvic) • Hypersensitivity reactions • Thrombocytopenia
Argatroban Bivalirudin Dabigatran	Direct thrombin inhibitors bind to circulating and clot-bound thrombin to block thrombus formation	• Bleeding • Pain (headache, back, abdominal, pelvic) • Nausea, vomiting, dyspepsia • Hypersensitivity reactions • Cardiac complications
Leperudin Desirudin Antithrombin-III (recombinant) Antithrombin-III (human)	Act primarily by inactivating thrombin, thrombin-III	• Bleeding • Pain (headache, back, abdominal, pelvic) • Nausea, vomiting, dyspepsia • Hepatotoxicity

*Does not include isolated and recombinant human factors, orphan drugs.

therefore determined by a change in the balance of thrombogenesis and fibrinolysis. If thrombogenesis is insufficient to match the vascular trauma of exercise, the risk of a bleed increases. In contrast, if exercise causes excessive thrombogenesis, the risk of an arterial or venous clot increases. Likewise, an insufficient increase in fibrinolysis would also increase the risk of a clot.

Despite recent advances in our knowledge of the exercise effects on various aspects of coagulation and fibrinolysis, the data are not sufficient to fully characterize the balance of these systems during exercise. Table 24.2 summarizes expert opinion regarding these independent effects.

After exercise, thrombotic pathway activity remains increased long after the fibrinolytic activity has returned to baseline. Interestingly, the incidence of myocardial infarction is increased in the hours shortly after physical activity. Many experts believe that vigorous exercise induces a transient increase in thrombogenesis and thereby increases risk of myocardial infarctions.

In addition to the effect of exercise on hemostasis, it is important to be familiar with the consequences of hemorrhages and thrombosis, as well as the effects of these complications on the exercise response.

About 20% of persons with hemophilia experience exercise-related intra-articular bleeds. The most commonly affected joints are the knees and hips. In addition, it is common for persons with hemophilia to have intramuscular bleeds, typically in the back and lower extremity. Both intra-articular and intramuscular bleeds cause contractures that lead to loss of range of motion, balance, and proprioception skills.

Another common circumstance occurs in a person who has suffered a DVT and subsequently resumes physical activity. Typically, DVT alters venous return in the affected limb, and the muscles distal to the DVT have altered venous drainage. This series of events increases venous pressure, reduces muscle pump effects,

and causes extravasation of fluid (edema), which is commonly painful or at least uncomfortable. This discomfort is frequently sufficient to limit exercise tolerance, with the patient or client stopping exercise due to pain.

Effects of Exercise Training

The effects of exercise training on the blood hemostasis system are complex and not well understood. It is well known that regular exercise protects against myocardial infarction and sudden cardiac death during physical activity, but this does not mean that clotting is causally related to this observation. Exercise training studies examining the balance of thrombosis and fibrinolysis have not yielded consistent results, in part because of different responses in the various study populations.

Markers of hemostasis appear to correlate with several risk factors for heart disease that are known to be altered with exercise training. Inflammation is an important aspect of atherogenesis, and recent data suggest that C-reactive protein (a marker of inflammation) is increased in coronary artery disease. Exercise training usually reduces resting blood pressure and also is usually associated with reduction of C-reactive protein, so the compound effects of exercise training on all the factors that influence hemostasis are extremely complex. The effects of exercise training on the thrombotic and fibrinolytic cascades in persons with disorders of these cascades have not been well researched.

Persons with hemophilia benefit from regular exercise, but exercise puts them at some risk of bleeding into joints and developing joint contractures. Neither aerobic nor strength training alters the underlying disorder of hemophilia, but non–weight-bearing aerobic exercise, as well as strength and flexibility training, can be of immense functional and psychological benefit.

Table 24.2 Hemostatic Effects of Acute Exercise

Exercise intensity	Platelet aggregation	Coagulation	Fibrinolysis
Light (<60% max heart rate [HR])	No effect	No effect	Increased
Moderate (60-80% max HR)	Decreased	No effect	Increased
Hard (>80% max HR)	Increased	Increased	Much increased

Recommendations for Exercise Testing

Use the ACSM Guidelines when instructing someone with bleeding or clotting disorders, and consider the following additional recommendations:

- Use range of motion testing to manage flexibility exercises for persons with hemophilia.
- No high-resistance strength testing should be performed by persons with low platelet counts due to the risk of intracranial bleeding.
- Presence of an acute DVT is a contraindication to exercise.
- Avoid exercise modes that risk falls or impacts in persons taking an anticoagulant.

Recommendations for Exercise Programming

In this very diverse and complex set of disorders, the special considerations for exercise programming must be considered according to the nature of the condition.

- Following the Basic *CDD4* Recommendations is probably quite safe in the vast majority of persons with platelet, coagulation, or fibrinolysis disorders.
- Otherwise, follow the ACSM Guidelines in all patients or clients, except as noted in the section Take-Home Message.

The consequences of a hemorrhagic stroke are devastating, often fatal or ending in prolonged institutionalization for mechanical ventilation, so erring on the conservative side seems prudent.

Thrombocytopenia

There is no consensus on what level of thrombocytopenia merits a contraindication for high-intensity exercise (of any form). The relationship between risk and platelet levels thus requires clinical judgment by the exercise specialist and medical team. These are the major mechanisms of

increased risk for exercise in persons with a low platelet count (or coagulation factor deficiencies):

- Bleeding from trauma
- Bleeding from high exertional blood pressure

Taking evidence from sedentary persons as a guide, bleeding from a low platelet count is rarely a problem in sedentary persons unless the count is well below $100,000/mm^3$. Occasionally, intracranial (inside the skull) bleeds occur spontaneously when the platelet count is markedly low ($< 20,000/mm^3$). There are no controlled data on exercise in clients with low platelets, but with an eye to being more conservative than in sedentary individuals, vigorous exercise is probably contraindicated when platelet counts are below $50,000/mm^3$. Since lifting heavy weights dramatically increases blood and intracranial pressures, one might consider increasing this to between 50,000 and $100,000/mm^3$ for high-intensity exercise. In persons with platelet counts between 20,000 and $50,000/mm^3$, exercise should probably be limited to use of elastic bands, stationary cycles, range of motion, and ambulation.

Coagulation Factor Deficiencies

Before the introduction of prophylactic clotting factor medication, children with hemophilia were discouraged from physical activity due to the risk of bleeds. Swimming or stationary cycling was historically recommended, but more and more sports physicians have been recommending all kinds of activity except for high-contact sports, coupling exercise with factor VIII replacement prophylaxis. Outdoor cycling is a non–weight-bearing activity on joints but risks trauma, though data have shown that up to 55% of Germans with hemophilia A commute by bicycle. It is worth noting that these recommendations are based on case report experience, not randomized trials.

In children and adolescents with hemophilia, vigorous physical activity is associated with a moderate increase in risk of bleeding. Indeed, about one in five youth with hemophilia report a history of hemarthrosis even with factor VIII therapy. However, the increased relative risk of bleeding is transient, and the absolute increase in risk of major bleeds associated with physical activity is thought to be small.

Accordingly, in persons with hemophilia or von Willebrand disease, the exercise mode should

- minimize joint trauma and weight bearing,
- emphasize moderate intensity to balance risk and benefit, and
- consider any limitations imposed by preexisting joint contractures from prior bleeds.

Of course, flexibility exercises may help restore joint mobility and proprioception in persons with hemophilia who are affected by contractures.

The optimal frequency and duration of training are not known, but most investigators have used common strength training paradigms (e.g., 30 min, three times a week, 8-15 repetitions maximum, emphasis on core strength and lower extremities). Cardiovascular and muscular adaptability are not known but are presumably normal.

Medical Anticoagulation

Medical anticoagulation is used in sufficiently different circumstances that they will be considered separately. In some conditions like a DVT, anticoagulation is short-term, but in others it is indefinite (presumably for the rest of the person's life).

Deep Venous Thrombosis

Exercise training should be curtailed in individuals who have had a recent DVT. The necessary duration of exercise restriction has not been thoroughly studied. Historically, resumption of training has been advised sometime after 3 to 9 months of treatment, with optional follow-up vascular studies to demonstrate resolution of the thrombus. Recent studies have challenged that notion and have resumed light walking as early as 4 weeks after diagnosis. Theoretical rationale for early return to walking includes the notion that increased flow may help with recanalization or growth of collateral vessels (or both) and thus reduced muscular edema and symptoms in the long term. Larger studies continue to examine this theory, and until these studies are complete the exercise therapist is advised to proceed with caution in order to avoid embolic complications.

Other Conditions of Medical Anticoagulation

In general, pharmacologic anticoagulation for purposes of cardiovascular disease (atrial fibrillation, coronary or cerebrovascular disease, prosthetic heart valves, arterial stents and other intravascular prostheses) is not a contraindication for exercise, most especially not for activities outlined in the Basic *CDD4* Recommendations. The reader is advised to see the relevant chapters of this book for information on these specific conditions.

Integration Into a Medical Home Model

Surveys indicate that the majority of people with bleeding disorders do not meet national physical activity guidelines. These individuals can experience significant gains in quality of life through enhancement of aerobic and muscular fitness. Although more studies are needed, professionally designed and supervised individualized exercise programs are feasible, safe, and beneficial for people with bleeding and clotting disorders. Comprehensive medical management should include referral to physical therapy or an exercise specialist who has expertise in people with these disorders.

Many individuals with bleeding disorders do not exercise adequately because of joint pain and swelling. Regular physical activity (e.g., swimming, a commonly prescribed exercise mode) can improve the ability to accomplish daily tasks through enhancement of cardiorespiratory function, muscle strength, balance, and proprioception.

Take-Home Message

The following are special considerations that one should take into account when working with persons who have bleeding or clotting disorders:

- Aspirin and other nonsteroidal anti-inflammatory drugs (e.g., ibuprofen) render platelets partially inactive, so their use is dangerous and should be avoided in individuals who have concomitant bleeding or platelet disorders.
- Persons who have any bleeding disorder or who are receiving medical anticoagulation are advised to avoid circumstances in which collisions are possible, for example bicycle racing and contact sports such as football, hockey, and basketball.
- Persons with low platelet counts are advised to avoid high-intensity exercise.
- Persons with hemophilia A are advised to use prophylactic factor VIII therapy as part of their exercise program.

Suggested Readings

Blamey G, Forsyth A, Zourikian N, Short L, Jankovic N, De Kleijn P, Flannery T. Comprehensive elements of a physiotherapy exercise program in hemophilia—a global perspective. *Haemophilia.* 2010;16(Suppl 5):136-145.

Broderick CR, Herbert RD, Latimer J, Barnes C, Curtin JA, Mathieu E, Monagle P, Brown SA. Association between physical activity and risk of bleeding in children with hemophilia. *JAMA.* 2012;308:1452-1459.

Coppola L, Grassia A, Coppola A, Tondi G, Peluso G, Mordente S, Gombos G. Effects of a moderate-intensity aerobic program on blood viscosity, platelet aggregation and fibrinolytic balance in young and middle-aged sedentary subjects. *Blood Coagul Fibrinolysis.* 2004;15:31-37.

Gomis M, Querol F, Gallach JE, González LM, Aznar JA. Exercise and sport in the treatment of hemophilic patients: a systematic review. *Haemophilia.* 2009;15:43-54.

Hilberg T, Herbsleb M, Puta C, Gabriel HHW, Schramm W. Physical training increases isometric muscular strength and proprioceptive performance in haemophilic subjects. *Haemophilia.* 2003;9:86-93.

Kahn SR, Azoulay L, Hirsch A, Haber M, Strulovitch C, Shrier I. Acute effects of exercise in patients with previous deep venous thrombosis. *Chest.* 2003;123:399-405.

Lockard MM, Gopinathannair R, Paton CM, Phares DA, Hagberg JM. Exercise training-induced changes in coagulation factors in older adults. *Med Sci Sports Exerc.* 2007;39(4):587-592.

Menzel K, Hilberg T. Blood coagulation and fibrinolysis in healthy, untrained subjects: effects of different exercise intensities controlled by individual anaerobic threshold. *Eur J Appl Physiol.* 2011;111:253-260.

Mulder K, Cassis F, Seuser DRA, Naraya P, Dalzell R, Poulsen W. Risks and benefits of sports and fitness activities for people with haemophilia. *Haemophilia.* 2004;10(Suppl 4):161-163.

Mulvany R, Zucker-Levin AR, Jeng M, Joyce C, Tuller J, Rose JM, Dugdale M. Effects of a 6-week, individualized, supervised exercise program for people with bleeding disorders and hemophilic arthritis. *Phys Ther.* 2010;90:509-526.

Shrier I, Kahn SR. Effect of physical activity after recent deep venous thrombosis: a cohort study. *Med Sci Sports Exerc.* 2005;37(4):630-634.

Souza JC, Simoes HG, Campbell CS, Pontes FL, Boullosa DA, Prestes J. Hemophilia and exercise. *Int J Sports Med.* 2012;33:83-88.

Van Sralen KJ, Le Cessie S, Rosendaal FR, Doggen CJM. Regular sports activities decrease the risk of venous thrombosis. *J Thromb Haemost.* 2007;5:2186-2192.

Wang J-S. Exercise prescription and thrombogenesis. *J Biomed Sci.* 2006;13:753-761.

Wittmeier K, Mulder K. Enhancing lifestyle for individuals with haemophilia through physical activity and exercise: role of physiotherapy. *Haemophilia.* 2007;13(Suppl 2):31-37.

Additional Resources

National Hemophilia Foundation (USA), www.hemophilia.org [Accessed November 8, 2015].

World Federation of Hemophilia, www.wfh.org [Accessed November 8, 2015].

NEUROMUSCULAR CONDITIONS

Medical textbooks commonly consider specific conditions, that is, diagnoses, as separate topics, segregated by the nature of the condition or, sometimes in neurology, by the region(s) of the nervous system and cell types that are affected. This is eminently sensible, since medical or surgical treatment and hope for a cure are specific to the diagnosis. Good medical care would not use an antibiotic for an infectious organism that is not sensitive to that antibiotic! But does a diagnosis-specific approach make the same sense when *exercise is the treatment*? In drafting *CDD4,* our team decided that it does not. The reason is that exercise is a very nonspecific therapy, affecting all tissues. The clinical power of exercise arises from its integrated nature, linking physiological systems into a unified function. We first addressed this in part II, emphasizing that many patients have multiple chronic conditions and that an exercise program needs to address all the conditions.

From a lifestyle-oriented perspective that integrates the person's health ecology, in which the diagnosis is more accurately expressed as *physical inactivity in the context of _____ [a health condition],* the medical diagnosis often has modest influence on how best to make use of exercise as a treatment. The challenges are more a function of the integrated physiological response to exercise or overcoming socioecological barriers that block participation in an exercise program (or both), and perhaps nowhere is this more apparent than in neuromuscular disabilities that affect mobility. People with severe neurological conditions most commonly die from cardiovascular disease, because neurological disability leads to adverse changes in body composition, energy metabolism,

and vascular function such that these populations take on a high-risk cardiometabolic pattern of prediabetes and type 2 diabetes. Research increasingly shows that the underlying reason for this metabolic shift is the effect of physical inactivity on skeletal muscle.

In conditions affecting the neuromuscular system, physical functioning is limited by the neural pathways involved in motor and autonomic control, often in a manner that is independent of how those pathways were damaged. Commonly, the mechanism of injury often doesn't make much difference because the integrated physiological barriers to exercise end up being

- flaccidity or hypertonicity,
- hyperreflexia and clonus,
- spasticity and loss of motor control,
- impaired proprioception and sensation,
- unstable stance and gait with risk of falls,
- chronic neurogenic pain,
- communication barriers, and
- seizure disorders.

These problems interact with the person's sense of identity and self-theory, often leading to complex psychosocial situations that should be addressed. Not the least of these is a societal tendency to protect such individuals, with everyone in the person's sphere accepting a very sedentary lifestyle.

In a patient or client with a seizure disorder who wants to exercise, does it affect the exercise program if the cause of his seizures was a stroke or a traumatic injury? In a person who has incomplete paralysis in her lower extremities, does it affect the exercise program if the cause was a spinal cord injury or polio? In the case of the man with seizures, if the seizures can be medically controlled, the mechanism of brain injury makes little difference; in the case of the paralyzed woman, it makes some difference, but most issues (e.g., spasticity, contractures) are much the same.

With this new perspective in mind, this part of the book outlines exercise management for conditions of the central and peripheral nervous systems, respectively. Using this type of perspective, we further discuss medical diagnosis-specific content for several neurological conditions, including Parkinson's disease, muscular dystrophy, cerebral palsy, and multiple sclerosis. Discussion of the integrated nature of exercise in patients with neurological conditions continues in part VIII. We hope you find this new approach illuminating.

Stroke, Brain Trauma, and Spinal Cord Injuries

Stroke, or cerebrovascular accident (CVA), is an injury to the brain from a critical reduction in blood flow to the area that is injured. About 80% of all strokes are due to thrombotic occlusion or embolism (ischemic stroke); the remainder are intracranial bleed (hemorrhagic stroke). Prevalence of stroke roughly doubles each decade beyond 50 years of age, representing the most common cause for serious disability in industrialized nations and an emerging public health concern in developing economies. Globally, ~15 million new strokes occur annually, with about 90 million stroke survivors worldwide. These figures are anticipated to double by 2050, due to aging populations on every continent except Africa.

Traumatic brain injury (TBI) is an injury to the brain caused by trauma, in which the head is hit by an object moving with high kinetic energy (momentum) or the head is moving with high momentum and strikes a mass (stationary or moving). Because the brain is tightly encased in bone and has very little mechanism for absorbing shock, such impacts can cause substantial trauma to brain tissues. Traumatic brain injury can be associated with intracranial bleeding but doesn't necessarily involve vascular injury. Only one in nine TBIs need hospitalization, and some are fatal. Globally, about half of all TBIs are caused by motor vehicle accidents involving cars, trucks, and motorcycles (as well as pedestrians and cyclists who are struck by a motor vehicle). Age distribution is bimodal, with persons <25 and >65 years of age at higher risk. Falls are a major cause of TBI in persons >65 years of age.

Spinal cord injury (SCI) is a third type of devastating injury to the central nervous system, affecting some 250,000 to 500,000 people per year worldwide. As with TBI, slightly less than half of all SCIs are due to motor vehicle accidents; about one in five are due to falls or to violence (mainly gunshot wounds), and another 15% are due to sport injuries. Males are the victims in four out of five cases.

Basic Pathophysiology of Stroke

Most authorities consider the fundamental pathophysiology of a stroke to be an infarction with resulting brain cell death. The residual clinical deficits depend on the lesion size and location. Ischemic strokes most commonly fall into the category of large vessel thrombotic or thromboembolic strokes with wedge-shaped cortical and often underlying subcortical white matter damage, or small vessel thrombo-occlusive disease producing lacunar infarcts (little holes) ≤2 cm in diameter in the subcortical white matter. Stroke can produce a myriad of neurological, neuropsychological, and systemic pathophysiological events.

At least two-thirds of stroke survivors are left with chronic neurological deficits that persistently impair function. Yet only 1 in 10 strokes produce overt clinical symptoms; the rest are so-called silent strokes. Because silent strokes go undetected, they take a cumulative toll and produce cognitive impairment with a ~10-fold increase in the rate of dementia. Stroke recurrence is most likely early after an initial stroke, generally ~10% over the first year, with two-thirds of events occurring within months. After the first year, recurrent strokes occur at a rate of about 3% to 5% annually. In contrast, dementia from silent strokes gradually increases over 3 years (affecting ~23% of persons who did not have dementia at baseline), so silent strokes behave more like a neurodegenerative disease in advancing age. Stroke survivors have very high rates of cardiovascular comorbidity, affecting 75% of people, with coronary artery disease in about half and a 3% to 5% annual rate of cardiac mortality. Hence, stroke represents a heterogeneous and multisystem vascular disease syndrome with multiple physiological, functional, and potentially cumulative cognitive impacts.

Basic Pathophysiology of Traumatic Brain Injury

Persons who have suffered a TBI have high rates of headaches, chronic pain, peripheral vestibular deficits, difficulty initiating and maintaining sleep, depression, and variable deficits in cognitive function (particularly executive control), which can influence multitasking and attention to self-care. These post-TBI problems can occur and persist whether or not the TBI was considered severe at the time of the injury. Even mild traumatic brain injury (MTBI or concussion) can be associated with persistent clinical issues. Historically, loss of consciousness was thought to reflect a severe injury, but research and experience have shown that loss of consciousness is not a predictor of severity or persistent symptoms.

Traumatic brain injury is the main cause of seizure disorders among adults. Traumatic brain injury can produce diffuse axonal injury, and this can in turn produce myriad neurological deficits depending on the severity, location, and number of TBI events. Damage to axonal tracts traveling near the base of the skull to frontal brain regions can contribute to impaired executive function, which is among the most prevalent categories of cognitive disturbance following TBI. Evidence in humans and animals shows that TBI of moderate or greater severity can activate ongoing inflammatory pathways in the brain. Animal studies suggest that exercise beyond the subacute recovery period can reduce the inflammation, but more research is needed in humans to learn whether exercise protects the brain tissue after clinically significant TBI. Because so little is understood about the pathophysiology of TBI, symptoms that persist are very difficult to manage (management consists mostly of medications to mitigate symptoms).

Basic Pathophysiology of Spinal Cord Injury

Spinal cord injury produces variable sensorimotor loss, depending on the lesion location and completeness. An SCI above the 1st thoracic vertebra is called quadriplegia (also tetraplegia) because nerve supply to all four limbs is affected, while an SCI below the 1st thoracic vertebra is called paraplegia because nerve supply to the upper extremities is not affected. The phrase "complete SCI" means that the entire cord is sufficiently damaged to block all neurological signals across the level of the injury, whereas incomplete SCI means that only part of the cord is damaged and that some neurological function remains across the level of the injury. The most common type of SCI is incomplete tetraplegia (30%), followed by complete paraplegia (25%), with complete tetraplegia and incomplete paraplegia both about 20%.

Disruption of autonomic innervation and SCI with an associated TBI deserve special mention. Traumatic brain injury is present in some SCI cases, particularly in persons who have injury to the cervical or high thoracic cord. Autonomic abnormalities, along with bowel and bladder control issues, are common in people who have had an SCI. In part, this is because autonomic innervation of the lower abdominal viscera is predominantly through the sacral plexus and therefore likely affected with an SCI at any cervical or thoracic level (and even lumbar injuries to the cauda equina).

Individuals should be evaluated by an examiner trained in the American Spinal Injury Association (ASIA) classification system (see figure 25.1), particularly if they will be referred to an exercise program.

Common Elements

From a medical perspective, brain and spinal cord injuries are quite different, but they have a number of common neurological elements that yield similar issues. These elements include the following:

- Upper motor neuron type of motor deficits:
  - Hemiparesis (half-paralysis) in cerebral hemispheric and lacunar infarcts
  - Bilateral and asymmetric motor deficits (some brain stem infarction, SCI, and TBI)

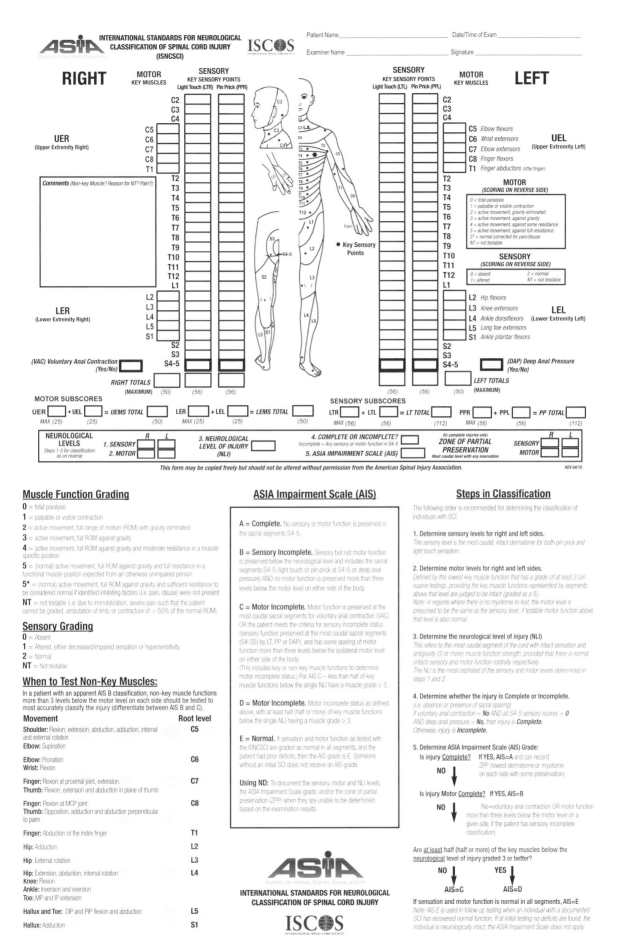

Figure 25.1 American Spinal Injury Association international standards for neurological classification of spinal cord injury.

American Spinal Injury Association: International Standards for Neurological Classification of Spinal Cord Injury, revised 2013; Atlanta, GA. Updated 2015.

- Sensory deficits:
  - Varying degrees of primary hemisensory loss
  - Hemimotor and hemisensory neglect
  - Visual deficits (hemianopsia or visual hemineglect)
  - Spinal cord injury level–dependent sensorimotor deficits
  - Dysesthetic pain and sensory processing abnormalities

In addition, people with all three types of these injuries can have a number of common health problems in areas of integrated neurological functioning, neuropsychological functioning, and mental health, as well as alteration-deterioration of systemic metabolism and body composition.

Common Neurological Problems

Chronic gait and balance deficits beyond 6 months

Fall risk (~70% of stroke survivors have a fall within the first year)

Hyperreflexia, spasticity, and clonus

Typically core body and leg problem for SCI, distal limb issue for stroke

Pathological reflexes that change the base of support at the ankle

Bilateral "scissoring" of leg tone that complicates mobility and may elicit pain

Cerebellar deficits, coordination difficulties affecting upper or lower extremity on same side as hemiparetic weakness; can be isolated in cerebellar strokes

Oculomotor deficits (less common), often accompanied by double vision

Central vestibular deficits (brain stem stroke)

Common Neuropsychological and Psychosocial Problems

Cognitive deficits—affect ~50% of stroke survivors (less commonly TBI or SCI)

Executive dysfunction—most prevalent with reduced attention, distractibility, and task sequencing that can impair motor learning

Communication difficulties due to cognitive deficits, aphasia, or both

Receptive aphasia a particularly difficult barrier

Affective disorders, including depression, in 25% or more of central nervous system injuries

Apathy, anxiety, and social phobia

Poor self-efficacy for balance—can lead to psychological paralysis

Systemic Effects of Central Nervous System Injury

The neurological injuries of CVA, TBI, and SCI all lead to gross muscular atrophy and increased intramuscular fat, with a shift from oxidative to nonoxidative muscle fibers that are insulin resistant. This change contributes to the metabolic syndrome that propagates glucose intolerance, hypertension, dyslipidemia, and increased body and organ fat and accelerates cardiovascular disease burden in persons with central nervous system (CNS) injuries.

The pathophysiologic bases for poor health outcomes in persons who have CNS injuries are increasingly clear:

- Impaired glucose tolerance and type 2 diabetes are seen in >75% of stroke survivors.
- Nearly all people with SCI develop metabolic syndrome (in proportion to deficit severity).
- Persons with TBI have increased insulin resistance, associated with decreased physical activity and endocrine problems (e.g., reduced growth hormone).
- In all three conditions, reduced bone density predisposes to pathologic fractures (a great concern in individuals with SCI who may not feel fractures).

Autonomic abnormalities are also very common, particularly dysfunction of sympathetic control of circulation and chronic pain:

- Autonomic dysreflexia—can dangerously raise blood pressure, triggered by pain or bowel–bladder dysfunction (mainly SCI)
- Exaggerated startle and pressor–chronotropic responses (mainly TBI)
- Chronic regional pain syndrome (formerly *reflex sympathetic dystrophy*) with limb pain syndromes sensitive to touch, with swelling and bone resorption

Sleep disorders, including obstructive sleep apnea and other disorders, are common after stroke and other CNS injuries. These factors commonly increase fatigue, blunt cognitive function, worsen diabetes, and increase risk for sudden death.

Management and Medications

Medical management of persons who have had a CNS injury with a residual deficit is largely focused on secondary prevention of recurrent vascular events and the management of postevent sequelae. Management techniques that are often overlooked are the use of diaphragmatic breathing exercises and mindful meditation, which can moderate dysautonomias and help with neuropsychosocial problems. Regular exercise is also helpful for these problems, as well as promoting better sleep.

Because all people who have a CNS injury are at markedly increased risk for cardiovascular disease, their care should include aggressive cardiometabolic disease prevention that may consist of some or all of the following:

- Antiplatelet medications (relative risk reduction of stroke by 20-32%)
- Evaluation and management of atrial fibrillation (see chapter 11), which carries a fivefold increased stroke risk (often treated with anticoagulation)
- Statin therapy for lipids and vascular function, even if cholesterol levels are not elevated
- Antihypertensive medications to manage blood pressure to guidelines
- Carotid endarterectomy of high-grade stenosis (e.g., >80%)—yields ~16% absolute stroke risk reduction (eight times superior to aspirin)
- Anticonvulsant medications for seizure control
- Tobacco cessation interventions for people who use tobacco

Tip: Avoid reductions in blood pressure due to orthostatic drops that could reduce blood flow across a critical carotid stenosis, or in the early weeks after stroke when cerebral autoregulation is most impaired.

Medical management of the complications that occur after a CNS injury may include the following:

- Physiatry, physical, occupational, language, speech, and swallow therapies—key to comprehensive stroke care
- Surveillance and treatment for depression (poststroke rates of 25-30%)
- Medical and multimodal management of neurogenic pain
- Braces and slings to reduce contractures and reduce pain (particularly for the hemiparetic upper extremity and shoulder that often produce pain syndromes)
- Antispasticity medications, less commonly botulinum toxin
- Ankle–foot orthoses (AFO) or functional electrical stimulation (FES) for foot drop and alignment
- Walkers, canes (single-point to quad), wheelchairs to meet mobility and safety needs

Stroke and other neurological disability conditions can change markedly over time and in response to structured exercise training. Vigilance must be maintained with regard to updating medical care and need for assistive devices.

Effects on the Exercise Response

Following a stroke or SCI, peak aerobic performance is ~50% below that of matched sedentary controls, whether exercise testing is done using an arm crank or lower-extremity ergometer. At the same time, the energy costs of walking with a hemiparetic gait are elevated 1.5- to 2-fold due to abnormal biomechanics. As a result of decreased aerobic abilities in the face of decreased aerobic economy, hemiparetic walking uses ~70% of the person's aerobic abilities, so walking can be sustained only for short periods. For this reason, most stroke survivors are unable to independently perform many activities of daily living. The situation is even worse for a person with an SCI, because the energy cost of locomotion can be 5- to 10-fold elevated after an SCI—a formidable metabolic barrier to mobility. After a CNS injury, skeletal muscle in the paretic limbs undergoes a shift from oxidative to nonoxidative fibers, in proportion to the neurological deficit, which leads to rapid onset of fatigue and exercise intolerance.

These reduced fitness levels combine with age-related declines in strength and fitness to compound neurological disability and reduce free-living physical activity. Physical activity monitoring reveals that stroke survivors take ~1200 to 2000 steps per day, well less than half the free-living physical activity reported in subjects who are older and have nonneurological chronic health conditions. This sedentary living, along with poor self-efficacy for balance, limits functional independence and worsens cardiometabolic risk profiles.

Most stroke survivors can achieve ~85% or greater of age-predicted maximal heart rate, suggesting that cardiac function is likely not the rate-limiting problem behind their low aerobic abilities. Rather, exercise tolerance is more likely a function of altered circulatory control, shift of skeletal muscle to a less aerobic quality, and the energy inefficiency of altered gait and neuromotor control. Fortunately, the mere presence of volitional movement provides a capacity for motor learning and increased motor strength, which has the potential to improve the aerobic qualities of muscle, physical functioning, and exercise capacity after CNS injury.

In persons who have had TBI, exercise response and physical function are contingent on severity and the constellation of associated medical and neuropsychological features in the syndrome. In general, individuals with TBI physiologically respond to exercise, but very little research on this has been completed (mostly in sport-related MTBI or concussion). More research is needed to understand the biology across the spectrum of TBI severity; current guidance is primarily coming from the Zurich guidelines on return to sport after a concussion.

For persons who have had an SCI, the exercise response and capacity for exercise are related to the level and completeness of the injury. The most commonly used nomenclature to describe an SCI is the American Spinal Injury Association's Impairment Scale (AIS), which grades the impairment by spinal level and degree of impairment, while the International Standards for Classification of SCI is useful to guide expectations. Lesions at or above the 4th thoracic level generally lead to a blunted chronotropic response to ≤130 beats/min, even in young adults. The exercise capacity may be limited in these circumstances by reduced blood flow in the affected musculature. Both central circulation and peripheral blood flow seem to be involved

in the reduced aerobic abilities, whether during volitional, electrically stimulated, or mechanically aided (e.g., robotics, partial weight suspension) exercise. Individuals with SCI can also have lower motor neuron (denervation) mixed in with the typical upper motor neuron type of injury, depending on the lesion and the nature of the injury. This can markedly reduce muscle activation and produce profound muscular atrophy.

Effects of Exercise Training

Exercise after CNS injury can improve multiple physiological and functional outcomes, even years after a disabling event. The approaches to exercise after stroke and traumatic brain and spinal cord injury are remarkably similar to the basic program for chronic disease and disability in nonneurological conditions. The major difference is an increased attention to neuromotor function and balance to safely adapt exercises to the person's neurological deficit. Despite the myriad of issues after a CNS injury, biological adaptations and a conditioning response do occur with a diversity of adapted exercise modes, improving glucose tolerance, improving blood flow to muscle and skin, and increasing bone structural integrity. Thus there is now a mandate for the safe and effective promotion of regular exercise to optimize fitness, physical function, and heart and brain health.

The majority of exercise training research in neurological populations has been in participants with mild to moderate deficits, which constitutes the vast majority of persons in these clinical populations. Approximately 80% of stroke survivors ultimately recover to this level of function, while ~50% of SCI cases have motor incomplete lesions and the vast majority of TBIs are graded as mild. That being said, mild to moderate deficits include a wide variety of motor and cognitive deficits.

Mild to moderate deficits can be operationally defined as those in individuals who have paretic mobility impairment yet can ambulate independently with an assistive device (e.g., cane, quad cane, walker). A second important element is the ability to do independent transfers, such as rising from and sitting back down to a chair. Having this ability generally renders sit-to-stand and many other multisegmental motor control exercises possible and safe, though a sturdy handrail or assistance from a coach or trainer may be needed.

Superimposing functional electrical stimulation on locomotor training may modestly enhance some outcomes in chronic SCI. However, meta-analyses mostly show that all modes of core loco-motor exercise provide benefit, without superiority of one mode over another. Adaptive exercise training methods warrant more study to better understand how to optimize dose and modality to improve health and function. More exercise research is especially needed for TBI.

Most information about the effects of exercise after stroke has been gathered from work with individuals who have mild- to moderate-severity paresis, implying some degree of volitional ambulatory capacity, albeit with an assistive device. The evidence base is larger for chronic stroke (>6 months), but earlier exercise is also effective. Randomized studies now show that regular exercise training improves the following physiological and functional outcomes in people who survive a CNS injury:

- *Fitness:* Aerobic exercise training in stroke survivors over 3 to 6 months increases peak fitness by 8% to 30%, a magnitude similar to that in age-matched controls.

- *Glucose tolerance:* Moderate-intensity aerobic training improves insulin sensitivity and glucose tolerance; some people (~60%) with impaired glucose tolerance or type 2 diabetes mellitus improve their glycemic status on glucose tolerance testing.

- *Cardiovascular disease risk:* Exercise training lowers blood pressure, improves fasting lipoprotein lipid profiles and body composition, and improves systemic and cardiac vasomotor function.

- *Brain blood flow:* Aerobic exercise training increases cerebral artery dilation capacity, which may protect against recurrent stroke.

- *Psychological factors:* Exercise training improves mood, serves as an adjunct treatment for depression, reduces anxiety, and in older populations can improve elements of cognitive function including memory and executive control.

- *Muscle function:* Aerobic exercise training can improve muscle metabolism and blood flow to enhance aerobic metabolism, while resistive training and functional electrical stimulation can improve strength and increase muscle mass without creating pathological changes in tone; mobility-oriented aerobic training can reduce paretic leg spasticity (hamstring).

- *Body composition:* Exercise training increases fat-free muscle mass and muscle quality and helps reduce intramuscular, organ, and central abdominal fat mass.

- *Skin:* Increased muscle integrity and blood flow improve body padding and resistance to skin breakdown and ulcers, which are often a big problem in SCI, severe TBI, and stroke.

- *Bone health:* Functional electrical stimulation and even lower-intensity weight-bearing exercise can improve bone density and trabecular bone content in hemiparetic limbs.

- *Mobility:* After mild- to moderate-severity stroke, exercise can improve walking velocity, walking endurance, balance (e.g., Berg Balance, Dynamic Gait Index), postural control, and simulated functional tasks (get-up-and-go, stair climb); after SCI this response depends on lesion severity and paresis level.

- *Bowel function:* Promising research suggests that some sorts of physical activity have the potential to improve bowel function, particularly upright activity, but more studies are needed to better understand this important and time-consuming issue in SCI.

- *Brain function:* Exercise can improve elements of cognitive function in people who are elderly and who do have not dementia, and possibly after stroke and other CNS injuries.

Recommendations for Exercise Testing

For individuals capable of higher-intensity exercise, use the ACSM Guidelines. For people with mild to moderate deficits, use the Basic *CDD4* Recommendations with individual modification as needed for efficacy and safety.

Low-intensity exercise rarely requires formal exercise testing, as this intensity is similar to elements of usual physical therapy and daily mobility. Some medically high-risk cases may need pretesting to define cardiopulmonary tolerance and safety even for low-intensity exercise.

Exercise testing and medical clearance depend on the planned intensity and goals of training, the neurologic deficits that influence exercise programming and safety, and individual medical risk profile.

- Before starting low-intensity training (e.g., ≤50% of heart rate reserve; community exercise classes), people should receive medical clearance from their providers.

- Neurological deficit profiles may necessitate adaptive devices and exercise testing protocols to fit capacities, optimize safety, and serve as metrics to document progress and outcomes across training.
- Six-minute walk test is also recommended, in addition to the tests in the Basic *CDD4* Recommendations.

Recommendations for Exercise Programming

For individuals capable of higher-intensity exercise programs, use the ACSM Guidelines. For those with mild to moderate deficits, use the Basic *CDD4* Recommendations with individual modification as needed for efficacy and safety.

By and large, anyone who is asymptomatic during activities of daily living can participate in low-level exercise rehabilitation programs. Individuals should not be cleared without further assessment if they have a high risk of cardiopulmonary decompensation (signs or symptoms of inadequate chronotropic or pressor responses, arterial desaturation) during the low-intensity exercise used in conventional physical rehabilitation (e.g., ~50% of heart rate reserve). Also, people need further assessment when they have postevent syndromes of pain or autonomic dysfunction, or if neuromotor control or balance are such that a particular exercise modality is unsafe. Caution is required for individuals in such situations, and further testing may be needed to establish cardiopulmonary or neurological safety and tolerability of exercise.

A plan for regular physical activity and prevention of cardiovascular deconditioning is recommended at the time of hospital discharge and across transitions in care for all individuals who have suffered a stroke or SCI. Because there is no outer limit to the timing of the benefits, exercise is considered a component of best care even decades after an index disabling event. People who have had a TBI require individual assessment before initiating an exercise program. If a mode or intensity of exercise exacerbates their neurological symptoms, they are probably not ready to do that activity.

The majority of medical literature states that exercise can be safely conducted without serious adverse events in all persons with CNS injury and stroke survivors, as long as exercise is done with proper oversight, attention to the guidelines, and awareness of the possible medical and physiological issues unique to these populations. The following issues must be considered:

- *Safety first:* Pay special attention to fall risk, clearance, and mobility with respect to getting on and off equipment (with vs. without assistance).
- *Therapist input:* Initial and follow-up assessments are used to grade neuromotor gains and to suggest specialized adjunct and assistive devices.
- *Adaptive equipment:* Safety devices (e.g., harnesses over treadmills) or specialized equipment that customizes training to neuromotor capabilities may be needed.
- *Mode:* Since a variety of aerobic and mixed modality exercise models are known to improve fitness, options for comfort and personal preference can be used to enhance compliance, including treadmills, modified cycle ergometry, group exercise, pool-based exercise, and steppers.
- *Warm-up:* While this has not been systematically studied, a slow warm-up is recommended to counter any paretic limb spasticity and reduced muscle blood flow.
- *Intensity:* Perceived exertion is always used in persons with disability to help judge exercise intensity, with periodic heart rate surveillance.
- *Pressor response:* Observe for signs of autonomic deconditioning that can cause low blood pressure (particularly for sustained upright exercise), as well as for elevated sympathetic response causing very high blood pressure (common in the first 6 months).
- *Body temperature:* Spinal cord injury can produce abnormalities of body temperature regulation in response to exercise, which warrants observation for safety and comfort.

High-intensity exercise programs are increasingly in vogue but should be approached cautiously in people after CVA, TBI, or SCI. Such exercises may be helpful for some, but current research is limited and inconclusive. Meta-analyses suggest that stroke survivors make greater gains in peak $\dot{V}O_2$ with higher-intensity training, though it is not clear whether this improves outcomes. Safety is paramount, and greater rigor with regard to cardiopulmonary clearance and monitoring is advised for individuals who are older or those with a history of stroke in whom there is a higher incidence of cardiac disease and dysrhythmias that can emerge at higher exercise

Steps in Exercise Programming in Persons With CNS Injury

1. Start Conservatively

   Deconditioned subjects who are elderly do well by starting with low-intensity interval training (e.g., walking at 75% of their self-selected 10 m walking velocity), consisting of three or four bouts that each last from a few to several minutes, with equivalent rest periods between bouts.

2. Increase Slowly

   Every 2 weeks, total exercise duration can be advanced by ~5 min and interval durations extended, as tolerated, toward an eventual goal of 30 min of low-intensity exercise on most days of the week. It may take ~3 months to reach goal.

3. Cross-Train

   Task practice relevant to walking and balance may aid the improvement in mobility skills. Exercises that safely challenge multisegmental balance control in the context of muscular endurance (e.g., sit to stand, lateral stepping, high stepping in place, all with appropriate handrail safety or standby aid) can improve selected balance outcomes and train core muscles.

4. Focus on Form

   Symmetry of posture and core is important to improve whole-body balance and motor learning across the kinematic chain while minimizing undesired compensatory movements. Slowing movements down can help reduce velocity-dependent pathologic reflexes (e.g., clonus, reflex agonist–antagonist spasticity) that can complicate exercise. Motor learning is influenced by the mental state of the learner; and for individuals with CNS injury, actively focused mindfulness on the movement therapy can augment sensorimotor gains.

5. Be Patient

   After CNS injury, improvements typically continue for at least 6 months, and progression tends to be slow and steady. Structured strength training should be considered an option in the exercise armamentarium, especially after stroke, in keeping with the Basic *CDD4* Recommendations.

6. Foster Mental Well-Being

   Disorders of executive function are common after CNS injury, and essentially must be remediated by the exercise professional's provision of external structure and oversight for success. Many people do not succeed if left to their own devices. Be aware of sleep disorders, and use exercise (with nutrition as appropriate) to enhance sleep hygiene. People should not exercise too close to bedtime to avoid activating alertness (especially individuals who are elderly who may tend to "sundown").

intensities. Cardiac disease is prevalent in persons with SCI as well, due to metabolic dysfunction in paretic muscle that greatly amplifies cardiovascular risk. Also, in higher-level spinal cord lesions, attention must be given to special conditions such as autonomic uncoupling of heart rate to physiological demands, which makes exercise programming more uncertain. In addition, TBI may produce long-term secondary neuroendocrine changes that increase cardiovascular risk, although this issue is less well understood than in stroke and SCI, where metabolic syndrome is the rule rather than the exception. Therefore, in individuals who have had a stroke, TBI, or SCI, careful clinical assessment with attention to risk–benefit assessment is advised before advancing to high-intensity training.

Special Recommendations for TBI

When to start: The time early after TBI, when brain inflammation is highest, should be a period of rest, including *cognitive rest,* to help facilitate recovery. The precise duration of needed rest must be individualized according to clinical features. Close supervision is recommended.

Start slowly: Exercise should be started at very low intensity after TBI, even for young healthy athletes, with attention to neurological symptoms that may be associated with or triggered by exercise. One useful technique is to attach a symptom diary to the exercise log.

Special Recommendations for SCI

In contrast to TBI, after SCI it is important to begin early mobilization and exercise training. Major declines in bone density and muscle mass occur rapidly, within months, after disabling SCI. Early strategies, including adaptive exercise and technologies (e.g., functional electrical stimulation), are important to preserve and maintain body composition.

Exercise should be customized based on the person's AIS score, as well as his individual capacities and interests. For example, in someone who has an incomplete SCI with paraplegia, assessment as AIS D (i.e., manual muscle testing below the level of the injury having muscle strength grade of 3 or higher [3 can lift against gravity; 4 can lift against some resistance, but with weakness; 5 shows no weakness]) may enable a more active volitional component to the exercise program. In practice, AIS D is a diverse group of paraplegic or quadriplegic deficit profiles with a wide range of functional capacity. Individuals who are more severely affected often benefit from the use of adaptive exercises to better match therapeutic exercise to muscle functioning, task-specific exercises aimed at improving neuromotor function such as gait or balance, or both.

In contrast, among individuals assessed as AIS C (i.e., at least half of muscles below the level of the lesion graded less than 3 on manual muscle testing), volitional movement against gravity may not be possible. Someone with AIS C may require specialized adaptive exercises from modalities such as functional electrical stimulation or robotic-aided exercises, which are still investigational. Close collaboration between exercise physiologists and rehabilitation providers may optimize health and function outcomes in such scenarios.

Nutritional tip: When planning exercise for individuals with neurological disease who are aging, it is important to pay attention not only to fluid, electrolytes, and caloric intake but also to protein intake. Exercise in older populations is enhanced by sufficient protein intake, including branched chain amino acids (e.g., leucine), when supplied within about 20 min after the completion of exercise. Food with a ratio of about 3 g carbohydrate to 1 g protein, consumed shortly after exercise, can reduce muscle injury and improve adaptive response

in terms of amino acid uptake by muscle. This need not be specialized supplements; regular foods that people like work fine. Nutrition timed around exercise especially warrants attention in the aging neurological population, but should be considered for all individuals who are recovering from a CNS injury.

Integration Into a Medical Home Model

Recovery from neurological injury is more rapid and complete when people receive well-trained guidance. Patient-centered medical home practitioners and staff should seek out and be familiar with local resources to help with this important service. Daily exercise is critically important in the long-term care planning for persons with a CNS injury. Individuals who have had a stroke, an SCI, or a TBI are all at risk for deconditioning and cardiovascular morbidity and mortality, which is true even for people with mild residual deficits. Best care recommendations for all of these conditions require referral for exercise rehabilitation by health professionals skilled in CNS injuries.

The persons affected, their families, and caregivers should all receive education conveying that profound cardiovascular deconditioning compounds disability and cardiovascular disease risk after any CNS injury or stroke, and that secondary changes rapidly occur in skeletal muscle that can be treated only with volitional physical activity or artificial adaptive activation of muscle (i.e., functional electrical stimulation). Brain health and function are critically dependent on keeping movement therapy going over the long term.

Spouses, caregivers, and family members are educational and social support for promoting healthy lifestyle after CNS injury. The concept is to keep moving and use the whole body for activities of daily living, in every room of the house or place of dwelling, despite the person's mobility disability. Ideally, these free-living home-based activities should expend more energy than structured exercise sessions themselves, adding to the cumulative healthiness of individuals with neurological mobility disability.

Take-Home Message

Keep in mind these points when working with people who have CNS injuries:

- The individual should participate in a program of regular physical activity, in a group setting or in an individually customized exercise program (or both), on most days of the week.

- For an individual with CNS injury to the spine or brain, regular exercise should become a lifelong goal, because the benefits of exercise therapy are rapidly lost with the cessation of exercise training.

- Progressive and functionally relevant motor task practice should be used to improve neuromotor control, avoid disuse atrophy of the muscles and brain, and optimize neurological function.

- Exercise is one of the most effective and cost-efficient medical treatments for persons who have had a CNS injury, who should be periodically reevaluated by rehabilitation providers to maintain continuity and upgrade assistive devices as needed.

- As with physical or occupational therapy, interactive coaching with log books on progress and progression of challenging cardiopulmonary, strength, balance, and neuromotor control exercises are key to long-term success.

Suggested Readings

Bland DC, Zampieri C, Damiano DL. Effectiveness of physical therapy for improving gait and balance in individuals with traumatic brain injury: a systematic review. *Brain Inj.* 2011;25(7-8):664-679.

Ivey F, Macko R. Prevention of physical deconditioning. In Stein J, Harvey R, Macko R, Winstein C, Zorowitz R, eds. *Stroke Recovery and Rehabilitation.* New York: Demos Medical; 2009.

Jacobs PL, Nash MS. Exercise recommendations for individuals with spinal cord injury. *Sports Med.* 2004;34(11):727-751.

Macko RF, Hidler J. Exercise after stroke and spinal cord injury: common biological mechanisms and physiological targets of training. *J Rehabil Res Dev.* 2008;45(2):323-328.

McCrory P, Meeuwisse WH, Aubry M, et al. Consensus statement on concussion in sport: the 4th International Conference on Concussion in Sport held in Zurich, November 2012. *Br J Sports Med.* 2013;47:250-258.

Mehrholz J, Kugler J, Pohl M. Locomotor training after spinal cord injury. *Cochrane Database Syst Rev.* 2012 Nov 14(11):CD006676.

Morawietz C, Moffat F. Effects of locomotor training after incomplete spinal cord injury: a systematic review. *Arch Phys Med Rehabil.* 2013 Nov;94(11):2297-2308.

Warms CA, Backus D, Rajan S, Bombardier CH, Schomer KG, Burns SP. Adverse events in cardiovascular-related training programs in people with spinal cord injury: a systematic review. *J Spinal Cord Med.* Epub ahead of print 2013 Nov 26.

Weightman MM, Bolgla R, McCulloch KL, Peterson MD. Physical therapy recommendations for service members with mild traumatic brain injury. *Head Trauma Rehabil.* 2010 May-Jun;25(3):206-218.

Wise EK, Hoffman JM, Powell JM, Bombardier CH, Bell KR. Benefit of exercise maintenance after traumatic brain injury. *Arch Phys Med Rehabil.* 2012 Aug;93(8):1319-1323.

Peripheral Neuropathy, Myopathy, and Myasthenia Gravis

Diseases affecting peripheral nerves, muscle, and the neuromuscular junction have not been well studied with regard to exercise. These conditions share common traits of fatigue, weakness, imbalance, and impaired mobility and dexterity and also have metabolic sequelae because people with these conditions tend to become physically inactive. This promotes deconditioning, compounds the functional decline associated with the neuropathy, and increases the risk of developing cardiometabolic disease.

In the past, medical practitioners have been extremely reluctant to prescribe exercise training for persons with neuropathy, myopathy, or myasthenia gravis, due to concerns that exercise would potentially worsen these conditions, lead to falls, cause muscle breakdown, or accelerate stress on the neuromuscular junction. Today, thinking is shifting toward the notion that a portion of weakness in these disorders results from disuse and deconditioning. Although exercise cannot reverse these conditions, it might be possible to strengthen weak and deconditioned muscles. Current management is therefore geared toward

helping individuals in these groups retain and maintain safe, functional independence.

Prior editions of this textbook focused on polio and postpolio syndrome. In this edition, we are addressing exercise in peripheral neuromuscular conditions by discussing them together, despite the varying nature of their pathophysiology. We did this because these conditions have common overarching issues that make them very similar from an exercise program perspective.

Basic Pathophysiology of Peripheral Neuropathy

Peripheral neuropathies are highly prevalent conditions. The causes are numerous and vary widely in age of onset, time course, secondary tissue changes, and pain, as well as effects on physical functioning and ability to tolerate exercise. A complete survey of etiologies is beyond the scope of this text, but a number of common causes are presented in table 26.1. The most common pattern is a length-dependent peripheral

Table 26.1   Common Causes of Peripheral Neuropathy

Category	Conditions
Hereditary	Charcot-Marie-Tooth syndrome, neurofibromatosis, schwannomas
Endocrine	Diabetes, thyroid dysfunction
Inflammatory and autoimmune	Vasculitis, lupus, Sjogren's syndrome, sarcoid, chronic inflammatory demyelinating polyneuropathy
Infectious	Polio, syphilis, Lyme disease, leprosy, tuberculosis, hepatitis, other viral illnesses
Vitamin deficiencies	Vitamin B_{12} deficiency
Toxins	Alcohol, heavy metals, organic solvents, acrylamide
Medications	Chemotherapy agents, antibiotics (nitrofurantoin, quinolones), amiodarone, excess pyridoxine (vitamin B_6)
Cancer	Lymphoma, monoclonal gammopathy, paraneoplastic syndromes (side effects of cancer)
Trauma	Nerve compression (carpal tunnel syndrome or sciatica), radiculopathy

neuropathy, in which the condition begins in the most distal of nerves, serving the toes or fingers, and progresses proximally over time. Neuropathy features vary according to the involvement of specific types of nerves—motor, sensory, or autonomic nerve dysfunction or more than one of these. Neuropathy symptoms may present as (italics indicate common symptoms among neuromuscular conditions)

- *fatigue,*
- *weakness,*
- *pain and discomfort,*
- *imbalance,*
- numbness,
- dizziness, and
- poor thermoregulation.

These symptoms can profoundly affect exercise capacity by impairing integrated sensorimotor and muscle–joint function and also reduce exercise safety by altering balance as well as skin integrity. Loss of trophic factors related to the neuropathy can secondarily impair the blood vessels, skin, sweat glands, muscle mass, energy metabolism, and fat distribution, including loss of fat pads on the sole of the foot that protect against compressive forces during walking.

Basic Pathophysiology of Myopathy

A myopathy is an *acquired* pathological condition that affects skeletal muscle. *Hereditary* skeletal muscle abnormalities are called muscular dystrophies, are somewhat different, and are considered in chapter 30. Myopathies typically present with symptoms of

- weakness of proximal muscles of the shoulders and hips;
- difficulties in lifting, climbing stairs, rising from a chair, and walking;
- muscular pain or aching (a common side effect of statin medications); and
- restrictive pulmonary disease (from weak respiratory muscles), more rarely.

Myopathies are usually secondary to some other acute or chronic condition, though some myopathies are primarily a disease of muscle.

Conditions known to cause myopathy include the following:

- Thyroid dysfunction
- Diabetes
- Kidney failure
- Various toxins (notably alcohol)
- Vitamin deficiencies
- Vasculitis, autoimmune, connective tissue disorders
- Cancer
- Prescription medications (e.g., statins and colchicine)

Autoimmune myopathies, such as polymyositis and dermatomyositis, can be secondary to an inflammatory condition or to cancer but also can occur as a primary condition.

Basic Pathophysiology of Myasthenia Gravis

Myasthenia gravis is an uncommon autoimmune disorder that causes weakness and fatigue and occurs at a prevalence of 14 to 20 in 100,000 persons. Myasthenia has a bimodal age distribution, tending to occur in women aged 20 to 30 years and men >50 years. Signs and symptoms include

- generalized fatigue,
- neck and limb weakness,
- ptosis (drooping eyelids),
- double vision,
- facial muscle weakness,
- slurred speech,
- difficulties chewing and swallowing, and
- shortness of breath.

Symptoms of myasthenia are often intermittent, sometimes increasing at the end of the day, and sometimes occurring with sustained or repetitive activities or in hot environments. Due to the low prevalence of this disorder and the intermittent and wide variety of symptoms and signs, there is often a significant delay in making the correct diagnosis of myasthenia. As with many autoimmune conditions, the course of myasthenia can be intermittent or gradually progressive, can take the form of episodic flares, or can culminate in a myasthenic crisis (which is a medical emergency of impending respiratory failure and severe dysphagia that increases the risk of choking and aspiration pneumonia).

Myasthenia is caused by an immune-mediated impairment and destruction of the postsynaptic acetylcholine receptor on skeletal muscle cells. Normally, stimulation of a motor nerve releases acetylcholine from the motor nerve terminal into the synaptic cleft, where the acetylcholine binds to receptors on the muscle end-plate membrane, depolarizing the muscle membrane to cause muscle contraction. In myasthenia, antibody binding and complement activation alter these acetylcholine receptors on the postsynaptic muscle membrane, impairing the muscle's reception of the neural impulse and thereby reducing the strength of contraction.

Management and Medications

Medical management of peripheral neuropathy, myopathy, and myasthenia (if not all neurological conditions) initially focuses on identifying and treating the underlying cause (see table 26.1). Subsequent care then focuses on improving physical functioning, maintenance of independent living, and treatment of pain or other problems that arise from having residual neurological deficits. Maintenance of physical functioning and independent living is partly about exercise, but much of it involves attention to proper joint positioning to prevent or exacerbate injuries, such as Charcot joints in the feet and radiculopathies.

For the purposes of exercise management, these supportive measures are very similar across all the conditions in this chapter, which is why they are grouped together in this textbook. Medication and surgical management may be unique to each condition, but for helping people remain active and living independently, the exercise management techniques are very similar.

Peripheral Neuropathy

Individuals with peripheral neuropathy face two major medical management issues—neuropathic pain and loss of sensation leading to tissue injuries (mostly feet and hands). When pain management is not a problem or is in good control, people can focus on skin and limb care. Emphasis is on educating and training individuals to inspect their hands and feet to detect minor injuries that they can't feel. Normally such injuries (even just a bruise) cause people to favor and protect an injury while it heals. Because people with peripheral neuropathy can't feel the injury, they inadvertently reinjure the tissue and can accumulate a great deal of tissue damage.

Good pain monitoring and management are essential for individuals with neuropathic pain, because pain causes them to not participate in exercise and daily activities at work and home. Poor management of pain medications can lead to addiction, and from a lifestyle perspective it leads to a cessation of physical activity because of fear avoidance (see chapter 7). Thus, one might think of good neuropathic pain management as preventing two kinds of withdrawal—that from narcotic addiction, which is highly feared, and that from active living, which is not as highly feared but may well be just as harmful to well-being. Medications commonly prescribed to treat neuropathic pain include the following:

- Antiepileptics (e.g., gabapentin, pregabalin, carbamazepine, valproic acid, topiramate)
- Antidepressants (e.g., duloxetine, tricyclic antidepressants)
- Topical analgesics (capsaicin, lidocaine) for localized pain

Opioid agents are useful but are used carefully to avoid tolerance, rebound pain, and addiction issues. Pain should be evaluated before initiating therapeutic exercise and should be rated and tracked using a subjective scale such as the Likert pain rating scale.

Loss of the normal protective ability to sense pain leads to multiple secondary complications of neuropathy. Lack of proprioception in the feet leads to more forceful and deliberate foot placement, resulting in higher-impact forces on the foot joints. When this is combined with a diminished ability to sense pain, individuals can develop arthritis, poor joint alignment, and even extreme bone and joint deformities (Charcot joints), which can culminate in complete destruction of the joint. In addition, imbalance from proprioceptive sensory loss coupled with weakness of the foot and ankle increases the risk of falls, and sensory loss increases the risk of skin breakdown and wounds that won't heal and become chronically infected, with risk of infection spreading to bone (osteomyelitis). These outcomes are horrific, are extremely expensive to treat, can lead to nontraumatic amputations (see chapter 6), and must be avoided.

The team approach among clinicians, podiatrists, and physical therapists focuses on inspection of hands and feet for skin integrity, abrasions, loss of protective fat pads under bony surfaces, compression points from shoes, and proper joint alignment, as well as maintaining good toenails. Footwear should be inspected for adequate absorptive cushioning, abnormal wear patterns on the soles, proper support and safe weight distribution across the base of support, and adequate room to accommodate bony changes of the feet.

Therapeutic exercise goals often emphasize balance training and awareness of proper body alignment, as well as strength and endurance training. Orthotics and adaptive shoes can cushion tissue and nerves to reduce sensitivity from compression, maintain a neutral position of joints to avoid stretch or compression of affected nerves, and improve functional independence by bracing weak muscles. Key considerations for physical or exercise therapy are to maintain safety and reverse deconditioning. Being alert to skin, joint, and bone abnormalities that could predispose to injury is a part of best care. Monitoring the foot (or other limb health) via visual inspection should be part of the "exercise calendar," and weight management is a crucial aspect of reducing foot pressures in neuropathy.

Myopathy

Identifying and treating the underlying cause is the first step in medical management of myopathy, which is achieved by considering exposure to causes of myopathy (e.g., use of a statin medication to lower low-density lipoprotein cholesterol) and using diagnostic testing to detect reversible or treatable conditions. Autoimmune muscle diseases such as polymyositis are treated with steroids and other immunosuppressant medications. Workup for myopathy can involve a search for an occult cancer, since polymyositis and dermatomyositis are sometimes a paraneoplastic phenomenon and the first indication that someone has cancer.

In the absence of a reversible or treatable cause, or when treatment doesn't fully cure the condition, medical management of myopathy is predominantly supportive. The mainstay of treatment is physical and occupational therapy to improve functional independence by adapting activities of daily living and paying attention to energy conservation for success at work, at home, and in leisure pursuits. Low-intensity exercise can maintain cardiovascular conditioning, and stretching can help preserve joint and muscle flexibility. During a flare of myopathy, exercise should be attenuated or put on hold until the flare subsides, with close medical oversight during the periods of disease activity or symptomatic flare-up.

Myasthenia Gravis

Diagnosis of myasthenia gravis is based on medical history and neurological examination and confirmed by several diagnostic tests. Evaluation to diagnose myasthenia can include blood tests for acetylcholine receptor antibodies and muscle-specific (MuSK) antibody; an intravenous edrophonium (tensilon) test, which transiently but rapidly reverses weakness; an ice test that cools muscles of the eye to improve their performance; measurement of muscle fatigue during electrical nerve stimulation; single-fiber electromyography to demonstrate tenuous neuromuscular transmission; and use of computed tomography or magnetic resonance imaging to detect a thymoma (a benign thymic tumor that can be the source of acetylcholine receptor antibodies).

Management of myasthenia includes use of medications to improve function of the neuromuscular junction and acetylcholinesterase inhibitors, which improve muscular strength and endurance for several hours before their effect wanes. Many individuals are treated with immunosuppressive drugs to reduce production of antibodies against the postsynaptic receptors. Medications to treat myasthenia include these:

- Anticholinesterase inhibitors (pyridostigmine)
- Immunosuppressants (prednisone, azathioprine, mycophenolate mofetil, cyclosporin, and tacrolimus)

These medications, especially the immunosuppressants, have major side effects (see chapter 6 on arthritis and content in chapter 20 related to organ transplantation). Surgical removal of the thymus gland in individuals with myasthenia gravis may also reduce the myasthenic symptoms. Myasthenia gravis crisis is an acute condition involving difficulties in breathing and swallowing; it is treated either with plasmapheresis to remove the high levels of the acetylcholine receptor antibodies from the blood or with high-dose intravenous immune globulin, which temporarily binds these antibodies to prevent them from binding to the acetylcholine receptors. After a myasthenic crisis has been successfully treated, more aggressive treatment is generally indicated.

Pulmonary function testing, as a measure of respiratory muscle function, is sometimes used to monitor the individual's status, particularly if she experiences dyspnea (shortness of breath). Information gained from this testing can be especially useful if the person has other comorbidities that cause dyspnea, since myasthenia and myopathies have a restrictive pattern whereas asthma and chronic obstructive lung disease have an obstructive pattern of respiratory impairment (see chapters 7 and 17). Sleep disturbances are common due to the weakness in the upper airway and respiratory musculature. Sleep hygiene is extremely important and influences the ability of people with myasthenia gravis to do daily activities.

A number of conditions can transiently worsen myasthenic symptoms, including infections, systemic inflammatory conditions, thyroid dysfunction, menstrual cycle, pregnancy, emotional stress, and sleep disturbances. Multiple medications can impair neuromuscular transmission in myasthenia, and symptoms can worsen with an elevated body temperature and ambient heat.

Effects on the Exercise Response

Objective exercise research on the conditions discussed here is sparse, in part because physicians and support systems for persons with these conditions have historically not encouraged people to be more active. As a result, much of our knowledge about the exercise response in these conditions has a limited perspective.

Individuals who have a chronic condition of nerve, muscle, or neuromuscular dysfunction often have reduced capacity for exercise that is partly due to the condition itself and partly due to physical deconditioning. Conditions that affect motor nerves, muscles, and neuromuscular junctions cause weakness and low endurance. Fear of falling or neuropathic pain can reduce the inclination to be physically active, often leading persons with these conditions to become quite sedentary.

Low muscular endurance in these conditions is generally believed to limit the individual's ability to exercise at a level that would produce cardiovascular conditioning, though exercise

may improve physical functioning in the short term. The following special considerations may affect the exercise response in neuromuscular conditions:

- People can mitigate sensory loss and impaired balance by assuming a wider stance.
- Canes, walkers, and hand railings can also aid balance and safety.
- Sensory loss in the hands may affect dexterity or reduce handgrip tightness.
- Orthotics can help reduce injury and optimize performance, and may be required to support the joint (e.g., an ankle–foot orthosis for foot drop).
- Exercise may need to be modified due to a weak muscle that cannot maintain joint position statically or dynamically.
- Heat intolerance may affect intensity, duration, mode, and participation.

Autonomic dysfunction can be a troublesome problem because it may cause wide fluctuations in blood pressure and pulse, leading to exaggerated pressor or chronotropic responses or to hypotensive responses, light-headedness, and orthostatic instability (usually mild). These problems can typically be improved by use of compression stockings, postural adjustments, proper hydration, and attention to electrolytes. Autonomic dysfunction may also affect sweating, thus altering thermoregulation. In addition, it can

Bear in mind that the autonomic and catecholamine responses to exercise are primarily driven not by the intensity of skeletal muscle functions, but by intensity of central command (intensity of command from the motor cortex). Systemic response and fatigue can be very high because of high-intensity central command, even though the patient is only able to produce a relatively modest amount of external work—people can be trying very hard, even though it looks as though they are not. For this reason, level of perceived effort and fatigue should always be closely monitored.

affect gastric emptying (so exercise should not be done soon after meals); and bowel (constipation or diarrhea) or bladder function (urinary urgency or urinary retention) can contribute to discomfort during exercise. Although exercise will not treat the autonomic dysfunction, it may in some cases help condition the autonomic nervous system.

Effects of Exercise Training

As with our understanding of the exercise responses in neuropathy, myopathy, and myasthenia, only recently have physicians started supporting higher levels of physical activity in individuals with these conditions. So while there are data on physical and occupational therapy as rehabilitation, less is known about more rigorous exercise training in this group. Nonetheless, some important findings and lessons likely apply to many of these individuals.

Core strength and endurance is particularly important in myopathies, which tend to affect proximal (shoulder and arm, pelvic and thigh) musculature. Unlike most people, who have sufficient reserves in strength to "get away with" bad technique and posture, people with weak proximal muscles cannot compensate for weak core musculature. For this reason, maintenance of core strength is very important even when core strength is not the main problem. People with these conditions need their core musculature to be strong, and they may feel as though they're working disproportionately on the core and not enough on their weaknesses. Exercises to facilitate improved respiratory mechanics can conserve energy also by improving body and respiratory posture and movements. People should be counseled to understand these nuances, lest they decide that these aspects are not as important and start to "cut corners" in their program.

Neuropathy

- Many neuropathies are only partial nerve injuries, rather than complete denervation, thus enabling exercise to produce beneficial adaptations to the neuromuscular apparatus.
- Even with complete denervation, electrical stimulation can produce muscle adaptations.
- Movement therapy helps to maintain flexibility, prevent secondary contractures, and avert skin and joint problems to improve the health of the limb.

Myopathy

- Exercise should be individualized to account for the affected muscles, as well as to facilitate compensation of weak and deconditioned muscles, while these muscles are slowly and gradually building strength and endurance.
- Training the core musculature is essential to maintaining proper structural support and prevent kyphosis and scoliosis as well as facilitating proper respiratory mechanics.

Myasthenia Gravis

- Breathing exercises can improve respiratory performance for persons with myasthenia.
- Resistive training can be performed safely and effectively in myasthenia gravis.
- Exercises that engage the core musculature while maintaining proper seated or sitting posture are an important first approach to exercise training in myasthenia gravis.

Recommendations for Exercise Testing

For individuals capable of higher-intensity exercise: Use the ACSM Guidelines. For individuals with mild to moderate deficits: Use the Basic *CDD4* Recommendations with modifications as needed for efficacy and safety.

The primary purpose of an exercise program for individuals with neuromuscular conditions is similar to that for normal individuals—to optimize and maintain long-term health and wellness while enhancing functional performance. Exercise testing may include functional assessment measures specific to neuropathy, myopathy, and myasthenia gravis (e.g., Myasthenia Gravis Activities of Daily Living Profile query of myasthenic symptoms and Quantitative Myasthenia Gravis with timed tests of muscle activities). Common elements of physical inactivity, physical deconditioning, and secondary changes in musculoskeletal tissue and skin constitute the rationale for regular exercise as a component of best care for these disorders.

Low-intensity exercise training is common in people with these conditions and does not require formal graded exercise testing with electrocardiogram (ECG) in all cases because the intensity of activities is similar to elements of usual physical therapy and daily mobility. Some medically high-risk cases may, however, benefit from graded exercise testing with ECG to define cardiopulmonary tolerance and safety even for a low-intensity exercise program. Exercise testing and medical clearance must thus be individualized, depending on the planned intensity and goals of training, the nature of the neurological deficits, and the medical comorbidities that will influence exercise programming and safety.

> • *Neurological deficit profiles may necessitate physical or occupational therapy evaluation (or both) for proper adaptive devices to optimize safety before exercise testing begins.*
>
> • *Exercise protocols may need to be adapted to match the individual's capacities. These modified exercise testing protocols can serve as metrics to document the individual's progress and outcomes across the exercise training.*

Recommended exercise or physical functioning assessments in this population include the following:

- Six-minute walk test
- Balance testing (Berg Balance and Dynamic Gait Index)
- Gait assessment (to determine the safety of walking)
- Timed get-up-and-go or timed five-repetition sit-to-stand for proximal lower-extremity strength
- One-repetition maximum strength in limb muscles with stable musculature
- Handheld dynamometer testing (must be performed in a standardized fashion with the same rater for more dependable test–retest reliability)
- Pulmonary function tests—provide useful measures of breathing function, including forced vital capacity, supine forced vital capacity, maximal inspiratory and expiratory pressures (MIP and MEP), negative inspiratory force, and maximal voluntary ventilation

For all exercise testing in neuromuscular conditions, generalized, muscular, and respiratory fatigue should monitored.

> *Individuals with neuropathy, myopathy, or myasthenia gravis should always obtain clearance from their medical provider before starting any exercise program, even low-intensity programs (e.g., ≤50% of heart rate reserve; community exercise classes). This clearance is necessary because the benefits of exercise training are predicated on the condition's being stable, and their provider should verify that they are indeed stable.*

Excessive levels of fatigue or perceived exertion or a rapid rise in fatigue or perceived exertion is an indication to terminate an exercise test.

Recommendations for Exercise Programming

Exercise programming recommendations are based on the severity of the disease:

- For individuals capable of high-intensity exercise: Use the ACSM Guidelines.
- For individuals with mild to moderate deficits: Use the Basic *CDD4* Recommendations with individual modification as needed for efficacy and safety.

A plan for regular physical activity and prevention of physical deconditioning is recommended for most individuals with neuropathy, myopathy, and myasthenia gravis. Supervision is recommended until the individual knows how to self-monitor for pain, muscular fatigue, and physical decompensation. To exercise independently, people must understand how to avoid the detrimental effects of excessive training of weak muscles. For severe disability, the emphasis should be placed on safe participation rather than the level of exercise.

Orthopedic and joint pain complaints of the knee, hip, and back are common concerns in individuals with neuropathy, myopathy, and myasthenia gravis. These pain issues should be referred for medical and physical therapy, as appropriate, before a formal exercise program is started.

Because of the need to individualize each client's exercise program, it is strongly advised that an exercise program begin with the supervision by trained health care professionals who are well versed in neuromuscular disorders. The exercise specialist should help the client find exercises that are specially selected to meet the individual's needs. The overarching goal is a gradual progression of client-centered exercise to prevent deconditioning and manage symptoms related to neuromuscular disorders. Exercise supervision is important because it helps to

- select the best exercises for the individual's needs,
- ensure that the exercises are performed properly and safely, and
- provide support and overcome fear avoidance of falls, pain, and fatigue.

Individuals with moderate to severe neuromuscular disorders should be managed carefully, and the exercise program should start very conservatively. Given the neuromuscular deficits and profound deconditioning, training should start with three or four discontinuous bouts of low-intensity aerobic training (e.g., walking back and forth at 75% of self-selected walk velocity over a 10 m distance). Each bout of walking can go for a minute or less to a few minutes, depending on the person's ability. Total time should be ~15 min of interval training. Such brief sessions, three times a week, confer beneficial physiological adaptations without overexertion of weak muscles and do not usually cause generalized fatigue. Every 2 weeks, total exercise duration can be advanced by about 5 min, either via adding another bout or via increasing the duration of each bout, toward a goal of 30 min of low-intensity aerobic exercise on most days. The intensity of the training can also be slowly advanced (e.g., walking speed may be increased by 0.25 to 0.5 mph [0.4-0.8 kph] each week). Most people achieve a walking velocity that cannot be safely increased due to abnormal neuromotor control that limits gait biomechanics, affecting safety. When this occurs, a way to increase exercise work rate is to increase the elevation for walking on a treadmill.

Patients may take ~3 months to reach the goal of 30 min of aerobic activity at a low to moderate intensity, a somewhat slower rate of progression than for most people with cardiopulmonary conditions. Nonetheless, this level of adaptation substantially increases the functional walking ability from room to room, at home, from building to building, improving community ambulatory capabilities that people need to sustain indefinitely.

Task practice relevant to walking and balance function may aid improvement in mobility skills. Exercises that safely challenge multisegmental balance control in the context of muscular endurance can improve selected balance outcomes and activation of core muscles, which are key to healthy metabolism. These exercises are examples (with appropriate handrail safety or standby aid):

- Sit to stand
- Lateral stepping
- High stepping in place

Participants should focus on form, maintaining neutral posture, and engagement of the core while minimizing undesired compensatory movements. Actively focused mindfulness on the movement pattern can augment sensorimotor gains and improve balance.

Weight-bearing activities and strength training should be considered in the exercise prescription, as long as balance is adequate and standing does not aggravate pain. These exercises will help address the high osteoporosis risk resulting from relative inactivity with reduced time in weight bearing and any prescribed steroids or immunosuppressant medications that accelerate osteoporosis.

A variety of aerobic, resistive, and mixed modality exercise models improve fitness and reverse deconditioning. These include treadmill, modified cycle ergometry, group exercise classes, pool-based exercise, and steppers. These different exercise modalities provide options for comfort, individual preference, and variety factors that will enhance compliance with an exercise program. Exercises also should incorporate skills that are important for activities of daily living.

Key Tips

- Neuropathic pain should be treated before starting exercise to lessen an individual's fear of performing movements that will induce or intensify pain.
- Physical therapy before, or in combination with, an exercise program can increase safety and serve as a bridge to a lifelong exercise program.
- Exercise planning must consider the medical, physical, and cardiopulmonary aspects of associated conditions causing neuropathy and myopathy.
- Sensory loss and impaired balance may require adaptations of certain standing

activities through provision of hand support or assuming a wider base of support.

- Sensory loss in the hands may affect dexterity or reduce handgrip tightness.
- Muscle weakness may require training at low absolute intensities or short durations.
- Start conservatively, progress slowly, and pace carefully—fatigue can come on quickly.
- Muscle and general fatigue should be closely monitored during exercise and recovery.
- Pulmonary musculature is amenable to training (dyspnea should be monitored).
- Mindful awareness of proper whole-body position improves posture, makes breathing easier, and reduces the likelihood of low back pain and fatigue.

Myopathy-Specific Tips

- Exercise should be monitored frequently to ensure that exercise training does not extend beyond the capacity of the muscle, which would result in muscle injury. Signs of muscle injury or failure include sudden or gradual loss of muscle power, activation of accessory muscles to compensate for fatigue in the target muscles, pain in the muscle or the surrounding joints, subjective report of fatigue or excessive effort, and postexercise weakness or fatigue that lasts hours or days.
- Myoglobinuria (muscle breakdown products in the urine, causing a tea-colored appearance of the urine) should be evaluated immediately by a medical provider, since rhabdomyolysis can cause kidney failure.

Myasthenia-Specific Tips

- A strong sense of body awareness will help people understand the impact of myasthenia on their day-to-day routine, timing of activities to maximize benefit from medications and fluctuating daytime fatigue, and their needs for rest.
- People should keep cool, drink cold beverages to hydrate, and wear breathable, wicking fabrics.
- They should exercise in a cool environment or while wearing a cooling vest.

Integration Into a Medical Home Model

Daily exercise is critically important for people with neuromuscular conditions in order to prevent decline of physical functioning, maintain joint and muscular flexibility, and promote overall health. Care of chronic neuromuscular conditions should include nutrition in combination with exercise, with the goal of optimal weight management since excessive weight increases the energy costs for all activities. Excessive weight also adds to loading on skin, bone, muscle and connective tissues, and persons with neuromuscular conditions have impaired neuromotor control that makes them vulnerable to overuse injury if they don't practice their movement technique and don't use proper biomechanics. The following are considerations relating to facilitation of mastery and maintenance of these essential skills:

- Individuals with neuropathy, myopathy, and myasthenia gravis should be referred to medical specialists and physical or occupational therapists (or both) for assistance.
- Investigate any new symptoms that develop as a result of exercise participation.
- Declines in physical performance might indicate a change in medical, psychological, or sleep status or a change in medications (or adherence to prescriptions).
- Log books help document the progression of conditioning, and success rates are higher in persons who keep a log.

Take-Home Message

Special considerations to keep in mind include the following:

- A regular and structured physical activity program is needed to avoid deconditioning and to optimize neurological function.
- The exercise therapist should establish appropriate goals that are achievable, starting conservatively with a long-term goal of maintaining lifelong regular exercise.
- Exercise therapists must be familiar with how to address fear avoidance.
- Individuals with neuropathy may be prone to muscle cramp and spasms, especially with changes in the exercise routine for muscles that have been relatively inactive

or pushed too far. Exercise should include muscle stretching following exercise and later in the day, with close attention to hydration and electrolyte balance.

- People with myopathy should be advised to report myoglobinuria (muscle breakdown products that give urine a tea-brown hue), which should be evaluated immediately by a medical provider since rhabdomyolysis can cause acute kidney failure.

Suggested Readings

Cass S. Myasthenia gravis and sports participation. *Curr Sports Med Rep.* 2013;12(1):18-21.

Colberg SR, Vinik AI. Exercising with peripheral or autonomic neuropathy: what health care providers and diabetic patients need to know. *Phys Sportsmed.* 2014;42(1):15-23.

Cup EH, Pieterse AJ, Ten Broek-Pastoor JM, Munneke M, van Engelen BG, Hendricks HT, van der Wilt GJ, Oostendorp RA. Exercise therapy and other types of physical therapy for patients with neuromuscular diseases: a systematic review. *Arch Phys Med Rehabil.* 2007;88(11):1452-1464.

Fregonezi GA, Resqueti VR, Güell R, Pradas J, Casan P. Effects of 8-week, interval-based inspiratory muscle training and breathing retraining in patients with generalized myasthenia gravis. *Chest.* 2005;128(3):1524-1530.

Kluding PM, Pasnoor M, Singh R, Jernigan S, Farmer K, Rucker J, Sharma NK, Wright DE. The effect of exercise on neuropathic symptoms, nerve function, and cutaneous innervation in people with diabetic peripheral neuropathy. *J Diabetes Complications.* 2012;26(5):424-429.

Lohi EL, Lindberg C, Andersen O. Physical training effects in myasthenia gravis. *Arch Phys Med Rehabil.* 1993;74(11):1178-1180.

Mermier CM, Schneider SM, Gurney AB, Weingart HM, Wilmerding MV. Preliminary results: effect of whole-body cooling in patients with myasthenia gravis. *Med Sci Sports Exerc.* 2006;38(1):13-20.

Quigley PA, Bulat T, Schulz B, Friedman Y, Hart-Hughes S, Richardson JK, Barnett S. Exercise interventions, gait, and balance in older subjects with distal symmetric polyneuropathy: a three-group randomized clinical trial. *Am J Phys Med Rehabil.* 2014;93(1):1-12.

Voet NB, van der Kooi EL, Riphagen II, Lindeman E, van Engelen BG, Geurts AC. Strength training and aerobic exercise training for muscle disease. *Cochrane Database Syst Rev.* 2013 Jul 9;7:CD003907. doi: 10.1002/14651858.CD003907.pub4.

Wong SH, Nitz JC, Williams K, Brauer SG. Effects of balance strategy training in myasthenia gravis: a case study series. *Muscle Nerve.* 2014 May;49(5):654-660. doi: 10.1002/mus.24054. Epub 2013 Dec 11.

Additional Resources

Foundation for Peripheral Neuropathy, www.neuropathy.org [Accessed September 14, 2015].

Myasthenia Gravis Foundation of America, myasthenia.org [Accessed September 14, 2015].

National Center on Health, Physical Activity, and Disability, www.ncpad.org [Accessed September 14, 2015].

Cerebral Palsy

Cerebral palsy (CP) is a group of common childhood-onset motor disorders that are due to permanent disturbances in the developing fetal or infant brain. Risk factors for CP include prematurity, hyperbilirubinemia, fetal or maternal infection, birth trauma, and intracranial hemorrhage, leading to about two children per 1000 live births diagnosed with CP. Although some children and adults with CP die prematurely, especially if they have severe or multiple comorbidities, most people with CP survive well into adulthood.

Basic Pathophysiology

The most common disturbances in the developing brain in CP are damage to white matter, basal ganglia lesions, cortical and subcortical lesions, and brain malformations. These conditions lead to various motor impairments, the most obvious of which are abnormalities in muscle tone:

- Bilateral spasticity
- Dyskinesia (athetosis or dystonia)
- Ataxia

The most common muscle tone abnormality is spasticity, which affects 80% of individuals with CP; the spectrum of motor impairments, however, varies by individual, depending on the type and site(s) of brain damage.

Motor impairments in CP are often accompanied by impairments to sensation, perception, cognition, communication, behavior, and the musculoskeletal system. These impairments are frequently seen musculoskeletal impairments:

- Muscle weakness
- Contractures
- Skeletal deformities

Epilepsy occurs in about 35% of those with CP and is more likely to occur in individuals who have other impairments.

Management and Medications

After the neonatal period, medical management is largely focused on addressing secondary impairments and limitations that interfere with activities of daily living (ADLs). The following are common medical interventions relevant to exercise:

- Anticonvulsant medication for seizure control
- Antispastic medication
- Selective dorsal rhizotomy surgery to reduce spasticity
- Orthopedic surgery to improve posture and balance (e.g., muscle lengthenings, tendon transfers, osteotomies)

Selective dorsal rhizotomy surgery (in which selected dorsal nerve rootlets are cut to reduce sensory input into neural pathways that contribute to spasticity) is typically done during childhood. Orthopedic surgeries are also typically performed during childhood, with revisions done in adulthood if necessary. Depending on the motor impairments, individuals with CP may use mobility-related equipment, for example:

- Ankle–foot orthoses (AFOs)
- Canes, crutches, or walkers
- Manual or powered wheelchairs

A person's use of this equipment can change, depending on the task or environment. For example, individuals may walk without support at home but use crutches at work.

In addition to medical and surgical management, individuals with CP may require the services of certain allied health care professions (e.g., nursing, physical or occupational therapy, recreation therapy, speech and language pathology, social work, psychology), depending on the needs of the individual, the family, and other caregivers.

Effects on the Exercise Response

People with CP show reduced peak aerobic capacity. This is hardly surprising given that for people to have a high aerobic capacity, they require strong muscles that are under smooth motor control in order to drive the cardiorespiratory system to high levels of oxygen uptake. Many people with CP do not have sufficient neuromusculoskeletal structure and function to do this. Since their *exercise capacity* is essentially limited by their *physical functioning,* this discussion focuses on physical functioning in CP.

About 70% of children with CP walk, but about 20% need support (e.g., a walker). From infancy, predictors of being able to walk include the following:

- Ability to roll, sit, or stand by the age of 2 years
- Absence of severe bilateral spasticity or blindness

For many people with CP who walk, maintaining these skills is a lifelong challenge. After age 8, children with CP who require support to walk are at risk for further decreases in walking ability over time. About 25% of adults with CP also lose at least some walking ability as they age. The following are risk factors for losing the ability to walk during adulthood:

- Being older
- Having severe motor impairments and walking limitations
- Easy fatigability
- Pain

To describe the extent of the mobility limitations in CP, clinicians commonly use the Gross Motor Function Classification System (GMFCS), summarized in table 27.1.

For people without disability, walking is an energetically efficient means of self-transport for ADLs. But energetically efficient walking, or bipedal locomotion, is neurologically complex and requires exquisite synchronization of muscle reflexes, activation, and relaxation. Complex groups of muscles must work in synergy from head to toe. One impact of the motor control problems associated with CP is that this synchronization is disrupted and as a result there is a reduction in walking efficiency. This means that the energy required to walk is increased and the aerobic economy of walking is decreased. As well, the aerobic economy of performing other mobility-related activities may also be decreased. Children and adults with CP therefore show two main alterations in the exercise response:

- An increase in energy expenditure for a given external work rate
- A reduction in peak aerobic capacity

The magnitude of these alterations is related to the magnitude of motor deficits: The greater the mobility limitations, the lower the aerobic economy, the peak aerobic capacity, and the level of regular physical activity.

Low aerobic economy (i.e., increased energy expenditure during walking) together with low peak aerobic capacity are believed to lead to early onset of fatigue, which is a common complaint of individuals with CP and may be one of the reasons they have lower levels of physical activity than their peers without disability.

Table 27.1   The Gross Motor Functioning Classification System (GMFCS)

Classification level	General description of physical functioning
I	Walks and runs without support Coordination, speed, and balance reduced
II	Walks without support but may use support in some situations Difficulty running
III	Walks with support Often uses wheeled mobility in the community
IV	Uses powered mobility or is pushed in a manual chair for most mobility needs
V	Is transported in a wheelchair for all mobility situations

Effects of Exercise Training

The preceding discussion of the effects of CP on the exercise response reveals that adults and children with CP are vulnerable to a downward cycle of the disuse syndrome (see chapter 1). This phenomenon can be characterized as follows:

- Physical activity is difficult due to motor impairments and easy fatigability.
- The individual becomes deconditioned (on a background of low muscle mass).
- Physical activity becomes increasingly harder.
- The individual is in a downward cycle of the disuse syndrome.

Data from girls with CP are consistent with this theory, as their peak aerobic capacity deficit appears to increase over time. As noted already and unless otherwise stated, the information that follows on exercise training pertains to those with CP who are ambulatory, either with or without support. The reason is that most of the exercise training studies in CP, like other exercise studies with this population, have targeted individuals who walk.

In children with CP, exercise training of short (≤3 months) or long duration (≥8 months) using functional activities such as walking, running, wheeling, and swimming (done two to four times per week for at least 30 min, at an intensity of about 60% to 75% peak heart rate [HR]), performed separately or in combination with strength training or anaerobic training, results in a significant increase in peak aerobic capacity of 15% to 40%. Greater improvements appear to be associated with longer-duration interventions and less severe mobility limitations, although this has not been directly evaluated.

For children in GMFCS levels I and II, long-duration combined aerobic and anaerobic lower-limb exercise training results in significant improvements in (1) functional anaerobic power and agility, (2) functional muscle strength, (3) participation in physical activities, and (4) health-related quality of life. In addition, long-duration training that is predominantly aerobic may lead to increases in the level of habitual physical activity, both for children with CP who walk (GMFCS I-III) and for those who do not walk (GMFCS IV). Short- and long-duration aerobic and combined training also result in improved mobility skills. The positive effects on mobility skills and other

functional activities of these interventions may be related to the functional character of the exercises. Lower-limb strength training alone has very little effect on functional mobility in children with CP despite its positive effect on muscle strength. The very limited experience from one long-duration study with children (GMFCS I and II) suggests that exercise interventions alone may not yield lasting increases in physical activity, as most of the immediate exercise-related gains the children made in this study were short-lived.

Very few data exist on the effects of aerobic exercise training in adults with CP. Preliminary evidence suggests that 8 weeks of cycling twice weekly for 30 min (with a gradual increase in intensity from 40% to 70% of peak aerobic capacity) results in about a 10% increase in peak aerobic capacity in young adults with CP who walk without support. Preliminary evidence also suggests that adults with CP, like children, do not continue to exercise on their own after the intervention program is completed. Adults with CP are also similar to the children with the condition in that those who perform lower-limb strength training alone do not improve mobility despite an increase in muscle strength.

Recommendations for Exercise Testing

These are the exercise testing recommendations for people with CP:

- The Basic *CDD4* Recommendations may be appropriate for individuals in GMFCS level I.
- People in GMFCS level II may need individualized adaptive measures.
- People in GMFCS levels III and IV will likely need individualized adaptive measures.
- Functional mobility testing, rather than classic exercise testing, will need to be done with individuals in GMFCS level V.

Because the exercise testing of children and adults with CP often requires adaptive techniques, the ACSM Guidelines for apparently healthy persons and even the Basic *CDD4* Recommendations are not always going to be appropriate. The reader is referred to the Suggested Readings at the end of the chapter for references that describe specific protocols for valid tests for individuals with CP. Screening questionnaires such as the Physical Activity Readiness Questionnaire or the Physical

Activity Readiness Medical Examination questionnaire are also useful to determine readiness for exercise testing and training and can guide the need for physician assessment before the exercise. For individuals with comorbidities such as epilepsy, cardiometabolic disease, or cognitive impairment, the reader is referred to the appropriate chapters of parts II, III, and VI.

Aerobic Capacity

Even if the ACSM exercise testing guidelines or the Basic *CDD4* Recommendations are not always suitable for people with CP, the ACSM Guidelines for client preparation for testing are appropriate and should be followed. Whether lower-limb orthoses are worn depends on the objectives of the test. Aside from issues related to comorbidities, the biggest challenge in testing those with CP is to choose a testing protocol that the individual can physically perform in a manner that is safe and results in valid data.

As noted previously, several aerobic exercise tests that yield valid results have been developed for both children and adults with CP. In general, the starting exercise intensity and the increases in intensity at each stage are individualized based on pretesting or based on the GMFCS level and in some cases also on body size.

Treadmill tests are feasible for those in GMFCS levels I and II, but care should be taken to avoid handrail support if possible, as this blunts the exercise response. The individual should also be comfortable walking on a treadmill before testing. Cycle ergometer tests are feasible for those in GMFCS levels I through III without severe lower-limb deformity or high levels of antagonist muscle coactivation (it is difficult to adapt a cycle ergometer to accommodate severe deformities, and a high level of antagonist muscle coactivation prevents a smooth pedaling motion). With use of a cycle ergometer, the feet are typically fixed to the pedals to help maintain contact on the pedal. Other positioning should be individualized to ensure optimal performance and clearance of the lower limbs during pedaling. The positioning should be documented to standardize repeated testing.

Arm ergometry is feasible for individuals in GMFCS levels I through III without severe antagonistic muscle coactivation or deformity in the upper limbs, and even some individuals in GMFCS IV are able to perform this type of test depending on their level of upper-limb motor impairment.

With use of an arm ergometer, the chair should be stabilized and have a back so the trunk can be supported and stabilized, particularly if peak aerobic power is being measured. As with the feet for cycle ergometry, in arm ergometry the hands may have to be fixed to the handles (usually with Velcro) depending on grip strength. Positioning should be individualized and documented to standardize repeated testing.

Valid and reliable CP-specific shuttle run tests have been developed for children in GMFCS levels I through III. These tests can be used with adults with CP who walk, although they have been validated only in young adults in GMFCS levels I and II. With use of a shuttle run test, peak aerobic capacity can be determined directly using a wearable metabolic device, but without such an apparatus one can obtain an estimate of aerobic locomotor function in very low, low, average, above average, and high levels. This can be used to track the person's cardiorespiratory fitness over time. One advantage of the shuttle run test is that it is the only locomotor-based test that yields valid data for those who can walk but cannot perform a valid treadmill test because they cannot walk on the treadmill unsupported. A second advantage is that there are reference values for children and young adults with CP in GFMCS levels I and II based on height and sex. This means that clinicians can compare the results for someone they are testing to those for a population of similar individuals with CP. The disadvantage of the shuttle run test is that if the laboratory doesn't have a mobile expired gas analysis device, inter-GMFCS level comparisons are not possible because no algorithms exist to estimate peak aerobic capacity from the number of shuttles completed.

Cardiopulmonary Exercise Testing

Graded exercise tests in people with CP generally end when the individual cannot maintain the desired exercise intensity, when he voluntarily stops, or when the evaluator stops the test for safety reasons (see the ACSM Guidelines for terminating a test, as they are applicable for this population). In both research and clinical practice, individuals with CP are considered to have reached their peak aerobic capacity when one observes

- HR ≥180 beats/min and respiratory exchange ratio (RER) >1 (children), and
- HR ≥95% estimated maximal HR and RER ≥1.1 (adults).

These values are somewhat lower than the maximum values seen in healthy individuals without disability because it may not be safe to push to maximum exertion persons who have motor control abnormalities, and cognitive or communication impairment may make it difficult to have them push themselves to their maximum. It is common to see lower values, denoted by "peak" aerobic capacity, in individuals who are particularly unfit or who have other comorbidities that preclude maximum testing.

Unless the individual has cardiometabolic or cardiorespiratory comorbidities, these tests can be used to determine the impact of the person's physical impairments on cardiorespiratory fitness, as well as changes over time (with or without interventions). Cardiopulmonary exercise data may also aid clinical reasoning to explain changes in health or a decreasing participation in ADLs or active leisure activities. The minimal data set, assuming no comorbidities and assuming that a metabolic cart is used, should consist of age, stature, body mass, GMFCS level, test modality and parameters, test duration, oxygen uptake, RER, HR, and the reason for the test termination. The clinical utility of other cardiorespiratory measures (e.g., ventilatory threshold) in individuals with CP has not yet been established.

It is worth noting here that motor control may also vary somewhat from day to day, in part because of variability in adherence to and timing of certain medications. If as a result the individual moves more excessively than what is normal for her during any exercise testing, or if she uses the handrails during treadmill testing, there could be an alteration in aerobic economy. If one is following the results longitudinally, an acute alteration in motor control during testing may need to be noted. The test may have to be repeated on a day when the person is demonstrating more typical motor control behavior.

Strength

Although strength training alone does not typically increase mobility, individuals with CP are weaker than their peers, and muscle strength is important to health and might be a useful measurement as part of a general health-related fitness assessment. Manual muscle testing is used for muscle groups in which full range of motion against gravity with resistance is not possible. For muscle groups with greater muscle strength, handheld dynamometry and repetition maximum testing can also be used,

as can functional testing of activities related to muscle strength (e.g., sit to stand).

Mobility

The 6-min walk test can be used to quantify walking capacity for those in GMFCS levels I through III. For those in GMFCS levels III and IV who use a manual wheelchair, the 6-min push test is appropriate.

Recommendations for Exercise Programming

These are the exercise programming recommendations for people with CP:

- The Basic *CDD4* Recommendations can be a useful starting point in persons with CP.
- However, adaptations are recommended depending on age and GMFCS level (see following list).

Most children and some adults with CP may find the Basic *CDD4* activities uninteresting, may not understand their importance, and thus may not be motivated to do them. Making the exercises into a game or a mild competition, having people do them with others, and charting progress are all ways to increase motivation. Interactive whole-body video games, such as Wii devices, may offer a path to adapting the Basic *CDD4* Recommendations into a more engaging and thus more acceptable physical activity program.

For children in GMFCS level I and some individuals in level II

The general pediatric physical activity guidelines and strategies to promote physical activity are appropriate (see the list of URLs at the end of this chapter for the link to this information).

For adults in level I and some individuals in level II

The Basic *CDD4* Recommendations are a useful starting point.

For children and adults in level III and some individuals in level II

The target of 150 min/week may include a combination of both moderate to intense and light to moderate physical activity.

For children and adults in level IV

The goal of 150 min/week will likely be mostly light-intensity physical activity. Some of this physical activity time should be spent not in the wheelchair, for example on a mat or in a swimming pool, to favor the use of as many muscle groups as possible. A swimming pool is an ideal environment for this subgroup, as they can usually move easily when their weight is partially supported in the water.

For children and adults in level V

The goal should be to spend 150 min/week not in the wheelchair, with appropriate positioning devices to support postures other than sitting (such as a side-lying or semisupine position, or supine in a heated pool). Any voluntary ability to move that does not cause pain or harm should be engaged as much as possible during this time.

Adaptations to the flexibility recommendation (chair sit-and-reach stretches on left and right)

Muscles in CP tend to be stiff, so flexibility exercises should be done every day.

For those in GMFCS levels III through V, this activity should be done in the wheelchair with the brakes on and the seatbelt attached for safety; the activity will require physical assistance for those in GMFCS level V (as well as some people in GMFCS levels III and IV), and care must be taken by the assistant not to stretch beyond tolerance or stress the shoulder joints.

Adaptations to the sit-to-stand recommendation for children and adults in level II

This activity should initially be supervised to see if the person is capable of safely performing a sit-to-stand activity without support.

For children and adults in level II who require support for the sit-to-stand activity, and those in levels III through V

The support should be a stable surface on which people can lean with their upper limbs.

Proper technique is to push down with the legs as the individual rises up, using support with the upper limbs only as much as needed for safety.

Supervision may be required.

For children and adults in level IV who can perform this activity (many cannot):

Support and proper technique should be used as explained for GMFCS level III.

Supervision and some physical assistance will likely be required.

For children and adults in level V:

Contraindicated; sit-to-stand exercise is not feasible.

Adaptations to the step-up recommendation for children and adults in levels II through V

People will need to use a handrail.

Those in GMFCS level III will most likely face sideways, holding the rail with two hands. They may not be able to ascend or descend with their more impaired leg first. It may also be more feasible and safe for the individual to climb up and down one or two steps repeatedly as tolerated, working up to 10 repetitions.

Supervision may be required.

For children and adults in levels IV and V:

Contraindicated, the step-up exercise is not feasible.

Adaptations to the upper-limb strengthening recommendation for children or adults in levels I through V

Those with low grip strength or other hand impairments may need a weight fixed to the hand.

Consider simply moving against gravity in weaker individuals.

For children and adults in level V:

Contraindicated; upper-limb strengthening is not feasible.

Integration Into a Medical Home Model

Many people with CP require the coordinated support of caregivers and health care providers throughout their lives to ensure that their health care needs are met. Children with CP in developed counties often have this type of coordinated care. With adults, however, the situation can vary greatly from optimal coordination of care to difficulty with access to care that is very fragmented. Integrating exercise into the clinical care plan is an important function that is missing for many people and should be an important goal and function of the medical home.

Physical activity and the ability to be mobile are among the central concerns of the long-term health care plan for people with CP. Cognitive, mental, sensory, communication, behavior, and seizure prevention or control are other concerns, depending on comorbidities. Besides addressing immediate health issues, the general goal of exercise programming in individuals with CP encompasses these aims:

- Maintain mobility.
- Increase physical activity.
- Counteract a sedentary lifestyle that predisposes to cardiometabolic disease.
- Reduce the physical and time burdens on caregivers.
- Improve quality of life.

The counseling aspects of exercise management for the person with CP are really no different than they are for anyone else but may seem so if the number of caregivers and health care professionals involved in the individual's care is large. Motivation may be somewhat more challenging for those with CP than for others. Lack of motivation can lead to frustration, and the person or her caregivers may be tempted to give up on overcoming problems that impair adherence to an exercise program. Motivation can be affected by cognitive deficits or simply because it is more difficult to exercise due to the physical impairments (that is, it may be more physically challenging to exercise, or it may be more challenging because of the lack of access to an appropriate place or exercise program). Because evidence is lacking on how to ensure optimal adherence to an exercise program for children and adults with CP, the team must become very skilled in the art of exercise medicine (see chapters 3 and 4).

Take-Home Message

When working with people who have CP, consider these important points:

- Support for behavior change should last at least 6 months.
- Support needs to be face-to-face as much as possible.

- More than one strategy should be used to increase physical activity.
- The individual needs support from significant others.

Suggested Readings

American Thoracic Society. ATS statement: guidelines for the six-minute walk test. *Am J Respir Crit Care Med.* 2002;166:111-117.

Bar-Or O, Inbar O, Spira O. Physiological effects of a sports rehabilitation program on cerebral palsied and post-poliomyelitic adolescents. *Med Sci Sports.* 1976;8:157-161.

Bredin SS, Gledhill N, Jamnik VK, Warburton DER. PAR-Q+ and ePARmed-X+: new risk stratification and physical activity clearance strategy for physicians and patients alike. *Can Fam Physician.* 2013;59:273-277.

Brehm MA, Balemans AC, Becher JG, Dallmeijer AJ. Reliability of a progressive maximal cycle ergometer test to assess peak oxygen uptake in children with mild to moderate cerebral palsy. *Phys Ther.* Epub ahead of print 2013 Nov 21.

de Groot S, Dallmeijer AJ, Bessems PJ, Lamberts ML, van der Woude LH, Janssen TW. Comparison of muscle strength, sprint power and aerobic capacity in adults with and without cerebral palsy. *J Rehabil Med.* 2012;44:932-938.

Hebert LJ, Maltais DB, Lepage C, Saulnier J, Crete M, Perron M. Isometric muscle strength in youth assessed by hand-held dynamometry: a feasibility, reliability, and validity study. *Pediatr Phys Ther.* 2011;23:289-299.

Maltais DB, Pierrynowski MR, Galea VA, Bar-Or O. Physical activity level is associated with the O2 cost of walking in cerebral palsy. *Med Sci Sports Exerc.* 2005;37:347-353.

Verschuren O, Bosma L, Takken T. Reliability of a shuttle run test for children with cerebral palsy who are classified at Gross Motor Function Classification System level III. *Dev Med Child Neurol.* 2011;53:470-472.

Verschuren O, Ketelaar M, de Groot J, Vila ND, Takken T. Reproducibility of two functional field exercise tests for children with cerebral palsy who self-propel a manual wheelchair. *Dev Med Child Neurol.* 2013;55:185-190.

Verschuren O, Ketelaar M, Takken T, Van Brussel M, Helders PJ, Gorter JW. Reliability of hand-held dynamometry and functional strength tests for the lower extremity in children with cerebral palsy. *Disabil Rehabil.* 2008;30:1358-1366.

Verschuren O, Takken T, Ketelaar M, Gorter JW, Helders PJ. Reliability and validity of data for 2 newly developed shuttle run tests in children with cerebral palsy. *Phys Ther.* 2006;86:1107-1117.

Additional Resources

How much physical activity do children need? www.cdc.gov/physicalactivity/everyone/guidelines/children.html [Accessed November 8, 2015].

National Center on Health, Physical Activity, and Disability, www.nchpad.org [Accessed November 8, 2015].

Multiple Sclerosis

Multiple sclerosis (MS) is a chronic, inflammatory neurological disease that is primarily diagnosed between the ages of 20 and 40, affecting a little fewer than 1 in 1000 people—about 2.5 million people worldwide. Multiple sclerosis is two to three times more prevalent in females than males, and, unfortunately, the cause is not known and there is currently no cure. Experts believe that a combination of environmental factors, including decreased vitamin D, the Epstein-Barr virus, and smoking, leads to the development of MS in genetically susceptible individuals. In some persons the effects are very mild, but in others MS can be a devastating disease. Due to its unpredictable nature and course, MS has multidimensional effects on the physical health and well-being of the individuals who have the condition.

Basic Pathophysiology

Multiple sclerosis is an autoimmune disorder affecting the central nervous system (CNS), in a process that primarily involves an immune-mediated attack on the myelin sheath of nerves. Myelin sheaths form a protective covering of nerve axons, providing electrical insulation to facilitate proper conduction of nerve impulses. Thus, when the myelin sheath is damaged or destroyed, nerve impulses are not properly conducted and neural signals are altered or lost. Additionally, MS leads to damage of nerve axons and the formation of hardened, sclerosed plaques within the white and gray matter.

The combination of plaque formation and demyelination leads to a multitude of signs and symptoms; these are the most notable:

- Decreased gait and balance
- Weakness and fatigue
- Spasticity
- Changes in sensation

- Visual disturbances
- Cognitive and emotional impairments
- Sleep disruption
- Loss of bladder and bowel control
- Depression and decreased functional ability

With this variety of symptoms, it is easy to see why MS is known among physicians as one of the "great imitators," and when MS first presents it can be quite difficult to make the diagnosis. MS can dramatically reduce quality of life, with an infinite combination of functional limitations that present a significant challenge to health care professionals.

Multiple sclerosis follows one of four specific courses, characterized by the pattern of relapses (flare-up of symptoms) and remissions:

- Relapsing-remitting (comes and goes, doesn't get worse)
- Primary progressive (never goes into remission, gradually gets worse)
- Secondary progressive (starts as relapsing-remitting, but becomes progressive)
- Progressive relapsing (progresses steadily, with severe episodes intermittently)

The impact is thus highly variable between individuals and even within an individual over time.

Management and Medications

Because MS is a great imitator, diagnosis of MS is usually made in a mix of confirmatory tests *and* the exclusion of other conditions or disorders. These are key features of MS that are necessary for the diagnosis:

- Two clinical attacks
- Affecting separate areas of the CNS
- Occurring at two different times

Magnetic resonance imaging (MRI) is very sensitive at detecting plaques and is typically used to confirm the presence of lesions and exclude other possible CNS conditions. Analysis of cerebral spinal fluid for the presence of monoclonal antibodies is often performed and is diagnostic of the immunological activity. Monoclonal antibodies found during the first attack are usually not sufficient to make the diagnosis, because to qualify as *multiple* sclerosis there must be either very good evidence of a second lesion (by MRI) or a second episode affecting a different area of the CNS. Some people have "silent" attacks, or more likely have had mild symptoms such as general fatigue that were not sufficient to cause them to seek medical attention. In such cases, the presence of multiple lesions in the CNS leads to the deduction that the individual has the relapsing–remitting form and that the current episode is actually a relapse and not the first attack.

Severity of MS is evaluated by clinical features more than by the number or magnitude of CNS lesions, most commonly using the Functional Systems Score or the Kurtzke Expanded Disability Status Scale (EDSS) (table 28.1). The EDSS is the most widely accepted measure of severity

Table 28.1 Expanded Disability Status Scale (EDSS) Scores and Approximate Functioning

EDSS score	Description of Functional Status (FS)
0	Normal neurological exam
1.0	No disability, minimal signs on 1 FS
1.5	No disability, minimal signs on 2 of 7 FS
2.0	Minimal disability in 1 of 7 FS
2.5	Minimal disability in 2 FS
3.0	Moderate disability in 1 FS or mild disability in 3 or 4 FS, though fully ambulatory
3.5	Fully ambulatory but with moderate disability in 1 FS and mild disability in 1 or 2 FS, or moderate disability in 2 FS, or mild disability in 5 FS
4.0	Fully ambulatory without aid, up and about 12 h a day despite relatively severe disability; able to walk without aid 500 m
4.5	Fully ambulatory without aid, up and about much of day, able to work a full day, may otherwise have some limitations of full activity or require minimal assistance Relatively severe disability Able to walk without aid 300 m
5.0	Ambulatory without aid for about 200 m Disability impairs full daily activities
5.5	Ambulatory for 100 m, disability precludes full daily activities
6.0	Intermittent or unilateral constant assistance (cane, crutch, or brace) required to walk 100 m with or without resting
6.5	Constant bilateral support (cane, crutch, or braces) required to walk 20 m without resting
7.0	Unable to walk beyond 5 m even with aid, essentially restricted to wheelchair, wheels self, transfers alone, active in wheelchair about 12 h a day
7.5	Unable to take more than a few steps, restricted to wheelchair, may need aid to transfer; wheels self, but may require motorized chair for full day's activities
8.0	Essentially restricted to bed, chair, or wheelchair but may be out of bed much of day; retains self-care functions, generally effective use of arms
8.5	Essentially restricted to bed much of day, some effective use of arms, retains some self-care functions
9.0	Helpless bed patient, can communicate and eat
9.5	Unable to communicate effectively or eat or swallow
10	Death due to MS

FS = Functional System, which includes pyramidal, cerebellar, brainstem, sensory, bowel and bladder, visual, cerebral, and other systems.

and provides a way to monitor progression. The EDSS scale ranges from 0.0 (normal) to 10.0 (death from MS) and increases in half-point increments. A person with an EDSS value of 4.0 can typically ambulate without aid for ~500 m and is mostly independent in activities of daily living (ADLs). Limitations in the completion of daily activities are typically observed at an EDSS of 5.0.

The disease process of MS is complex and unique for every individual, and treatment is typically focused on both disease and symptom management. Consequently, a multidisciplinary approach to treatment and management of the disease is crucial.

The medical management for an acute attack of MS typically consists of high-dose intravenous glucocorticoids to reduce inflammation within the CNS. For medical management between relapses, several disease-modifying therapies now exist to reduce the rate of relapse and slow disease progression (table 28.2). Other medications may help alleviate a broad range of symptoms such as depression, fatigue, spasticity, weakness, pain, bowel and bladder dysfunction, and endocrine

Table 28.2   FDA-Approved Disease-Modifying Medications Approved for MS

Medication	Mechanism of action	Adverse effects
Interferonβ-1b	Binds to multiple receptors to modulate immune-mediated activity, MS-specific effects not established	• Flu-like symptoms (headache, myalgia, nausea, fever, chills, asthenia, arthralgia) • Depression • Dizziness • Urinary tract infections • Upper respiratory infections
Interferonβ-1a	Binds to multiple receptors to modulate immune-mediated activity, MS-specific effects not established	• Flu-like symptoms (headache, myalgia, nausea, fever, chills) • Abdominal pain • Abnormal blood cell counts • Hepatotoxicity
Natalizumab	Monoclonal VCAM-1 (vascular cell adhesion molecule) antibody blocks cell surface integrins to inhibit leukocyte adhesion to target ligands	• Flu-like symptoms (headache, myalgia, nausea, fever, chills, asthenia, arthralgia) • Hepatotoxicity • Gastroenteritis, abdominal discomfort • Urinary tract infections • Upper respiratory infections
Mitoxantrone	Topoisomerase II inhibitor that impairs DNA repair and interferes with RNA to impair translation	• Headache, nausea, alopecia • Cardiotoxicity • Diarrhea, constipation • Urinary tract infections • Upper respiratory infections • Acute myeloid leukemia
Glatiramer acetate Dimethyl fumarate	Mechanisms not known	• Nausea, vomiting, dyspepsia, asthenia • Dyspnea, palpitations, chest pain • Back pain • Flushing, edema, rash
Fingolimod	Sphingosine 1-phosphate receptor antagonist, believed to reduce ability of leukocytes to translocate and migrate into the central nervous system (CNS)	• Hepatotoxicity • Headache, dizziness • High blood pressure • Upper and lower respiratory infections
Teriflunomide	Pyrimidine synthesis inhibitor, thought to reduce leukocyte activation in CNS	• Hepatotoxicity • Headache, diarrhea, nausea • Anxiety, high blood pressure • Upper and lower respiratory infections

disorders. Individuals sometimes supplement symptom medical management with alternative methods including cannabis, herbal remedies, acupuncture, massage therapy, and vitamins.

In addition to pharmacotherapy, people with MS may require adaptive equipment to enable mobility and participation in physical activities. Numerous assistive devices are available, including these:

- Canes, walkers, and crutches
- Orthotics
- Functional electric stimulation
- Wheelchairs and motorized scooters

People with MS are typically sensitive to climate changes and heat, so additional aids such as cooling vests may be recommended.

Effects on the Exercise Response

Individuals with MS tend to live sedentary lifestyles, with little to no participation in light physical activity, much less strenuous exercise. In part, this is because the progressive course and episodic exacerbations of MS make sustained participation a challenge. The consequence is that it is very common for people with MS to be deconditioned, with a reduced muscular and cardiovascular fitness as compared to the general population (though the general population doesn't meet physical activity guidelines!). Fatigue is a common and debilitating symptom seen in many of those living with MS, and a large majority of individuals experience heat sensitivity. Thus, people with MS commonly have an atypical exercise response, including these:

- Low or very low physical functioning
- Dyspnea on exertion
- Exaggerated pressor and chronotropic responses
- Weakness and easy fatigability
- Low thermal stress tolerance

Effects of Exercise Training

Experts previously thought that physical activity was relatively contraindicated in MS, but recent investigations have demonstrated that aerobic exercise can increase peak oxygen uptake ($\dot{V}O_2$peak). Additionally, endurance and resistance training may increase strength and range of motion and help reduce the negative effects of MS on movement and physical functioning. It is worth noting, however, that people may see a temporary worsening of symptoms upon initiating an exercise program and typically report an increase in fatigue. This exacerbation of symptoms generally resolves within 30 min of completing the exercise session. Unfortunately, science hasn't clarified what type of exercise and at what intensity and frequency may be the most beneficial for individuals living with MS, as well as how exercise may affect individuals with more severe impairments (EDSS of 7 and above).

Like patients who have other autoimmune disorders, persons with MS are best served by starting low, progressing slowly, and emphasizing enjoyment and fun activities that help promote physical functioning. To this end, light-to-moderate activities, "listening to their body" and erring on the side of doing too little is likely to pay dividends to individuals with MS.

Recommendations for Exercise Testing

Follow the Basic *CDD4* Recommendations for persons with MS.

Due to the wide range of neuromuscular and musculoskeletal impairments (including high-dose corticosteroids that increase the likelihood of osteoporosis), maximal testing may not be safe for these individuals and is not recommended. Additionally, assessments may be adversely influenced by medications, secondary conditions, and disease severity.

Submaximal physical functioning tests, as in the Basic *CDD4* Recommendations, are a good choice for people with MS. Balance, sensation, and flexibility are important to assess, since these measures are typically compromised and may guide the elements of a training program. Individual performance may be diminished if testing occurs later in the day or in hotter climates.

Recommendations for Exercise Programming

Follow the Basic *CDD4* Recommendations for persons with MS.

No one specific exercise program is appropriate for all individuals living with MS. Individuals with an EDSS of 6.0 or lower may respond well to exercise training but should avoid overexertion and exercising in hot temperatures. Additionally, resistance exercise may be more beneficial for those who are sensitive to heat since core body temperature is not raised as significantly as with endurance training. The sit to stand and step-ups are extremely useful exercises to try to maintain lower-extremity function.

Because of the need to individualize each person's program, it is strongly advised that an exercise program begin with supervision of a trained professional who is well educated about MS. Depending on the person's individual circumstances and the skills of the professionals who are locally available to provide guidance, these options are the most suitable:

- Physical therapist
- Occupational therapist (with more debilitated individuals)
- Registered Clinical Exercise Physiologist (RCEP)

Supervision is needed for these reasons:

- To select the best exercises for the person's individual needs
- To ensure that the exercises are performed properly and safely
- To offer counseling for support and to help overcome fear avoidance

A key goal is for the person to become independent in a home program. Since boredom is a key factor in decreased participation, it may be beneficial to offer a variety of activities. Finally, individuals with MS tend to isolate themselves, and group activities may help to increase socialization and exercise adherence. Conversely, some people are very self-conscious and reluctant to participate in a group setting. A key function of the exercise specialist is to help people find what is right for them. The overarching goal is a gradual progression of client-centered exercise to prevent deconditioning and manage symptoms related to MS.

Integration Into a Medical Home Model

Regular exercise will not stop the course of MS, but routine engagement in physical activity can have neurorehabilitative and neuroprotective properties that will help persons with MS better cope with their symptoms and limitations. A well-designed exercise program will enhance their independence and quality of life, and everyone with MS should be referred for evaluation by an exercise therapist trained to manage persons with MS.

People with MS are more likely to lead sedentary lifestyles, and the combination of inactivity and disease progression can lead to a downward spiral that increases risk for diabetes, osteoporosis, obesity, and hypertension. A secondary function of the exercise program is to combat the tendency to develop these secondary cardiometabolic conditions.

Medical home staff, physicians, and therapists should work together to develop a regimen that is tailored to the individual.

- Exercise should begin at low levels using a familiar activity.
- Exercise should be gradually increased as tolerated.
- Exercise should focus on increased engagement rather than continuous intensive exercise.
- Alternative exercise programs should be offered, with an emphasis on fun.
- Try multiple short bouts of exercise during the day, varying intensity and duration.
- Try yoga, aquatics, and safe but adventurous activities like therapeutic horseback riding.
- Most especially, people should learn to "listen to their body," backing off or skipping a day when they are not feeling well, if they are experiencing a relapse, or maybe just because they have started or stopped a medication and need to adjust to the side effects.

Take-Home Message

When working with people with MS, remember these special considerations:

- Supervision is to ensure client safety and ensure that exercises are performed properly.

- Advise caution with exertion in warm conditions (e.g., heated yoga rooms and pools).

- Strength training is less likely to increase the body's core temperature and may be better tolerated in persons living with MS, especially those who are heat sensitive.

Suggested Readings

Dalgas U, Stenager E, Ingemann-Hansen T. Multiple sclerosis and physical exercise: recommendations for the application of resistance-, endurance-, and combined training. *Mult Scler.* 2008;14:35-53.

Finlayson M. *Multiple Sclerosis Rehabilitation: From Impairment to Participation.* Boca Raton, FL: CRC Press; 2013.

Gappmaier E. The submaximal clinical exercise tolerance test (SXTT) to establish safe exercise prescription parameters for patients with chronic disease and disability. *Cardiopulm Phys Ther J.* 2012;23(2):19-29.

Kuspinar A, Andersen RE, Teng SY, Asano M, Mayo NE. Predicting exercise capacity through submaximal fitness tests in person with multiple sclerosis. *Arch Phys Med Rehabil.* 2010;91:1410-1417. doi: 1016/j.apmr.2010.06.005.

Mayo NS, Bayley M, Duquette P, Lapierre Y, Anderson R, Bartlett S. The role of exercise in modifying outcomes for people with multiple sclerosis: a randomized trial. *BMC Neurol.* 2013;13:69.

McAuley E, Motl RW, Hu L, Doerksen SE, Elavsky S, Konopack JF. Enhancing physical activity adherence and well-being in multiple sclerosis: a randomized controlled trial. *Mult. Scler.* 2007;13:652-659.

Stuifbergen AK, Blozis SA, Harrison TC, Becker HA. Exercise, functional limitations, and quality of life: a longitudinal study of persons with multiple sclerosis. *Arch Phys Med Rehabil.* 2006;87:935-943. doi: 10.1016/j.apmr.2006.04.003.

White LJ, Dressendorfer RH. Exercise and multiple sclerosis. *Sports Med.* 2004;34(15):1077-1100. www.direct-ms.org/pdf/GeneralInfoMS/ExerciseAndMS.pdf.

Widener GL. Multiple sclerosis. In Umphred DA, Lazaro RT, Roller ML, Burton GU. *Umphred's Neurological Rehabilitation.* St. Louis: Elsevier Mosby; 2013:585-600.

Additional Resource

National Multiple Sclerosis Society, www.nationalmssociety.org/index.aspx [Accessed November 8, 2015].

Parkinson's Disease

Parkinson's disease is a progressive condition involving the extrapyramidal part of the nervous system, causing impairment in motor function. The most common form of Parkinson's disease is an idiopathic neurodegenerative disorder that usually occurs after the age of 50, is found slightly more frequently in men than in women, and is less prevalent in African blacks and Asians. There is no known cause of Parkinson's disease; however, both genetics and environment (e.g., exposure to toxins) are risk factors.

Individuals with Parkinson's disease have problems with many aspects of movement. Tremors are common, both at rest and with action, as is rigidity of the spine, trunk, and extremities. The ability to rapidly move fingers, hands, arms, or legs is drastically reduced (bradykinesia), while standing posture becomes kyphotic and flexed. People often have difficulty initiating the first step in walking (known as start hesitation) and have a gait that is typically slow and shuffling, with shortened steps that are involuntarily hurried (festinating gait). Episodes of freezing can occur during walking, and postural righting reflexes cause problems with frequent falls. Fine motor functions can also be severely affected, which can make it very difficult to communicate and perform activities of daily living (ADLs) that require manual dexterity. Many people require assistance with dressing, bathing, and preparing and eating meals. All of these abnormalities are independent of cognitive functions, which are unaffected by Parkinson's, making these problems very frustrating to someone with the condition.

Basic Pathophysiology

Parkinson's disease is associated with a reduction in the neurotransmitter dopamine, primarily in the substantia nigra, a component of the basal ganglia. The reduction of dopamine results from the death of dopaminergic cells that live within the basal ganglia and produce dopamine, but symptoms do not occur until loss of these dopaminergic cells is greater than 80%. Other factors that may contribute mechanisms to Parkinson's are aging, autoimmune disorder, and mitochondrial dysfunction. The following are the symptoms caused by the loss of dopamine:

- Resting and action tremors
- Rigidity
- Bradykinesia
- Standing kyphosis with flexed extremities
- Start hesitation with festinating gait
- Dynamic postural instability

The motor symptoms that occur with Parkinson's disease affect many aspects of gross motor function. Tremors can be evident both at rest (spending energy needlessly) and with action (impairing coordination). Rigidity often begins in the neck and shoulders and spreads to the trunk and extremities, and rigidity combined with bradykinesia causes many individuals to have difficulty rising from a chair, or getting into or rising from bed or both. Ambulation can be severely compromised by start hesitation, festinating gait, freezes, and difficulty navigating doorways and narrow spaces. Individuals who have a Hoehn and Yahr disability level of 3.0 are unable to recover balance on a pull test, increasing fall risk.

Fine motor control is also affected. Many people develop micrographia (minute, illegible handwriting), are unable to cut food or handle utensils, and have difficulty swallowing food. Many have trouble speaking loudly or understandably enough to communicate, a difficulty that can be exacerbated by a loss of facial expression (hypomimia).

Management and Medications

Parkinson's disease is clinically classified in a number of categories related to the gross features of the person's condition:

- Age at onset (<40 [juvenile], between 40 and 70, or >70)
- Predominant symptom (tremor, akinetic rigidity, postural instability–gait difficulty)
- Mental status (dementia present, absent)
- Clinical course (benign, progressive, malignant)
- Level of disability (Hoehn and Yahr stages 1.0-5.0)

These classifications also have prognostic implications. For example, a malignant clinical course is the situation in which symptoms have been evident for less than 1 year but disability has progressed rapidly. The prognosis is poor in this instance.

Medications are the most successful way to treat the symptoms of Parkinson's disease. Drug management is aimed at correcting or preventing neurochemical imbalances in relation to dopa-mine, epinephrine, and norepinephrine deficiencies and the relative increase in acetylcholine. The most common antiparkinson medications are listed in table 29.1.

A key drug for Parkinson's disease is levodopa, which is a metabolic precursor to dopamine that can pass through the blood–brain barrier. In the brain, levodopa is metabolized into dopamine and thereby increases the amount of dopamine available in the basal ganglia. Levodopa is also metabolized in peripheral muscle, which reduces the bioavailability of levodopa for the brain. Carbidopa, which is a modified form of levodopa, is often administered with levodopa because carbidopa is less rapidly metabolized by skeletal muscle and thereby has more bioavailability for crossing the blood–brain barrier.

Antiparkinson medications have both peripheral and central side effects; the following are the most common (see also table 29.2):

- Gastrointestinal upset
- Confusion
- Delusional states
- Hallucination
- Insomnia
- Changes in mental activity

Table 29.1   Medications for Parkinson's Disease

Drug class	Examples	Mechanism of action
Dopaminergic	Levodopa Levodopa, carbidopa	Increase dopamine levels Carbidopa inhibits breakdown of levodopa
Dopamine agonists	Bromocriptine Pramipexole Ropinrole Rotigotine	Activate dopamine receptors in basal ganglia
Antiviral	Amantadine	Increases presynaptic release of dopamine
Anticholinergics	Benztropine Trihexyphenidyl	Increase breakdown of acetylcholine
Monoamine oxidase type B (MAO-B) inhibitors	Deprenyl Selegiline Rasagiline	Inhibit breakdown of dopamine by mitochondrial oxidation of dopamine
Catechol-O-methyltransferase (COMT) inhibitors	Entacapone Tolcapone	Inhibit breakdown of dopamine
Discontinued in United States	Pergolide	Dopamine and serotonin agonist

Long-term use of drug therapies, particularly dopaminergics and monoamine oxidase type B (MAO-B) inhibitors, often leads to movement disorders (e.g., dyskinesias), dystonias, and clinical fluctuations of motor disability. More than 50% of all those who have taken medication for Parkinson's disease for more than 5 years have reduced responses to their medications and experience fluctuation of motor disability.

Medical management of individuals with Parkinson's is often difficult because medications really are the main form of treatment, but they decline in effectiveness over time and the variability in individual responses to the drugs is very great. Most people take multiple medications, with the combination often achieved by trial and error to determine what works best for the particular individual. In general, there is not good understanding of how drug absorption, metabolism, and effectiveness combine in a particular person, and there is very little knowledge on how exercise training affects drug efficacy. In theory, strenuous exercise can reduce absorption or interact with factors such as anticholinergic drugs, autonomic dysfunction, recent food intake, amount of protein in the diet, nutrient supplementation, and many other factors (see table 29.3 for additional special considerations). Little of this has been carefully studied with regard to impact on physical functioning in Parkinson's disease.

Table 29.2   Adverse Effects of Parkinson's Disease Medications

Medication	Adverse effects
Levodopa, carbidopa	Can produce exercise bradycardia and transient peak-dose tachycardia and dyskinesia
Pramipexole Ropinirole Pergolide (discontinued in the United States)	May lower blood pressure and exacerbate dyskinesia
Selegiline	Associated with dyskinesia; produces mood elevation

Table 29.3   Other Considerations in Parkinson's Disease

System	Considerations
Physical functioning	Walking may be a problem with balance deficits. Some patients can jog safely.
Exercise programming	Increase exercise dose very slowly (about once a month). Duration and intensity should receive equal priority.
Orthopedic	Joint dysfunction is common in Parkinson's patients over 50 years of age. Caution against overuse syndromes from exercise training. Dystonia or dyskinesia can exacerbate joint degeneration.
Sleep	Painful nocturnal dystonias can interfere with sleep. Vivid dreaming may occur.
Psychological	Depression is common. Mask-like facial features make it difficult to read emotions and feelings. People may feel vulnerable when off meds and invincible on them.
Metabolic	High resting metabolic rates are common. Exercise training may alter medication bioavailability.
Gastrointestinal	Discomfort is a common side effect of medication.
Autonomic	Sweating may be absent but improve with exercise training (increased risk of hyperthermia).

Effects on the Exercise Response

The effect of Parkinson's disease on exercise is quite difficult to characterize because no two people with the disease are alike and even the same person can be different from day to day. Symptoms may fluctuate from hour to hour, day to day, and week to week. All of this makes it very difficult to generalize about exercise response in this population. The motor control abnormalities adversely affect many aspects of movement and result in poor physical functioning:

- Episodes of decreased movement or freezing during walking
- Increased difficulty getting through doorways or narrow spaces
- Difficulty getting into and rising from bed or getting out of a chair
- Need for assistance with dressing and bathing

The exercise response in Parkinson's can be unpredictable and seemingly illogical. Sometimes walking is severely impaired, but other tasks such as stair climbing are easily accomplished. Freezing and start hesitation may make certain activities more difficult and inefficient than others. During a single session of activity there can be a significant loss of upright posture (increasing kyphosis).

In addition to these motor control difficulties, autonomic nervous system dysfunction is common in Parkinson's disease. This dysfunction can cause problems with thermal regulation, as well as altered heart rate and blood pressure responses with postural changes and during physical activity. The cardiovascular system is often adversely affected by the movement disorders and muscular rigidity, because this decreases exercise efficiency and results in higher heart rates and oxygen uptake for any absolute level submaximal exertion. The combination and variability of all these abnormalities mean that the difficulties in physical functioning experienced by people with Parkinson's disease cannot be characterized as a simple trait.

Effects of Exercise Training

Knowledge about exercise training in individuals with Parkinson's disease is modest but growing. Aerobic training can improve function, can fail to affect function, or can reduce function in individuals with Parkinson's. From the preceding description of the impact of Parkinson's disease on the ability to do ADLs, it should be apparent that in someone with mild and reasonably stable disease, strengthening and motor control exercises may be of great benefit, but for someone with severe and high daily variability, every day can present a different challenge.

Individuals can experience direct, indirect, and what are called composite effects of Parkinson's disease. Direct effects are those that occur directly as a result of Parkinson's disease—for example, tremor and rigidity. Indirect effects, such as aerobic deconditioning or loss of range of motion from inactivity, occur along with the disease. Composite effects may be a combination of the direct central nervous system changes and compensatory musculoskeletal symptoms, such as changes in axial mobility and balance problems. Exercise interventions could have a minimal effect on the symptoms resulting directly from the disease process, but appropriately designed interventions may alter the indirect or composite effects of musculoskeletal and cardiovascular or metabolic deconditioning.

Because exercise training is a chronic activity that is predicated on a predictable dose–response relationship in order to yield anabolic effects and improvements in function, one cannot assume that exercise training will always prove beneficial to a given individual with Parkinson's. The complexity of the problem, the progressive nature of the disease, and the impact of medications on the condition introduce uncertainty. In effect, each person is in a trial-and-error situation, so the therapist must

Exercise management for persons with Parkinson's disease can be very challenging. Each patient has unique burdens of the condition, their response to medication affects the exercise response and fluctuates daily, persons with Parkinson's are likely to have other co-morbid conditions and the effect on facial expression can make it difficult to interpret their reactions to a situation. Exercise specialists working with such individuals should have skill at interpreting all of these factors and with improvising a program that suits each patient's situation.

have a great deal of mastery in the art of exercise management.

Recommendations for Exercise Testing

Recommendations for exercise testing for people with Parkinson's disease are based on Hoehn and Yahr class:

- People in Hoehn and Yahr class 1 and 2: follow the ACSM Guidelines.
- People in Hoehn and Yahr class 3 or higher: use the Basic *CDD4* Recommendations, with modifications for Parkinson's.

Because rigidity and gait-balance are so commonly affected and have adverse impacts on physical functioning, these are often the most important evaluations to do in someone with Parkinson's. And because the spine and proximal joints have great impact on function of extremities, evaluation of neck, trunk, and more proximal joint flexibility is also important. Therefore, evaluations to be considered as part of an exercise assessment include these:

- Pull test
- Timed tandem stand and 360° turn
- Functional reach or chair sit and reach
- Reaction time

Gait and physical functioning should be evaluated to determine competency with ADLs; assessment can include these measures:

- Gait speed
- Timed walk
- Walking cadence and step length
- Sit to stand

Together, these measures, which are a minor modification of the Basic *CDD4* Recommendation, establish a baseline to determine progress or decline, classify Hoehn and Yahr disability level, define existing balance deficits, reveal quadriceps weakness or poor motor control, and help determine whether driving competence may be questionable.

With regard to cardiovascular risk, persons with Parkinson's disease are typically at an age that is at high risk for latent cardiovascular disease, so follow the ACSM Guidelines to evaluate for coronary artery disease. But the need for graded exercise testing also depends on the severity of Parkinson's, the person's age, the other traditional cardiovascular risk factors, and the individual's general robustness (i.e., ability to do higher-intensity activities). Persons in Hoehn and Yahr stage 2 typically have aerobic capacities comparable to those of healthy age-matched controls, so following the ACSM Guidelines is probably reasonable for individuals in both stages 1 and 2. This procedure would be particularly pertinent if people seek to participate in a moderate to vigorous exercise program.

People in Hoehn and Yahr stages 3, 4, and 5, in addition to having lower functional capacities, have greater inter- and intraindividual variability in physical functioning. If an individual in one of these groups mainly seeks a low to moderate level of physical functioning, focused more on maintaining ADLs, graded exercise testing with an electrocardiogram (ECG) may not be indicated. If the person exhibited very low exercise efficiency, however, and seemed to have inappropriately high exertional heart rate and pressor responses, one would have more rationale to perform an exercise test. The decision on whether to test should be made clinically by the exercise specialist in conference with the person's physician.

There are a number of condition- and medication-related concerns to be aware of during exercise in people with Parkinson's:

- Use a cycle ergometer or treadmill with a safety harness if balance is poor.
- Prevalence of cardiac dysrhythmias is high (drug side effect and age of population).
- Autonomic dysfunction is common (poor temperature, orthostatic, and heart rate regulation).
- All exercise should be started 45 to 60 min after medication has been taken.
- Some individuals demonstrate a brief, intense tachycardia or dyskinesia at peak drug levels.
- Caution should be used after a change in medications; the impact may be unpredictable.
- Individuals who fluctuate dramatically may need to be evaluated on and off medications.
- Face masks are recommended if there is a need to do respired gas analysis (many people have difficulty sealing lips on mouthpiece).

There are a number of sometimes surprising phenomena that affect feasibility and safety. People with Parkinson's disease who have balance deficits or freezing episodes should use treadmills with a safety harness, though many individuals with advanced Parkinson's can pedal a cycle ergometer or perform an arm crank exercise even if they are in an "off" period and unable to walk without assistance. Frequently, nonroutine motor behaviors are easily accomplished, so don't make assumptions about what an individual can do. Experienced therapists typically help the client try an activity as long as he is safe from injury.

Recommendations for Exercise Programming

Like recommendations for exercise testing, exercise program recommendations for people with Parkinson's disease are based on the Hoehn and Yahr classification:

- Persons in Hoehn and Yahr class 1 and 2: follow the ACSM Guidelines.
- Persons in Hoehn and Yahr class 3 or higher: use the Basic *CDD4* Recommendations, with modifications for Parkinson's.

The approaches to exercise training for Parkinson's disease fall into five categories: flexibility, aerobic training, functional training, strengthening, and neuromuscular coordination-skill. Any exercise prescription should keep the goals of the exercise clearly in mind, recognizing that different interventions will have different expectations and outcomes. The Basic *CDD4* Recommendations are applicable for people with Parkinson's, particularly for maintaining

- strength and flexibility,
- gait and balance,
- physical functioning (sit to stand and transfer exercises), and
- orofacial and manual dexterity (speech and occupational therapy).

Table 29.4 suggests specific exercises that may be suitable for people with Parkinson's disease.

Parkinson's disease can interfere with motor planning and motor memory. Repeated demonstrations along with written and visual cues are needed to ensure adherence. In some instances, supervision may be necessary for participation in an exercise program.

Most of the concerns regarding exercise testing are equally applicable to an exercise training situation; however, some additional considerations apply to people with Parkinson's:

- Some individuals have an impaired chronotropic response and have difficulty achieving target heart rates.
- Heart rate responses to the same activity may vary greatly from day to day.

Table 29.4   Suggested Training Activities for Parkinson's Disease

Type of training	Suggested exercises
Resistance	Shoulder shrugs, shoulder squeezes, trunk twists, hip bridges, straight-leg raise, leg kicks, toe–heel raise, standing against the wall, and wall push-up
Flexibility	Head turns and tilts, chin tucks, arm stretches, wrist circles, finger–thumb circles, back extension, hamstring stretch, bringing the knees to the chest, hip rolls, calf stretch, and ankle circles
Gait	Use attentional strategies: reminding about larger steps, using visual or auditory cues to improve step length, and walking with alternate arm swing
Balance	Tandem walking, sideways walking, backward walking, standing on one limb, alternate standing on toes and heels, and turning the head and looking over the shoulders while standing
Transfers	Bed turns (sequential trunk rolling and pushing up) and getting up (leaning forward, shifting weight, and standing up quickly), sit to stand
Orofacial	Chin tucks (in and out), eyebrow lifts, deep breathing, sticking the tongue in and out, saying "u-k-r" loudly, lip presses, whistling, circling the tongue inside the lips, long blowing through the mouth, big smiles, face squeezes, and cheek stretches

- Heart rates should be carefully observed for evidence of this variability.
- Training benefit may depend on consistently exercising after the same period of time following a dose of medication.
- The time to peak drug effect may thus be useful for exercise training, and should be noted.

Observe for changes in Parkinsonian symptoms and signs during training (e.g., increases or decreases in dyskinesias, bradykinesia, dystonias, freezing, tremor), which might indicate changes in drug absorption or metabolism.

Integration Into a Medical Home Model

Parkinson's disease is a very complex neurodegenerative disease that causes both motor and nonmotor problems. Staff in the medical home should follow these individuals closely to help them and their families deal with medications and changes in severity of their condition, and most especially to provide guidance on living with a frustrating condition that can last 20 to 30 years. Many people need encouragement and coaching on how to maintain an active lifestyle and achieve a high quality of life despite the substantial challenges that Parkinson's disease presents.

Individuals with Parkinson's disease can exercise safely and achieve many beneficial results, especially through participation in regular physical activity consistent with the Basic *CDD4* Recommendations. Beneficial outcomes they should strive for include the following:

- Increased physical fitness in Hoehn and Yahr stages 1 and 2
- Increased physical functioning in higher Hoehn and Yahr stages
- Improved gait and balance
- Generally better function in ADLs

Many people prefer to participate in group exercise activities for the additional benefit of socializing with others who have Parkinson's disease as well as their families. Regular contact, such as frequent phone calls, may improve compliance with an exercise program.

Take-Home Message

These are important considerations to keep in mind when working with people with Parkinson's disease:

- All exercise should be started about 45 to 60 min after medication has been taken.
- If balance is poor, always use an ergometer with a seat or a treadmill with a safety harness.
- Heart rate responses and exercise efficiency may vary greatly from day to day.
- Because of variable chronotropic response to exercise, use rating of perceived exertion for estimating intensity.
- People may have poor temperature regulation and orthostatic intolerance.
- Caution is urged after a change in medications, as the impact may be unpredictable.
- Monitor for increases or decreases in Parkinsonian symptoms and signs (dyskinesias, bradykinesia, dystonias, freezing, tremor).
- Arthritis is common in people over 50 years old and exercise may help, but dyskinesia and dystonia can aggravate degenerative joint disease.
- Painful dystonias may interfere with exercise and sleep.
- Mask-like facies and communication difficulties can make it difficult to interpret a person's reaction or perceived exertion.
- Depression is a common comorbidity.

Suggested Readings

Allen NE, Sherrington C, Suriyarachchi GD, Paul SS, Song J, Canning CG. Exercise and motor training in people with Parkinson's disease: a systematic review of participant characteristics, intervention delivery, retention rates, adherence, and adverse events in clinical trials. *Parkinsons Dis.* 2012;2012:854328. Epub 2011 Nov 16.

Corcos DM, Robichaud JA, David FJ, Leurgans SE, Vaillancourt DE, Poon C, Rafferty MR, Kohrt WM, Comella CL. A two year randomized controlled trial of progressive resistance exercise for Parkinson's disease. *Mov Disord.* 2013 Mar 27. doi: 10.1002/mds.25380.

Ellis T, Cavanaugh JT, Earhart GM, Ford MP, Foreman KB, Fredman L, Boudreau JK, Dibble LE. Factors associated with exercise behavior in people with Parkinson disease. *Phys Ther.* 2011 Dec;91(12):1838-1848.

Goodwin VA, Richards SH, Taylor RS, Taylor AH, Campbell JL. The effectiveness of exercise interventions for people with Parkinson's disease: a systematic review and meta-analysis. *Mov Disord.* 2008 Apr 15;23(5):631-640.

Schenkman M, Ellis T, Christiansen C, Barón AE, Tickle-Degnen L, Hall DA, Wagenaar R. Profile of functional limitations and task performance among people with early- and middle-stage Parkinson disease. *Phys Ther.* 2011 Sep;91(9):1339-1354. Epub 2011 Jul 21.

Schenkman M, Hall DA, Bar-on AE, Schwartz RS, Mettler P, Kohrt WM. Exercise for people in early- or mid-stage Parkinson disease: a 16-month randomized controlled trial. *Phys Ther.* 2012;92:1395-1410.

Tomlinson CL, Patel S, Meek C, Clarke CE, Stowe R, Shah L, Sackley CM, Deane KH, Herd CP, Wheatley K, Ives N. Physiotherapy versus placebo or no intervention in Parkinson's disease. *Cochrane Database Syst Rev.* 2012 Aug 15;8:CD002817. doi: 10.1002/14651858.CD002817.pub3.

Trail M, Protas EJ, Lai E. *Neurorehabilitation in Parkinson's Disease: An Evidenced-Based Treatment Model.* Thorofare, NJ: Slack; 2008.

van Nimwegen M, Speelman AD, Overeem S, van de Warrenburg BP, Smulders K, Dontje ML, Borm GF, Backx FJ, Bloem BR, Munneke M; ParkFit Study Group. Promotion of physical activity and fitness in sedentary patients with Parkinson's disease: randomised controlled trial. *BMJ.* 2013;346:f576. doi: 10.1136/bmj.f576.

Additional Resources

National Institute of Neurological Disorders and Stroke, www.ninds.nih.gov [Accessed November 8, 2015].

Parkinson's Disease Foundation, www.pdf.org [Accessed November 8, 2015].

WebMD, Parkinson's Disease Health Center, www.webmd.com [Accessed November 8, 2015].

Muscular Dystrophy

Diseases of skeletal muscle—*myopathies*—are surprisingly common; many of them are *acquired conditions* due to chronic disease such as kidney failure, to toxins or malnutrition such as cardiomyopathy of vitamin B_{12} deficiency (so-called wet beriberi), or to autoimmune polymyositis or many other diseases. Some diseases of striated muscle (i.e., not smooth muscle in the gastrointestinal tract and vasculature) are congenital, for example the muscular dystrophies. Muscular dystrophies are a very diverse group of myopathies, mostly affecting skeletal muscle, but some muscular dystrophies can also affect the heart. Most muscular dystrophies become apparent in youth, but some are not manifest until young adulthood.

Historically, people afflicted with myopathies were often advised that they were disabled and that they should use assistive devices to make it easier to do activities of daily living. In recent decades, however, more people with muscular dystrophies have pushed the boundaries of their limitations, seeking independence and a better quality of life. Because of the very broad spectrum of diseases of muscle, and the fact that only a small percentage of people are afflicted, there is disappointingly little knowledge about exercise in the management of these conditions (especially with specificity for each of the unique diseases). It seems paradoxical that knowledge of how to exercise—the very function of skeletal muscle—is something we don't know much about with regard to diseases of muscle, but that is the state of the science today. This chapter focuses primarily on two specific conditions, Becker and Duchenne muscular dystrophies. One barrier to both research and clinical programming with muscular dystrophies is that the individuals are often minors, and deeply involved parents have difficulty weighing the risks and benefits to their child.

Basic Pathophysiology

Becker muscular dystrophy (BMD) and Duchenne muscular dystrophy (DMD) are two of the more common muscular congenital diseases of skeletal muscle. These conditions are closely related X-linked diseases that affect, respectively, about 1 in 17,000 and 1 in 3600 live male births. Both conditions lead to a progressive loss of muscle strength and functional abilities during childhood and adolescence. They are caused by deficiency or reduced expression of muscle protein dystrophin due to deletions or point mutations in the dystrophin gene. This leads to an absence of or defect in muscle dystrophin, causing a progressive weakness of the proximal muscles in the upper and lower extremities. People with DMD lose their ability to ambulate independently by the age of 13 years or earlier. Becker muscular dystrophy is a milder variant of DMD because BMD is an in-frame mutation, which therefore results in some functional dystrophin protein. Persons with Becker therefore generally show better functional prognosis than those with Duchenne and may not be diagnosed until adolescence.

Because muscular dystrophy is a condition of striated muscle, about 10% of persons with dystrophin mutations have a phenotype isolated to cardiac muscle. Cardiomyopathy is present in about 90% of people with BMD and DMD, and is the cause of death in about 50% of those with BMD and 20% of those with DMD. The cardiac involvement in BMD may be disproportionate to the skeletal involvement, which is generally less affected than in DMD.

Management and Medications

The management of BMD and DMD is primarily supportive because there currently is no technology to correct genetic deficiencies in structural muscle proteins. As a result, treatments are mainly physical and occupational therapy.

Glucocorticoid therapy is used in pharmacologic treatment in an attempt to slow the decline in muscle strength and functional capacity. Other pharmacologic interventions, in particular exon-skipping drugs for DMD, are under investigation.

Effects on the Exercise Response

For people with muscular dystrophy, there is much anxiety regarding the role of exercise in disease progression, due to the possible negative effects on muscle and cardiac function. People with muscular dystrophy are thus somewhat trapped between the need and desire to do more and improve, to maintain status quo or even just slow the progression of their condition, and the risk that muscular activity will provoke delayed-onset muscle soreness (DOMS) and initiate the normal remodeling process of muscle in someone who has a faulty ability to rebuild muscle. It's easy to imagine, whether in skeletal or cardiac muscle, how provoking breakdown without having an ability to rebuild might accelerate the overall decline.

As a result of such fears, much exercise research in DMD is in rodent models of the disease. In murine models to test pharmacologic interventions, mice genetically engineered to have deficiency of the dystrophin gene undergo intensive and forced exercise regimens that will provoke muscle damage. The response of mdx mice (the DMD rodent model that lacks dystrophin) to exercise testing and training protocols has been studied in more detail.

In forced running models of mdx mice, exercise is clearly detrimental to skeletal and cardiac muscle, justifying concerns about exercise in people with muscular dystrophy. Nonetheless, human studies exist; and in human adults with mild BMD (and according to anecdotal observational data in children), the cardiopulmonary exercise response is low-normal for $\dot{V}O_2$peak and peak heart rate. Whatever the muscle dysfunction consequences of muscular dystrophy, in people with mild BMD the integrated physiological response to exercise and circulatory control mechanisms are intact.

The situation with the response to high-intensity exercise (e.g., >85%) is less clear. Studies of high-intensity exercise showed elevated levels of serum creatine kinase, supporting the concept that high doses of exercise (duration or intensity) may risk adverse effects. Maximal cardiopulmonary exercise testing (i.e., very brief maximal exercise) does not appear to increase serum creatine kinase in BMD.

Effects of Exercise Training

In studies on mdx mice, voluntary wheel exercise (in contrast to forced exercise) had beneficial effects on muscle mass and walking distance without signs of exercise-induced muscle damage. Recent intervention research in adults with BMD showed that it was safe to participate in endurance training at 65% of $\dot{V}O_2$peak, and low-intensity assisted bicycle training seemed safe in children with DMD as well. These studies suggest that one can expect humans with muscular dystrophy to show sensitivity with regard to exercise dose-response, such that a little bit of activity is beneficial but that it is probably unwise to push individuals with BMD or DMD too far.

Research is needed to evaluate the effects of training on functional deterioration in BMD and DMD. Limited knowledge regarding the exercise response makes it difficult to define parameters for exercise training, though the available data suggest that light to moderate activities, terminated well in advance of fatigue, do not cause immediate or short-term harm. When new pharmacologic strategies such as the exon-skipping medications are available, people with DMD and BMD may have more exercise training options to possibly prevent early functional decline.

Recommendations for Exercise Testing

Exercise testing recommendations include the following:

- In children and adults with muscular dystrophy: use the recommendations from the TREAT-NMD network (see www.treat-nmd.eu), adapting protocols to suit children.

- When maximal exercise testing is indicated: use an adapted cycle ergometer protocol with small increments between stages.

- For children (BMD or early DMD): for 6-min walk test <400 m, use increments of 5 W/min; otherwise use increments of 10 W/min.
- In adults (BMD only): use the Basic *CDD4* Recommendations.

Higher-intensity or maximal exercise testing should be performed only after careful consideration of the person's overall status; there must be no present signs of rhabdomyolysis or cardiac risks, and testing should be approved by the treating physician.

There is increasing interest in maximal exercise capacity as an outcome measure for interventions in people with BMD and DMD, but use of maximal exercise testing has been limited by the fear of exercise-induced muscle damage or cardiac side effects (especially in minors). European guidelines recommend a number of functional tests such as the North Star Ambulatory Assessment (NSAA) to assess motor function, quantitative muscle testing, and the 6-min walk test (6MWT) to evaluate physical functioning. The advantage of obtaining maximal cardiopulmonary exercise testing with respired gas analysis and 12-lead electrocardiogram (ECG) is the possibility of differentiating between various factors that limit exercise capacity in persons with muscular dystrophy, that is, muscular or cardiac versus pulmonary limitation. This testing protocol could be particularly important in an individual who has cardiac manifestations and who has been sedentary because of the muscular dystrophy. While cardiac problems are common, cardiomyopathy should not be regarded as an absolute contraindication for testing. Given the prevalence of cardiac mortality in this population, however, exercise testing should always be performed in collaboration with a cardiologist.

Recommendations for Exercise Programming

These are exercise programming recommendations for people with muscular dystrophy:

- In children with DMD: use light-intensity dynamic exercise, preferably in a warm setting. Types of activities should be in line with the child's interests and abilities (ambulatory or wheelchair dependent),

with a maximum of 30 min (divided over arm and leg exercise) per session 3 or 4 days a week.

- In children and adults with BMD: use the Basic *CDD4* Recommendations, limiting exercise intensity to ≤65% of $\dot{V}O_2$peak, with a maximum of 30 min on 3 or 4 days a week.

Despite the lack of clear guidelines for persons with DMD and BMD, low- to moderate-intensity exercise may help to preserve physical functioning and prevent the development of secondary deconditioning. The progressive course and characteristics of muscular dystrophy make participation and recommendations for exercise challenging:

- Continuous change in equipment and type of activity may be needed.
- Begin a program only in agreement with the treating specialist(s) (neurologist, cardiologist, or both).
- Rule out contraindications for exercise, for example recent rhabdomyolysis or cardiac dysfunction.
- Carefully monitor for signs of muscle breakdown or cardiac events.

Integration Into a Medical Home Model

Maintaining and preventing the decline of physical functioning is the essence of care planning for people with muscular dystrophy. Most people with muscular dystrophies are children, and therefore care should be taken to foster as normal a developmental and social environment as possible. Very little is known about exercise and physical activity programming for individuals with muscular dystrophies, especially the rarer conditions.

Specialists in neurology and cardiology are likely to be closely involved in the medical management of persons with muscular dystrophy and need to be comfortable with the physical functioning and quality-of-life objectives of the exercise management care team. Key principles to remember are:

- Refer people to physical or occupational therapists (or both) with expertise in muscular dystrophy.
- In adults, beware of cardiometabolic conditions related to a sedentary lifestyle.
- Watch for cardiomyopathy and symptoms and signs of heart failure.

- Monitor for DOMS with a rhabdomyolysis questionnaire.
- Measure serum creatine kinase as indicated by symptoms or questionnaire or functional decline.

Take-Home Message

One should take these special considerations into account when working with people who have muscular dystrophy:

- Advise light- to moderate-intensity activities.
- Avoid eccentric exercise.
- Avoid exercise to the point of fatigue.
- Adaptive technologies may be necessary in those who are more severely affected.

Suggested Readings

Andersen SP, Sveen ML, Hansen RS, Madsen KL, Hansen JB, Madsen M, Vissing J. Creatine kinase response to high-intensity aerobic exercise in adult-onset muscular dystrophy. *Muscle Nerve.* 2013 Dec;48(6):897-901.

Bartels B, Takken T, Blank AC, van Moorsel H, van der Pol WL, de Groot JF. Cardiopulmonary exercise testing in children and adolescents with dystrophinopathies: A pilot study. Pediatr Phys Ther. 2015;27(3):227-34. doi: 10.1097/PEP.0000000000000159.

de Groot JF, Bartels B, Blank CA, van der Pol WL, Takken T. Cardiopulmonary exercise testing in ambulatory children with Duchenne and Becker Muscular Dystrophy: a pilot study. *Med Sci Sports Exerc.* 2013;45(5):S362.

Eschenbacher W, Mannina A. An algorithm for the interpretation of cardiopulmonary exercise tests. *Chest.* 1990;97(2):263-267.

Jansen M, de Groot IJ, van Alfen N, Geurts ACH. Physical training in boys with Duchenne Muscular Dystrophy: the protocol of the No Use is Disuse study. *BMC Pediatr.* 2010 Aug 6;10:55.

Jansen M, van Alfen N, Geurts AC, de Groot IJ. Assisted bicycle training delays functional deterioration in boys with duchenne muscular dystrophy: the randomized controlled trial "no use is disuse." *Neurorehabil Neural Repair.* 2013 Nov-Dec;27(9):816-827.

Markert CD, Case LE, Carter GT, Furlong PA, Grange RW. Exercise and Duchenne muscular dystrophy: where we have been and where we need to go. *Muscle Nerve.* 2012;45(5):746-751.

McDonald CM, Henricson EK, Abresch RT, Florence J, Eagle M, Gappmaier E, Glanzman AM; PTC124-GD-007-DMD Study Group, Spiegel R, Barth J, Elfring G, Reha A, Peltz SW. The 6-minute walk test and other clinical endpoints in duchenne muscular dystrophy: reliability, concurrent validity, and minimal clinically important differences from a multicenter study. *Muscle Nerve.* 2013 Sep;48(3):357-368.

Sockolov R, Irwin MA, Dressendorfer RH, Bernauer EM. Exercise performance in 6 to 11 year old boys with Duchenne muscular dystrophy. *Arch Phys Med Rehabil.* 1977;58(5):195-201.

Sveen ML, Andersen SP, Ingelsrud LH, Blichter S, Olsen NE, Jønck S, Krag TO, Vissing J. Resistance training in patients with limb-girdle and Becker muscular dystrophies. *Muscle Nerve.* 2013 Feb;47(2):163-169.

Sveen ML, Jeppesen TD, Hauerslev S, Køber L, Krag TO, Vissing J. Endurance training improves fitness and strength in patients with Becker muscular dystrophy. *Brain.* 2008 Nov;131(Pt 11):2824-2831.

Additional Resources

Duchenne and Becker Muscular Dystrophy, www.medicalhomeportal.org/living-with-child/diagnoses-and-conditions---faqs/duchenne-and-becker-muscular-dystrophy [Accessed December 2, 2015].

TREAT-NMD, www.treat-nmd.eu/dmd/overview [Accessed December 2, 2015].

VII

COGNITIVE AND PSYCHOLOGICAL DISORDERS

Another group of conditions that medicine has historically left out of exercise programming is the diseases of the mind. For purposes of this discussion, it is useful to define the mind as integrative cerebral functioning, which provides individuals the ability to interpret their relationship to their surroundings and then provides them with adaptive cognition and the ability to function appropriately.

Although physicians extending back to antiquity—indeed, to Hippocrates—have believed that exercise is beneficial to the mind, this perspective was somewhat lost from modern medicine until quite recently. In the late 1800s and early 1900s, medical education and technologies put an increasing emphasis on the molecular model of disease. Modern psychiatry and psychology did not, until recently, develop a particular interest in the role of exercise in functions of the mind. But research is increasingly showing that exercise is in fact very beneficial for promoting health of the mind, which makes part VII an important section of this book.

As with the other noncardiopulmonary, nonneuromusculoskeletal conditions, health insurance in the United States typically does not cover any exercise programming benefits for persons with dementia, depression, or anxiety. Perhaps even more than in the other disorders, the purpose of physical activity programming in disorders of the mind is to preserve not only physical functioning, but also cognitive functioning. Both are needed for independent living and maintaining quality of life. Clinical staff and exercise specialists, again, need to be creative and supportive of persons with cognitive and psychological disorders.

Dementia and Alzheimer's Disease

he term *dementia* refers not to a disease per se, but rather to a set of symptoms that are caused by various diseases or conditions. Dementia symptoms include deterioration in memory, thinking, or behavior that is severe enough to interfere with daily life. Numerous forms of dementia exist, with Alzheimer's disease the most prevalent. Alzheimer's accounts for approximately 50% to 80% of dementia cases and is the sixth leading cause of death in the United States. The second most common form of dementia is vascular dementia, followed by frontotemporal dementia and dementia with Lewy bodies. More than one type of dementia may also be simultaneously present; this condition is termed *mixed dementia*. There are approximately 24 million people with dementia of multiple etiologies worldwide, and this number is predicted to double every 20 years.

Basic Pathophysiology

Dementia results from the death of or damage to neurons. When neurons are impaired, they cannot effectively function and communicate with one another. Thinking, behavior, and mood can all be affected as a consequence. The cognitive abilities affected depend on the specific region of the brain that has been damaged. For example, Alzheimer's begins with the destruction of neurons in the hippocampus. Because this brain region is critical to memory formation, one of the most common early symptoms of Alzheimer's is difficulty remembering newly learned information. Vascular dementia is often caused by a stroke or a series of strokes that obstructs a brain artery and deprives local neurons of oxygen and nutrients. Symptoms of this type of dementia are highly variable since the damage can occur in a number of arterial regions of the brain.

Dementia symptoms vary further depending on the cause, stage of disease, and brain region affected, but these are common signs:

- Memory loss
- Difficulty communicating
- Impaired reasoning and judgment
- Problems with attention and focus
- Difficulty with complex tasks
- Confusion and disorientation and getting lost
- Mood and personality changes
- Agitation
- Hallucinations

Although the greatest known risk factor for dementia is increasing age, the pathological changes causing dementia are not normal aging. Not only persons who are elderly are affected. In fact, approximately 5% of people with Alzheimer's develop the disease before age 65. This early-onset form of the disease often runs in families and has a stronger genetic path of inheritance than the more prevalent, late-onset Alzheimer's disease. Familial inheritance is a known risk factor for all forms of dementia, but potentially controllable risk factors for averting or delaying dementias (especially vascular and alcoholic forms) include the following:

- Alcohol abuse
- Atherosclerosis
- Hypertension
- High blood levels of low-density lipoprotein (LDL)
- Depression
- Obesity

The exact etiology of Alzheimer's remains unclear, but amyloid plaques and neurofibrillary tangles in the brain are hallmarks of the disease. Most people develop some plaques and tangles as they age, but those with Alzheimer's have considerably more, and these tend to develop in predictable patterns in the brain. Plaques and tangles are thought to cause synaptic and neuronal cell death by blocking communication among nerve cells, disrupting processes that neurons depend on for survival. This process begins in the hippocampus and typically proceeds in a rostral fashion, toward frontal regions of the brain. As this occurs, cognitive regions are often affected, leading to increasingly severe symptoms related to functions such as language and reasoning. In mid- to late-stage Alzheimer's, people may experience major personality and behavioral changes, including paranoia, delusions, severe irritability, and extreme emotional outbursts. The motor cortex is relatively spared until the final stages of Alzheimer's, at which point people may have difficulty speaking, sitting without support, or holding their head up. Muscles grow rigid, and swallowing can become impaired.

Management and Medications

The degenerative changes in the brain responsible for dementia are typically progressive and cannot be stopped or reversed, but other conditions can cause problems with memory and thinking that resemble dementia. These conditions include severe infections and immune disorders, nutritional deficiencies (typically B vitamins), substance abuse, depression, and endocrine abnormalities. Because these conditions can often be reversed with treatment, early detection of dementia followed by an accurate diagnosis is critically important in getting the individual proper treatment.

Persons with dementia commonly present with symptoms of depression and anxiety, two conditions that have a bidirectional relationship (see chapter 32). The progressive decline in cognitive abilities can cause feelings of depression or anxiety, and there is evidence to suggest that late-life depression may also contribute to dementia. This may stem from the notion that high cortisol levels, often seen with depression, have adverse effects on the hippocampus. Depression can also worsen behavioral symptoms in people with dementia, exacerbating aggression, problems with sleeping, or resistance to eating.

The basic pharmaceutical treatment strategy is to improve symptoms, slow disease progression, and enable people to maximize function and maintain independence. Approaches to this can include antidementia medications, supplements to reduce inflammatory response and free radical formation, and psychiatric medications aimed at treating particular aspects of brain degeneration that lead to these behavioral symptoms. Because the antidementia drugs focus on preserving mental function, they are often used to treat all types of progressive dementias.

Anti-Alzheimer's Drugs

Unfortunately, there are no medications today that treat dementia. Two classes of Food and Drug Administration (FDA)-approved prescription medications are used to treat Alzheimer's disease:

- Acetylcholinesterase inhibitors
- N-methyl-D-aspartate receptor (NMDA) antagonists (see table 31.1)

Table 31.1 Anti-Alzheimer's Medications

Generic name	Mechanism of action	Adverse effects
Donepezil Rivastigmine Galantamine Tacrine	Prevent the breakdown of acetylcholine, important for learning and memory, and thought to benefit communication among nerve cells	• Nausea • Vomiting • Loss of appetite • Increased frequency of bowel movements • Liver toxicity
Memantine	N-methyl-D-aspartate receptor antagonist that appears to regulate glutamate activity	• Headache • Constipation • Confusion • Dizziness (increases fall risk)

Acetylcholinesterase inhibitors prevent the breakdown of acetylcholine, an important neurotransmitter for learning and memory. By keeping acetylcholine levels in the brain high, this class of drugs is thought to benefit communication among nerve cells. The acetylcholinesterase inhibitors are approved for mild to moderate Alzheimer's disease. The NMDA receptor antagonists appear to regulate the activity of glutamate, which is a neurotransmitter involved in information processing, storage, and retrieval. Memantine is the only NMDA antagonist that is FDA approved for moderate to severe Alzheimer's disease.

Vitamin E

The antioxidant vitamin E, while not FDA approved as a dementia treatment, helps slow the progression of Alzheimer's disease in some people. Vitamin E can interfere with medications prescribed to lower cholesterol or prevent blood clots, which are also common medications in people who are frail or elderly and have dementia. Due to the advanced age of typical onset, many clients with dementia take various medications to control comorbid conditions such as arthritis, heart disease, hyperlipidemia, and hypertension. Hypertension and high cholesterol are considered risk factors for vascular dementia, so medicines to treat these conditions are often prescribed with the intent of preventing future stroke.

> *Vitamin E is sold over the counter as a supplement, but it has many effects and potential drug interactions. The decision whether or not to use vitamin E in someone with Alzheimer's should be referred to the physician managing the condition.*

Psychiatric Medications

It is not uncommon for people also to be prescribed medications to help manage the behavioral and psychiatric symptoms of dementia. Antidepressants may help improve affect, while anxiolytics and antipsychotics may be needed for symptoms such as restlessness, anxiety, severe agitation, delusions, or hostility. (See chapter 32 for information on antidepressant and anxiolytic medications.) Antipsychotic medications have more varied effects on the central nervous system,

and little is known about antipsychotic medications and exercise. Many of these medicines cause increases in fat mass, which could affect exercise participation.

Effects on the Exercise Response

In addition to their deleterious cognitive, emotional, and behavioral effects, dementias also have an adverse effect on physical functioning. People with dementia have an increased risk of falls and fractures, a more rapid decline in mobility, and more undernutrition than those without the condition. Dementia also leads to deterioration in activities of daily living (ADLs). A decline in the ability to bathe, feed oneself, and dress reduces autonomy and is a major cause of institutionalization. Long-term care costs are particularly high for dementia, creating a financial burden on society similar to that for heart disease and cancer.

Dementia in early stages should not have a measurable impact on physical functioning or the physiological response to exercise. It is primarily the more advanced stages of dementia that cause decline in physical functioning. However, several drug classes used to treat dementia—particularly medicines prescribed for behavioral or psychiatric symptoms—can affect perception and coordination and increase fatigue. Another notable medication is memantine, which can cause dizziness and pose a concern for balance and stability, particularly among people who are severely deconditioned or frail.

Effects of Exercise Training

Regardless of the stage of dementia, exercise has been shown to improve physical functioning, increase cardiovascular fitness, and slow or reverse disability in basic ADLs. The extent to which exercise training helps physical functioning varies based on the age of onset, stage of disease, and other comorbidities. For many people, physical activity may extend independent mobility and enhance quality of life despite disease progression. In more advanced stages of dementia, in persons with a particularly late age of onset, or in those with other neuromusculoskeletal impairments (e.g., vascular dementia in stroke survivors), the potential to improve mobility may be highly limited.

Risk of falls depends on the cause of falls. In situations in which atrophy, weakness, and loss of coordination increase the risk of having a fall, exercise training to preserve muscle mass and lower-body strength while enhancing balance and mobility can reduce the risk. Furthermore, physical activity programs, or even just getting "out and about," creates valuable opportunities to socialize with others, which benefits well-being and reduces feelings of isolation. The combined effects of exercise training may ultimately serve to delay institutionalization. Multicomponent interventions that combine endurance, strength, and balance appear to benefit gait speed and functional mobility more than progressive resistance training alone.

Due to the high prevalence of depressive symptoms in these populations, exercise should also be viewed as a potential means to improve affect. In elderly persons without dementia, exercise reduces depression symptoms. The same benefit was found in people with Alzheimer's when exercise was combined with behavior management, though other exercise interventions in persons with dementia have shown only a statistical trend toward improving depression. Thus, the use of exercise to treat depression in people with Alzheimer's remains uncertain, but exercise might help, and given the low risks of supervised exercise, it's likely worth trying.

Use It Before You Lose It

As with many physiological functions, as people age and functioning declines, it becomes harder and harder to reconstitute physical and cognitive functioning that has been lost. This seems to particularly be true for those who have Alzheimer's, so interventions to preserve existing physical functioning, which have the side benefit of helping cognitive functioning, should be implemented as soon as the individual is diagnosed with Alzheimer's.

With regard to cognition, exercise training appears to play a stronger role in delaying the onset of dementia symptoms than it does in slowing or reversing cognitive decline after dementia is established.

With few exceptions, population-based studies have shown an inverse association between regular physical activity or cardiorespiratory fitness levels and the risk of dementia and Alzheimer's disease. These data are similar to those showing a relationship between physical activity and fitness versus cardiovascular and all-cause mortality. Physical activity may be a reasonable prescription to delay the onset of or even prevent certain dementias. It is hypothesized that exercise helps build a cognitive reserve such that the brain can better tolerate the effects of cognitive disease.

Whether the protective effects of exercise on neural integrity, as well as memory function, extend to progressive dementias remains to be determined. Research in healthy older adults offers promise, as the benefits of exercise training on brain function are particularly robust in the hippocampus—the very region predominantly affected early in the progression of Alzheimer's disease. A recent aerobic exercise intervention increased hippocampal volume in healthy older adults by 2% and reversed age-related volume loss in the region by 1 to 2 years. These physical changes were also manifest in memory benefits.

Recommendations for Exercise Testing

Exercise testing recommendations for people with dementia depend on physical function:

- Very early, in people with no limitation in physical functioning: follow the ACSM Guidelines.
- In debilitated individuals with limited functioning: use the Basic *CDD4* Recommendations.

Exercise testing may be best suited for clients in the early stages of dementia, but decisions should be made on a case-by-case basis. Many clients are prescribed a low-intensity physical activity program comparable to their usual activities of daily living. In such cases, exercise testing is unnecessary. Testing may also be deemed unreliable, particularly for those in later stages of disease. Mental capacity or balance may be compromised such that testing on a treadmill or bike ergometer is too demanding or risky. In more advanced dementia, the individual may not understand the value of testing or may become confused once a test has begun. High levels of agitation and distress can also affect the ability to tolerate a lengthy testing session. If a client becomes agitated or disoriented during a test, it should be stopped and scheduled for another day. Using another test modality or foregoing exercise testing altogether should be considered.

Recommendations for Exercise Programming

Like the recommendations for exercise testing, exercise programming recommendations are based on level of physical function:

- Very early, in people with no limitation in physical functioning: follow the ACSM Guidelines.
- In debilitated individuals with limited functioning: use the Basic *CDD4* Recommendations.

The primary goal of an exercise program in persons with dementia is to increase or maintain physical functioning and independent mobility. As with any exercise program, consideration of people's abilities, needs, and preferences is paramount. While some may have participated in regular exercise before, others may find exercising an unfamiliar concept. Because the majority of people with dementia are elderly, age-related comorbidities such as osteoarthritis, cardiovascular disease, and other inactivity-related conditions are common and should be taken into consideration (see part II).

Establishing a regular routine of physical activity is especially important in early stages of dementia. The main focus of any program should be low-intensity exercise that is pleasurable and easily accomplished. Simple exercises with a high degree of repetition and familiarity (e.g., walking, stationary cycling, and basic strengthening) are more likely to attract participation than more complex routines. Especially if the client is exercising at home, a daily walk with an exercise partner may be the optimal means of establishing a structured routine. This also creates important social engagement.

Exercises that target leg strength, balance, and range of motion are important to maintaining a client's ambulatory abilities. Weight-bearing and functional strength exercises in the Basic *CDD4* Recommendations help minimize the loss of muscle mass, which is heavily linked to loss of independence. The Basic *CDD4* Recommendations may also serve to counteract fear avoidance and risk of falls. Referral to physical or occupational therapy is a good idea for someone who has had a fall or is at high risk of falling.

As memory and cognitive abilities deteriorate, people become more susceptible to depressive symptoms, which can result in withdrawal from an exercise program. Patience is paramount, as clients may forget to come to sessions, may not remember what exercises have been completed in a session, or may need to be reminded how to perform certain exercises.

As dementia progresses, memory loss, disorientation, and the risk of wandering eventually necessitate vigilant and constant exercise supervision. Even in small gyms or centers, clients can become disoriented and get lost, which is highly distressing. Language skills and comprehension may also be greatly affected, so regular communication and collaboration with the primary caregiver is important at every stage.

Behavioral symptoms pose a different set of challenges for the exercise leader. Emotional instability, severe irritation, and extreme verbal or physical outbursts can occur in the intermediate and late stages of the disease. Keep in mind that these are symptoms of the disease and therefore to be very forgiving and tolerant of things that would be socially inappropriate in someone else. Change in a person's surroundings or new events may trigger behavioral symptoms, likely reflecting the stress associated with trying to make sense out of an increasingly confusing world. Exercise leaders may need to recruit the support of caregivers to work through severe agitation.

Integration Into a Medical Home Model

Exercise should be part of the care plan to delay progression of dementia, particularly if the cause of the dementia is Alzheimer's or cerebrovascular disease. Exercise is essential to retard or reverse the progression of cardiometabolic conditions that affect cognition (notably heart disease, diabetes, and stroke).

Refer people to a local program or a personal trainer with expertise in clients who have dementia (many nursing homes and assisted living facilities have such personnel).

Useful tips regarding behavioral management in dementia include these:

- Having a caregiver present or "on call" can be helpful for clients prone to severe agitation.
- Exercise may be better tolerated during morning hours. People with Alzheimer's often have a higher level of restlessness, fatigue, and agitation at the end of the day (this is termed *sundowning*).
- Create an environment that is calm and as familiar as possible. Avoid loud noises and excess background noise and distraction (including television).
- Soothing or familiar music can help clients relax and reduce anxiety or distress.
- Avoid being confrontational or arguing about facts.

In late stages of Alzheimer's and other dementias, deconditioning may be such that ambulation is severely limited. At this point, it is critical to maintain range of motion and strength. Seated exercises are less strenuous than those in a standing position and are a suitable option for individuals who are frail or experience problems with balance.

Take-Home Message

When working with people who have dementia or Alzheimer's disease, keep in mind these special considerations:

- Constant supervision may be necessary.
- Use visual cues when possible. Exercise instructions with figures, or numbers at exercise stations, can help clients' memory and allow them to feel more proficient.
- A medical alert bracelet with identification and emergency contact should be considered in advanced stages, as the individual may tend to wander.

Suggested Readings

Blankevoort C, Heuvelen M, Boersma F, Luning H, Jong J, Scherder E. Review of effects of physical activity on strength, balance, mobility and ADL performance in elderly subjects with dementia. *Dement Geriatr Cogn Disord.* 2010;30:392-402.

Diniz B, Butters M, Albert S, Dew A, Reynolds C. Late-life depression and risk of vascular dementia and Alzheimer's disease: systematic review and meta-analysis of community-based cohort studies. *Br J Psychiatry.* 2013;202:329-335.

Erickson KI, Voss MW, Prakash RS, Basak C, Szabo A, Chaddock L, et al. Exercise training increases size of hippocampus and improves memory. *Proc Natl Acad Sci U S A.* 2011;108:3017-3022.

Heyn P, Abreu B, Ottenbacher K. The effects of exercise training on elderly persons with cognitive impairment and dementia: a meta-analysis. *Arch Phys Med Rehabil.* 2004;85:1694-1704.

Larson E, Wang L, Bowen J, McCormick W, Teri L, Crane P, Kukull W. Exercise is associated with reduced risk for incident dementia among persons 65 years of age and older. *Ann Intern Med.* 2006;144:73-81.

Rovio S, Kareholt I, Helkala E, Vitanen M, Winblad B, Tuomiehto J, et al. Leisure-time physical activity at midlife and the risk of dementia and Alzheimer's disease. *Lancet Neurol.* 2005;4:705-711.

Teri L, Gibbons LE, McCurry SM, Logsdon RG, Buchner DM, et al. Exercise plus behavioral management in patients with Alzheimer disease: a randomized controlled trial. *JAMA.* 2003;290:2015-2022.

Vreugdenhill A, Cannell J, Davies A, Razay G. A community-based exercise programme to improve functional ability in people with Alzheimer's disease: a randomized controlled trial. *Scand J Caring Sci.* 2012;26:12-19.

Additional Resources

Alzheimer's Association, www.alz.org [Accessed November 8, 2015].

Alzheimer's Foundation of America, www.alzfdn.org [Accessed November 8, 2015].

National Institute on Aging, Alzheimer's Disease Education and Referral Center, www.nia.nih.gov/alzheimers [Accessed November 8, 2015].

Depression and Anxiety Disorders

epression and anxiety are among the most common disorders seen in medical practice. Almost everyone experiences symptoms of depressed mood and anxiety at times, because these emotions serve an important role in facilitating appropriate responses to life's stressors. These adaptive responses can develop into a maladaptive disorder, however, when symptoms become chronic and severe. For many people, the frequency and intensity of affective symptoms can become overwhelming and interfere with daily functioning.

Basic Pathophysiology

The pathophysiology and etiology of these illnesses are diverse and complex. A combination of environmental factors and genetic predispositions is believed to contribute to both anxiety and depressive disorders. Traumatic or stressful life events can also trigger episodes. While depression and anxiety are distinct clinical conditions, there is a high degree of concurrence between them. It is common for someone with an anxiety disorder to also suffer from depression and vice versa. The biological underpinnings and clinical presentations of the conditions frequently overlap, suggesting a common neural pathophysiological basis. Altered levels of neurotransmitters (including norepinephrine, serotonin, dopamine, and gamma-aminobutyric acid [GABA]) have long been implicated and are the primary target of pharmaceutical treatments for both conditions. However, it is not clear if neurotransmitter imbalances are a fundamental cause or instead a consequence of these conditions.

The burden of these disorders is further increased in that comorbidities for other chronic illnesses are common (see part II). These relationships are bidirectional: Just as depression and anx-iety are associated with increased risk of chronic diseases, numerous chronic diseases increase the risk of depression or anxiety. Studies suggest a high prevalence of modifiable risk factors as a major contributor to the increased morbidity in people with these conditions. Behaviors such as smoking, alcohol abuse, sleep disturbances, and physical inactivity are all common with symptoms of depression and anxiety. Clinical and epidemiological studies suggest that physical inactivity may even be associated with the development of a variety of mental disorders, including depression and anxiety.

Depressive Disorders

Four main types of depressive disorders are pertinent to this text:

- Major depressive disorder
- Bipolar disorder
- Disruptive mood dysregulation disorder (for minors <18 years of age)
- Persistent depressive disorder (formerly called dysthymia)

Major Depressive Disorder

Also known as clinical or major depression, major depressive disorder (MDD) is the leading cause of disability among adults in the United States and throughout the world. MDD is predicted to be the second-largest contributor to the global burden of disease by the year 2020. In the U.S. population, approximately 16% of adults will meet criteria for MDD at some point in their life.

Diagnosis of MDD is characterized by at least five of the following symptoms, occurring on most days, for at least 2 weeks. According to the *Diagnostic and Statistical Manual of Mental Disorders*

(DSM V), effects of medication or substance abuse should not account for these symptoms:

- Depressed mood most of the day
- Diminished interest in or pleasure from most activities
- Significant weight loss or gain, or change in appetite
- Altered sleep patterns (insomnia or hypersomnia)
- Psychomotor agitation or retardation observable to others
- Fatigue or loss of energy
- Feelings of worthlessness or excessive guilt
- Diminished ability to think or concentrate, indecisiveness
- Recurrent thoughts of death or suicide

Under *DSM V,* bereavement is no longer an exclusionary criterion for MDD as it had been under *DSM IV.* Current understanding of a prolonged recovery from grieving remains incomplete and controversial.

Bipolar Disorder

Also referred to as manic–depressive illness, bipolar disorder causes atypical shifts in mood, energy, and behavior. Symptoms are severe and affect the ability to carry out day-to-day responsibilities. People with bipolar disorder experience extreme emotional states that occur in distinct periods or "episodes." Extreme euphoria, hyperactivity, and impulsivity may accompany a manic episode, whereas intense sadness and a feeling of hopelessness characterize a depressive episode. These cycles of mood state may last anywhere from a few days to several months and correspond to extreme changes in energy, activity, sleep, and behavior. Between episodes, people may be symptom free, but episodes of mania and depression typically come back over time.

Some minors (under the age of 18) have symptoms consistent with bipolar disorder, and under *DSM V* these have been reclassified as disruptive mood dysregulation disorder (DMDD).

Persistent Depressive Disorder

Persistent depressive disorder (PDD, formerly known as dysthymia) is a chronic type of depression characterized by a consistently low mood, though symptoms are not as severe as with MDD. People present with many of the same symptoms as with MDD: low energy, sleep disturbances, changes in appetite, and so on. Individuals may be more irritable, stress easily, or experience anhedonia, which is the inability to derive pleasure from activities once found enjoyable. The name was changed from dysthymia in *DSM V* because it was concluded that there is no clear factor differentiating the disorder from MDD other than intensity or severity of symptoms. The renaming from dysthymia blurs the distinction between PDD and MDD, but the clinical impact of this change remains to be seen. PDD is common in people with chronic conditions (see chapter 9).

Anxiety Disorders

Anxiety disorders encompass a group of illnesses that include social phobias, panic disorder, obsessive–compulsive disorder (OCD), posttraumatic stress disorder (PTSD), and generalized anxiety disorder (GAD). Collectively, these disorders affect approximately 18% of U.S. adults and cost an estimated $42 billion per year in the United States. Yet anxiety disorders are considered to be among the most undertreated health problems and often go unrecognized.

There is considerable individual variability in both the severity and type of symptoms exhibited with these conditions. Symptoms typically include excessive worry, apprehension, fear, and uneasiness. Physiological symptoms stem from hyperactivation of the sympathetic nervous system, known as the fight-or-flight response, and are manifest as increases in the following:

- Blood pressure
- Resting heart rate
- Perspiration
- Blood flow to the major muscle groups

At the same time, immune and digestive symptoms may be inhibited. Under nonpathological conditions, the fight-or-flight response is adaptive and helps the body prepare to handle perceived threats or stressors. But in anxiety disorders, chronic stimulation of the physiological stress response exacerbates and even causes chronic conditions such as cardiovascular disease, immunosuppression, and MDD. Thus, symptoms of anxiety may manifest in the following ways:

- Heart palpitations
- Tachycardia
- Muscle weakness or tension
- Nausea

* Chest pain
* Dizziness
* Shortness of breath

Management and Medications

Treatments for depression and anxiety are diverse and have varying degrees of effectiveness. Management traditionally consists of pharmacological intervention, cognitive behavioral therapy, dialectical behavior therapy, or a combination of pharmacological and behavioral therapies. Cognitive behavioral therapy is highly variable, somewhat depending on the training, skills, and biases of the counselor, but is designed to help people modify their thought patterns and the way they react to stress. Dialectical behavior therapy is similar, though perhaps more clearly focused on increasing the awareness of one's own emotions, being more accepting of them, and responding in a more favorable and adaptive pattern. Treatment for depression can help not only manage symptoms but also prevent recurrence. A wide variety of anxiolytic and antidepressant treatments are currently available; some that are notable are discussed next.

Depressive Disorders

Direct-to-consumer marketing of medications by pharmaceutical companies, as well as compression of the time that primary care physicians can devote to patient visits, has driven an increase in the use of antidepressant medications. There are a number of proven treatments for depression (including exercise) that may be appropriate in a particular individual's situation.

Antidepressants

Most classes of antidepressants work by changing the level of one or more neurotransmitters in the brain, mostly serotonin and norepinephrine. By blocking their absorption (reuptake) from the synaptic cleft—the space between neural cell synapses—more of the transmitter is available for stimulation. See table 32.1 for a listing of antidepressant medications.

Electroconvulsive Therapy

Electroconvulsive therapy (ECT) is typically used for people with severe MDD who are unresponsive to all other antidepressant treatments, may be at high risk of suicide, or may not be able to take pharmaceuticals for health reasons or pregnancy. The mechanisms of action are unclear, but the electric currents that pass through the brain are hypothesized to reduce depressive symptoms by affecting neurotransmitter levels. The most common side effect is short-term confusion. Electroconvulsive therapy is often very effective and does not deserve the stigma associated with ECT because of historical use. Modern ECT is done under general anesthesia with short-acting paralytic medications to prevent muscle contractions.

Mood Stabilizers

Bipolar disorder can be difficult to manage and is commonly treated with mood-stabilizing drugs. Lithium is usually one of the first medicines prescribed for bipolar disorder, and many people are stable on lithium alone. Others may need a second-line treatment or combination of agents during a manic episode or to prevent mood cycling between mania and depression.

Exercise as a Primary Treatment for Affective Disorders

Exercise should not be overlooked as a potential first-line treatment for affective disorders. The efficacy of exercise compares favorably with that of antidepressant medications for mild to moderate depression, and depressive symptoms are further improved when exercise is used as an adjunct treatment to antidepressant medications. Motivation to exercise can be especially challenging for someone with depression, but exercise adherence is comparable to that in medication trials, at 60% to 80%. Because of the risks of suicidal ideation in persons with depression, the decision to try exercise as first-line therapy should always be referred to the physician who is managing the person's depression.

Anxiety Disorders

Most selective serotonin reuptake inhibitor (SSRI) and selective serotonin–norepinephrine reuptake inhibitor (SNRI) antidepressants (see table 32.1) also have anxiolytic properties. They are usually the first-line pharmacological treatment for the majority of anxiety disorders, though they often are prescribed at higher doses than for depression.

Benzodiazepines are the other main anxiolytic class of medications and are believed to diminish

Table 32.1 Medications Used to Treat Depression

Generic name	Mechanism of action	Adverse effects
Citalopram Escitalopram Fluoxetine Paroxetine Sertraline	Selective serotonin reuptake inhibitors (SSRIs) block the reuptake of serotonin in the neural synapse, increasing serotonergic stimulation by reducing the recycling of serotonin.	• Stomach upset, weight change • Decreased sexual desire • Fatigue, dizziness • Insomnia, headaches
Desvenlafaxine Duloxetine Venlafaxine	Selective serotonin–norepinephrine reuptake inhibitors (SNRIs) increase serotonergic and norepinephrine stimulation by reducing the recycling of serotonin and norepinephrine.	As for SSRIs, plus: • Sweating • Dry mouth • Fast heart rate • Constipation
Bupropion	Norepinephrine–dopamine reuptake inhibitors (NDRIs) increase dopaminergic and norepinephrine stimulation by reducing the recycling of dopamine and norepinephrine.	• Lowers seizure threshold (high doses)
Amitriptyline Nortriptyline Protriptyline Amoxapine Desipramine Doxepin Imipramine Trimipramine	Tricyclic antidepressants (TCAs) act through nonselective serotonin and norepinephrine stimulation by reducing the recycling of serotonin and norepinephrine.	• Generally more adverse effects than SSRI, SNRI, or NDRI medications • Stomach upset, dizziness, dry mouth • Changes in blood pressure • Changes in blood sugar levels • Constipation and nausea • Confusion • Life-threatening cardiac conduction abnormalities in an overdose
Mirtazapine Trazodone Vilazodone	These are atypical antidepressants.	• Primary or adjunct medication
Tranylcypromine Phenelzine Selegiline Isocarboxazid	Monoamine oxidase inhibitors (MAOIs) inhibit the deamination of serotonin, melatonin, epinephrine, norepinephrine, or phenylethylamine.	• Oldest class of antidepressants • Adverse effects can be severe • Extreme hypertension with some medications • Side effects similar to those for SSRIs and SNRIs

neural hyperactivity by increasing the inhibitory neurotransmitter gamma-aminobutyric acid (GABA). Benzodiazepines have more side effects than SSRIs or SNRIs and thus are usually second-line agents (see table 32.2). Common side effects are related to the sedating and muscle-relaxing action of these drugs: drowsiness, dizziness, and decreased alertness and concentration. Such side effects may impair coordination and increase risk of falls and injuries, especially in persons who are elderly or frail.

Exercise also appears to be an effective alternative treatment for certain anxiety disorders. However, exercise alone does not reduce anxiety to the same extent as pharmaceuticals. The wide range of symptoms and disorders associated with anxiety complicates the study and interpretation of exercise-related benefits.

Effects on the Exercise Response

Depression and anxiety disorders do not typically alter physiological changes to exercise. However, psychomotor retardation, most commonly seen with depression, can cause a visible slowing of motor movements and reaction times. Benzodiazepines, often prescribed for anxiety, are also associated with profound psychomotor slowing and impaired coordination. This is especially true following initial administration or with a sudden

Table 32.2   **Medications Used to Treat Bipolar Disorder and Anxiety Disorders**

Generic name	Mechanism of action	Adverse effects
BIPOLAR DISORDER		
Lithium	Monovalent cation, mechanism of action not currently known	• Requires frequent blood test monitoring • High white blood cell count • Tremors, weakness, twitching • Nausea, vomiting, headache • Confusion, decreased memory
Valproate Divalproex Carbamazepine Lamotrigine	Anticonvulsant medications with mood-stabilizing effects Likely have GABAergic effects (acting on gamma-aminobutyric acid pathway)	• Requires frequent blood test monitoring • Increased liver enzymes • Decreased white blood cell count • Gastric irritation
ANXIETY DISORDERS		
Escitalopram Fluoxetine Paroxetine Sertraline Sitalopram	Mechanism of action not known for improvement of anxiety	• Stomach upset, weight change • Decreased sexual desire • Fatigue, dizziness • Insomnia, headaches
Desvenlafaxine Duloxetine Venlafaxine	Mechanism of action not known for improvement of anxiety	As for SSRIs, plus: • Sweating • Dry mouth • Fast heart rate • Constipation
Lorazepam Oxazepam Diazepam Alprazolam Chlordiazepoxide	Benzodiazepines (BDZs) increase the activity of gamma-aminobutyric acid (GABA) at the receptor level	• Drowsiness, dizziness • Decreased alertness • Decreased concentration • Impaired coordination, fall risk

dosage increase. Motor responses and reaction times may be significantly slowed on these medicines.

Other concurrent pharmacological interventions may affect the exercise response. Thus, gathering a thorough medical history and listing of current medications is important. Several drug classes used for treatment in these populations may reduce functional aerobic capacity, affect perception and coordination, or reduce the desire to be physically active.

Effects of Exercise Training

Research on anxiety, depression, and exercise indicates that the psychological and physical benefits of exercise can help reduce anxiety and alleviate symptoms of depression. These effects do not appear to be limited to aerobic forms of exercise.

Depression

Exercise and physical activity are effective in significantly reducing symptoms of depression, and regular aerobic exercise is equal in effectiveness to some pharmacotherapy treatments in mild to moderate depression. Furthermore, a relapse of depression is less likely with exercise as compared to antidepressant treatment. When used as an adjunct to antidepressant medicines, exercise appears to further improve symptoms compared to medication alone.

The benefits that can be attained from exercise show dose–response effects with both aerobic and resistance exercise. One study concluded that total energy expenditure is the main determining factor for the reduction and remission of symptoms, but this needs further study. The amount of exercise necessary to achieve antidepressant efficacy is consistent with the Health and Human Services and Basic *CDD4* Recommendations.

High-intensity progressive resistance training (80% maximal load) is believed to be more effective than low-intensity training (20% maximal load) in reducing depression, because a direct relationship seems to exist between strength gain and reduction in depressive symptoms. While more research is needed, current understanding leans toward the quantity and intensity of exercise as more important than the mode of exercise.

The mechanisms underlying exercise-related improvements in depression are not known. Proposed psychological factors include increased self-efficacy, gaining a sense of mastery, distraction from negative thoughts, and enhanced self-concept. Exercise-related improvements in sleep quality and the circadian sleep cycle may also reduce symptoms. Biological pathways through which exercise may exert its effects include the hypothalamic–pituitary–adrenocortical axis, central norepinephrine neurotransmission, and serotonin synthesis and metabolism. One hypothesis is that major depression may be linked to decreased hippocampal volume and a decrease in the synthesis of new neurons in the hippocampus. This region is critical to functions such as long-term memory and spatial navigation. Antidepressants and exercise both enhance the synthesis of hippocampal neurons. The mechanism of these exercise-induced changes occurs, in part, through increased expression of several key brain growth factors.

Anxiety

The impact of exercise on clinically diagnosed anxiety disorders has not been as extensively investigated as that of depression, but findings support the utility of exercise in the treatment of anxiety. Exercise training significantly reduces symptoms of anxiety when compared to no treatment or cognitive behavioral therapy. Unlike the situation with depression, exercise alone does not appear to reduce anxiety to the same extent as pharmaceuticals.

The optimal intensity and duration of acute exercise still need to be characterized across people suffering from different anxiety disorders. There is likely a high degree of individual variability, but anxiety symptoms may be most alleviated by exercise training sessions of at least 30 min, though antianxiety benefits may be of limited duration.

Recommendations for Exercise Testing

Exercise testing recommendations for people with depressive or anxiety disorders depend on the impact of the condition on the individual's daily functioning:

- High functioning without comorbid conditions: follow the ACSM Guidelines.
- Low functioning due to comorbid conditions: use the Basic *CDD4* Recommendations.

Any contraindications to exercise testing are primarily related to comorbidities and other underlying conditions. Note whether or not the person to be tested is experiencing any of the medication-associated side effects (tables 32.1 and 32.2). If a client has been diagnosed with a specific medical condition, see the corresponding chapter in this book.

Recommendations for Exercise Programming

Just as with the recommendations for exercise testing, the recommendations for exercise programming when working with people who have depressive and anxiety disorders depend on the individual's level of functioning:

- High functioning without comorbid conditions: follow the ACSM Guidelines.
- Low functioning due to comorbid conditions: use the Basic *CDD4* Recommendations.

People should be encouraged to achieve at least the minimum recommended levels of 150 min/week or more of moderate-intensity physical activity. For mild to moderate depression, total energy expenditure appears to be a key consideration, with activity levels below 150 min/week having no significant effect.

Integration Into a Medical Home Model

Exercise should be part of the care plan to help decrease anxiety and symptoms of depression. Exercise compares favorably to antidepressant medications for mild to moderate depression and supplements the benefits of pharmaceuticals.

Exercise may appeal to people because it is seen socially as positive, whereas many people perceive counseling in a negative light. Furthermore, exercise does not trigger the more severe and longer-lasting side effects sometimes seen with medication. Anxiolytic and antidepressant medications can also grow expensive, and people reluctant to try exercise for affective disorders should bear that in mind. Exercise may not only be more attractive to people but may also ultimately decrease their dependence on psychopharmaceuticals.

- Prescribe exercise to all persons with depression, and recommend that people with anxiety try exercise training.
- Refer people to a local program or personal trainer with expertise in clients who have depression and anxiety (if you know of none, cultivate such resources).
- There is likely a high degree of individual variability, so client feedback and monitoring are critical in maximizing adherence to an exercise program.

Take-Home Message

Remember these key points when working with people who have depressive or anxiety disorders:

- Keep goals achievable, but consider encouraging the process of change and growth rather than achievement of goals. Many individuals with depression lack the motivation to exercise, and setting goals risks failure and the fostering of a fixed self-theory (see chapter 4).
- Rejoice in the person's path of self-discovery as a way to boost confidence and provide positive feedback to foster a sense of growth and mastery of something new rather than a specific accomplishment.
- Emphasize interest and enjoyment. Depression can fuel apathy toward activities once viewed pleasurable, which can further worsen symptoms. To break this cycle, seek the person's feedback in order to plan activities that are enjoyable, interesting, relaxing, or satisfying.
- Encourage leisure-time physical activity (gardening, shopping, golf, walking, and so on).

- Completing an exercise session tends to be self-fulfilling and usually provides a small sense of satisfaction and feeling good about oneself. Leverage that tendency, but acknowledge it in a subtle fashion, letting the person's persuasive internal voices be heard and not drowned out by the nagging of the health care team.
- Consider encouraging opportunities to interact with others. Social interaction can boost levels of well-being and confidence, especially with depression. However, some people may find that social interaction elevates stress and anxiety and wish to avoid social environments and group exercise sessions.
- Exercise environment is important. Note any aspects or characteristics that may help or worsen symptoms in your client. Crowded gyms or overstimulating environments could trigger stress or anxiety in some. The home environment may be more comfortable for some people.

Suggested Readings

Babyak M, Blumenthal JA, Herman S, Khatri P, Doraiswamy M, Moore K, Craighead WE, Baldewicz TT, Krishnan KR. Exercise treatment for major depression: maintenance of therapeutic benefit at 10 months. *Psychosom Med.* 2000;62:633-638.

Blumental JA, Babyak MA, Doraiswamy MP, Watkins L, Hoffman BM, Barbour KA, Herman, Craighead WE, Brosse AL, Waugh R, Hinderlitter A, Sherwood A. Exercise and pharmacotherapy in the treatment of major depressive disorder. *Psychosom Med.* 2007;69:587-596.

Carek P, Laibstain S, Carek S. Exercise for the treatment of depression and anxiety. *Int J Psychiatry Med.* 2011;41(1):15-28.

Craft L, Perna F. The benefits of exercise for the clinically depressed. *Prim Care Companion J Clin Psychiatry.* 2004;6(3):104-111.

Dunn A, Trivedi M, Kampert J, et al. Exercise treatment for depression: efficacy and dose response. *Am J Prev Med.* 2005;28(1):1-8.

Dunn A, Trivedi M, O'Neal H. Physical activity dose-response effects on outcomes of depression and anxiety. *Med Sci Sports Exerc.* 2001;33:S587-S597.

Herring MP, O'Connor PJ, Dishman RK. The effect of exercise training on anxiety symptoms among patients: a systematic review. *Arch Intern Med.* 2010;170:321-331.

Herring MP, Puetz TW, O'Connor PJ, Dishman RK. Effect of exercise training on depressive symptoms among patients with a chronic illness. *JAMA.* 2012;172(2):101-111.

Meyer T, Broocks A. Therapeutic impact of exercise on psychiatric diseases: guidelines for exercise testing and prescription. *Sports Med.* 2000;30(4):269-279.

Rethorst C, Wipfli B, Landers D. The antidepressive effects of exercise: a meta-analysis of randomized trials. *Sports Med.* 2009;39(6):491-511.

Additional Resources

American Institute of Stress, www.stress.org [Accessed November 8, 2015].

Anxiety and Depression Association of America (ADAA), www.adaa.org [Accessed November 8, 2015].

Beyond Blue, www.beyondblue.org [Accessed November 8, 2015].

Black Dog Institute, www.blackdoginstitute.org [Accessed November 8, 2015].

National Institute of Mental Health (NIMH), www.nimh.nih.gov [Accessed November 8, 2015].

VIII

CASE STUDIES

I n the first edition of this textbook, only a few sample cases were provided. The sample cases were expanded in the second and third editions such that a case accompanied each condition. It is common for clinicians who present cases in rounds or team discussions to try to highlight their care management expertise with a "perfect case." The sample cases in *CDD1* through *CDD3* were just this kind of material, even though the knowledge base was often scanty for many conditions and the meaning of "perfect" was more the best that the authors could offer.

During the 20 years that have transpired since the writing of the first edition, much has been learned about exercise in chronic disease, altering how we look at sample cases. Many young clinicians worry that the case they want to present is not "good enough," or not interesting because it seems rather ordinary (even so-called textbook!). Quite the contrary, every case is interesting because there is just about always some twist that makes the story unique and requires the clinician to improvise based on her particular skills. From this perspective, what makes a case valuable is not how well the case follows the textbook. Rather, what is interesting is how the story exposes what is not known, either because some data are missing for that patient or, quite often, because the situation simply goes beyond the scientific knowledge and evidence base. Going beyond the evidence base is particularly the situation for exercise in most chronic conditions, especially when the clinician must try to accommodate the patient's socioecological situation (which is rarely

embedded in exercise training studies). In such moments, the exercise specialist and care team find themselves having to use bits of knowledge from disparate parts of their training, clinical experiences, and even their life experiences in order establish a theory of how to help the particular patient. Their key question thus is this:

How do I improvise with my scientific knowledge to design a safe and sound solution?

The sample cases in *CDD4* are presented in the SOAP note format that was used in prior editions of this book. This time, we have included comments of the chief editor—not as a critique of the care plan or case management, but more to illustrate where and how the care team must improvise, either filling in gaps or going beyond published knowledge. Such gaps can provide the inspiration for the next generation of good research ideas.

Patients come to clinicians in need of help, and it's very unsatisfying to turn someone away just because there is not enough scientific evidence. In the case of exercise and physical activity, there is very rarely much risk in giving a trial of light to moderate physical activity, particularly in people who have lost a lot of their physical functioning and are at risk of losing their independence. But often the program must be carefully crafted, with great skill (skill that is sometimes more in the area of counseling than of exercises in the gym). It is our great hope that the cases and commentary provided in this section lead to your greater enlightenment and growth, because that is what our patients need from us.

Abdominal Aortic Aneurysm

PRESENTERS Holly Fonda, MS, and Jonathan N. Myers, PhD, FACSM

S

"I feel OK, but I know I have a lot of health problems and I'd like to try to become healthier."

A 75-year-old male with a history of heavy smoking, hypertension, hyperlipidemia, renal insufficiency, obesity, and abdominal aortic aneurysm was referred for an exercise test as part of a study on the effects of exercise therapy on abdominal aortic aneurysm expansion. An ultrasound revealed a distal aortic aneurysm of 4.2 cm diameter. He was asymptomatic but reported doing very little activity.

O

Height: 5 ft 10 in. (1.78 m)

Weight: 215 lb (97.5 kg)

BMI: 30.8 kg/m^2

HR: 69 beats/min

BP: 136/86 mmHg

Pertinent Exam

Abdominal bruit; femoral pulses intact and non-delayed

Medications

Atenolol, amlodipine, atorvastatin, aspirin

Graded Exercise Test

Individualized ramp treadmill protocol 2.7 METs (40% of age predicted)

Peak HR: 138 beats/min

Peak BP: 182/100 mmHg

Terminated due to shortness of breath

No significant ST change during exercise or recovery; occasional PVCs

A

Low exercise tolerance secondary to

- History of smoking
- Vascular disease
- Deconditioning

Abdominal aortic aneurysm (no vascular compromise)

P

Initiate an exercise program

Refer to smoking cessation program and provide supportive counseling

Monitor risk factors including smoking, lipids, body weight, and hypertension

Consider more aggressive hypertension management

Reassess aneurysm size in 6 months

Goals

Adopt and sustain a more physically active lifestyle

Improve physical functioning with emphasis on walking

Reduce risk factors including smoking, lipids, body weight, and hypertension

Exercise Program

Basic *CDD4* Recommendation with special considerations for patients with aneurysms:

Aerobic exercise 4-5 days a week

- Walk at an RPE <13 (out of 20), or use talk test
- Begin with 15-20 min (rest stops allowed)
- Build ~5 min a week up to 40 min/session
- May increase to RPE <16 as tolerated, using interval design
- May decrease session to 30 min if concurrent with strength training

Resistance exercises 2 days a week

- Sit-to-stand exercises
- Arm curls using a 2 L or 4 L jug filled with water
- Build up to a minimum of 2 sets of 8 reps

Flexibility exercises 3 days a week

- Chair sit-and-reach stretches daily on L and R
- Hold for 30 s on each side, goal of 0 cm/in. from toes
- Additional stretches as desired, 30 s per stretch below discomfort point

Warm-up and cool-down phases @ RPE <11/20 for ~10 min

Special Considerations

Limit intensity to no more than a 20 mmHg increase in systolic pressor response

Monitor blood pressure and changes in aneurysm size

Senior Editor's Comment

To me, the most notable thing about this patient is his very low physical fitness, with peak exercise capacity of <3 METs. Peak exercise capacities this low are associated with increased risk of cardiovascular and all-cause mortality, and this "warning sign" is often overlooked. In counseling such patients, I tend to note that if functioning slips any further, the patient will lose his independence. Most patients understand that exercise training is what will keep them out of the nursing home and keep them from being a burden to their family, and this often is very motivating. With his medical history, physical inactivity, and smoking history, this patient almost surely has widespread atherosclerosis even though no clinically significant coronary artery disease was detected during the graded exercise test. With a history of heavy smoking, he likely has a component of obstructive chronic bronchitis, if not COPD, and termination of the test due to dyspnea is not surprising. I would expect his work of breathing to be increased and to perhaps limit his ability to improve. A spot check of his SaO_2 during exercise would be worthwhile. Most patients like this can build up their endurance fairly quickly, but I would not be surprised if his peak exercise capacity doesn't improve much.

Amyotrophic Lateral Sclerosis

PRESENTER Heather Hayes, PhD, DPT, NCS

S

"I've been getting weakness in my right hand; it gets tired and sometimes cramps. It's making it hard to write, and I want to know if there are any exercises that can help."

A 58-year-old Caucasian woman with a diagnosis of ALS 2 years earlier came in with problems of progressive weakness. Her ALS began with left lower-extremity weakness, which was progressing in her left hand as well as left leg, and she was now presenting with cramping in her right hand, which was affecting her work. She continued to work as an RN but had cut her hours to part-time. She was using a four-wheeled walker and an AFO on her left foot in order to walk at work and in the home and reported no falls. She was using a manual wheelchair for longer distances. She was anticipating having to stop working in the near future because of increased difficulty with writing and standing, and the demands of job were fatiguing. She planned to travel to France in the next week for a vacation with her supportive husband and had ordered a power wheelchair and an elevator for the home, which she anticipated to arrive in the next 3 weeks.

O

Height: 63 in. (1.6 m)

Weight: 147 lb (66.7 kg)

BMI: 26 kg/m^2

HR: 87 beats/min

BP: 137/70 mmHg at rest

Pertinent Exam

Bulbar: normal, good appetite, no respiratory concerns

Upper-extremity motor examination: shoulder abduction bilaterally 5/5, elbow flexion bilaterally 5/5, elbow extension bilaterally 5/5, finger intrinsic 3+/5 bilaterally, obvious thenar wasting L > R

Lower-extremity motor examination: hip flexion L 3/5, R 4/5; hip extension 2/5; knee flexion L 3/5, R 4-/5; knee extension L 2/5, R 3/5; dorsiflexion L 0/5, R 3/5

Medications

Gabapentin 300 mg × 2 at bedtime as needed for fasciculations and sleep (takes 2-3×/week)

Lisinopril 20 mg daily

Paxil 20 mg daily

Clonazepam 0.25 mg at bedtime and as needed for anxiety

Trazodone 25 mg at bedtime and additional 25 mg if unable to sleep

Rilutek 50 mg twice a day on empty stomach

Lovaza 2 capsules daily

Ibuprofen 200-400 mg for body ache, headache (rarely)

Multivitamin daily

TUMS, 1 tablet as needed for heartburn

Caltrate-D occasionally

Pulmonary Labs

O_2 saturation: 94% on room air

FVC: 89% of predicted

ALS Functional Rating Scale

Speech 12

Occupational therapy 9

Physical therapy 3

Respiratory therapy 12

A

ALS, primarily affecting LE and progressing to UE in hand and fine motor control

Fall risk; has good adaptive equipment for transition in mobility status

P

Education on energy conservation/management, safety, and adaptive equipment

Range of motion and strengthening exercises

Goal

Decrease fall risk during functional activities

Exercise Program

The Basic CDD4 Recommendations are not appropriate for this patient. A PT student went to the patient's home 2 days/week to assist with range of motion exercise. The patient was educated on the goal of PROM/ROM versus strength training for weak muscles. She had been going to the pool 1 day/week for range and walking, and this was continued. Due to difficulty with stairs, she was going up/down only once a day.

Exercise Program

Exercise	Reps	Possible changes (3 months) due to progression of disease
Hip extensors	PROM, hold 60 s daily Can be performed supine	May need to be assisted
Hip flexors	PROM, hold 60 s daily Lying prone in bed; with assistance	With progression and difficulty vs. gravity, may do in side-lying position
Knee flexors	PROM, hold 60 s daily Can be performed seated	May need to be assisted
Knee extensors	PROM, hold 60 s daily Lying prone in bed; with assistance	With progression and difficulty vs. gravity, may do in side-lying position *This is a key stretch as patient progresses into increased time in power wheelchair.*
Dorsiflexion	PROM, hold 60 s daily Standing calf stretch	May need to be assisted May need a night splint for left lower extremity
Shoulder flexion/abduction	AROM Overhead reach, may use 1-2 lbs × 10 reps May need ankle weight secured to wrist to avoid overfatiguing weakening grip	Transition to performing without use of hand weight
Hand	PROM, hold 60 s daily	Adaptive equipment for grip May need a night splint
Walking	As needed for functional activities, work, home with walker Limit use of stairs (fall risk)	Transition to power wheelchair for work and half-day home activities Eventually will need power wheelchair for all mobility
Pool	Once or twice a week for water walking and ROM May cause fatigue and require more use of wheelchair on these days (75-100%)	Continue pool exercise indefinitely

Senior Editor's Comment

This case demonstrates how some patients have a chronic condition such that the best we can hope for, at least with today's medical knowledge, is to make their decline more gentle and humane. Likely, things will ever be thus, because our existence on Earth is finite. Health care workers—perhaps most especially exercise professionals—must acknowledge that exercise training does have a limit in what it can do. All too often, I hear or read of exercise professionals or physicians blindly extolling the virtues of exercise and physical activity as if it's always a good thing. This patient shows that for some patients, it's very easy to have too much exercise and for professionals to do harm by just urging everyone to do more at higher intensities or durations. It would not be the least bit difficult, by encouraging this woman to do too much to resist the natural course of her illness, to cause her to have a fall and suffer a fracture or head injury.

At the same time, this patient also shows that with supportive family and good exercise management, one can go down the path of progressive disease and disability gracefully. Patients need support to continue being the person they are, fulfilling their spiritual being, doing what is important to them, and contributing to the richness in the lives of people they touch. For this woman, that meant continuing to work until she simply was unable to do so and continuing to travel and explore the world with her family. Learning how to help a patient go down the path of inevitable decline with grace and dignity may be the most important art an exercise professional can learn.

Asthma

PRESENTER **Kenneth W. Rundell, PhD, FACSM**

S

"I would like to be able to have this under control for the World Championships and I need a therapeutic exemption to take a prerace medication."

A 21-year-old female elite cross-country runner complained of wheeze during exercise and presented with postexercise cough and chest tightness. She had previously seen her family doctor, who had prescribed preexercise albuterol as needed. However, she was using albuterol daily and was still somewhat symptomatic. No indication of childhood asthma was presented with her history. By rules governing international competition, in order for her to obtain therapeutic exemption to use a short-acting bronchodilating agent (SABA), she must test positive to a bronchial challenge test.

O

Height: 5 ft 9 in. (1.75 m)
Weight: 120 lb (54.4 kg)
BMI: 17.8 kg/m^2

HR: 62 beats/min
BP: 116/68 mmHg

Pertinent Exam

Atopic, normal-appearing elite runner

Medications

Albuterol as needed

Lab Tests

Skin testing positive to dust mite, cat dander, ragweed, and mold spores

Exercise Test Results

$\dot{V}O_2$max: 73 mL · kg^{-1} · min^{-1}

Peak HR: 198 beats/min

RPE: 10/10

Termination reason: plateau in oxygen uptake and heart rate despite increasing treadmill elevation at 9 mph (14.5 km/h)

Clinical symptoms: no symptoms of bronchial hyperresponsiveness

Pulmonary Lab Tests

Resting FEV1: 120% of predicted, with no increase in FEV1 after administration of SABA

Eucapnic voluntary hyperventilation (EVH) bronchial challenge test, in which the patient ventilates compressed medical-grade air with 5% CO_2 for 6 min at 30× FEV$_1$ (~85% maximal voluntary ventilation):

- The patient did not have clinically significant posttest bronchoconstriction.

A more detailed history was obtained, which revealed that symptoms occurred only toward the end of long-duration events or training.

Additional lab tests were performed:

- Methacholine test: positive
- Treadmill test to mimic race conditions revealed a significant postexercise decrease in FEV_1

A

Allergies to multiple environmental allergens

Symptomatic asthma during training runs and races 5K and longer

Tolerance to short-acting bronchodilators from daily use

P

Inhaled corticosteroids for daily use

Short-acting bronchodilators for use as needed

Avoid allergens: do laundry at >54 °C (130 °F), recommend noncarpeted floors, and avoid cats, ragweed, and mold

Regular visits to her allergist, who monitored her condition

Goals

Training and competing as per normal schedule, without respiratory symptoms

Exercise Program

This patient is an elite athlete, so neither the Basic *CDD4* Recommendation nor the Guidelines are applicable to her program. Her exercise training program was custom designed by her coach.

Follow-Up

After completing college and her competitive career, the patient became less adherent to her medications, and the asthma worsened. She now has moderate to severe asthma requiring higher-dose inhaled corticosteroids and long-acting bronchodilators.

Senior Editor's Comment

This case is most interesting because many experienced exercise physiologists and sports physicians are quite familiar with the problem of athletes having pain or symptoms during practice or competition but then not being able to replicate the symptoms in the clinic or exercise laboratory. This phenomenon represents an interaction of technological inadequacies and the ability of athletes to achieve a higher level of exercise intensity during competition, and it is seen in a number of exercise-related problems (e.g., exertional compartment syndrome and exercise-associated collapse as two examples). Elite endurance athletes exercising at very high intensities can often breathe for hours at nearly 100 L/min of minute ventilation, and this creates an enormous thermal and osmotic load on the airways, particularly at altitude or in cold dry air. This situation is very difficult to replicate in the lab. In this case, the runner was treated for asthma, though it remained possible (but not probable) that she was having exercise-induced laryngeal obstruction. It is interesting that she was ultimately diagnosed with asthma.

Atrial Fibrillation

PRESENTERS Jonathan N. Myers, PhD, FACSM, and J. Edwin Atwood, MD

S

"I'd like to be able to get through each day without feeling so fatigued."

A 68-year-old male had a 5-year history of intermittent AF before it became chronic. After several failed attempts at electrical cardioversion, a decision was made to accept the fact that he had AF and to control the rate pharmacologically. He had no history of other cardiovascular disorders, but had been taking medication for hypertension for approximately 20 years. He had a 20-year history of smoking, but had stopped smoking approximately 25 years ago. He reported an inability to perform daily physical activities at the level he used to do.

O

Height: 5 ft 9 in. (1.75 m)

Weight: 171.3 lb (77.7 kg)

BMI: 25.3 kg/m^2

HR: 88 beats/min, irregular

BP: 136/90 mmHg, irregular

Pertinent Exam

Middle-aged male in no distress

Cardiovascular: irregular rhythm; no rubs, gallops, or murmurs

Medications

Warfarin, sotalol

Labs

INR (International Normalized Ratio): 2.12

ECG: Atrial fibrillation with a ventricular response of 88 beats/min, but otherwise normal; cardiac exam unremarkable

Echocardiogram: Mildly enlarged left and right atria, normal left ventricle, normal wall thicknesses, and normal ejection fraction (55%); aortic and mitral valves show mild thickening; mild mitral and tricuspid regurgitation; findings similar to an echocardiogram performed 1 year earlier.

Graded Exercise Test (Naughton Protocol)

Peak exercise: 5.5 METs (70.5% of predicted)

Peak RPE: 18/20

Peak HR: 166 beats/min

Peak BP: 178/94 mmHg

Termination from leg fatigue

Chronotropic response to exercise: normal

No significant ST changes during exercise or recovery

No report of chest discomfort

A

Chronic AF

Low exercise tolerance for age

No evidence of significant underlying heart disease

P

Improve resting and exercise ventricular rate pharmacologically

Begin cardiovascular rehabilitation program

Goal

Improve cardiorespiratory endurance

Exercise Program

Aerobic exercise 4-5 days a week
- Walk at an RPE <13 (out of 20), or use talk test, limiting intensity to ~145 ventricular beats/min
- Begin with 15-20 min (rest stops allowed)
- Build ~5 min a week up to 40 min/session
- May increase to RPE <16 as tolerated, using interval design
- May decrease session to 30 min if concurrent with strength training

Resistance exercises 2 days a week
- Sit-to-stand exercises
- Arm curls using a 2 L or 4 L jug filled with water
- Build up to a minimum of 2 sets of 8 reps

Flexibility exercises 3 days a week
- Chair sit-and-reach stretches daily on L and R
- Hold for 30 s on each side, goal of 0 cm/in. from toes
- Additional stretches as desired, 30 s per stretch below discomfort point

Warm-up and cool-down phases @ RPE <11/20 for ~10 min

Special Considerations

Monitor ventricular rate response to exercise

Monitor pressor response to exercise

Consult cardiologist for any changes in these 2 parameters

Senior Editor's Comment

This patient has good cardiovascular function, despite the atrial fibrillation, as evidenced by the high ventricular heart rate (166 beats/min) during exercise along with the ability to generate good blood pressure (178/94 mmHg). He shows relatively poor functional capacity, however, with a peak exercise capacity of only 5.5 METs at a high perceived exertion (18/20). The combination of good cardiovascular function but low functional capacity is suggestive of deconditioning with good prospects for improvement. Some of the disappointingly low work rate at max may be that cardiac output is not benefitting from atrial kick. It is unclear whether or not his smoking history plays much of a role, but that seems unlikely. As his fitness and ability to redistribute blood flow to skeletal muscles improve, it will be interesting to see if this affects his exertional blood pressure. As long as his physical functioning is robust with good venous return and preload, this likely will not have a clinically significant effect but might alter the dose of sotalol. Prospects for this patient to improve his physical functioning are good.

Becker Muscular Dystrophy

PRESENTERS Janke de Groot, PhD, PT, and Bart Bartels, PT, MSc

S

"I have problems keeping up with my friends—do you have advice for training?"

A 15-year-old boy with Becker muscular dystrophy and subsequent dilated cardiomyopathy was referred to the exercise lab after his yearly checkup to get some advice regarding physical activity. While considered "normal ambulatory" and not having many difficulties in activities of daily living, he increasingly found it difficult to keep up with his peers when running and jumping, which was limiting his participation in sports. Yet he still biked to and from school daily (45 min one way) using an electrically assisted bike, and he enjoyed weekly scouting activities. In general, he reported low levels of fatigue or muscle soreness except on days that he was more active than usual (e.g., a day of work instead of school or when helping out to prepare for camp). During those days he reported considerable muscle pain (6/10) and fatigue (5/10).

O

Height: 5 ft 6 in. (1.68 m)
Weight: 118.8 lb (53.9 kg)
BMI: 18.0 kg/m^2

HR: 89/min
BP: 120/80 mmHg

Pertinent Exam

Handheld dynamometry

- Reduced hip abductor strength (Z score = –2.3)
- Reduced knee extensor strength (Z score = –2.5)
- Other muscle groups: within normal range

Functional tests

- North Star Ambulatory Assessment: 100%
- Gower's sign (getting up from the floor): negative

Exercise Tests

- 10 m sprint test: normal
- 6-min walk test: 470 m (66% of predicted using the formula by Geiger et al. *J Pediatr.* 2007 Apr;150(4):395-9)

Cardiopulmonary cycle ergometry test with 10 W/min increments and 12-lead ECG

- Resting SBP: 138 mmHg
- Max SBP: 202 mmHg
- Max HR: 167 beats/min (81% of predicted)
- Max work: 112 W
- RPE at max: 9/10
- W/kg: 2.1 (55% of predicted)

- Test duration: 12.4 min
- Reason for termination: clear signs of intense effort (e.g., unsteady biking, sweating, facial flushing) and unwillingness to continue despite encouragement
- ECG: no rhythm or conduction abnormalities

- $\dot{V}O_2$peak: 29.4 mL · min^{-1} · kg^{-1} (58% of predicted)
- RER @ max: 1.42
- $\dot{V}E$peak/$\dot{V}CO_2$peak: 19.2 (normal)
- Ventilatory threshold @ 22 mL · kg^{-1} · min^{-1} $\dot{V}O_2$
- $\Delta\dot{V}O_2/\Delta W$: 9.8 mL · kg^{-1}/W · min^{-1}
- O_2 pulse: 9.52 mL/beat

A

Normal pulmonary and peripheral circulatory function

Mildly blunted chronotropic response with normal pressor response

Decreased strength in hip abductors and knee extensors

Mildly reduced functional ambulation

Moderately reduced $\dot{V}O_2$peak, with signs of mild cardiac limitation

P

Maintain current physical activity level

Continue monitoring for fatigue and muscle pain

If symptoms or weakness increase, contact physician

Yearly follow-up recommended to monitor physical and cardiac function

Exercise Program

This patient's lifestyle seemed sufficient for optimizing fitness levels and prevention of secondary deconditioning, so no changes were recommended. His typical exertion correlated with the recommended limits for exercise in patients with Becker muscular dystrophy. $\dot{V}O_2$peak equals 8.6 METs, while his daily cycling to school probably averages around 5.5 METs, which reaches the recommended intensity of 65% of $\dot{V}O_2$peak.

Special Considerations

Limit intensity to moderate (65% for BMD)

Terminate session for extreme or long-lasting muscle soreness or fatigue during or after exercise

Chief Editor's Comment

This case reflects a common situation: Exercise testing often helps characterize the nature of limitations in physical functioning, but for many patients with chronic conditions, the best one can hope for from exercise training is to retard the rate of deterioration. One cannot generalize from this patient's story alone, but it is likely that his regular cycling to school and participation in scouting and other games are maintaining his physical functioning, though he's beginning to fall behind his peers. He's from the Netherlands, where cycling is a very common commuting mode, thereby providing a supportive environment (unlike what is seen in most of the United States, where he would spend 45 min commuting on a school bus instead of cycling).

It would be tempting to try to strengthen his hip abductors and knee extensors, but he's probably doing OK with a more functionally oriented life. Strength training might induce muscle soreness and rhabdomyolysis. Here, his exercise team elected to err on the side of "good enough" as more prudent. If he asked to do something to strengthen his legs, one suggestion might be to just ride some of the way to and from school without the electric motor assist—maybe turn it off except when going uphill or when he's getting fatigued. Because of the electrical assist, it's not fully clear to me that estimates of the cycling work rates are accurate, so it might be worth investing in a heart rate monitor to wear a few times. This would give a better estimate and would also give the patient some additional tools if he gets to a bad patch and develops symptoms.

Breast Cancer Survivor

PRESENTER Kathryn Schmitz, PhD, FACSM

S

"I feel like I have aged 10 years during one year of treatment."

Mrs. J., a 59-year-old breast cancer survivor, presented with a history of diabetes, hypertension, and an anxiety disorder and seeks to regain her functioning. She had completed curative treatment 6 months earlier for stage IIB ER+/PR+/HER– breast cancer, which consisted of breast-conserving surgery with tram flap reconstruction, removal of 8 lymph nodes, external beam radiation to the chest wall and axilla, and combination chemotherapy (Adriamycin, cyclophosphamide, and Taxotere).

As a home health care aide, she has not been able to return to work, which involves lifting and providing physical care to patients, due to weakness in her upper body and general deconditioning. She cannot carry her groceries inside from the car anymore because of a sharp stabbing pain in her shoulder that appeared 3 months ago and seems to be limiting her range of motion and causing weakness. She reports that her affected side feels heavy and thick. She is anxious about money and eager to be back to work but is afraid that this is her "new normal" and that she will have to apply for disability and stop working. She just wants her old life back.

Symptoms include fatigue, anxiety, upper-body pain, weakness, swelling, and altered range of motion.

O

Height: 5 ft 6 in. (1.68 m)

Weight: 170 lb (77 kg)

BMI: 27.4 kg/m^2

HR: 74 beats/min

BP: 140/85 mmHg

Pertinent Exam

Appearance of some hardened or thickened soft tissue in the armpit on the side affected by breast cancer. Affected side seems puffy.

Medications

Celexa 20 mg/day, hydrochlorothiazide 25 mg/day

Lab Tests

Hgb 9 g/dL

Normal resting ECG, no signs of cardiomyopathy, and normal ejection fraction on resting echocardiogram

Exercise Test

6-min walk test distance 312 m (~20% below age predicted)

A

Decreased endurance

Upper-body weakness, swelling, pain, and loss of range

Anxiety

Low hemoglobin (borderline anemia)

P

Goals

Improve 6-min walk distance to age-matched median

Increase strength and pain-free range of motion

Reduce anxiety by referral to a breast cancer survivorship support group

Treat anemia: iron supplementation/iron-rich diet and consider erythropoietin

Exercise Program

Basic *CDD4* Recommendation with special considerations:

For fatigue and loss of endurance
- Basic *CDD4* aerobic and strength training components
- Ask her to log her exercise and return in 3 weeks to reevaluate

For the upper-body problems
- Refer for evaluation and treatment for axillary web syndrome (cords of irritated fascial tissue treatable through myofas- cial massage) and lymphedema with a certified lymphatic therapist; then, as the axillary web syndrome improves, initiate a slowly progressive supervised weight-lifting program to gradually increase strength, function, and range of motion in the upper extremities.

Aerobic exercise 4-5 days a week
- Walk at an RPE <13 (out of 20), or use talk test
- Begin with 15-20 min (rest stops allowed)
- Build ~5 min a week up to 40 min/session
- May increase to RPE <16 as tolerated, using interval design
- May decrease session to 30 min if concurrent with strength training

Resistance exercises 2 days a week
- Sit-to-stand exercises
- Arm curls using a 2 L or 4 L jug filled with water
- Build up to a minimum of 2 sets of 8 reps

Flexibility exercises 3 days a week
- As for axillary web syndrome

Warm-up and cool-down phases @ RPE <11/20 for ~10 min

Special Considerations for Adaptative Approaches

Needs treatment for axillary web syndrome before strengthening and full upper-body program

Will likely experience fear avoidance and needs supportive counseling for adherence to recommendations

Senior Editor's Comment

This case illustrates that the consequences of cancer and impact on physical functioning depend heavily on the nature of treatment and not just the type of cancer. In this patient's case, the unaffected side is not noted, but it is likely there is a substantial difference between the affected and unaffected sides. This can be especially important if the swelling involves her dominant hand. It's also very likely that if she hadn't had chemotherapy, she probably would not be anemic. So, as noted in the chapter on cancer, all patients with cancer are different and need a program uniquely suited to them.

Cerebral Palsy

PRESENTER Désirée B. Maltais, PhD, PT

S

"It's more and more difficult to get around outside of my apartment. I was much more active when I was in university and used to walk from class to class and go out with friends using public transportation. Now I drive everywhere and I don't feel as physically capable as I used to be. I feel like I am aging prematurely. I want to stop this."

This 24-year-old male has bilateral spastic cerebral palsy and functions at GMFCS level II. He has worked as an IT consultant since finishing university a year ago. He lives in an apartment with a friend and drives to and from work. He has no history of seizures, and had bilateral Achilles tendon and hamstring lengthenings at age 7.

O

Height: 5 ft 1 in. (1.55 m)
Weight: 140 lb (63.5 kg)
BMI: 26.4 kg/m^2

HR: 85 beats/min
BP: 117/78 mmHg

Medications

None

Cycle Ergometer Test

Test parameters: 60 rpm, stage 1 20 W, increases of 10 W/min

Test duration: 8 min

Peak HR: 175 beats/min (~90% of predicted)

Peak $\dot{V}O_2$: 28.2 mL $O_2 \cdot$ kg$^{-1} \cdot$ min^{-1}

Peak RER: 1.14

Termination reason: unable to maintain 60 rpm

6-Min Walk Test

335 m

Functional Strength

30 s maximum test, left 18, right 20

For this test, the scores are a composite sum of the number of repetitions from the following tests: sit to stand, lateral step-up, and half-knee standing, where the score for each test is the number of repetitions completed in 30 s. Sit to stand is done once, and the lateral step-up and half-knee standing tests are done twice, once each leading with the left and right foot. To calculate the left and right side scores, the step-up and half-knee standing scores are taken from the respective sides and added to the sit-to-stand score.

A

Low cardiorespiratory fitness
Low 6-min walk test distance
Poor lower-limb functional strength
Overweight

P

Exercise program as outlined below
Refer to nutritionist for weight management counseling

Goals (Over the Next 6 Months)

Improve peak exercise capacity and 6-min walk test distance by 15%

Improve functional strength by 20%

Reduce weight by 10 lb (4.5 kg)

After the 6-month program, follow *CDD4* Recommendations modified for GMFCS level II to maintain gains

Exercise Program

Aerobic exercise 4-5 days a week
- Walk at an RPE <13 (out of 20), or use talk test
- Begin with 15-20 min (rest stops allowed)
- Build ~5 min a week up to 40 min/session
- May increase to RPE <16 as tolerated, using interval design
- May decrease session to 30 min if concurrent with strength training

Resistance exercises 2 days a week (all major muscle groups)
- Circuit lifting: 1 set, 50-70% of 1RM, 8-12 reps to failure
- Build up over ~8 weeks to a minimum of 2 sets of 8 reps

Functional exercises 2 days a week (all major muscle groups)

- Step-ups (leading with each foot): 1 set of 10 reps *as fast as safely possible*
- Build up over ~8 weeks to a minimum of 2 sets of 8 reps

Flexibility exercises 3 days a week

- Chair sit-and-reach stretches daily on L and R (hold for 20-30 s on each side, goal of 0 cm/in. from toes)
- Additional stretches as desired, 20-30 s per stretch, below discomfort point

Warm-up and cool-down phases @ RPE <11/20 for ~10 min

Special Considerations

Patient should hire a personal trainer for support during initial 6 months

Trainer should be educated in the issues and safety needs specific to CP

Recommend finding an exercise partner to help maintain motivation

Chief Editor's Comment

This is an interesting case on many levels. The patient is an example of someone whose peak $\dot{V}O_2$ is not "low" by our *CDD4* definition of <20 mL $O_2 \cdot$ $kg^{-1} \cdot min^{-1}$, or about the level of New York Heart Association class I heart failure (see chapter 5). But as a 24-year-old male, he is just below the 5th percentile, which is not enough for him to keep up with his peers. Holding up one's friends because you can't keep up is frustrating, no matter how old you are. Intriguing that a statistical definition of "low" and his subjective feelings corroborate, since *statistically significant* and *clinically significant* do not always have the same meaning.

His 6-min walk and functional strength tests, however, seem lower than I would expect from someone with 28 mL $O_2 \cdot kg^{-1} \cdot min^{-1}$ of aerobic capacity (I've seen very few patients with CP, but have seen many older patients with this range of $\dot{V}O_2$ and think most of them would outperform this patient). Also, his maximal heart rate was ~90% of age predicted, which is also just below the lower limit of normal, and is coupled with a peak RER of 1.14 that is just shy of qualifying as a "true max" in most laboratories (RER >1.15). Putting these findings together, one cannot decide if he's deconditioned, or if his CP impairs his neuromotor control to dramatically reduce his aerobic economy and thereby produce his low physical functioning. If one were to measure his lower-limb muscle strength with electrically stimulated contractions, the contractile ability of his muscles might be more normal than it appears from this patient's data.

What makes the case so interesting is this question: To what extent is his functional strength low because of intrinsic muscle weakness (i.e., inability to generate force), to what extent is this because his neuromotor control limits his energy economy and peak power, and to what extent is it from deconditioning from becoming more sedentary in the prior year? How much comes from deconditioning, and how much from his inability to go harder to drive his cardiovascular function to its maximum potential? At this point, there isn't enough science to provide the answer. As to answering *his* question, a functional exercise trial is really the only way. Because of his neuromotor issues, I would prefer for his personal trainer to have some skills in occupational or physical therapy.

I've seen an awful lot of IT specialists who are very sedentary, who do all their work sitting at a computer terminal. It can be useful to have people who sit most of the time they're at work try using a pedometer, with the goal of increasing their awareness of physical activity. Working on this patient's weight is important, because with his intrinsically low aerobic economy and sedentary job, excess weight will only further diminish his low physical functioning, and he needs help to avoid that outcome (and the cardio-metabolic conditions that would ensue).

Chronic Fatigue Syndrome
PRESENTER Stephen P. Bailey, PhD, PT, FACSM

S

"I just feel out of it all the time. I used to be very active; now I can't do anything. I try to get outside and work in the garden, but I work for 10 to 15 min-utes and then I have to stop and rest for an hour. I seem to end up sitting in front of the television or computer and not really thinking about anything."

A 28-year-old Caucasian woman was referred to an exercise specialist by a psychiatrist because she had become deconditioned secondary to chronic fatigue syndrome. She stated that her fatigue started dramatically, 14 months ago, after she had been involved in an auto accident. Since the time of the accident, she has lived with her husband in a two-story home and leaves the house only occasionally (approximately two times a week) due to profound fatigue and a fear of not being able to return. She believes that starting an exercise program will help her "get her life back," but when she does it on her own she gets profound fatigue for several days. Before her accident she worked as an intensive care nurse.

O

Height: 5 ft 3 in. (1.60 m)
Weight: 148 lb (67.1 kg)
BMI: 26.3 kg/m^2

HR: 82 contractions/min
BP: 102/52 mmHg

Pertinent Exam

Caucasian female with normal muscle tone and mass, otherwise unremarkable; mental status—difficulty concentrating on immediate question

Medications

Prozac, Advil, Ritalin-SR, CORTEF

Routine Blood Tests

Within normal limits

Graded Exercise Test

Low-level protocol at self-selected pace
Duration: 6 min 15 s
Peak work rate: 2.5 mph @ 1% grade
Peak HR: 114 contractions/min
Peak BP: 128/58 mmHg
$\dot{V}O_2$peak: 21 mL · kg^{-1} · min^{-1}
RPE: 19/20

Flexibility: normal
Declined to perform other tests because of fatigue

A

Chronic fatigue syndrome
Borderline low exercise tolerance and physical functioning

P

Continue psychiatric treatment, including trial of Florinef
Prescribe an exercise program

Goal

Resume normal activities

Exercise Program

Basic *CDD4* Recommendation with special considerations for chronic fatigue syndrome:

Aerobic exercise 2-3 days a week
- Walk at an RPE <13 (out of 20), or use talk test
- Try treadmill walking RPE <13 (out of 20) for 15-20 min (rest stops allowed)
- Build ~5 min a week up to 40 min/session

Resistance exercises 2 days a week
- Sit-to-stand exercises
- Arm curls using a 2 L or 4 L jug filled with water
- Build up to a minimum of 2 sets of 8 reps

Flexibility exercises 3 days a week
- Chair sit-and-reach stretches daily on L and R
- Hold for 30 s on each side, goal of 0 cm/in. from toes
- Additional stretches as desired, 30 s per stretch below discomfort point

Warm-up and cool-down phases @ RPE <11/20 for ~10 min

Special Consideration

Limit intensity to low-intensity activities, with very slow progression

Follow-Up

The client frequently canceled appointments and discontinued participation after only 3 sessions.

Chief Editor's Comment

I have known only a few patients with CFS who achieved long-term adherence to an exercise program. The vast majority drop out, though most of them last for more than 3 sessions. Anecdotally, the keys to success seem to have been the interpersonal relationships between the exercise therapist and the patient, in which a strong bond was formed by a gentle and caring therapist who was understanding, accepting, and very forgiving.

In those few patients, the rate of progress was truly glacial in comparison to what is seen in normal individuals, literally taking 1-2 years to achieve exercise training benefits that normal untrained individuals can expect to achieve in 4-6 months.

The ability of such patients to succeed seems driven more by the personality of the therapist than by any other factor. Perhaps one should even discard the entire notion of a "program" for such patients?

Chronic Heart Failure With Mild COPD

PRESENTERS Peter H. Brubaker, PhD, FACSM, and Jonathan N. Myers, PhD, FACSM

S

"I get easily fatigued and winded when I'm doing things around the house or yard."

A 70-year-old male complained of increasing difficulty sustaining recreational activities and household chores. The patient had a 10-year history of reduced left ventricular function from ischemic heart disease and had bypass surgery performed 5 years earlier. He was no longer smoking but had a 40 pack/year history of smoking. Other risk factors included a sedentary lifestyle, history of high blood pressure (controlled pharmacologically), and slightly excessive weight. He carried nitroglycerin for chest pain and an albuterol inhaler for bronchitis, but he rarely used either one.

O

HR: 55 contractions/min

BP: 130/65 mmHg

Pertinent Exam

Elderly male, normal appearance; chest shows mildly prolonged I:E ratio, lungs clear without rales or wheezes; no S3 gallop, no edema

Labs

ECG: sinus bradycardia

Echocardiogram:

LVEF: 30%, mild ventricular hypertrophy, posterior wall dyskinesis, inferior wall akinesis, mild mitral valve thickening and moderate regurgitation, mild tricuspid valve regurgitation

Pulmonary Function Tests

FVC: 2.84 L (60.6% of expected)

FEV_1: 70.4% of normal (low)

Graded Exercise Test

Peak exercise: 4.6 METs (estimated), with peak $\dot{V}O_2$ = 15.3 mL · kg^{-1} · min^{-1} (measured, 62% of age predicted)

Peak HR: 95 contractions/min

Peak BP: 160/70 mmHg

Peak rate–pressure product: 15,200

ECG: No significant ST changes during exercise or recovery; occasional PVCs

Terminated due to shortness of breath @ RPE 17/20, no chest discomfort

A

Congestive heart failure, NYHA class III

Coronary artery disease, status/post-CABG surgery

Mild COPD

P

Introduce home-based exercise to increase physical activity level

Ask him to keep an exercise log and follow up in 2 weeks

Reevaluate in 6 months

Goals

Improve functional capacity and decrease symptoms of fatigue and dyspnea

Monitor for symptoms of worsening CHF

Assist with lifestyle changes to decrease cardiovascular risk

Exercise Program

Basic *CDD4* Recommendation with special considerations for chronic heart failure:

Aerobic exercise 4-5 days a week

- Walk at an RPE <13 (out of 20), or use talk test
- Begin with 15-20 min (rest stops allowed)
- Build ~5 min a week up to 40 min/session
- May increase to RPE <16 as tolerated, using interval design
- May decrease session to 30 min if concurrent with strength training

Resistance exercises 2 days a week

- Sit-to-stand exercises
- Arm curls using a 2 L or 4 L jug filled with water
- Build up to a minimum of 2 sets of 8 reps

Flexibility exercises 3 days a week

- Chair sit-and-reach stretches daily on L and R
- Hold for 30 s on each side, goal of 0 cm/in. from toes
- Additional stretches as desired, 30 s per stretch below discomfort point

Warm-up and cool-down phases @ RPE <11/20 for ~10 min

Special Considerations

Take these precautions when working with people with congestive heart failure:

- Check for rales, an increase in 3rd heart sounds, and edema prior to an exercise session when the daily weight gain is >2 kg (fluid retention), and consult cardiology if this decompensation is not explained by dietary indiscretions or if these findings persist.
- Reduce intensity/duration or postpone session on days when the patient is more fatigued than usual (also perform cardiovascular screening exam as above).

Senior Editor's Comment

This man's case is an altogether too frequently told story. The combination of COPD and CHF may create a narrow range of opportunity for him to improve his functional capacity, because the COPD increases intrathoracic pressures that impair venous return/cardiac preload, and also increase the work of breathing such that a relatively large fraction of his cardiac output must be distributed to respiratory muscles rather than muscles for locomotion. Focusing on increasing strength and perhaps higher-intensity interval-type exercise might improve his ability to maintain his activities (see chapter 7 and COPD case study).

Chronic Kidney Disease: Stage 4, Renal Insufficiency

PRESENTER Patricia L. Painter, PhD, FACSM

S

"I have a lot of fatigue just getting around. It's a major effort to keep up with any regular exercise, I'm just barely able to get through the day at work."

This 69-year-old Hispanic male was diagnosed with chronic kidney disease with an estimated glomerular filtration rate of 16 mL · min^{-1} · 1.74 m^2. He was not yet requiring renal replacement therapy and had been referred for a transplant evaluation. He continued to work full-time but was getting fatigued during the day. He had been physically active in the past; however, his activity levels had deteriorated with the decline in kidney function. He was interested in knowing what he could and should do for physical activity, wanting to prevent further deterioration in physical functioning.

O

Height: 5 ft 6 in. (1.68 m)
Weight: 194.4 lb (88.2 kg)
BMI: 30.9 kg/m^2

HR: 88 beats/min
BP: 122/78 mmHg

Pertinent Exam

Normal elderly-appearing male, slight pitting edema below the shins, normal cardiovascular exam without abnormal heart sounds

Medications

Amlodipine, atenolol, lisinopril, furosemide, Aranesp, calcitriol, glipizide, Lipitor

Labs

Creatinine: 4.4 mg/dL

Blood urea nitrogen: 39 mg/dL

Hematocrit: 38%

Total cholesterol: 168

Hemoglobin A1c: 7%

Exercise Test Results

Peak exercise: ~7 METs

Peak HR: 139 beats/min

Peak BP: 147/80 mmHg

Peak RPE: 17

Reason for termination: leg fatigue/shortness of breath

ECG-LVH, no rhythm, conduction, or ST-segment abnormalities

Physical Function

6-min walk distance: 314.5 m

4 m normal-gait speed: 0.9 m/s

A

Chronic kidney disease (stage 4) nearing end stage that will require renal replacement therapy

Hypertension (controlled with medication)

Type 2 diabetes

Deterioration in physical functioning

Outcome will depend on progression of renal disease and consequences of losing renal function

P

Initiate Basic *CDD4* Recommendations for aerobic, resistance, and flexibility exercises

Goals

At a minimum, maintain physical function; at best, improve physical function

Maintain strength and range of motion

Maintain participation in physical activity at least 3 days/week

Exercise Program

Basic *CDD4* Recommendation with special considerations for persons with chronic kidney disease:

Aerobic exercise 4-5 days a week

- Walk at an RPE <13 (out of 20), or use talk test
- Begin with 15-20 min (rest stops allowed)

- Build ~5 min a week up to 40 min/session
- May increase to RPE <16 as tolerated, using interval design
- May decrease session to 30 min if concurrent with strength training

Resistance exercises 2 days a week

- Sit-to-stand exercises
- Arm curls using a 2 L or 4 L jug filled with water
- Build up to a minimum of 2 sets of 8 reps

Flexibility exercises 3 days a week

- Chair sit-and-reach stretches daily on L and R
- Hold for 30 s on each side, goal of 0 cm/in. from toes
- Additional stretches as desired, 30 s per stretch below discomfort point

Warm-up and cool-down phases @ RPE <11/20 for ~10 min

Special Considerations

The program will be followed as tolerated, and progression may be a bit slower than in other clinical populations due to fatigue/progression of renal dysfunction. Exercise tolerance may diminish with continued loss of renal function.

Senior Editor's Comment

Renal failure strikes lower socioeconomic groups disproportionately, because they have less access to preventive services for hypertension and diabetes that would avert or delay end-stage renal disease. Many renal patients therefore live in a built environment that is not amenable to a home-based walking program, but they also don't have discretionary funds to invest in gym memberships or fitness programs. Combined with the progressive weakness and fatigue, many patients with progressive renal insufficiency do not participate in any structured fitness and physical activity plan, even one as simple as the Basic *CDD4* Recommendation.

Chronic Kidney Disease: Stage 5, Treated With Hemodialysis

PRESENTER Patricia L. Painter, PhD, FACSM

S

"I lost so much muscle mass and endurance, I have no idea where to start. I feel better now that I'm on dialysis, but have lost so much."

This 69-year-old Hispanic male with chronic kidney disease had progressed to the point of requiring hemodialysis. He had been on hemodialysis treatments for 2 months, and was feeling better and interested in starting to regain his fitness. He had been physically active in the past; however, his activity levels had deteriorated with the progressive decline in kidney function. Due to the unavailability of evening dialysis sessions, he had to take leave on medical disability and was hoping to go back to work when he found a dialysis clinic that offered evening sessions. His goal was to regain strength and endurance and get in shape for a kidney transplant.

O

Height: 5 ft 6 in. (1.68 m)
Weight: 188.2 lb (85.4 kg)
BMI: 30.3 kg/m^2

HR: 80 beats/min
BP: 148/78 mmHg

Pertinent Exam

Hemodialysis fistula on L forearm, otherwise normal physical exam

Medications

Carvedilol, amlodipine, Aranesp, calcitriol, atorvastatin

Labs

Creatinine: 11.1
Blood urea nitrogen: 26
Hematocrit: 31.2%
Total cholesterol: 142
Hemoglobin A1c: 6%

Exercise Test Results

Peak exercise: 5.1 METs
Peak HR: 132 beats/min
Peak BP: 188/82 mmHg
Peak RPE: 17
Reason for termination: leg fatigue/shortness of breath
Clinical endpoints: ECG-LVH, no rhythm, conduction, or ST-segment abnormalities
Symptoms: shortness of breath, leg fatigue

Physical Function

6-min walk distance: 269 m
4 m normal-gait speed: 0.8 m/s
Sit to stand: able to perform only one repetition without using arm assist

A

Chronic kidney disease stage 5, requiring hemodialysis 3×/week for 3-4 hours per treatment
Hypertension
Type 2 diabetes
Significant deterioration in physical function and muscle strength
Outcome will depend on any complications and hospitalizations associated with the dialysis treatment

P

Initiate Basic *CDD4* Recommendations for aerobic, resistance, and flexibility exercises

Goals

At minimum, maintain physical function; at best, improve physical function
Maintain strength and range of motion
Maintain participation in physical activity at least 3 days/week

Exercise Program

Basic *CDD4* Recommendation with special considerations for patients with hemodialysis:

Aerobic exercise 4-5 days a week

- Walk at an RPE <13 (out of 20), or use talk test

- Begin with 15-20 min (rest stops allowed)
- Build ~5 min a week up to 40 min/session
- May increase to RPE <16 as tolerated, using interval design
- May decrease session to 30 min if concurrent with strength training

Resistance exercises 2 days a week

- Sit-to-stand exercises
- Arm curls using a 2 L or 4 L jug filled with water
- Build up to a minimum of 2 sets of 8 reps

Flexibility exercises 3 days a week

- Chair sit-and-reach stretches daily on L and R
- Hold for 30 s on each side, goal of 0 cm/in. from toes
- Additional stretches as desired, 30 s per stretch below discomfort point

Warm-up and cool-down phases @ RPE <11/20 for ~10 min

Special Considerations

The timing of this patient's exercise sessions may be determined by his physiological response to dialysis. Often there is extreme fatigue after dialysis, so exercise may be better accommodated on nontreatment days. If at all possible, exercise during the dialysis treatment would be ideal, although the dialysis providers may not support (or even allow) such programs. The patient needs encouragement from all health care providers to maintain an active lifestyle, so outreach to the dialysis care staff may help them know what the patient can do and how to provide encouragement. A given patient's adoption of and adherence to exercise recommendations often depend on the complications and hospitalizations associated with the dialysis treatment (how many, how severe, how frequent). Problems with the fistula are very common, and dialysis patients have many other intercurrent illnesses that cause them to be admitted to the hospital.

Senior Editor's Comment

Here is an example of a patient who was given a home exercise recommendation but was unable to benefit from it. His situation is likely to go from bad to worse, however, now that he's on dialysis. That's because, as his subjective history notes, he's unable to hold down a job while receiving the dialysis treatments. The combination of weakness, fatigue, chronic malaise, and three 1/2 to 3/4 days lost each week to the dialysis treatments makes it very difficult for dialysis patients to hold down a normal job. In the United States, almost all dialysis patients are on full disability, whereas some countries do a better job of providing dialysis services to try to help patients maintain a more normal life. The consequences for this patient are that he is very likely to find himself less and less able to maintain his level of physical functioning and, like most of his peers, he probably will do well if he just stops deteriorating as rapidly. Because the recommended exercise program is focused on physical functioning, the advice he is given is very similar to what he was told before. Having a supportive dialysis center with a proactive perspective on exercise is enormously beneficial.

Chronic Kidney Disease: Status/Post–Renal Transplantation

PRESENTER Patricia L. Painter, PhD, FACSM

S

"I feel so much 'clearer' and have more energy since I had my transplant! I want to get back to my normal life and make the most of this."

This 69-year-old Hispanic male presented again 3 months after kidney transplantation. He had an uneventful posttransplant course and had been able to reduce his immunosuppression medications to minimal doses of prednisone. He thought he'd gained too much weight, because of a good appetite and no restriction on his diet. He still felt like his legs were weak and his muscles felt "mushy," but he had started walking a little bit more without fatigue. He likes being active, and had been frustrated in his attempts to exercise because of progressive fatigue and frequent hospitalizations. He got very discouraged by the setbacks he encountered. He was looking to get back to work and his normal life.

O

Height: 5 ft 6 in. (1.68 m)

Weight: 210.4 lb (95.4 kg)

BMI: 33.9 kg/m^2

HR: 80 beats/min

BP: 152/78 mmHg

Pertinent Exam

Bright-eyed and showing more energy, no dependent edema, heart and lung exams normal

Slightly Cushingoid facial appearance; L arm hemodialysis fistula clotted off; scar in his lower left quadrant at the site of the donor kidney

Medications

Carvedilol, amlodipine, atorvastatin, prednisone, tacrolimus, glipizide

Labs

Creatinine: 1.2 mg/dL

Blood urea nitrogen: 16 mg/dL

Hematocrit: 39.2%

Total cholesterol: 212 mg/dL

Hemoglobin A1c: 7.5%

Exercise Test Results

Peak exercise: ~7 METs

Peak RPE: 17

Peak HR: 152 beats/min

Peak BP: 198/82 mmHg

Reason for termination: leg fatigue

ECG-LVH, no rhythm, conduction, or ST-segment abnormalities

Physical Function

6-min walk distance: 589.3 m

4 m normal-gait speed: 1.3 m/s

A

Deconditioning

S/P renal transplant

P

Gradual return to functioning and quality of life prior to end-stage renal failure

Goals

Perform instrumental activities of daily living without assistance

Play with grandchildren

Exercise Program

Basic *CDD4* Recommendation with special considerations for kidney transplant recipients:

Aerobic exercise 4-5 days a week

- Walk at an RPE <13 (out of 20), or use talk test
- Begin with 15-20 min (rest stops allowed)
- Build ~5 min a week up to 40 min/session
- May increase to RPE <16 as tolerated, using interval design
- May decrease session to 30 min if concurrent with strength training

Resistance exercises 2 days a week

- Sit-to-stand exercises
- Arm curls using a 2 L or 4 L jug filled with water
- Build up to a minimum of 2 sets of 8 reps

Flexibility exercises 3 days a week

- Chair sit-and-reach stretches daily on L and R
- Hold for 30 s on each side, goal of 0 cm/in. from toes
- Additional stretches as desired, 30 s per stretch below discomfort point

Warm-up and cool-down phases @ RPE <11/20 for ~10 min

Special Considerations

There are no restrictions on type of activity unless musculoskeletal discomfort occurs due to excess weight. Prednisone may slow progression and adaptation to weight training compared to what is seen in persons not taking prednisone; however, strength gains are achievable. Contact sports may be contraindicated (e.g., basketball) due to risk of damage to the transplanted kidney.

Senior Editor's Comment

This patient was, like so many, unable to participate in an exercise program while he was on dialysis because he couldn't muster the resilience on his own and the dialysis center had no program for him. He was very fortunate to receive a kidney transplant, however, and also to not have any substantial problems with rejections, postoperative infections, or other adverse events related to the transplantation procedure and posttransplant management. His renal status became essentially normal, which was reflected in his subjective history; but he remains a challenging problem because of the need for potent immunosuppression drugs and their adverse effects. Patients who don't have rejection issues and remain stable on modest doses of immunosuppression (especially prednisone) do remarkably well. Higher doses of prednisone, particularly, increase the risk of weight gain, diabetes, and osteoporosis, though these problems pale in comparison to life on dialysis. Whether stage 5 chronic kidney disease is treated with dialysis or renal transplantation, the costs of care are so substantial that it really makes no sense to invest such resources in renal replacement therapy but not bother to help these patients maximize their physical functioning to maintain their independence and quality of life.

Chronic Obstructive Pulmonary Disease

PRESENTERS Brett A. Dolezal, PhD, and Christopher B. Cooper, MD

S

"I need help getting around."

A 60-year-old male was diagnosed 4 years earlier as having COPD with emphysema. He had a 75 pack/year smoking history but quit smoking a year after his diagnosis. His exercise capacity varied considerably from day to day, but he generally was able to walk about half a city block and climb 15 steps at home, although he had some difficulty climbing stairs. He was in a motor vehicle accident 20 years ago in which he sustained 5 fractures to his left leg with a residual varus deformity of his left foot. Consequently, he often walked with a cane. He used supplemental oxygen overnight at a flow rate of 2 L/min.

O

Height: 5 ft 11 in. (1.8 m)
Weight: 176 lb (80 kg)
BMI: 24.7 kg/m^2

HR: 80 beats/min
BP: 140/90 mmHg

Pertinent Exam

Moderately dyspneic at rest, pursed-lip breathing using accessory respiratory muscles with thoracoabdominal paradox; chest hyperinflated and hyperresonant to percussion; breath sounds considerably diminished throughout; heart sounds regular but indistinct; no right ventricular heave, no increased jugular venous pressure or peripheral edema; truncal obesity and mild degree of muscle wasting affecting all limbs; left foot has a 45° varus deformity.

Medications

Tiotropium inhaler 18 mcg daily, albuterol inhaler (2 puffs several times daily), prednisone (intermittent courses)

Labs

Chest X ray: bilateral emphysematous changes prominent in the upper lobes

Pulmonary Function Tests

FVC: 45% of predicted
FEV_1: 16% of predicted (13% improvement after inhaled bronchodilator)
FEV_1/FVC: 35%
DL_{CO}: 52% of predicted

Arterial Blood Gas (Breathing Room Air)

pH: 7.43
PCO_2: 41 mmHg
PCO_2: 75 mmHg

Graded Exercise Test (Treadmill)

Speed: 0.8 mph, 0% grade
Duration: 10 min
Peak RPE: 16/20
Reason for termination: fatigue, shortness of breath
Breathlessness: 92/100 (visual analog scale)
Arterial PO_2 decreased to 59 mmHg

A

Severe COPD with emphysema secondary to tobacco smoking

Left leg deformity

Hypoxemia during low-intensity exercise

Moderate deconditioning with marked symptomatic limitation during exercise

Excess fat weight with loss of lean body mass

P

Add indacaterol 75 mcg puffs daily

Avoid systemic corticosteroids

Prescribe portable O_2 with a flow rate of 1 L/min during exercise

Initiate a multidisciplinary pulmonary rehabilitation program

Goals

Improve functional capacity to increase and maintain ADLs

Reduce sensation of dyspnea

Exercise Program

Basic *CDD4* Recommendation with special considerations for chronic obstructive pulmonary disease:

Aerobic exercise 3 days a week

- Walk at an RPE <11 (out of 20), THR ~110 beats/min
- Begin with 15-20 min (rest stops allowed)
- Build ~5 min as tolerated up to 30 min/session
- May increase to RPE <16 as tolerated, using interval design
- May decrease session to 20 min if concurrent with strength training

Resistance exercises 2 days a week

- Resistance circuit training machines
- Build up to a minimum of 2 sets of 12 reps maximum

Flexibility exercises 3 days a week

- Chair sit-and-reach stretches daily on L and R
- Hold for 20-60 s on each side, goal of 0 cm/in. from toes
- Additional stretches as desired, 30 s per stretch below discomfort point

Warm-up and cool-down phases @ RPE <10/20 for ~10 min

Special Considerations

Supervised aerobic exercise program

Use portable O_2 with a flow rate of 1 L/min during session

Senior Editor's Comment

Pulmonary rehabilitation will help this man improve his physical functioning, but he really needs to maintain this type of program indefinitely. His physical state, reflecting a prednisone-induced Cushingoid body habitus, left foot deformity with need for cane, and use of supplemental O_2 during activity, creates an image of appearing much older than his stated age. With truncal obesity, limb muscle atrophy, and a 75 pack/year history of smoking in combination with a sedentary lifestyle, he's at high risk for cardiovascular disease or diabetes. He's closer to a nursing home than he may realize, but he can still do something about these issues. His health care team really needs to help him understand that his pulmonary rehabilitation program is not just about his breathing.

Coronary Artery Disease and Dyslipidemia, Status/Post-Angioplasty With Stent Placement

PRESENTERS Benjamin Gordon, PhD, RCEP, and J. Larry Durstine, PhD, FACSM

S

"I don't want to have a heart attack."

A 51-year-old man with hypertension presented to the emergency department with an acute coronary syndrome. His cardiovascular risk factors were hypertension, tobacco abuse, and a sedentary lifestyle. His ECG showed 2 mm of anterior ST-segment depression, so he underwent cardiac catheterization and received angioplasty with placement of a stent in his left anterior descending coronary artery. The day after placement of the stent, an echo showed normal wall motion and his ejection fraction was preserved, so he was discharged from the hospital on aspirin, clopidogrel, a β-adrenergic blocker, and a statin and was told to go to cardiac rehabilitation. At the intake session for cardiac rehab, the patient stated that he had quit smoking when he was in the hospital and had been walking daily since he was discharged. His wife was watching his diet and he had lost 5 lb (2.4 kg). They came to cardiac rehab 2 weeks after the stent was placed, saying that they were motivated to do more.

O

Height: 6 ft 1 in. (1.85 m)
Weight: 236 lb (107.05 kg)
BMI: 31.1 kg/m^2

HR: 72 contractions/min
BP: 110/60 mmHg

Pertinent Exam

Well-appearing obese male, no distress, heart and lung exam normal

Labs

Total cholesterol: 163 mg/dL
Triglycerides: 148 mg/dL
LDL-C: 104 mg/dL
HDL-C: 29 mg/dL
Fasting glucose: 117 mg/dL
HbA1c: 5.9%

Graded Exercise Test (2 Weeks Post–Stent Placement)

Resting HR: 66 contractions/min
Resting BP: 120/80 mmHg
Total treadmill time (Bruce protocol): 6 min 30 s, ~8 METs
Peak HR: 155 contractions/min (92% of age predicted)
Peak BP: 180/76 mmHg
Peak rate–pressure product: 27,900
No ECG evidence of cardiac ischemia, no conduction or rhythm abnormalities

A

Coronary artery disease, status/post–percutaneous coronary intervention with stent placement

Dyslipidemia with low HDL-C
Tobacco use
Fair exercise tolerance with good cardiac work indices
Obesity

P

Exercise program
Tobacco cessation
Weight loss intervention

Goals

Consistent participation in exercise/physical activity
40-60 min of daily exercise

Exercise Program

Aerobic exercise 4-5 days a week
- Walk or cycle at RPE <13 (out of 20), use talk test or THR (60-75% HRR)
- Begin with 20 min (rest stops allowed)
- Build ~5 min a week up to 45 min/session
- Increase intensity ~5% every 2 weeks, up to 75% HRR; may use interval training
- May decrease session to 30 min if concurrent with strength training

Resistance exercises 2 days a week
- 1 set of 10-15 repetitions of 8-10 different resistance training exercises
- Build up to a minimum of 2 sets of 8 reps at a comfortable resistance

Flexibility exercises 3 days a week
- 20-30 s/stretch (all major muscle groups)

Warm-up and cool-down phases @ RPE 7-9/20 for ~10 min

Follow-Up

He attended cardiac rehab and walked at home on the alternate days, remaining tobacco free without participating in a tobacco cessation group or using nicotine replacement. He was placed on niacin therapy because of his low HDL-C and remained motivated to increase his exercise duration as well as adding resistance training. Repeat evaluations revealed a gradual and profound improvement in key biometric cardiovascular risk factors.

After 20 weeks, he underwent repeat treadmill testing (Bruce protocol) and exercised for 13:45 (~15 METs), achieving peak HR of 169 contractions/

Trend of Risk Factors

Risk factor	Program entry	8 weeks	20 weeks
Weight (lb)	236 (107.3 kg)	224 (101 kg)	193 (87.7 kg)
Resting HR (beats/min)	72	70	66
Resting BP (mmHg)	110/60	105/65	90/60
Total cholesterol (mg/dL)	163 (4.2 µmol/L)	120 (3.1 µmol/L)	124 (3.2 µmol/L)
LDL-C (mg/dL)	104 (2.7 µmol/L)	68 (1.8 µmol/L)	58 (1.5 µmol/L)
HDL-C (mg/dL)	29 (0.8 µmol/L)	39 (1.0 µmol/L)	57 (1.5 µmol/L)
Triglycerides (mg/dL)	148 (1.7 µmol/L)	65 (0.7 µmol/L)	44 (0.5 µmol/L)
Blood glucose (mg/dL)	117 (6.5 µmol/L)		95 (5.3 µmol/L)

min, peak BP of 140/76 mmHg, and peak RPP = 23,660. He was asymptomatic, and his ECG showed <1 mm of upsloping ST-segment changes.

Chief Editor's Comment

This patient, almost surely due in part to the support of his wife, represents the best that cardiac rehabilitation has to offer. Only a small percentage of patients "put it all together" as well as this one did (especially the very big increase in physical fitness), but some patients do. The case doesn't detail his wife's involvement, but in my experience the patients who do really well have a spouse who is very involved, comes to the educational classes, and works hard on promoting a physically active lifestyle and a much heart-healthier diet (the patient couldn't have succeeded like this without a big change in his food shopping and cooking habits). Both partners have to participate willingly for this to succeed (as is the case for tobacco cessation, too). With many patients having a standard American diet (sometimes called "SAD") and a sedentary lifestyle, there is potential for many patients to make big improvements. In today's era of primary angioplasty and thrombolysis, a

substantial number of patients survive their acute coronary syndrome with relatively little cardiac damage and essentially normal cardiac function. Most such patients could have an outcome like this if they and their spouse (or partner) were willing to invest the time, money, and energy in lifestyle change.

As a note on medical management, this case predated the most recent lipid and blood pressure management recommendations. There are no data indicating that attempting to raise the HDL by a means other than exercise (i.e., niacin) is associated with reduced cardiovascular risk (and there are few data to support better outcomes via raising HDL through exercise training). Also, new blood pressure guidelines would probably have this patient stop his β-adrenergic blocker (many physicians already would have anyway), and perhaps such a patient will now be started on an angiotensin-converting enzyme inhibitor right after the angioplasty. Cardiovascular physiologists tend to favor the ergomimetic qualities of afterload reduction inherent in angiotensin-converting enzyme inhibitors and angiotensin receptor blockers over the β-adrenergic blockers, because β-blockers tend to blunt the exercise response.

Cystic Fibrosis

PRESENTERS Erik Hulzebos, PhD, PT, RCEP;
Maarten S. Werkman, PhD, PT; Bart C. Bongers, PhD; Tim Takken, PhD

S

"I get tired and can't breathe during field hockey."

A 16-year-old female with cystic fibrosis was known to have CF-related liver disease, pseudomonas aeruginosa and mycobacterium abscesses, and allergic bronchopulmonary aspergillosis

(ABPA). In the 3 months prior to coming to the clinic, she lost 4 kg of weight (~8% of her body weight). Pulmonary function testing revealed a decline in lung function since her last annual checkup, as measured by forced expiratory volume in 1 s (FEV_1) and by the ratio of FEV_1 to forced vital capacity (FVC). Both had decreased significantly, the FEV_1 decreasing from 73% (2.12 L) to 49% (1.41 L) and FEV_1/FVC from 83% to 63%. She also had developed more air trapping, as measured by residual volume (RV) as a fraction of total lung capacity (TLC), with a RV/TLC of 51% and increased airway resistance of 149% of predicted. Her physicians suspected a pulmonary infection leading to an exacerbation of her pulmonary function. Before starting antimicrobial treatment, a maximal cardiopulmonary exercise test (CPET) was performed according to the ATS Guidelines.

O

Height: 5 ft 5 in. (163.5 cm)
Weight: 108.5 lb (49.2 kg)
BMI: 18.4 kg/m^2

Pertinent Exam

None provided, but presumably consistent with lower respiratory infection in a lean adolescent girl and significant weight loss since her prior exam

Cardiopulmonary Exercise Test

Peak work (WRpeak): 163 W (76% of predicted); 3.3 W/kg (86% of predicted)

Maximal HR: 171 beats/min (89% of predicted)

Respiratory rate: 51 breaths/min

She desaturated (SpO_2 <90%) after 6.30 min of exercise, at a heart rate of 150 beats/min; SpO_2 reached a nadir of 83% at maximal exercise

Reason for termination: leg fatigue (RPE 9/10), dyspnea scale (9/10)

$\dot{V}O_2$peak: 31.3 mL · kg^{-1} · min^{-1} (79% of predicted)

$\dot{V}E$peak: 50 L/min

Tidal volume: 0.98 L

RERpeak: 1.18 (96% of predicted)

Ventilatory threshold (VT): 50% of predicted $\dot{V}O_2$peak

Ventilatory reserve, VR = ([1 − ($\dot{V}E$peak / maximal voluntary ventilation)] × 100%): −3% (VR of >15% is normal by ATS criteria)

End-tidal CO_2 partial pressure (PETCO$_2$): 4.0 kPa at rest; 6.6 kPa at max

Ratio of dead space to tidal volume (VD/VT): 18% at rest; 27% at max

Efficiency of ventilation: $\dot{V}E/\dot{V}O_2$ = 26.5; $\dot{V}E/\dot{V}CO_2$ = 30.0

Steep Ramp Test (Cycle Ergometry)

After 3 min of unloaded pedaling, the work rate was increased by 25 W/10 s until she could not maintain 60 rpm despite verbal encouragement

Maximum short-time exercise capacity (MSEC): 225 W

A

Pneumonia with presumed pseudomonas/mycobacterium

Dynamic hyperinflation and an inability to adequately expire CO_2

Exertional arterial desaturation limiting aerobic exercise capacity and ventilatory reserve

P

Home intravenous antibiotics × 2 weeks

Ceftazidime and tobramycin

High-intensity interval training (HIIT) × 12 weeks

Goals

Increase aerobic exercise capacity

Use a training regimen with the least strain on her respiratory system (HIIT)

Achieve physiological benefits of aerobic exercise in her locomotor muscles

Exercise Program

High-intensity interval training

- Interval training 3× each week for 12 weeks
- Continue with field hockey on other days

Start

- 10 intervals, 30 s each at 50% MSEC with 60-s recovery at 25% MSEC
- Pedaling cadence 60 and 80 rpm; RPE scales to assess peripheral fatigue and dyspnea

Reassess

- MSEC every 2 weeks to adjust training intensity

Progression

- Every 1-2 weeks, increase intensity by 10% of the current MSEC, and add 2 more intervals to each session

Hardest HIIT workout

- 20- × 30-s high-intensity intervals at 90% MSEC, with 60-s recovery @ 25% MSEC

Follow-Up

After 12 weeks, attendance rate was 90%; all absences were for holidays. She reported that her physical activity (field hockey) had increased back up to 90 min a day

Weight increased from 49.3 kg to 51.8 kg

Pulmonary function increased +19% for FEV_1, +14% for FVC, and −16% for RV/TLC.

Cardiopulmonary Exercise Testing Was Repeated

WRpeak (+37 W; +16%), and WRpeak/kg (+0.56 W/kg; +14%)

$\dot{V}O_2$peak/kg (+7.5 mL $O_2 \cdot kg^{-1} \cdot min^{-1}$; +11%)

$\dot{V}E$ increased to 75 L, primarily due to increased respiratory rate 57 breaths/min

VT: 68% of $\dot{V}O_2$peak predicted

O_2 pulse ($\dot{V}O_2$peak/HRpeak): +10.3%

$\dot{V}E/\dot{V}O_2$ and $\dot{V}E/\dot{V}CO_2$ were unchanged

$\Delta\dot{V}O_2/\Delta WR$: +4%

VR decreased from −3% to −14%.

SpO_2 decreased from 98% at rest to 90% at max (11 min)

RPE and dyspnea ratings were unchanged

Chief Editor's Comment

This is another provocative and eye-opening case as to what can be done with persons who have a severe underlying chronic condition and who develop an acute illness on top of that. Here, a young girl, fairly severely affected by cystic fibrosis (as evidenced by low BMI and impacts on multiple organs—lung, liver, and skeletal muscle mass), developed an acute pneumonitis with weight loss and quite reduced pulmonary function. Despite this, her desire to play field hockey was great enough that she was willing to use a high-intensity interval training approach to get back in shape.

What remains unclear in this case study is how much of her improvement is secondary to the exercise program and how much is related to recovery from her pneumonitis and subsequently improved ventilatory and cardiorespiratory functions. Much of it is likely due to the antibiotics and resolution of pulmonary function, allowing her body to spend less energy on the cost of breathing and more on functions like increasing lean mass (which the HIIT surely helped promote). If pulmonary function didn't improve and the work of breathing remained very high, would lean mass still increase? Would cardiorespiratory fitness still improve ~10%, or less than that? While answers to these questions await randomized trials that will be very complicated to perform, at the least this case shows that in selected patients it's possible and appears to be safe to do an aggressive exercise trial to see what happens.

Lastly, this patient's story illustrates one point that I've found in my own patients: Athletes (and former athletes) usually understand that some of the things I've asked them to do will call for sacrifice and discomfort (but not pain). Athletic patients, seemingly more than nonathletic patients, are usually not intimidated by such a challenge and are able to muster the motivation to meet it. In the case of this young girl, field hockey seems to have been strongly motivating, and her high adherence to "practice" reflects that. In appropriate patients, appealing to their athletic desires can be very helpful in motivating them.

Deep Venous Thrombosis

PRESENTER Michael Lockard, MA

S

"My leg is swollen."

A 20-year-old female track athlete presented to the emergency department after 5 to 6 days of swelling and tightness in the left thigh and calf. She had a history of a synovial sarcoma in the left hip 3 1/2 years prior to admission, which was surgically excised and treated with 6 months of radiation therapy. A CT scan 3 weeks before admission had showed no recurrence of cancer. She did not use tobacco or oral contraceptives and did not recall a recent injury. She denied family history of bleeding or clotting disorders.

O

HR: 53 contractions/min

BP: 110/62 mmHg

Temp: 37.3 °C (99.1 °F)

Pertinent Exam

Left leg and thigh were warm and red, with 2+ nonpitting edema, good posterior tibial pulses, neurologically intact in both lower extremities.

Labs

Doppler ultrasound: DVT from iliac to popliteal vein in left leg

She was started on Lovenox and warfarin with a transition to long-term warfarin, advised on the bleeding risks of anticoagulation therapy, and told to do no running or other form of heavy exertion. Six days later, her symptoms had improved and she had no bleeding.

A

History of DVT in a track athlete

History of synovial sarcoma, in remission after surgical excision, lymph node dissection, and radiation therapy

DVT likely secondary to interaction of minor sport trauma and prior treatment

P

Gradual resumption of exercise

Coagulation studies were followed weekly to biweekly. Repeat Doppler ultrasound revealed that the DVT had resolved. She was maintained on warfarin for >6 months, monitored via INR (International Normalized Ratio). Evaluations for coagulopathies were negative.

Goals

Resume active lifestyle

Consider risk/benefit of competitive athletics against the risk of recurrent DVT

Exercise Program

Basic *CDD4* Recommendation transitioning to higher-intensity exercise, with special considerations when working with people with bleeding or clotting disorders:

Aerobic exercise daily

- Start with non–weight-bearing activities (e.g., cycling, pool exercise) at an RPE <13 (out of 20), or use talk test
- Begin with 30 min (rest stops allowed)
- Build ~5 min a week up to 40-60 min/session
- May increase to RPE <16 as tolerated, using interval design
- May decrease session to 30 min if concurrent with strength training

Resistance exercises 2 days a week

- Nonballistic weight training, starting at 1/2 the weight of preinjury
- 2 sets of 10 reps, gradually building up to resistance of 10RM

Flexibility exercises 3 days a week

- Regular stretching program, 30 s per stretch below discomfort point

Warm-up and cool-down phases @ RPE <13/20 for ~10 min

Special Considerations

Take these precautions when working with people with bleeding or clotting disorders:

- Gradually progress to slow jogging over next 6 months
- Consider changing sport to one that doesn't involve pounding footfalls

Chief Editor's Comment

A story that is not rare, though more usually occurring in young women taking oral contraceptives and therefore needing some counseling about considering an alternative method. With only a very small number of studies examining the optimal period of rest after the diagnosis of DVT, nobody knows how long to rest the athlete or even the optimal duration of anticoagulation; but almost all athletically minded patients want to get back to their sport. At this point, these decisions must be individualized according to the judgment of the care team. The possibility of recurrent DVT or pulmonary thromboembolism (PTE) instills great fear among physicians. As much as I favor allowing people to live the life they choose, and given that the available data on early resumption of activity are positively encouraging, it is hard to resist delaying until one is certain that the risk of DVT/PTE seems to be low. The management of DVT/PTE in athletes leads to a very difficult judgment call.

Dementia and Frailty

PRESENTERS Jessica S. Oldham, MS, and Jo B. Zimmerman, MS

S

"I don't even know why I'm here. My daughter brings me—she says this business is good for me."

A 71-year-old female with mixed dementia had become increasingly frail over the prior year, and her family sought assistance out of concern for her health. They were also hopeful that an exercise program could help her gain more functioning in her daily life. As the dementia progressed, her disorientation had made basic activities such as grocery shopping and gardening increasingly difficult. Her family also reported that she had had several falls in the preceding months, which had shaken her confidence, and her gait had become slow and measured. The patient herself was highly irritable and quite explicit in indicating that she had no real interest in starting an exercise program but consented to doing so upon the urging of her children.

O

Height: 5 ft 4 in. (1.63 m)
Weight: 103 lb (46.72 kg)
BMI: 17.7 kg/m^2

HR: 64 contractions/min
BP: 128/86 mmHg

Pertinent Exam

Elderly-appearing woman, minimally kyphotic, of tiny stature and marked sarcopenia, alert and oriented to person with questionable orientation to place and time; mild osteoarthrosis

Medication

Donepezil, 10 mg/day

Physical Functioning

Five-times sit-to-stand (time necessary to stand from a chair 5× unassisted): 19.5 s

Gait speed: 0.89 m/s

Stride length: 101.5 cm

Unipedal stance time: 12 s balance without support

Sit-and-reach: 23 cm (fair)

6-min walk: 300 m, estimated $\dot{V}O_2$peak 13 mL · kg^{-1} · min^{-1} (poor; <10th percentile)

A

Frailty (by Fried criteria) with generalized weakness

Slow gait speed and short stride length

Increased fall risk; may reflect loss of automaticity of gait

Cardiorespiratory deconditioning

Osteoarthritis in knees and hands

P

Gait training

Initiate a daily exercise program

Goals

Improve functional independence

Increase stride length and frequency

Improve strength and balance

Increase steps/day and provide client with simple pedometer for monitoring

Exercise Program

Basic *CDD4* Recommendation with special considerations for fall risk:

Aerobic exercise 4 days a week
- Walk or stationary cycle at an RPE 10-14 (out of 20), or use talk test
- Begin with 10-15 min (rest stops allowed)
- Build ~5 min a week up to 30-40 min/session
- May decrease session to 30 min if concurrent with strength training

Resistance exercises 2-3 days a week
- Sit-to-stand exercises
- Arm curls using a 2 L jug filled with water
- Build up to 2-3 sets of 10-12 reps

Flexibility exercises 3 days a week
- Chair sit-and-reach stretches daily on L and R
- Hold for 20-30 s on each side, goal of getting closer to 0 cm/in. from toes

Warm-up and cool-down phases @ RPE <11/20 for ~10 min

Special Consideration

Adaptive precaution: Supervision strongly recommended.

328

Chief Editor's Comment

There are lots of "little old ladies" like this patient, and it's disconcerting that she's this way but only 71 years old. Her very poor physical functioning parameters and meeting 4/5 Fried criteria for frailty imply a poor prognosis. Her social history is not clear from this brief vignette, but her living situation is a concern—it sounds as though she is either still living independently or with family or close friends (she's still gardening and shopping). It's not explicitly stated in the plan of care, but one of the ulterior objectives is surely to keep her in this environment with as little assistance as necessary so that her family can continue their own lives with minimal disruption. Otherwise, she is destined for assisted living or a nursing home.

In aging or frail patients with chronic conditions who understand why their physical functioning is so important, exercise in a gym as a workout may be seen as means to an end. In such patients, the social aspects of the environment help with adherence to the program. In a patient with dementia, however, the socioecological aspects may play a stronger role in maintaining her interest. To the extent that the program can be transposed to a gardening environment (since she likes to garden), I would be inclined to have some of her program medically supervised in a gym and some of it at home with family in the garden. Hopefully, the family members can make some time to "garden with Grandma" and view that less as a burden and more as a precious opportunity to share life with her.

Fibromyalgia

PRESENTER **Stephen P. Bailey, PhD, PT, FACSM**

S

"I'm tired and everything hurts."

A 40-year-old woman complained of a 3-year history of pain, including aches and pains in the neck, upper and lower back, and both arms and legs. Her pain was usually worse later in the day. She usually awoke unrefreshed and occasionally experienced morning stiffness. She had been taking an antidepressant for 5 years.

O

Height: 5 ft (1.52 m)
Weight: 125 lb (56.7 kg)
BMI: 24.5 kg/m^2

HR: 80 contractions/min
BP: 120/80 mmHg

Pertinent Exam

Healthy-appearing middle-aged woman, no notable joint inflammation

Medications

Paroxetine daily

Graded Exercise Test (Modified Balke Protocol)

Completed 4 stages (12 min)

Peak exercise capacity: ~5.5 METs

Peak HR: 140 contractions/min (78% of age predicted)

Normal pressor response

Test terminated secondary to fatigue

ECG: no abnormalities, nondiagnostic study due to inadequate maximum heart rate

Questionnaires

Widespread Pain Index (WPI): 12

Symptom Severity Scale (SS): 7

A

Fibromyalgia

Low exercise capacity/physical functioning (decreased tolerance for ADLs)

P

Consider switching medication to duloxetine or tricyclic antidepressant

Low-impact exercise program

Goals

Maintain participation

Decrease symptoms

Improve ability to perform ADLs

Exercise Program

Basic *CDD4* Recommendation with special considerations for fibromyalgia:

Aerobic exercise 2-3 days a week

- Walk at an RPE <13 (out of 20), or use talk test

- Begin with 15-20 min (rest stops allowed)
- Build ~5 min a week up to 40 min/session
- May eliminate session on strength training day
- Avoid eccentric and ballistic forms of exercise
- May wish to try aquatic exercise class

Resistance exercises 1-2 days a week

- Sit-to-stand exercises
- Arm curls using a 2 L or 4 L jug filled with water
- Build up to a minimum of 2 sets of 8 reps

Flexibility exercises 3-4 days a week

- Chair sit-and-reach stretches daily on L and R
- Hold for 30 s on each side, goal of 0 cm/in. from toes
- Additional stretches as desired, 30 s per stretch below discomfort point

Warm-up and cool-down phases @ RPE <11/20 for ~10 min

Special Consideration

May self-select down to 2 days/week (all modes included)

Chief Editor's Comment

Maintaining motivation is often the most difficult problem for patients with fibromyalgia, similar to the situation in those who have chronic fatigue. As in the case of chronic fatigue, it is very helpful to have a good social component of the exercise program and an exercise specialist who connects well with the patient. Forgiveness and understanding for many skipped/missed sessions are important to try to avert a high likelihood of dropping out. Light yoga and other mindful meditation/alternative medicine approaches can help maintain engagement. Counseling and focusing on participation more than on achieving specific goals may help avert negative thinking.

Hearing Impairment

PRESENTER Jo B. Zimmerman, MS

S

"I am fed up with being overweight and feeling sluggish and I worry about what might happen if I don't start taking better care of myself. I'd like to enjoy my retirement in a few years."

A 55-year-old woman wanted to begin some general exercise for fitness. She had had a sedentary occupation and very little recreational activity for most of her adult life. She was feeling the effects of creeping weight gain and was growing concerned for her long-term health. She had developed hearing impairment bilaterally after long-term noise exposure in her 20s and 30s and wore two hearing aids to assist her communication. She reads lips very well, but people tell her that her own speech is very hard for others to understand. She has very little exercise experience and never did any sports.

O

Height: 5 ft 1 in. (1.55 m)
Weight: 165 lb (74.8 kg)
BMI: 31.2 kg/m^2

HR: 76 beats/min
BP: 138/88 mmHg

Pertinent Exam

55-year old Caucasian woman, mildly obese and not well muscled; bilateral hearing impairment, with no vestibular impairment

Laboratory Data

Audiogram: significant bilateral deficit in the 250- to 1000-Hz range

Submaximal Exercise Test

Rockport Walk Test (patient was not comfortable on a cycle ergometer)

1-mile walk time: 16 min 35 s

Peak HR: 136 beats/min

Estimated $\dot{V}O_2$max at 23.4 mL · kg^{-1} · min^{-1}

A

Severe mixed sensorineural auditory impairment

Prehypertension with class 1 obesity

Aerobic fitness very poor for her age

P

Use a personal trainer/program to help get started

Exercise programming per ACSM Guidelines, though the Basic *CDD4* Recommendations may be a good starting point if she struggles with intensity or duration

Refer to Registered Dietitian for additional support toward BMI and blood pressure goals

Goals

Reduce BMI and blood pressure

Find a form of exercise/physical activity that she likes

Encourage her to try swimming or group exercise (if bass component of music is perceptible) for additional variety and social support

Exercise Program

ACSM Guidelines with special considerations for hearing impairment:

Aerobic exercise 5 days a week

- Walk at an RPE <13 (out of 20), or use talk test
- Begin with 10- to 20-min blocks (rest stops allowed)
- Build ~5 min a week up to 30-60 min/session
- May increase to RPE <16 as tolerated, using interval design; vary duration and intensity as time and energy allow
- May decrease session to 30 min if concurrent with strength training

Resistance exercises 2 days a week

- Selectorized or plate-loaded machines, free weights OK if gym has mirrors
- Light to moderate intensity (not to failure), emphasis on higher repetitions
- Full body routine, building up to a minimum of 2 sets of 8 reps
- Emphasis on safety and not doing Valsalva maneuver during lifts

Flexibility exercises 3 days a week, for 5-10 min

- Chair sit-and-reach stretches daily on L and R
- Hold for 30 s on each side, goal of 0 cm/in. from toes
- Additional stretches as desired, 30 s per stretch below discomfort point

Warm-up and cool-down phases @ RPE <11/20 for ~10 min

Special Considerations

Adaptive precautions: Provide clear written instructions for her, and have a notepad or whiteboard available for questions and answers as they arise.

Resources for This Client

National Center on Health, Physical Activity, and Disability (NCHPAD), www.ncpad.org [Accessed September 20, 2015].

National Institute on Deafness and Other Communication Disorders (NIDCD), www.nidcd.nih.gov/Pages/default.aspx [Accessed September 20, 2015].

USA Deaf Sports Federation, www.usdeafsports.org [Accessed September 20, 2015].

Senior Editor's Comment

In many cultures, at the time of this writing, women who were middle-aged and older had never been involved in any kind of exercise program. I myself have held an elderly woman's feet to the pedals of a cycle ergometer and used my hands to make the pedals go around in order to help her learn how to pedal—she'd never ridden a bicycle in her life. Even in the United States, it's been only since the 1960s and 1970s that women were able to participate on par with men in high-exertion sports events, and still today, women's sport organizations are arguing their case to do the same events as men.

This woman's auditory impairment may indeed have led her to a more sedentary existence than she might otherwise have had, but many women come to a cardiac rehab program or to the clinic who really have a very low exercise tolerance because they've never done anything to be hardy and fit. One can approach their situation by following the ACSM Guidelines and getting a stress test with 12-lead ECG, but in many cases the process works fine to start such patients out slowly and increase gradually. Furthermore, such a protocol might actually be a better use of clinical resources to allow them to build themselves up a little before sending them for a stress test, to ensure that they don't give out at a nondiagnostic work rate. This decision needs to be individualized, and the wise provider will get an exercise specialist's recommendation on formal testing, since the exercise specialist is the health care team member who really sees the patient in action.

Heart Transplant

PRESENTERS Audrey Borghi Silva, PhD, PT, and Gerson Cipriano Jr., PhD, PT

S

"I had a heart transplant about one month ago. I am having difficulty walking long distances, going up and down stairs, and getting in and out of a chair. My family is very supportive of my rehabilitation."

This 45-year-old Caucasian male underwent heart transplantation because of ischemic heart failure following a myocardial infarction 2 years earlier. Before the transplant surgery, he was on an implantable left ventricular assist device (LVAD) for 6 months. Two months after the LVAD was implanted, he started participating in a cardiac rehabilitation program (three times per week) and maintained that program until the heart transplantation. His past medical history includes non–insulin-dependent diabetes, dyslipidemia, and hypertension. One month following successful transplantation he was referred to outpatient cardiac rehabilitation for a reconditioning program and cardiovascular risk reduction.

O

Height: 5 ft 9 in. (1.75 m)
Weight: 185 lb (83.9 kg)
BMI: 27.5 kg/m^2

HR: 95 contractions/min
BP: 136/84 mmHg

Pertinent Exam

Overweight male post–heart transplant, well-healed surgical scars, lungs clear, no abnormal heart sounds, no edema, good peripheral circulation

Medications

Tacrolimus, prednisone, isosorbide dinitrate, pravastatin, furosemide, enalapril, glyburide, aspirin, magnesium

Labs

Baseline echocardiogram: normal left ventricular ejection fraction (55%), mild mitral and tricuspid regurgitation

Estimated pulmonary artery pressure: 30 mmHg

Total cholesterol: 215 mg/dL

Fasting blood glucose: 110 mg/dL

HDL: 40 mg/dL

LDL: 149 mg/dL

Triglycerides: 130 mg/dL

Minnesota Living With Heart Failure Questionnaire: 55 out of 105 (higher is worse)

Duke Activity Status Index: 6 METs

Graded Treadmill Test

Conservative ramping protocol (~0.5 METs/30 s)

Peak exercise: 4.0 METs

Peak HR: 140 contractions/min

Peak BP: 189/90 mmHg

$\dot{V}O_2$peak: 14.0 mL $O_2 \cdot kg^{-1} \cdot min^{-1}$

Peak RER: 1.14

$\dot{V}E/\dot{V}CO_2$ slope: 35.0

$\dot{V}O_2$ at ventilatory threshold: 9.8 mL $O_2 \cdot kg^{-1} \cdot min^{-1}$

No ECG changes or ischemia indications

Test terminated by patient primarily due to lower-extremity fatigue (RPE =17/20)

Dyspnea also reported (2/4)

Decline in HR during recovery: 7 contractions/min @ 1 min; 18 contractions/min @ 3 min

Note: Improved from 1 month pretransplant exercise test: $\dot{V}O_2$peak = 9.8 mL $\cdot kg^{-1} \cdot min^{-1}$, peak RER= 1.19, $\dot{V}E/\dot{V}CO_2$ slope = 49.0.

A

Status post–cardiac transplant, low exercise tolerance (~NYHA Class III)

Reduced exercise tolerance, although improved from pretransplant level

Appropriate hemodynamic response, as expected post–cardiac transplant; ECG response to exercise within normal limits

Hypertension and dyslipidemia persist, although pharmacologically managed

Osteoporosis risk due to immunosuppressive agents

Type 2 diabetes, fasting blood glucose above desired threshold (<100 mg/dL)

P

Reinstitute supervised aerobic conditioning initiated in the pretransplant stage and progress as tolerated

Assess muscle force production and initiate resistance exercise training program after 1 month of aerobic exercise training is completed

Introduce diabetic, dyslipidemia, hypertension, and dietary education

Closely monitor resting and exercise HR, BP, and ECG

Progress patient to unsupervised home exercise program over the next 3 months

Goals (12 Weeks)

Independently perform activities of daily living with peak exercise capacity of 5 METs

Increase muscle force production by 10%

Decrease BMI to 25 kg/m^2

Improve glucose and lipid values

Decrease in SBP and DBP by 10%

Improve perceived quality of life as measured by MLWHF (Minnesota Living with Heart Failure) instrument

Independently manage exercise program, dietary monitoring, and risk factor control

Reverse skeletal muscle weakness and insulin resistance, secondary to deconditioning from CHF/LVAD limitations

Attenuate bone density effects of tacrolimus and prednisone

Exercise Program

Basic *CDD4* Recommendation with special considerations for cardiac transplant recipients:

Aerobic exercise 4-5 days a week
- Walk at an RPE <13 (out of 20), or use talk test
- Begin with 15-20 min (rest stops allowed)
- Build ~5 min a week up to 40 min/session
- May increase to RPE <16 as tolerated, using interval design
- May decrease session to 30 min if concurrent with strength training

Resistance exercises 2 days a week
- Sit-to-stand exercises
- Arm curls using a 2 L or 4 L jug filled with water
- Build up to a minimum of 2 sets of 8 reps

Flexibility exercises 3 days a week
- Chair sit-and-reach stretches daily on L and R
- Hold for 30 s on each side, goal of 0 cm/in. from toes
- Additional stretches as desired, 30 s per stretch below discomfort point

Warm-up and cool-down phases @ RPE <11/20 for ~10-15 min

Special Considerations

Adaptive approaches: A slower and more prolonged warm-up is needed, since sinoatrial node in the transplanted heart is denervated and thus will not respond to decreasing vagal tone. Increase in exertional heart rates is secondary to sympathomimetic amines in cardiac transplant recipients.

Senior Editor's Comment

This case study is a characteristic example of the course many transplant recipients follow (cardiac and other organs), with a prolonged waiting period before transplantation that leads to deconditioning. If at all possible, patients waiting for transplant should participate in a "prerehabilitation" program, which costs little compared to the cost of the transplant itself but begins the reconditioning phase. Regarding the pretransplant exercise test when the patient was on the LVAD, his performance was surprisingly good. The LVADs available today are constant-flow devices that cannot increase pump output to match demand (the output of the LVAD can be varied a little, but once the flow is set it does not change). The patient's native heart does continue to be rate-responsive, but obviously it is not working effectively as a pump (and perhaps less so because of the mechanical alteration of the heart by the LVAD). Thus, the resting and exercising cardiac output with an LVAD in place is likely ~5 L/min, which would nominally deliver an oxygen output of ~1 L/min of O_2 (12 mL $O_2 \cdot kg^{-1} \cdot min^{-1}$). Thus, the measured peak $\dot{V}O_2$ of 10 mL $O_2 \cdot kg^{-1} \cdot min^{-1}$ actually represents pretty good conditioning—he's able to use all of the cardiac output that he can deliver! Moreover, the fact that he demonstrated a ventilatory threshold at such low cardiac outputs suggests that his muscles still have ability to generate metabolic demand and that his circulatory control is adequate. I would expect this patient to do well in his posttransplant training program.

Human Immunodeficiency Virus

PRESENTERS Gregory A. Hand, PhD, FACSM;
G. William Lyerly, PhD; Wesley D. Dudgeon, PhD

S

"I would like to get in better shape, feel more energetic about doing daily activities, and feel like my old self."

A 37-year-old man has been infected with HIV for 5 years. He typically had a CD4 count of 300 cells/mm³ that had been slowly dropping, so he began a highly active antiretroviral therapy (HAART) regimen of dapsone, tenofovir/emtricitabine, and lopinavir/ritonavir. Before starting HAART, he had been having frequent diarrhea and muscle weakness, but after the HAART he had less GI distress and put on weight in his abdomen, neck, and chest. After finding out that he was HIV+, he became more sedentary due to its being "hard to do the things I used to do" and had developed osteoporosis. He complained of being tired more often and was "having trouble keeping up with my friends like I used to."

O

Height: 5 ft 10 in. (1.78 m)
Weight: 170 lb (77 kg)
BMI: 24.38 kg/m²

HR: 72 contractions/min
BP: 138/90 mmHg

Pertinent Exam

Central adiposity, with waist circumference 34.1 in. (86.6 cm)

Medications

Dapsone, tenofovir/emtricitabine, lopinavir/ritonavir, metoprolol, and atorvastatin

Labs

CD4 count: 300 cells/mm³
Viral load: 5500 RNA copies/mL
Hemoglobin: 14.4 g/dL
Hematocrit: 41%
Fasting glucose: 96 mg/dL
Total cholesterol: 290 mg/dL
Triglycerides: 180 mg/dL

Graded Exercise Test (Modified Bruce Protocol)

Peak work rate: 2.5 mph @ 12% grade
Total treadmill time: 6.5 min
Peak HR: 183 contractions/min
Peak BP: 190/95 mmHg
$\dot{V}O_2$peak: 15.84 mL · kg^{-1} · min^{-1}
Peak RPE: 18 out of 20
Reason for test termination: volitional exhaustion
ECG: Sinus rhythm at rest and throughout exercise and recovery
No dysrhythmias observed or reported
No report of chest discomfort

Body Composition (DEXA)

22.7% body fat
Bone density: 1.27 g/cm²

A

Low physical functioning with easy fatigability
Increased cardiometabolic risk from hyperlipidemia, hypertension, and sedentary lifestyle
Osteoporosis

P

Increase physical functioning
Increase lean body mass and reduce body fat
Short-term goals (1-2 months):

- Attend ~90% of exercise sessions
- Sustain aerobic activity for 30 min at a moderate intensity without terminating due to fatigue
- Increase muscular strength

Long-term goals (6 months):

- Sustain aerobic activity for 40-60 min at a moderate intensity without terminating due to fatigue
- Increase lean tissue mass and decrease central obesity
- Improve subjective QOL

Exercise Program

Basic *CDD4* Recommendation:

Aerobic exercise 4-5 days a week

- Walk at an RPE <11 (out of 20) or ~40% of HRR, or use talk test
- Begin with 10-15 min (rest stops allowed)
- Build ~5 min a week up to 40 min/session
- May increase to RPE <13 or ~60% of HRR, as tolerated, using interval design
- May decrease session to 20-30 min if concurrent with strength training

Resistance exercises 2 days a week

- Sit-to-stand exercises
- Arm curls using a 2 L or 4 L jug filled with water
- Build up to a minimum of 2 sets of 8 reps

Flexibility exercises 3 days a week

- Chair sit-and-reach stretches daily on L and R
- Hold for 30 s on each side, goal of 0 cm/in. from toes

- Additional stretches as desired, 30 s per stretch below discomfort point

Warm-up and cool-down phases @ RPE <11/20 for ~10 min

Senior Editor's Comment

This patient illustrates that being HIV+ often brings with it multisystem chronic conditions, involving lipid/glucose metabolism, cardiovascular risk, and bone health on top of being HIV+. As in many patients with chronic conditions, there is a gradual deterioration toward frailty that should be resisted with aerobic and strength training exercise. The patient's comments hint at the phenomenon that chronic disease can lead to social isolation, noting that he can't keep up with his friends. Exercise training, with a little bit of counseling to help the patient and his support system gain tolerance regarding his dwindling physical functioning, can go a long way toward helping such a patient avoid becoming isolated.

Hypertension, Dyslipidemia, and Obesity

PRESENTER Benjamin Gordon, PhD, RCEP

S

"I think I should start an exercise program."

A 56-year-old male was interested in starting an exercise program. He had no health complaints but did have hypertension, dyslipidemia, and obesity, meeting the criteria for the metabolic syndrome. He had been sedentary for many years, did not smoke cigarettes, and had a family history of premature atherosclerosis.

O

Height: 5 ft 10 in. (1.78 m)
Weight: 210 lb (95.3 kg)
BMI: 30.2 kg/m^2

HR: 56 contractions/min
BP: 146/94 mmHg

Pertinent Exam

Moderately obese male, no distress

No vascular changes on funduscopic exam, peripheral pulses nondelayed, no bruits

Normal heart sounds, nondisplaced ventricular apical impulse

Medications

Metoprolol SR 50 mg daily, rosuvastatin 10 mg daily

Graded Exercise Test (Bruce Protocol)

Peak work rate: 6 min (end of stage 2) ~7 METs

Peak HR: 134 contractions/min

Peak BP: 192/96 mmHg

Peak RPP: 25,728

RPE: 18/20

Reason for termination: leg fatigue

ECG: no significant ST changes; occasional PVCs

No chest discomfort reported

A

Hypertension

Hyperlipidemia

Obesity

Metabolic syndrome

Sedentary lifestyle/deconditioning

P

Initiate an exercise program, with primary emphasis on aerobic activity and calorie expenditure

Goals

Improve blood pressure and pressor response to exercise

Manage weight

Control cholesterol

"Reverse" the diagnosis of metabolic syndrome

Exercise Program

Use ACSM Guidelines.

Aerobic exercise, start at 3 days a week

- Walk at an RPE <13 (out of 20), use talk test or 40-59% of HRR (87-102 contractions/min)
- Begin with 15 min/session (rest stops allowed)
- Build ~5 min a week up to 30-60 min/session
- May increase to RPE <16 as tolerated, using interval design
- Increase over a few months to 5-7 days/week

Resistance exercises 2-3 days a week

- 1 set of 10-15 repetitions of 8-10 different resistance training exercises
- Build up to 2 sets of 8RM for each lift

Flexibility exercises 3 days a week

- 20-30 s/stretch (all major muscle groups)
- Progress as tolerated to at least 4 repetitions per muscle group

Warm-up and cool-down phases @ RPE <11/20 for ~10 min

Chief Editor's Comment

Here is another extremely common scenario today, applicable to nearly one-third of the population in most countries. This patient is someone who would benefit from gradual progression to as much exercise as he will do. If he builds to the full program in this care plan, he will accumulate 300+ minutes of moderate to vigorous activity each week. The exercise program would have great benefits on his blood pressure and dyslipidemia, but he will likely need to use a dietary intervention to reduce and sustain his weight much under 200 lb (90.9 kg). He probably needs to get down to ~190-195 in order to "reverse" the diagnosis of metabolic syndrome, though lower is likely better for reducing his risk for osteoarthosis and other musculoskeletal overload syndromes.

Interstitial Lung Disease (Chronic Restrictive Lung Disease)

PRESENTER Connie C. W. Hsia, MD

S

"I feel tired and short of breath when walking."

A 45-year-old African American female complained of episodic fatigue, night sweats, and skin lesions that had been going on for several years. She was found to have skin nodules on the face and extremities, enlarged liver and spleen, and lymph nodes in the chest. Biopsy of skin and lung tissue revealed pathology compatible with the diagnosis of sarcoidosis, a multisystem granulomatous inflammatory disease. Initially she had no respiratory complaints, but she gradually developed worsening shortness of breath on exertion to the point that she could walk only 800 m on a flat surface or climb just one flight of stairs without stopping. She was easily fatigued at work, and her lung volumes and diffusing capacity (DL_{CO}) had markedly decreased. She has required oral prednisone therapy and as a result developed weight gain, hypertension, and hyperglycemia.

O

Height: 5 ft 4 in. (1.63 m)

Weight: 180 lb (81.6 kg)

BMI: 30.9 kg/m^2

HR: 70 beats/min

BP: 140/90 mmHg

Pertinent Exam

Enlarged parotid glands and cervical lymph nodes; lung sounds were clear; cardiovascular examination was normal with no peripheral edema; spleen and liver enlarged; there were several flesh-colored nonitchy skin nodules on her face and legs.

Medications

Prednisone, insulin, lisinopril, calcium and vitamin D supplement

Labs

Normal blood counts and electrolytes, with mildly elevated liver enzymes

Serum markers of systemic inflammation (angiotensin-converting enzyme and gamma globulins) elevated

Pulmonary Function Testing

Vital capacity 60% of predicted, DL_{CO} 55% of predicted

CT Scan of the Chest

Enlarged lymph nodes in central mediastinum, and reticular infiltrates in both lungs

Exercise Test Results

Peak exercise: 5.0 METs

Peak HR: 150 beats/min

Peak BP: 170/95 mmHg

Transcutaneous oximetry: oxygen saturation 97% at rest and 93% at end of exercise

Peak RPE: 18/20

Termination reason: shortness of breath and leg fatigue

Symptoms: no chest discomfort

Chronotropic response: normal

A

Interstitial lung disease due to sarcoidosis

Restrictive pulmonary physiology, with adequate exertional oxygen saturation (>90%) while breathing room air

Increased cardiometabolic risk, secondary to reduced physical activity and prednisone

P

A second-line anti-inflammatory and immunosuppressive treatment added

Oral prednisone tapered, with a goal of facilitating weight loss

Follow-up clinic visits and lung function testing to monitor response to treatment

Exercise program to facilitate and maintain weight loss, reduce cardiometabolic risk, and improve physical functioning

Goals

Improve ventilatory endurance

Reduce the sensation of dyspnea

Improve respiratory muscle strength and endurance

Exercise Program

Basic *CDD4* Recommendations with special considerations for chronic restrictive lung disease:

Aerobic exercise 4-5 days a week
- Walk at an RPE of 8-13/20
- Begin with 20 min (rest stops allowed)
- Build ~5 min a week up to 40 min/session
- May gradually increase to RPE <16 as tolerated, using interval design
- May decrease to 30 min if strength training is on a day concurrent with aerobic exercise

Resistance exercises 2 days a week
- Sit-to-stand exercises
- Arm curls using a 2 L or 4 L jug filled with water
- Build up to a minimum of 2 sets of 8 reps

Flexibility exercises 3 days a week
- Chair sit-and-reach stretches daily on L and R
- Hold for 30 s on each side, goal of 0 cm/in. from toes
- Additional stretches as desired, 30 s per stretch below discomfort point

Warm-up and cool-down phases @ RPE <11/20 for ~10 min

Special Considerations

Anti-inflammatory medications should be taken regularly to obtain the best exercise performance

Fear avoidance of dyspnea is a common barrier to maintaining regular exercise in lung disease

Worsening hypoxemia during exertion may induce chest pain and/or arrhythmias

If needed, supplemental O_2 flow rate should be adjusted to maintain O_2 saturation >90%

Senior Editor's Comment

This patient is likely following a downhill course, but at this time she still has potential to improve her physical functioning and reduce her sensation of dyspnea. While exercise may likely help her to lose some weight, she should combine a weight loss diet intervention with resistance training that is as intensive as she can muster. Recent studies suggest that high-intensity interval and strength training might provide a myogenic stimulus without provoking her dyspnea symptoms. The natural course of sarcoidosis would be for her to get gradually worse, with increasing need for O_2 supplementation and increased difficulty in physical functioning, in part because the work of breathing will gradually increase as her condition gets more severe. Building up a reserve of muscle strength and mass may be her best defense to prolong her quality of life.

Major Depressive Disorder

PRESENTERS Jessica S. Oldham, MS, and Jo B. Zimmerman, MS

S

"I used to be an athlete and want to get back into an exercise routine, but I'm just tired all the time and can't get motivated to go to the gym after work. I'd like to race a 10K with my wife, but I'm not sure I could do it."

A 35-year-old male presented with a several-month history of major depressive disorder and wanting to be more active. He had been taking fluoxetine prescribed by his physician but does not think that it is helping. Many of the activities he previously enjoyed with his family are no longer appealing to him, and he worries that his relationships with his wife and children are suffering. He has irregular sleep habits and reports frequently waking for long periods in the middle of the night and then having trouble getting up for work in the morning. He notices that he is having an increasingly hard time concentrating at work, and this is making him concerned about his job security. For physical health reasons, he has a strong desire to start exercising again but cannot get motivated to do so. He denied suicidal ideation.

O

Height: 6 ft 2 in. (1.88 m)
Weight: 190 lb (86.2 kg)
BMI: 24.4 kg/m^2

HR: 78 beats/min
BP: 135/87 mmHg

Medications

Fluoxetine 50 mg/day

Pertinent Exam

Middle-aged healthy-appearing male, with depressed mood and flat affect

Cardiopulmonary Exercise Test (Modified Bruce Protocol)

Reason for termination: volitional exhaustion. No symptoms reported.

Max HR: 187 beats/min

Max BP: 188/89 mmHg

RPE at max: 19/20

$\dot{V}O_2$max: 46 mL · kg^{-1} · min^{-1}

ECG: sinus rhythm at rest and through exercise testing and recovery, no ectopy, conduction abnormalities, or ST changes

Body Composition

18% body fat (by skinfold)

A

Major depression
Sedentary male
Prehypertension

P

Combine moderate-intensity aerobic activity and resistance training exercise programs

Reevaluate in 6-8 weeks

Register for a running race as motivation

Goals

Attend majority (~90%) of exercise sessions

Walk or jog for 30 min 4-5 days a week

Gradually increase duration and percent of time spent jogging

Increase lean tissue mass

Exercise Program

Follow ACSM Guidelines for exercise prescription.

Aerobic exercise 4-5 days a week

- Walk/jog at an RPE 8-13 (out of 20)
- Begin with 20 min (rest stops allowed)
- Build ~5 min a week up to 40 min/session at RPE 11-16/20
- May decrease session to 20 min if concurrent with strength training

Resistance exercises 2 days a week

- All major muscle groups
- 1 set, 8-12 reps using 50-70% of 1RM
- Continue each exercise all the way to failure
- Increase to 2-3 sets over 8 weeks

Flexibility exercises 3 days a week

- All major muscle groups
- Hold for 20 s per stretch
- Keep below discomfort point (at gradually higher range of motion over days/weeks)

Warm-up and cool-down phases @ RPE <11/20 for ~10 min

Chief Editor's Comment

In many ways this is a garden-variety case of a middle-aged male with no evidence of cardiometabolic or musculoskeletal conditions that would limit his participation in physical activity, and reflects a typical course of treatment for a patient with depression. From the history given, one cannot tell if the fluoxetine is helping or not (likely so, since the doctor hasn't changed the medication), but for sure the fluoxetine alone isn't an adequate solution. Another common scenario is seen in a patient who's been on antidepressants for a long time and is doing better, but everyone is reluctant to stop the medication. I myself would hesitate to diagnose prehypertension at a single visit, especially as I am unfamiliar to the patient, but would note that he has either prehypertension or a bit of anxiety. Designing his exercise program is relatively straightforward.

The patient is an ex-runner who says he'd like to get back into it, and I'm not sure if these are his true feelings or if he is grasping for a solution. Whichever, an exercise program is a good idea because it's likely to help, and after several months he needs to do something more than take fluoxetine. In such patients, I typically ask how they feel about counseling, since bringing in a clinical psychologist would a good idea. If he were resistant, I'd be patient and allow the exercise specialist or personal trainer to become more of a trusted resource, and see how the exercise program was helping. If the patient had a high PHQ-9 (depression screening questionnaire) score or did have suicidal thoughts, I would press harder for counseling (assuming no suicidal intentions were present).

As far as motivation, in a former runner who says he wants to run again, it's logical to set a goal of participation in a 5K or 10K event, but I would do so warily until I knew the patient well. I always want to know the socioecological issues in the patient's life before I set any goals: what he wants to be, how his relationships with his wife and kids are going, how his work life is going, if money or alcohol or drugs or other issues and stressors are lurking. If he's the kind of person who is inclined to judge himself negatively, I would approach the running event conservatively, where participation is enough, and would resist his setting some performance-oriented goal (as a formerly competitive runner is likely to do). I like the plan to pursue participation in the exercise program as the first goal, and would advise using those opportunities to learn more about what's going on in his life.

Multiple Sclerosis

PRESENTERS Tara Patterson, MEd, PhD, and Jill Seale, PhD, PT, NCS

S

"I want to be able to continue working and to improve my walking and balance so that I can return to my usual activities."

This 42-year-old female, who had been diagnosed with relapsing–remitting multiple sclerosis (MS) at age 38, was recovering from a recent exacerbation that began 7 days before her clinic visit. Prior to her MS diagnosis, the patient had no significant medical history. Since diagnosis, she has experienced nine significant exacerbations, and after the last two exacerbations she did not return to her baseline functioning. Her physician was concerned that the patient might be transitioning to secondary progressive MS. The patient reported that she had to stop cycling and playing golf due to fatigue and balance dysfunction. She was a vice president of a technology company and had been forced to work at home several days each week due to fatigue, impaired walking, and diminished balance.

O

Height: 5 ft 8 in. (1.73 m)
Weight: 145 lb (65.8 kg)

HR: 76 beats/min
BP: 110/70 mmHg
RR: 16 breaths/min

Physical Exam

Normal unless noted otherwise

Mental status/cognition: alert and oriented × 4; recent and remote memory intact

Communication: clear speech, no aphasia

Musculoskeletal: lower-extremity weakness, left > right

Neuromuscular: ambulatory with single-point cane; alterations in kinematics and temporospatial parameters; increased muscle tone in left lower extremity; impaired standing balance

Range of motion: R foot dorsiflexion 0°, L dorsiflexion lacks 8° to neutral

Sensation: bilateral lower legs—impaired kinesthesia, vibratory sense, and light touch

Deep tendon reflexes: brisk in bilateral patellar and Achilles tendons

Coordination: normal

Medications

Baclofen, Avonex, and Effexor

Manual Muscle Testing

Bilateral hip flexors and knee extensors: 4+/5; bilateral hip extensors, hip abductors, and knee flexors: 4/5; R plantarflexors: 3/5; L plantarflexors: 2/5; R dorsiflexors: 3+/5, L dorsiflexors: 3/5

Muscle Tone (Modified Ashworth Scale [MAS])

L hip extensors, knee extensors, and plantarflexors: 2 MAS

R knee extensors and plantarflexors: 1+ MAS

All other LE muscles with MAS = 0

Functional Mobility

Modified independent (MOD I) in bed mobility, sit-to-stand from car, toilet, bed, and standard-height chair. Needs increased time, uses single-point cane, and appears mildly unstable.

Gait Analysis

Ambulates over level surfaces with MOD I using single-point cane. Observed gait deviations include R LE, flat foot initial contact, decreased dorsiflexion in mid and terminal stance, decreased knee flexion in swing; L LE, forefoot initial contact, decreased knee flexion in loading response, decreased dorsiflexion in mid and terminal stance, knee extensor thrust in midstance, decreased hip extension in mid and terminal stance, and decreased hip, knee, and dorsiflexion throughout swing.

Balance

Timed up-and-go = 22 s (fall risk)

Functional Gait Assessment (FGA)

14/30 (fall risk)

Gait speed: 0.8 m/s

Gait endurance: 623 ft (190 m) in 6-min walk test

Evaluation Scales

Expanded Disability Status Scale (EDSS) score: 3.0

Modified Fatigue Impact Scale 5-item version: 12/20

A

Secondary progressive multiple sclerosis

Predominantly lower-extremity impairments

Fall risk

Identified Problems

Decreased strength, sensation, and PROM in bilateral LEs

Increased muscle tone in bilateral LEs

Impaired balance

Multiple gait deviations, with decreased gait speed and endurance

Increased fatigue

Anticipated Problems

Progressive weakness and sensory changes

Decreased PROM due to weakness and spasticity

Eventual need for adaptive equipment

Decreased cardiovascular fitness due to progressing disability

P

Improve balance and educate on fall risk

Increase lower-extremity strength and PROM

Improve gait speed and efficiency

Provide education on energy conservation and activity modification to lessen fatigue

Exercise Program

Modified *CDD4* Recommendation with special considerations for multiple sclerosis; balance exercises added (see text in italics):

Aerobic exercise 3-4 days a week

- Walk at an RPE <13 (out of 20), or use talk test
- Begin with 10 min (rest stops allowed)
- Gradually increase by 5 min each week over 4 weeks, maintain RPE 11-16/20
- May decrease session to 30 min if concurrent with strength training

Resistance exercises 2 days a week

- Sit-to-stand exercises
- Arm curls using a 2 L or 4 L jug filled with water
- Add other exercises as tolerated to strengthen all major muscle groups
- Build up to a minimum of 2 sets of 8 reps, over ~8 weeks

Flexibility exercises 3 days a week

- Chair sit-and-reach stretches daily on L and R
- Hold for 30 s on each side, goal of 0 cm/in. from toes
- Additional stretches as desired, 30 s per stretch below discomfort point

Balance exercises 2-3 days a week

- *Start balance exercises on stable surface with eyes open*
- *Progress to balance exercises on unstable surface with one or both eyes closed*

Warm-up and cool-down phases @ RPE <11/20 for ~10 min

Special Considerations

Terminate session if symptoms are provoked by exercises

Adaptive precautions: Supervise activities that have fall risk until patient is skilled with the activities

Chief Editor's Comment

This patient seems to be on a cusp in the time course of her MS, and may soon need to face some difficult decisions. She is obviously an accomplished individual but is only 42 years old and seems to be facing a decline in her physical functioning. Cognitively she is unimpaired, but she is losing physical functioning at an alarming rate for a 42-year-old woman. The role of stress as an inciting agent for her exacerbations may need to be evaluated. With nine exacerbations over 4 years, or about one every 4-5 months, she should be able to look back and decide whether work or family or social stressors were associated with the exacerbations. If there is such a pattern, she may need to think about "backing off" in one or more areas of her life and reranking her priorities. Initiating a discussion about a patient's dreams, goals, and priorities in life can be very difficult, but sometimes it is unavoidable. My own preference is to raise generalized questions about life and allow the patient to go home and ponder them. Patients need time to figure out that they've come to a point in life when they have to scale back on their aspirations, and that they may need to spend more of their time, money, and energy on taking care of themselves. Many parents have great difficulties with putting their own needs ahead of their children's needs, as this presents a severe challenge to their identities and how they conceive of parenthood. Be warm, caring, and patient in this discussion, allowing people time to figure out what they need to do, and help them understand that tending to their own health needs is not a mark against their abilities to be a parent. Reading this case, my sense is that these discussions loom for this young woman. If she were my patient, I would be thinking about my approach with her and waiting for the right moment.

Myasthenia Gravis

PRESENTER Charlene Hafer-Macko, MD

S

"I would love to run a marathon again and I need to lose weight."

This 36-year-old female, who had been diagnosed with myasthenia gravis (MG) 8 years earlier when she presented with double vision and problems swallowing and talking, had developed new troubles with raising her arms overhead to wash her hair and climbing stairs without using the railing. Since her diagnosis, her myasthenia has been treated with pyridostigmine, an acetylcholine esterase inhibitor, taken three times a day with meals. She has also been treated with prednisone, which has led to a 25-lb (11 kg) weight gain, as well as the development of hypertension and impaired glucose tolerance. These medications stabilized her myasthenia, but she still complains of fatigue after 3 p.m., in the heat, and during her menstrual period. Before her diagnosis, she was a marathon runner and would like to resume exercise. This seems out of reach, though, since she can barely complete her daily activities and work in human resources at the university.

O

Height: 5 ft 2 in. (1.57 m)
Weight: 175 lb (79 kg)
BMI: 32

HR: 92 beats/min
BP: 144/90 mmHg
RR: 18 breaths/min

Pertinent Exam

Obese with flexed truncal posture, facial acne with flushing, and Cushingoid appearance; cognition normal, clear speech, no aphasia, no slurred or nasal speech; drooping of the right eyelid, double vision develops with sustained up-gaze due to left lateral rectus weakness; smile is not full, neck flexion is weak; skin bruises with mild lower-extremity edema; no obvious deformities, + proximal muscle weakness (upper-extremity abduction), details as below; short stride length with poor postural balance; no muscle atrophy, pain to palpation, or fasciculations, normal tone; sensation intact, all modalities; deep tendon reflexes normal throughout; coordination normal in upper and lower extremities, standing tandem balance intact with eyes closed.

Medications

Pyridostigmine, prednisone

Review of Systems (Positive)

Dyspnea on exertion, bruising

Manual Muscle Testing

Bilateral deltoids 4/5; triceps 4/5, finger extension 4/5, first dorsal interossei 4/5, bilateral hip flexors 4/5, and right tibialis anterior 4/5; all other muscles 5/5

Physical Functioning Tests

Timed up-and-go (TUG): 9 s

6-min walk: 1500 ft (457 m)

30 ft (9.1 m) self-selected walking speed: 0.85 m/s

Daily steps: 1750 steps/day

Exercise Testing

Fitness level (VO$_2$peak) 21 mL O$_2 \cdot$ kg$^{-1} \cdot$ min^{-1}; calculated peak work rate 6.7 METs

Aerobic economy of 13 mL O$_2 \cdot$ kg$^{-1} \cdot$ min^{-1} @ 0.7 m/s

Forced vital capacity: 2.2 L

A

Myasthenia gravis, inadequate response to therapy, as evidenced by:

Proximal muscle weakness and fatigue (sustained arm abduction and TUG/sit to stand)

Mild restrictive pulmonary pattern from respiratory muscle weakness (low FVC, supine FVC 15% lower than seated FVC, with maximal voluntary ventilation)

Deconditioning secondary to myasthenia, as evidenced by:

Slow gait speed, reduced 6-min walking distance, and low fitness level

Mildly impaired balance

Fatigue affecting daily function (low daily steps)

Adverse metabolic effects of prednisone (Cushingoid traits)

P

Trial of better dose timing of pyridostigmine (vs. additional treatment)

Exercise training to counteract elements of deconditioning/inactivity and prednisone

Education on strategies for better coping with myasthenia

Goals

Provide education on energy conservation and activity modification to lessen fatigue

Provide education on timing of acetylcholine esterase inhibitors in relation to exercise, and avoidance of activity at the end of day when fatigue is greatest

Raise awareness of body position, using exercises to improve core stability

Improve respiratory control with breathing exercises

Increase extremity strength

Improve gait speed and efficiency

Exercise Program

Basic *CDD4* Recommendation with special considerations for myasthenia gravis; exercises for balance and breathing added (see text in italics):

Aerobic exercise 4 days a week

- Walk at an RPE <8-11 (out of 20); start at a low intensity

- Begin with 10-15 min (using rest intervals to avoid fatigue)

- Build ~5 min other weeks up to 30 min/session

- Gradually increase to RPE <11-15 as tolerated, using interval design

- May decrease duration by 5-10 min if concurrent with strength training

Resistance exercises 2 days a week

- Start with functional balance/core stability exercises (e.g., sit to stand)

- Arm curls using a 1 L or 2 L jug filled with water; emphasis on proper body mechanics

- Start with 1 set, 8-15 reps at 50-70% of 1 rep max

- Increase to 2 sets over ~8 weeks

Balance exercises 2-3 days a week

- *10-15 min/session*

- *Start with exercises on stable surface with eyes open*

- *Progress to exercises on unstable surface with one or both eyes closed*

Flexibility exercises 3 days a week

- Chair sit-and-reach stretches daily on L and R

- Hold for 30 s on each side, goal of 0 cm/in. from toes

- Additional stretches as desired, 30 s per stretch below discomfort point

Pulmonary exercises 3 days a week

- *Diaphragmatic and pursed-lip breathing exercises*

- *1 set of 10 increasing to 2 sets*

Warm-up and cool-down phases @ RPE <11/20 for ~10 min

Special Consideration

Terminate session for excessive fatigue

Follow-Up After 3 Months of Exercise Program

The patient had increased her regular aerobic exercise from 15 min with 2 seated rests of 5 min to 30 continuous min, progressing from 1.8 mph (0.8 m/s) and 0% grade to 3 mph and 3% grade, raising her exercising heart rate from 110 to 135 beats/min.

Functional Testing After 3-Month Intervention

Timed up-and-go: 7 s

6-min walk: 1850 feet (564 m)

30 ft (9.1 m) self-selected walking speed: 1.41 m/s

Daily steps: 2950 steps/day

Fitness level ($\dot{V}O_2$peak) 22 mL $O_2 \cdot kg^{-1} \cdot min^{-1}$; calculated peak work rate 8.9 METs

Aerobic economy of 10 mL $O_2 \cdot kg^{-1} \cdot min^{-1}$ @ 0.7 m/s

Forced vital capacity: 2.3 L

Her exercise program produced demonstrable gains in timed up-and-go and walking measures, which translated into greater home, work, and community participation with higher daily steps.

Peak oxygen uptake did not change, but estimated peak work rate and peak minute ventilation both increased ~25%, reflecting her improved aerobic economy and suggesting better ability to use nonoxidative energy metabolism (glycolytic pathway with lactate accumulation).

Senior Editor's Comment

One of the most dreaded discussions a sports medicine doctor has is about advising athletes to end their competitive career, because with many athletes you feel as though you are tearing at their soul. With so much time, money, effort, and spiritual energy invested in their sport, many athletes have their sense of identity bound up with the sport. There comes a time, however, in every athlete's life when this is a discussion that must be faced. Most of the time, such discussions are not that difficult because athletes have figured it out themselves, and what they really need is celebration of their time as an athlete, confirmation that it's OK to let go of that dream, and—most important—help with refocusing their life to pursue a new dream. The process is not unlike mourning the loss of a loved one, because in a real sense that's what is going on inside the patient.

Sadly, there is a version of this discussion in patients with a severe chronic health condition. This young woman provides an example, because encouraging this woman to think about running a marathon is not good advice. Even if her myasthenia could be managed to the point where she had the strength and endurance to complete a marathon (even just walking), she—alas—cannot escape the biomechanical issues of her added weight and risk of pathological stress fractures from the prednisone.

I think it would be a mistake, however, to confront her directly with this opinion. Sometimes a confrontation is necessary, but the only reason to rush into that is when a patient is imminently going to do something unwise and risk doing herself harm (say, if she said she was going to do a marathon this weekend . . . which a new patient sometimes comes in and says!). In this woman's case, a more tactful approach would be to leverage the education and exercise interventions—use this time to build the patient–provider relationship and win her trust. Like most athletes at the end of their career, most patients who have unrealistic notions of what they can do end up figuring it all out, and I suspect this woman would do so as well. She is much more likely to succeed at this if she has developed a supportive relationship with her providers, is more able to reveal her fears (a necessary step to overcoming her denial about her marathon running), and can begin to let go of her old vision of herself and dream anew.

Realistic use of exercise is going to be central to this woman's quality of life, for the rest of her life. Getting her to this level of insight and cooperation is one of the most important roles her health care team can serve, but try to let your patients figure it out for themselves. Use your authority to "prohibit" any notions they have that might cause harm, but otherwise be patient, using time and your relationship to facilitate internal spiritual growth. The education and training program are the easy part!

Note: This patient's improvement in aerobic economy and submaximal measures of performance is a very common finding in patients with severe chronic conditions, illustrating the benefit of the Basic *CDD4* Recommendation of starting low, going slow, and being alert for signs and symptoms.

Myocardial Infarction

PRESENTER Benjamin Gordon, PhD, RCEP

S

"I had a heart rhythm that felt extremely rapid, like a fluttering in my chest, and they told me I had a heart attack."

A 59-year-old college professor presented to the emergency department about 4 h after experiencing symptoms, including a rapid heart rate and mild throat tightness and discomfort. His coronary risk factors were generally unremarkable but included a slightly elevated low-density lipoprotein cholesterol that had ranged 140-190 mg/dL and a slightly low high-density lipoprotein cholesterol that had been 32-36 mg/dL. He walked regularly, was normotensive, did not smoke cigarettes, and was not diabetic but was slightly overweight. He had been very busy, working 70-80 h per week, attempting to meet several deadlines.

O

Height: 5 ft 9 in. (1.75 m)

Weight: 183 lb (83.0 kg)

BMI: 27.1 kg/m^2

HR: 150-180 beats/min, rapid and irregular

BP: 146/86 mmHg

Pertinent Exam

Middle-aged male in distress, normal lung sounds, rapid irregularly irregular rate with variable pulse strengths in radial arteries, trace of pitting edema below the ankles

Medications

Metoprolol, aspirin, clopidogrel, atorvastatin

Labs

ECG: atrial fibrillation with rapid ventricular response, inferolateral ST-segment depression (~1.0-1.5 mm)

Cardiac enzymes (from emergency department): 1st blood sample was normal; the second blood test levels of CK-MB enzyme and troponin I had risen to >5% of total CK and >0.5 ng/mL

Hospital Course

The patient underwent emergent cardiac catheterization:

- Total occlusion of the proximal right coronary artery; otherwise normal vessels
- Ejection fraction: 50%
- An emergent percutaneous transluminal coronary angioplasty with stent placement was successful.

Submaximal Exercise Test (Post-PTCA, Day 3)

Peak exercise: 4-5 METs

Peak HR: 112 contractions/min

Patient had reverted to normal sinus rhythm at rest

No ECG changes, threatening dysrhythmias, or symptoms of myocardial ischemia with exercise

A

Acute inferior wall MI

CAD with successful PTCA

Overweight

History of lipid/lipoprotein abnormalities

High stress/workaholic

P

Refer to cardiac rehabilitation program for comprehensive lifestyle modification

Reduce body weight/fat stores (goal weight = 160-165 lb [72.5-75 kg])

Dyslipidemia therapy: 40 mg atorvastatin, once per day; niacin 500 mg twice daily; and omega-3 fish oil, 1200 mg capsules twice daily

Counseled to refrain from work for 4-6 weeks and to then resume and maintain a reduced work load/schedule

Initiate a program of weight reduction, dietary modification, stress reduction

Goals

Peak exercise tolerance 8.9-10.9 METs

Increase functional capacity

Reduce body weight/fat stores

Improve lipid/lipoprotein profile

Stress reduction, reduce work-related activities

Reinforce lifestyle changes/adhere to prescribed medications/regular exercise/medical management

Exercise Program

Follow ACSM Guidelines for exercise prescription with special considerations after myocardial infarction.

Aerobic exercise 3 days a week

- Walk at an RPE <13 (out of 20)
- Initial target HR: 84-96 contractions/min
- Begin with 15-20 min (rest stops allowed)
- Build ~5 min a week up to 40 min/session, using interval design
- May increase to RPE <16 or target HR of 60-70% of max as tolerated
- May decrease session to 30 min if concurrent with strength training
- RPE: 11-13 (fairly light to somewhat hard)
- Distance walking training: ≥30 min/session, 3-5 days/week

Resistance exercises 3 days a week

- Circuit-type exercise on machines
- Start 1-10 reps @ 50-70% of max
- Build up over 8-12 weeks to 2 sets of 10-12 reps

Flexibility exercises 3 days a week

- Chair sit-and-reach stretches daily on L and R
- Hold for 20-30 s on each side, goal of 0 cm/in. from toes
- Additional stretches as desired, 20-30 s per stretch below discomfort point

Warm-up and cool-down phases @ RPE <11/20 for ~10 min

Special Consideration

Observe for A-fib or other rhythm/conduction abnormality (these usually resolve after revascularization)

Follow-Up

Approximately 14 weeks after his heart attack, he completed 11 min on a Bruce protocol (4.2 mph, 16% grade, 10-11 METs). Peak heart rate was 132 contractions/min. The test was terminated due to volitional fatigue, without significant ST-segment depression or anginal symptoms. Isolated PVCs were noted during and after exercise. The exercise prescription was updated, and he continued his home-based rehabilitation program.

Chief Editor's Comment

This case study is commonly seen in the modern era of primary coronary interventions with emergent angioplasty and stent placements. As far as the stress component, today it doesn't seem to matter whether it's a college professor, a businessman or -woman, a laborer, a working parent or homemaker; everyone has lots of stress and many competing demands. Successful and sustained lifestyle interventions depend much less on health education than on how well patients are able to reprioritize their life stressors and invest the time, money, and energy needed to reduce their risk factors.

Parkinson's Disease

PRESENTERS Elizabeth J. Protas, PhD, PT, FACSM, and Rhonda K. Stanley, PhD, PT

S

"I'm afraid of freezing and falling."

A 62-year-old woman had bilateral total knee replacements and was referred to physical therapy for knee rehabilitation after the surgery. The program included strengthening of the knee and hip muscles, gait training, ROM exercises, general exercise for endurance, and pain reduction. After completing the rehabilitation program, however, she still had difficulty walking. She showed shortened step lengths and was referred to a neurologist, who made a diagnosis of idiopathic Parkinson's disease. Antiparkinsonian medications were prescribed, and she was referred to physical therapy. She had difficulty performing activities of daily living (e.g., bathing, grooming, rising from a chair and bed, turning over in bed), and complained of getting tired easily. She also lacked confidence in making conversation and had a fear of choking.

O

Pertinent Exam

Elderly-appearing woman in no distress; masked facies and neck flexed with stooped shoulders; cranial nerves intact; slurred speech; poor finger-to-nose and heel-to-shin coordination

Romberg's sign: positive; slight flexion of knees while standing; reduced ROM in neck, shoulders, trunk, hips, knees, and ankles; normal strength; festinating gait with reduced arm swing

Gait Study

Slow, shortened steps, narrowed base of support and instability (festinating gait)

50-ft (15.2 m) walk time: 16 s

Gait velocity: 0.73 m/s

Step length: left 36.6 cm (14.4 in.); right 38.4 cm (15.1 in.)

Physical Function

Pull test: unable to recover balance (Hoehn and Yahr stage 3)

360° turn: 6.7 s

Functional reach: 8.5 in. (21.6 cm)

A

Status post–bilateral knee replacement surgery

Parkinson's disease

Gait and balance impairment

Difficulty in transfers (turning in bed, getting up)

Decreased flexibility

Decreased endurance

Impaired orofacial functions

P

Sinemet/anticholinergic medication

Initiate a physical rehabilitation program

Goals

Be able to ambulate with confidence

Improve balance

Self-sufficiency with transfers

Improve endurance

Exercise Program

Basic *CDD4* Recommendation for persons with Parkinson's disease (modification in italics) with special considerations:

Aerobic exercise 3 days a week

- Walk *or swim* at a self-selected pace, within the talk test
- Begin with 15-20 min (rest stops allowed)
- Build as tolerated up to 30 min/session

Resistance exercises 2 days a week

- Strengthening exercises (as below)
- Build up to 3 sets of 10 reps (or holding position for 10 s)
- *Resistance exercises include shoulder shrugs, shoulder squeezes, trunk twists, hip bridges, straight-leg raises, leg kicks, toe/heel raises, standing against the wall, and wall push-ups*

Flexibility exercises 3 days a week

- 20 s per stretch, 3 sets, do 30 min total stretching
- *Stretching exercises include head turns and tilts, chin tucks, arm stretches, wrist circles, finger/thumb circles, back extensions, hamstring stretches, bringing the knees to the chest, hip rolls, calf stretches, and ankle circles*

Gait training and/or orofacial exercises 1-5 days a week, 15-30 min as needed

- *Gait program requires an attentional strategy: reminding about larger steps, using visual or auditory cues to improve step length, and walking with alternate arm swing. Balance training includes tandem walking, sideways walking, backward walking, standing on one limb, alternate standing on toes and heels, and turning the head and looking over the shoulders while standing*

Orofacial exercises

- *Chin tucks in and out, eyebrow lifts, deep breathing, sticking the tongue out and in, saying "u-k-r" loudly, lip presses, whistling, circling the tongue inside the lips, long blowing through the mouth, big smiles, face squeezes, and cheek stretches*

Functional training 1 day a week

- *Transfer training includes bed turns (sequential trunk rolling and pushing up) and getting up (leaning forward, shifting weight, and standing up quickly)*

Special Considerations

Observe for fall risks, changes in neuromotor control during session, exaggerated/blunted cardiovascular response

Adaptive precautions: Perform during peak drug levels (start 45-60 minutes after dose)

Chief Editor's Comment

This patient shows how chronological age can be quite different from biological age, as this patient is only 62 years old but now has serious decline in physical functioning. Her story shows how particular combinations of conditions can be very challenging, with the reconditioning after bilateral knee replacements being far more challenging in the context of Parkinson's. As far as cardiovascular risk, her age and postmenopausal status does put her in a group of increasing risk, but at Hoehn and Yahr stage 3 she is not likely to provoke cardiovascular symptoms with her exercise program focused on physical functioning. If the medications and exercise program lead to dramatic improvements and she becomes Hoehn and Yahr stage 2 or 1, her therapists should shift from the Basic *CDD4* Recommendations to the ACSM Guidelines for apparently healthy normals. She might still not need a graded exercise test, however, if her exercise increased to 150 min or more per week of moderate to vigorous activity with no cardiovascular symptoms. She and her therapists should be aware of the potential for cardiovascular symptoms as her functioning improves, but if she did well and had no symptoms, she probably would not need further workup.

Peripheral Artery Disease

PRESENTER Shane A. Phillips, PhD, PT

S

"The cramps in my left calf and buttocks interfere with my ability to work."

A 65-year-old man with history of hypertension, hyperlipidemia, and tobacco use presented with complaints of cramping and fatigue in his left calf and buttocks during ambulation, which was interfering with his ability to work as a mail carrier. The patient had noted a significant decrease in his walking ability over the prior year and was frequently having to stop to relieve the pain. He denied pain at rest, and the pain was crampy in nature. He had also developed erectile dysfunction. The patient was sent to the exercise laboratory for evaluation of intermittent claudication.

O

Height: 5 ft 10 in. (178 cm)
Weight: 220 lb (100 kg)
BMI: 31.6 kg/m^2

RHR: 72 contractions/min
BP: 150/90 mmHg

Pertinent Exam

Vascular examination:

Neck: carotid bruits, bilaterally

Cardiac: regular rate and rhythm

Abdomen: no abdominal bruits, no pulsatile mass

Extremity: 1+ dorsalis pedis (DP)/posterior tibialis (PT) pulses on the left, 2+ DP/PT pulses on the right, evidence of hair loss over the left lower extremity, no evidence of ulcerations

Medications

Lisinopril 20 mg once daily, simvastatin 40 mg once daily

Labs

Blood lipids: total cholesterol 213 mg/dL, high-density lipoprotein cholesterol 31 mg/dL, triglycerides 165 mg/dL, low-density lipoprotein cholesterol 149 mg/dL

Left ankle/brachial index: 0.70

Right ankle/brachial index: 0.89

Graded Exercise Test

Continuous treadmill at 2.0 mph, 0.0% grade increasing 2% every 2 min:

- Time to onset of claudication pain: 510 s
- Time to maximal claudication pain: 956 s

Physical Function

Free-living daily energy expenditure as measured by accelerometer: 300 kcal/day

6-min walk distance: 231.6 m

A

Peripheral arterial disease with intermittent claudication

Decreased functional capacity caused by early-onset claudication

P

Initiate a treadmill exercise program

Consider starting cilostazol and antiplatelet medications as medically appropriate

Consider increasing statin dose or using a more potent agent

Stop smoking

Goal

Increased time to claudication onset and maximal pain

Increased free-living energy expenditure

Increased functional capacity and endurance

Associated heart disease risk factor management

Exercise Program

Basic *CDD4* Recommendation (modifications in italics) with special considerations for persons with peripheral artery disease:

Aerobic exercise 5-7 days a week

- *Walk* at 60% of $\dot{V}O_2$peak, *or elicit 3 (out of 4) claudication pain*
- Begin with 15-20 min (rest stops allowed)
- Build ~5 min a week up to 40 min/session
- May increase to RPE <16 as tolerated, using interval design

- May decrease session to 30 min if concurrent with strength training

Resistance exercises 3 days a week

- Sit-to-stand exercises
- Arm curls using a 2 L or 4 L jug filled with water
- Build up to a minimum of 2 sets of 8 reps

Flexibility exercises 3 days a week

- Chair sit-and-reach stretches daily on L and R
- Hold for 30 s on each side, goal of 0 cm/in. from toes
- Additional stretches as desired, 30 s per stretch below discomfort point

Warm-up and cool-down phases @ RPE <11/20 for ~10 min

Special Consideration

Walking specifically recommended for improving claudication symptoms

Follow-Up

The 6-month rehabilitation program increased the time to onset of claudication pain by 5 min, and the patient no longer terminated the test due to maximal claudication pain. His free-living daily activity increased from 300 kcal/day to 550 kcal/day. One year after completing the formal rehabilitation program, he maintained a 4-min improvement in time to onset of claudication and did not experience maximal (4/4) claudication pain. The patient was able to resume his job as a mail carrier.

Chief Editor's Comment

This patient has widespread vasculopathy, and aggressive medical management with intensive lifestyle intervention is indicated. Fortunately, his life as a mail carrier facilitates the physical activity component of lifestyle intervention, which is going to be essential for him to maintain his occupation. Good counseling would help him come to understand this and maintain his motivation for these changes.

Pulmonary Hypertension

PRESENTER Kelly Chin, MD

S

"I want to be able to be able to go shopping and carry things without getting totally out of breath."

A 59-year-old female was diagnosed with severe idiopathic pulmonary hypertension at the age of 55. She was on combination therapy of intravenous epoprostenol and tadalafil, and was also taking furosemide 40 mg twice daily for lower-extremity edema. Over the preceding 2 years she had worsening exercise tolerance and intermittent lower-extremity edema. Her oxygen saturation at rest was 90%, and it fell with exertion into the mid-80s, which was partially related to a patent foramen ovale.

O

Height: 4 ft 11 in. (1.50 m)
Weight: 81 lb (36.7 kg)
BMI: 16.4 kg/m^2

HR: 107 beats/min
BP: 92/64 mmHg

Pertinent Exam

Mild elevation in jugular venous distention, right ventricular heave, no lower-extremity edema

Medications

Epoprostenol, tadalafil, furosemide

Labs

BNP (brain natriuretic peptide) level 193

Exercise Test Results (Naughton Protocol)

Peak exercise: 4 METs

Peak HR: 145 beats/min

Peak BP: 135/80 mmHg

Peak RPE: 18 of 20

Termination reason: dyspnea

Chronotropic response: normal

No significant ST changes during exercise or recovery

No chest pain during testing; desaturation to 80% on room air

A

Pulmonary hypertension

Low exercise tolerance for age

Exertional hypoxia

P

Continue medical therapy

Begin cardiopulmonary rehabilitation program

Optimize oxygen therapy at rest and with exertion

Goals

Improve cardiovascular response to exertion

Monitor oxygen levels during exertion

Refer to PH specialist to optimize PH therapy

Exercise Program

Basic *CDD4* Recommendation with special restriction for pulmonary hypertension:

Aerobic exercise 4-5 days a week

- Walk at an RPE <13 (out of 20), or use talk test
- Begin with 15-20 min (rest stops allowed)
- Build ~5 min a week up to 40 min/session
- May increase to RPE <16 as tolerated, using interval design
- May decrease session to 30 min if concurrent with strength training

Resistance exercises 2 days a week

- Sit-to-stand exercises
- Arm curls using a 2 L or 4 L jug filled with water
- Build up to a minimum of 2 sets of 8 reps

Flexibility exercises 3 days a week

- Chair sit-and-reach stretches daily on L and R
- Hold for 30 s on each side, goal of 0 cm/in. from toes
- Additional stretches as desired, 30 s per stretch below discomfort point

Warm-up and cool-down phases @ RPE <11/20 for ~10 min

Special Consideration

No Valsalva maneuver allowed

Chief Editor's Comment

This is a patient for whom I would encourage endurance over fitness. She can generate a reasonably robust arterial blood pressure during exercise, and no estimates of pulmonary artery pressure are provided. Presumably these are 50-55 or more, because it is "severe" pulmonary hypertension. With arterial desaturation during exercise going down to 80% on room air, I would obtain an opinion from a cardiologist on whether or not the patent foramen ovale should be addressed.

Refractory Angina

PRESENTERS Sherry O. Pinkstaff, PhD, PT, and Abraham Samuel Babu, MPT, FCR

S

"I have pain in the chest when I walk."

A 48-year-old male was admitted with complaints of chest pain for 6 months and recent onset of angina at rest. The pain was aggravated during activity and relieved with rest. The pain was over the chest and was diffuse in nature, with radiation to the neck and back. His ECG showed signs of ischemia in the precordial leads, and his cardiac catheterization revealed a diseased left anterior descending coronary artery, on which angioplasty was performed and a stent was placed. Two weeks following his discharge, he continued to experience angina on exertion and had thereby reduced his activities, thus limiting his work. He was referred to cardiac rehabilitation to improve his functional capacity.

O

Height: 5 ft 6 in. (1.68 m)

Weight: 170 lb (77.1 kg)

BMI: 27.4 kg/m^2

HR: 74 beats/min

BP: 155/95 mmHg

RR: 16 breaths/min

Pertinent Exam

Lungs: clear

Heart sounds: no murmurs or S3 gallop

Normal peripheral pulses; no edema

Medications

Aspirin, clopidogrel, metoprolol, ramipril, simvastatin, and isosorbide dinitrate

Labs

Resting ECG: normal

Graded Exercise Test (Naughton Protocol)

Peak METs achieved: 7.3

Max HR: 146 beats/min

Max BP: 173/98 mmHg

Peak rate–pressure product: 25,258

Reason for termination: fatigue (RPE: 16/20), did not experience angina

ECG: nonsignificant ST/T changes

A

Exertional angina, possibly caused by CAD

P

Goals

Reduce or eliminate exertional chest discomfort

Improve anginal threshold

Increase functional capacity

Educate about managing angina symptoms

Reinforce lifestyle changes

Promote return to work and active lifestyle

Monitor fasting lipids, blood pressure, and adherence to recommendations

Exercise Program

Refer to supervised cardiac rehabilitation program, following ACSM Guidelines for angina.

Aerobic exercise 3-4 days a week

- Aerobic exercise machines @ *40-50% of heart rate reserve (below angina)*
- RPE 11-12/20 for 15-20 min as tolerated initially, with rest stops allowed
- Build ~5 min a week up to 30-40 min/day, 4 days/week

Resistance exercises 2 days a week

- Resistance circuit exercise machines 2 days a week
- 1 set of 10 reps with 40-50% of 10RM

Flexibility exercises 3 days a week

- Stretching every session, before and after as desired
- 30 s per stretch below discomfort point

Warm-up and cool-down phases @ RPE <11/20 for ~10 min

Special Consideration

Limit intensity to below angina threshold (by symptoms and/or rate–pressure product; see text in italics)

Follow-Up

"I've gone back to work and I don't have any more chest pain."

No formal exercise test performed.

Indirect estimation from exercise time suggests benefits in exercise capacity.

Aerobic exercise duration at end of program: 45 min at 50% to 60% heart rate reserve.

Senior Editor's Comment

Most patients who have angioplasty with or without a stent have a resolution of their exertional angina. This patient was found to have diffuse disease and thus may have been experiencing a component of vasospasm (i.e., Prinzmetal's angina) and perhaps anxiety, but the graded exercise test was nondiagnostic. The cause of his persistent angina was thus uncertain, but coronary artery disease seems a likely culprit. The fact that the angina resolved after exercise training would be consistent with the known benefits of exercise on angina, and he would be wise to maintain his exercise regimen to promote good vascular health.

Spinal Cord Injury

PRESENTERS Sangeetha Madhavan, PhD, PT, and Tanvi Bhatt, PhD, PT

S

"I was part of a professional tennis league before my injury, was very athletic, and played other recreational sports. After my injury I continued to train and play wheelchair tennis and competed in several recreational tournaments. I would now like to compete in the Paralympics."

A 27-year-old female suffered a T10 ASIA Impairment Scale, Type A, injury (a complete spinal cord injury at the T10 level) in a skiing accident at the age of 21 years. She is in graduate school and is active in wheelchair tennis and other recreational activities such as skiing and racing. She lives with her fiancé in a wheelchair-accessible home. She takes baclofen to control spasms and is currently on ciprofloxacin for recurrent urinary tract infections (UTIs), which have been occurring frequently. Since her injury, she has maintained a healthy lifestyle, competing in several recreational wheelchair tennis tournaments. She would like to compete in the Paralympics, so she has started more rigorous training. Since increasing the intensity of practice, she has noticed right shoulder (dominant side) pain and fatigue that has limited her performance, especially her ability to serve. She has developed an irregular practice schedule because of the shoulder pain, decreased endurance, and fear of injury. She rates the right shoulder pain at 8/10 with overhead lifting and 3/10 at rest, and was using ibuprofen for pain.

O

Height: 5 ft 6 in. (1.68 m)
Weight: 139 lb (63.2 kg)
BMI: 22.4 kg/m^2

HR: 66 beats/min
BP: 114/65 mmHg
RR: 14 breaths/min

Pertinent Exam

Neurological and musculoskeletal

Neurological assessment reveals AIS A T10, sensation intact to T10, deep tendon reflexes 3+ below T10; integument reveals skin intact throughout.

Right shoulder exam:

ROM: terminally restricted in all planes; + Neer test for shoulder impingement (pain at 90° of forward flexion)

Motor: shoulder abductor, internal and external rotators and flexors –4/5; all other shoulder muscles –5/5

Left upper-extremity ROM and strength all within normal limits

Mobility and ADL Checklist/Evaluation

Independent with bed mobility, transfer to all surfaces including wheelchair to floor, pressure relief, bowel and bladder care, dressing and driving, and manual wheelchair maneuvering with high-level wheelchair skills

Medications

Baclofen, ciprofloxacin, ibuprofen

Exercise Test

Wheelchair ergometry with respired gas analysis:

Test parameters: wheel handrim speed equivalent to 3 km/h, stage 1 starting @ 5 W with increases of 10 W every 5 min

Test duration: 8 min

Peak HR: 160 beats/min

Peak BP: 190/75 mmHg

Peak RPE: 18/20

Reason for termination: unable to maintain speed of 3 km/h at 15 W

$\dot{V}O_2$peak: 19.7 mL $O_2 \cdot kg^{-1} \cdot min^{-1}$

A

Right shoulder impingement syndrome

Low aerobic capacity for a young Paralympic tennis athlete with a T10 AIS A injury

Decreased endurance and fatigue, possibly secondary to chronic UTI

P

Shoulder physical therapy

Refer to primary physician for evaluation of persistent UTI

Provide education on prevention of UTI in patients with a spinal cord injury

Goals

Decrease right shoulder pain and improve right shoulder ROM and strength

Increase cardiovascular fitness and endurance

Exercise Program

Modified *CDD4* Recommendation with special considerations for spinal cord injury (modifications in italics).

Rotator cuff/scapular stability exercises every day

- Internal/external/abduction/elevation/scaption exercises with elastic bands
- Start with 20 reps each exercise, both shoulders, increasing to 50 reps as tolerated by discomfort and fatigue
- Scaption exercises to be done pushing out and pulling back (arm straight)
- Increase to 2 sets, both sides, as tolerated
- Advanced shoulder exercises (tennis specific) per physical therapist

Aerobic exercise 4-5 days a week

- Arm cranking or wheelchair propulsion at an RPE <13 (out of 20), or use talk test
- Begin with 15-20 min (rest stops allowed)
- Build ~5 min a week up to 40 min/session
- May increase to RPE <16 as tolerated, using interval design
- May decrease session to 30 min if concurrent with strength training

Resistance exercises 2 days a week *(trunk and upper-extremity muscle groups)*

- *Circuit lifting: 1 set, 50-70% of 1RM, 8-12 reps to failure*
- Build up over ~8 weeks to a minimum of 2 sets of 10 reps
- *No overhead lifting or any other activities that cause right shoulder pain (mild discomfort allowed)*

Flexibility exercises 3 days a week

- Chair sit-and-reach stretches daily on L and R
- Hold for 30 s on each side, goal of 0 cm/in. from toes
- Additional stretches as desired, 30 s per stretch below discomfort point

Warm-up and cool-down phases @ RPE <11/20 for ~10 min

Special Considerations

Encourage patient to adhere to her Paralympic training schedule

Consider hiring a personal trainer for 6 months to minimize risk of shoulder reinjury

Shoulder pain management:

- Apply warm moist heat to shoulder before stretching and strengthening activities
- Apply cold packs to relieve pain and inflammation after activities that aggravate symptoms

Senior Editor's Comment

This young athlete's story demonstrates the mix of challenges faced by patients who have a spinal cord injury. She is comparatively fortunate in having a low, albeit complete, cord injury, which allows her to remain reasonably active and engaged in a lifestyle consistent with the individual she was prior to the injury. Undoubtedly, her continuing to pursue sports relates to her innate emotional being and her fiancé's and family's support, but, having known many athletes and patients, I sense an incongruity that deserves deeper exploration.

Athletes who rise to the level of becoming a professional are just not the kinds of people who give up easily or back off, yet she had been allowing her shoulder pain to interrupt her training. That would be very uncharacteristic of an elite athlete, whether without disability or Paralympic. Were I her physician, I would want to explore this, because it could be very important to her long-term health and well-being. Was she a gifted high school tennis star who never had to practice, or was she someone who worked hard to get so good?

If she was the gifted type, then what she needs to be a successful Paralympic athlete is additional support, encouragement, and advice conveying that it's not going to be as easy as it was when she was not disabled. Patients with spinal cord injuries face many issues—this patient's chronic UTI and upper-extremity overuse syndrome are examples. With the support she has, she can surmount these problems and perhaps become a Paralympic champion, but not by giving up easily. She shouldn't follow a "no pain, no gain" mantra, but she does need a persevering attitude and insight into when something isn't right and she needs some help (e.g., with her shoulder).

If she was the kind of athlete who worked hard to get where she was, on the other hand, then she seems to be faltering. Her fitness level is far too low for an elite Paralympic athlete. Does that represent fatigue from a chronic UTI, or did she not try very hard? (*NB:* On an exercise test, this requires some judgment by the person supervising the test, on assessing the test data, how the patient reacts to the exercise, and—most important—body language and facial expression. These are all things one must see in person that don't come across well in a SOAP note.) If it's the chronic UTI, overcoming that problem takes on added importance as a

medical issue. If she's giving up, that would more be a sign of succumbing to the many burdens and challenges of life as a woman with paraplegia. It would be a great loss if the heart of a champion were to accept a lesser fate because of a skiing accident. Notwithstanding the substantial medical expertise that is helping this young woman, nurturing her dreams of fulfillment is the most important element in her care plan.

The spirit becomes diminished for too many patients with severe neurological conditions, who face formidable life challenges. Most of them cannot ever become a Paralympic athlete, but our health care system needs to do a much better job of helping these patients find happiness and fulfillment.

Stroke

PRESENTER Richard Macko, MD

S

"I'm not able to do what I want to do and want to get back to being myself."

A pleasant 74-year-old African American female with a history of stroke complained of getting tired easily, which limited her ability to be on her feet, participate in church activities, and do her shopping and her usual activities around the house. She also felt that she did not have very good balance, needed her cane, and had to move around with caution because of concern about falling. Six years earlier she had a left carotid artery hemispheric stroke with residual mild to moderate hemiparesis of the right arm and leg, and she completed a year of poststroke rehabilitation. She prefers to use a single-point cane when leaving the house but can walk for short household distances without a cane.

Comorbid conditions: poststroke seizure disorder (well controlled), hypertension, mild anemia, B_{12} and vitamin D deficiency, osteoporosis R > L

O

Height: 5 ft 3 in. (1.60 m)
Weight: 152 lb (69 kg)
BMI: 27 kg/m^2

HR: 86 contractions/min
BP: 128/88 mmHg

Pertinent Exam

Moderate R-sided hemiparesis, reduced primary sensation in the R arm and leg compared to the L, but still capable of feeling moderate finger touch; mild spasticity in the R leg and arm indicated by increased resting tone throughout the range of motion, but she had full ROM, no clonus at the foot or wrist

Medications

Levetiracetam 1000 mg twice daily, phenobarbital 30 mg twice daily, hydrochlorothiazide 12.5 mg daily, pravastatin 20 mg daily in evening, aspirin 81 mg daily, folate 1 mg daily, B_{12} injections monthly

Labs

Mild anemia

Cardiovascular screening: ECG normal, no history of coronary artery disease, carotid duplex <50% blockage bilaterally, cleared by primary care for aerobic exercise

Physical Functioning

Rises out of a chair only once

Uses the L (less affected) > R armchair handrails for assistance

10 m self-selected gait velocity: 0.6 m/s with single-point cane

Stance and gait:

- Hemiparetic asymmetry, reduced stance time

- Increased swing with circumduction of the R leg due to spasticity across the knee and reduced dorsiflexion

A

Status/post-CVA for 6 years

Deconditioning, with adequate neuromotor capacity for low-intensity aerobic exercise

Limitations: hemiparetic gait and impaired balance with moderate fall risk

High risk for recurrent stroke and cardiometabolic disease

Osteoporosis, hemiparetic side > unaffected side

Need for increased social support

P

Low-intensity group exercise class with functional mobility training

Add cardiovascular disease prevention medications (statin, aspirin)

Goals

Improve balance and endurance

Reduce cardiometabolic risk

Exercise Program

Exercise component	Weeks 1-5	Weeks 6-16	Weeks 17-24
Walking warm-up	Slow 5 min	5-7 min	5-7 min
Standing weight shift, then step in place Attention to posture, breathing	1 min each Full grip handrail	2 min each Light handrail	3 min each Face handrail
Forward step, back step, side step, paretic and nonparetic leg	5 reps each Handrail contact	8 reps Light handrail	10-12 reps Light handrail
Partial squat	5 reps of 1/4 squat Grip handrail	8 reps Light handrail	10-12 reps Light handrail
Leg raise to front, rear, side, both sides Done slowly, concentrating on form	5 reps Full grip handrail	8 reps Light handrail	10-12 reps Light handrail
Arm raises and arms reach across midline; focus on supination Use paretic arm independently as much as possible	5 reps each Seated, less affected arm helps paretic arm	8 reps each Standing in front of handrail	10-12 reps Hold ball between hands
Sit to stand Emphasize form: forward weight shift, up to symmetrical balanced stand	5 reps Use both arms to assist	8 reps Minimize arm push-off	10-12 reps No arm support
Walking obstacle course with slightly S-curved rope, hula hoop step in and step out, around cones, 4-in. step-up Close observation by trainer	6 min Self-paced, slow	8-12 min S rope more curved, head up and wave at bends (dual task)	12-15 min Faster pace, tight S curve, sequential cones, hoops, and steps

Progression formula following the University of Maryland School of Medicine and Rehabilitation Institute.

Follow-Up at 24 Weeks

Outcome measure	Baseline	24 weeks
6-min walk	287 m	327 m
Berg Balance	42 points	49 points
Dynamic Gait Index (DGI)	18 points	23 points
Daily step count (pedometer)	2891	3012
Cardiovascular fitness ($\dot{V}O_2$peak)	14.5 mL $\cdot$ kg^{-1} $\cdot$ min^{-1}	16.2 mL $\cdot$ kg^{-1} $\cdot$ min^{-1}

The patient reported feeling more comfortable on her feet. She described being a little tired after the classes initially, but by about 3 months of training she was feeling more active overall. The Berg Balance and DGI showed clinically significant improvement in standardized balance metrics. The gains in cardiopulmonary fitness are consistent with those expected to improve glucose tolerance and reduce insulin resistance, which may help reduce risk of recurrent stroke. Her daily step count remained at the level of sedentary elderly persons, so she was advised to build an additional 30 min of low-intensity regular physical activity into her daily habits on the days that she did not have exercise class.

Senior Editor's Comment

This patient provides a good example of someone who is dwindling away after a major medical event, in this case a stroke. Her 5-year history reflected a gradual decline, with objective evidence of disuse having metabolic impact—her osteoporosis being greater on the side affected by her stroke. Because her functioning was nearly to the level of needing some assistance, her gains in physical functioning of ~15% over 4 months are modest in absolute magnitude but provide a clinically meaningful buffer of functional capacity that allows her to do activities of daily living without assistance. That noted, she has precious little excess capacity to lose should she have another medical event such as a heart attack. For this reason, medical management of a patient in her situation requires aggressive cardiovascular risk treatment.

This patient also reflects a common phenomenon, which is a mismatch between her subjective report of being more active and the pedometer reports showing only 120 more steps per day. Surely this gain is heavily attributable to the thrice-weekly exercise class, implying no change in her usual daily activities. Her subjective sense thus may be more related to ease in her physical functioning and not being quite as close to her limits than to her actual life activities. This illustrates why it can be very important for individuals who are elderly and have a disability to find a program in which they can participate in group activities every day. Her experience shows why the social support aspects of her exercise program remain critically important for her future, helping to protect her from decline in physical, cognitive, and social functioning, as well as getting her closer to meeting the recommended physical activity guidelines. Medical home practitioners need to connect patients like this woman to the resources available in the community.

Type 2 Diabetes and Disability From Morbid Obesity With Multiple Chronic Conditions

PRESENTER Geoffrey E. Moore, MD, FACSM

S

"I lost a lot of weight [at the nursing facility], but since I left there I've gained 15 pounds of fluid in two weeks."

A 65-year-old woman had a long history of type 2 diabetes, morbid obesity with a BMI of ~80 kg/m^2, vasculitis, and depression. She was disabled from her weight and multiple comorbidities and often required prolonged treatment for vasculitis with high doses of prednisone, which exacerbated her weight issues and made control of diabetes difficult. She decided to have bariatric surgery, but prior to the surgery she developed an infection that resulted in a prolonged hospitalization and several months in a rehabilitation facility. As a result, she never had the surgery. She was sleeping OK and feeling fairly positive about where she'd come from. She said it had been a long way back to living in her apartment again and was feeling that she could improve even more with PT. She had become very appreciative of her family members, who had helped her cope with stress.

After discharge from the rehab facility, she went from 427 to 442 lb. Despite this, her blood sugars had nearly all been under 200 in the prior 2 months (measured four times a day). At that time she was ranging from 140 to 185 (and rarely to <100).

Exercises: At the time of admission to rehab, she was only able to lift her left arm, but at the office visit she was able to do arm exercises with light hand weights, as well as supine kicking, marching, and abduction exercises for core strengthening. She was doing them only once a day. She had not fully settled down in her apartment, which was making adherence to the exercises more difficult. She walked with a walker in the hallway and with a physical therapist following behind with a wheelchair, but PT could come only twice a week. She was hoping to get help with walking, but the fluid gains were making it difficult to walk. She had a pending appointment with her internist to discuss the fluid weight gains.

O

Height: 5 ft 3 in. (1.60 m)
Weight: 438 lb (198.7 kg)
BMI: 77.6 kg/m^2

HR: 80 beats/min
BP: 120/70 mmHg

Pertinent Exam

General: normal development, morbidly obese body habitus, no deformities—looks her usual self

Chest: normal respiratory effort; I:E ratio = 1.5-2:1; clear to auscultation, distant lung sounds

CV: regular rhythm with normal S1 and S2, without rubs, gallops, or murmurs; distal pulses 1+ (radial); 3-4+ dependent edema in lower extremities

Musculoskeletal: in wheelchair; digits and nails without clubbing, cyanosis, inflammatory conditions, petechiae, ischemia, infections, or nodes; normal alignment and symmetry, no crepitation, defects, tenderness, masses, or effusions; limited ROM secondary to obesity, but without pain, crepitation, or contracture; adequate joint stability without subluxation or laxity; 4-5/5 muscle strength and tone without atrophy or abnormal movements

Skin: bruises from Coumadin; peau-d'orange look on legs, secondary to edema

Psych: appropriate judgment and insight; normal mood; normal affect

Medications

Lantus 38 U subcutaneous once daily

Humalog insulin, subcutaneous sliding scale

Lisinopril 5 mg once daily

Potassium Cl 20 mEq once daily

Baclofen 10 mg once daily

Celexa 40 mg once daily

Synthroid 75 µg once daily

Furosemide 20 mg once daily

Coumadin 5 mg once daily

Amiodarone 150 mg once daily

Senna-S 2 tabs as needed for constipation

Flexeril 10 mg as needed for muscle cramps

Most Recent Lab Tests

A1c: 6.8

Sedimentation Rate: 22

Hb: 13.1; Hct: 39

WBC: 6.4; Platelets: 273

Physical Functioning

Sit to stand (modified)—able to rise from wheelchair once in 30 s

Wheelchair—needs help pushing the chair on carpet

Can do upper-extremity lifting and moving objects up to 2 kg

Chair sit and reach (modified)—can reach just below knees

A

Looked surprisingly good considering what she'd been through

Fluid gain of uncertain cause, likely differences in diet/medication adherence

She was on new medicines (amiodarone/Coumadin) for uncertain reasons

Diabetes well controlled

Doing exercises, but had marginal physical functioning

P

Seeing primary care physician in 2 weeks for follow-up on amiodarone, Coumadin, sleep apnea, fluid gain, prednisone taper

Continue daily home exercises with PT supervision 2 times weekly

Continue family-supervised walking on other days as tolerated by breathing

Goal of returning to center-based aquatic exercise

Follow-Up

This patient did well with home PT, while her diet and diuretics were adjusted to reduce the excess fluid weight. The amiodarone was discontinued, and she subsequently joined a supervised aquatic exercise program that she attended until her available funds and financial aid were exhausted.

Senior Editor's Comment

There are patients who just rarely seem to get a lucky break with their health, and this patient is one of them. She experiences burdensome, unpredictable, and frequent intercurrent illnesses that make sustained progress hard to achieve. To her great credit, she did fight to pull herself along through her problems, and when she was last seen she was stable and doing OK. Unfortunately, her situation was like that of many patients disabled with chronic illness, and she ran out of funds and financial aid for the health maintenance programs she was doing. One does wonder at what point there comes such a severe burden of chronic conditions and polypharmacy that the best health promotion efforts and exercise programs do little more than delay deterioration. This patient is an example of someone who seemed to be in that situation, but was doing a pretty good job of holding her own against all odds.

Type 2 Diabetes and Obesity With Osteoarthrosis

PRESENTER Geoffrey E. Moore, MD, FACSM

S

"I intend to live into my 90s but I can't get there unless I'm dancing."

A 52-year-old woman with type 2 diabetes was referred for help on lifestyle. She had a family history of diabetes and tended toward a centripetal/abdominal fat distribution pattern associated with high CV risk. Her recent A1c values had been 6.5-6.6, and her a.m. fasting glucose was recently 102. She qualified as T2DM by A1c criteria but wasn't working on her diet or weight and had refused to start taking metformin, and her primary care physician was increasingly concerned.

She expressed a desire to be more active through dance, which she had loved since childhood and felt was a part of her culture (having immigrated to the United States as a child). She expressed little concern about her weight and was more worried that the pain in her L shoulder kept her from dancing. She had first experienced the shoulder problem 10 years earlier while swimming backstroke. She had been to physical therapy (PT) twice, which made it worse. She also had received subacromial cortisone injections, but the relief didn't last. Her shoulder is worse at night (rolls over on her left side); the pain was rated 8/10 and was daily and constant except for sharp pain that radiated to the side of the neck when she moved the shoulder too much. She tried not to take pain or anti-inflammatory medications because she said they "wreck her stomach." Acupuncture (cupping) helped.

She also noted a history of her L knee "giving out," causing falls, and of having 2 knee arthroscopic surgeries (4 and 10 years earlier). When she overdoes weight-bearing activity, her left knee swells "like a watermelon." She did PT after the surgeries and gets to her normal daily activities but can't dance.

O

Height: 5 ft 4¾ in. (1.65 m)

Weight: 234.4 lb (106.3 kg)

BMI: 39 kg/m^2

HR: 65 contractions/min

BP: 188/98 mmHg

Pertinent Exam

General: normal development, good nutrition, obese body habitus, no deformities

CV: regular rhythm with normal S1 and S2, without rubs, gallops, or murmurs; distal pulses/circulation normal; no edema

Musculoskeletal:

- Normal gait and station, muscular build especially of lower extremities
- Valgus alignment of both knees, no tenderness but the L knee has a small effusion
- Normal ROM without pain, substantial crepitus in both knees

Left shoulder-specific tests:

+ Painful arc

+ Glenohumeral laxity, with crepitus

- + Neer test, +/- Hawkins test, +/- Speed's test
- + Painful with resisted external rotation, – pain with resisted internal rotation
- + AC joint tenderness

X Rays

Osteoarthrosis in the acromioclavicular joint; severe osteoarthrosis in the glenohumeral joint with advanced sclerosis, subchondral cysts, and spurring; subluxed position of the humeral head at rest

Medications

Coreg 10 mg once daily

Diovan HCT 320/12.5 once daily

Multivitamin once daily

A

Increased CV risk: HTN, hyperlipidemia, diabetes by A1c criteria, sleep apnea, obesity

Advanced glenohumeral osteoarthrosis and reduced function due to pain, status post–bilateral partial meniscectomies

Her shoulder markedly diminishes quality of life, impairing her ability to sleep and to recreate. Dance is her preferred form of physical activity, which she wants to do as her approach to improve the blood glucose and reduce her CV risk profile.

P

L shoulder OA: refer to PT shoulder specialist

It was explained that her shoulder would take a long time to get better, but she was likely to have some gains in reduced pain and better function after 3-6 months of PT

Ibuprofen 500 mg every 8-12 hours and prior to bedtime as needed for pain

Dance for exercise, but any dances that cause pain should be avoided

Review on the difference between pain and discomfort

High cardiovascular risk: She was not interested in addressing her cardiometabolic risk at the time of the first visit.

Goals

Pain-free near-normal physical functioning of L shoulder

Dance, including tango, for recreation and physical activity

Physical therapy focused on rotator cuff and scapular stabilizer strengthening, postural control, manual therapy, and education on pain avoidance

Gentle dancing that avoids painful arm movements started, daily for 30 min

Dancing and daily activities to be increased as tolerated (i.e., shoulder discomfort)

Exercise Program

Physical therapy twice weekly

Home exercises daily, as advised by therapist

Dancing that does not cause pain allowed as tolerated

Follow-Up

L shoulder OA

Pain-avoidance techniques gave rapidly improved physical functioning of L upper extremity.

Physical therapy made good progress over 3 months to near-normal physical functioning.

Dance for aerobic exercise was very successful in getting her motivated.

High cardiovascular risk

After 2 months of PT, she was motivated to work on weight management and diabetes prevention. She enrolled in a partial meal-replacement lifestyle intervention plan, lost 25 lb (11.3 kg) over the subsequent 3 months, and her fasting blood glucose/A1c returned to normal.

Senior Editor's Comment

This patient's story illustrates some of the art of exercise medicine, revealing the importance of working with patients to meet their emotional expectations and needs. Her primary care team was concerned about the diabetes and had been advising her about diet, weight loss, and taking metformin, but this was less important to the patient than the loss of her ability to dance. Dancing was the only type of physical exercise she enjoyed, so she needed to be able to dance as part of her pathway to improving insulin sensitivity. Close liaison with the physical therapist facilitated the process, because the therapist provided her with tips on how to avoid painful movements and begin to dance immediately. By accepting the need to dance as her top priority and then working to help overcome her barriers to dancing, the lifestyle intervention team gained her faith in them and she began to follow their advice on diet and weight. This approach seemed unusual to many who were involved in her care, as it appeared to put the diabetes problem on hold while focusing on a painful shoulder. But from another perspective, the painful shoulder was the most important problem to the patient and thus a significant barrier that needed to be overcome if exercise (in the form of dance) was to become her most important medicine.

Valvular Heart Disease
PRESENTER Matthew W. Parker, MD, FACC

S

"I can't do anything for more than 10 or 15 minutes."

A 49-year-old man had been told for years that he had a murmur. Many years ago, a family doctor told him they would "follow it." Over the past year he had become increasingly breathless during his daily activities. He was having dyspnea with exertion, which kept him from running for more than 10-15 minutes at a time for the past several months. He was occasionally having chest pain with exercise but had not experienced episodes of syncope or presyncope. He described a sense of "pounding" in his chest when running, but not at other times. He had previously run 2.5 miles (4 km) every morning and occasionally ran in road races and half-marathons.

O

Height 6 ft 1 in. (1.85 m)

Weight 194 lb (88.0 kg)

BMI: 25.7 kg/m^2

HR: 74 contractions/min

BP: 118/59 mmHg

RR: 12 breaths/min

Pertinent Exam

Athletic male in no distress

Lungs clear to auscultation bilaterally

Heart: IV/VI systolic murmur along the right second interspace and also a II/IV diastolic murmur along the left sternal border; a faint ejection click heard at the apex

Extremities: no edema, pulses slightly reduced

Labs

ECG: left ventricular hypertrophy by voltage, otherwise normal

Echo: normal systolic function, LVEF 65%; bicuspid aortic valve; peak velocity 4.5 m/s, mean gradient 45 mmHg, and calculated valve area 0.9 cm^2; moderate aortic regurgitation

A

Bicuspid aortic valve with severe aortic stenosis and moderate aortic regurgitation

P

Left heart catheterization and magnetic resonance aortography prior to aortic valve replacement surgery

Cardiac rehabilitation with exercise training

Avoid contact sports but encourage other activity following surgery

Exercise Program

Initially a Basic *CDD4* Recommendation, advancing to ACSM Guidelines with special considerations for prosthetic valves:

Aerobic exercise 4-5 days a week
- Walk at an RPE <13 (out of 20), or use talk test
- Begin with 15-20 min (rest stops allowed)
- Build ~5 min a week up to 40 min/session
- May gradually increase RPE as tolerated, using interval design

Resistance exercises 2 days a week
- Lower extremity and core exercises as tolerated
- Delayed initiation of upper-extremity exercises as sternum heals
- Build up to a minimum of 2 sets of 8 reps maximum (RM)

Flexibility exercises 3 days a week
- Stretching as desired, daily on L and R

- Hold for 30 s on each side, below the discomfort point

Warm-up and cool-down phases @ RPE <11/20 for ~10-15 min

Special Consideration

Contact sports for prosthetic valve recipients are not allowed due to bruising/hematoma risk in persons who are fully anticoagulated

Senior Editor's Comment

This patient should be an easy participant to have in cardiac rehabilitation. As a middle-aged recreational runner, he's likely to be eager to get back to his usual routine. Sometimes patients like this need a little bit of restraint because of the prolonged deconditioning they experience in the period leading up to valve replacement. Having a discussion to establish reasonable expectations is very worthwhile, as many patients are eager, but rushing doesn't really serve a good purpose. Needing to start back up again and the convalescence from the thoracotomy to do the valve replacement are reasons to start off a little bit slowly, but once the chest wall is healed, such patients rebound pretty quickly and do well.

Visual Impairment

PRESENTER Jo B. Zimmerman, MS

S

"I want to stay healthy and manage my other health risks, and I really miss running. Can I ever be an athlete again?"

A 27-year-old African American male, who had been diagnosed with retinitis pigmentosa (RP) at age 17, wanted to return to general fitness activity. He quit playing football and running track in high school after he was diagnosed with RP. His rapidly diminishing vision affected his physical activity participation substantially, as he had little confidence in his personal safety or abilities outside of football and track. In college, he learned that his studies could be adapted to his needs and increasingly found himself wanting to be more physically active. While he knew that team sports that relied on visual interaction and passing balls between players were not viable, he had begun to wonder what other activities might take their place.

O

Height: 74 in. (1.88 m)
Weight: 205 lb (93 kg)
BMI: 26.3 kg/m^2

HR: 68 beats/min
BP: 128/84 mmHg

Pertinent Exam

Detects light/dark, general shapes are blurred, substantial night blindness; visual acuity 20/500 in both eyes (stable for 2 years); otherwise healthy male with no other health problems

Family history of diabetes and hypertension

Submaximal Exercise Test

Peak work rate: 600 kgm/min

Peak heart rate: 104 beats/min

Reason for terminating test: local fatigue in his legs

Estimated $\dot{V}O_2$max: 48.5 mL $O_2 \cdot kg^{-1} \cdot min^{-1}$

A

Severe visual impairment, legally blind

Aerobic fitness above average for age/sex, with excellent fitness potential

Increased familial risk for cardiometabolic diseases of sedentary lifestyle

P

Engage a personal trainer/find a program in adapted exercise for the visually impaired

Exercise programming per ACSM Guidelines

Goals

Ensure sense of safety

Maintain normal BMI and blood pressure

Explore sports options as interests and opportunities unfold

Find a partner for outdoor running; try swimming and adapted sports

Exercise Program

Follow ACSM Guidelines, adapted with safety precautions for visually impaired.

Aerobic exercise 30-40 min 3-5 times per week

Strength training using circuit machines, 2 sets of 10 reps, 2 times per week

Stretching as needed or desired

URL Resources

International Blind Sports Federation, www.ibsasport.org [Accessed September 23, 2015].

National Center on Health, Physical Activity, and Disability (NCHPAD), www.ncpad.org [Accessed September 23, 2015].

National Federation of the Blind, https://nfb.org [Accessed September 23, 2015].

Vision Aware, www.visionaware.org [Accessed September 23, 2015].

Senior Editor's Comment

This case study illustrates how easily patients with visual impairment drift away from a physically active lifestyle, in part because of their own internal challenges and in part because our society reduces its expectations. It's bad enough that, in the United States, most adolescents don't meet the U.S. Department of Health and Human Services physical activity guidelines, but this young man's sports career was terminated because of his RP. Did he get guidance that he couldn't play sports anymore (he might have been able to compete on the track team in running and throwing events)? And did his university provide support for either individual or intramural sporting activities? Perhaps these institutions did offer support but he and his family were able to cope only with learning the new skills of living while visually impaired and perceived that his studies were the most important defense against becoming dependent. It may be that he hasn't been emotionally/spiritually open to remaining active and exploring that aspect of his life but is now growing into this as a new phase of his individuality. These background issues need to be explored because they may influence the counseling he needs for health promotion.

His family history indicates that it is important for him to do everything he can to avert the development of hypertension and diabetes. It would be all too easy for his health care providers to watch him remain sedentary because he's blind, watch him gradually gain weight and go on medications for hypertension and diabetes, and gradually see his kidney function deteriorate until he had to go on dialysis. In the United States, the largest group of patients with end-stage kidney failure on hemodialysis is African Americans with hypertension or diabetes. His health care providers need to make it crystal clear that one thing he most certainly needs to run away from is being a blind, diabetic, hypertensive African American male on dialysis 3 days a week. His counseling on lifestyle is about more than opening doors to his spiritual fulfillment in sports; it's also about not letting his visual disability lead to a life that would make it exceedingly difficult to maintain his independence. Why does our health care system not provide him that advice or the resources he needs to make it happen?

Common Medications

Appendix A is a listing of common medications that exercise and health care professionals are likely to encounter among their clients or patients who are, or are soon to be, physically active. It includes the name of each drug, the brand names, and indications for drug use. This listing is not intended to be exhaustive or all-inclusive and is not designed for the determination of pharmacotherapy or medication prescription for patients by clinicians or physicians. Rather, this listing should be viewed as a resource to further clarify the medical histories of research study participants, patients, and clients encountered by exercise professionals nationally and internationally. To this end, some brand names, although recently discontinued (i.e., only generic formulations available) or no longer marketed in the United States, are included for reference. For a more detailed informational listing, the reader is referred to the American Hospital Formulary Service (AHFS) Drug Information (American Society of Health-System Pharmacists, 2014) or the U.S. Food and Drug Administration, U.S. Department of Health and Human Services website from which the following listings were obtained.

Cardiovascular

β-Blockers

Indications: Hypertension (HTN), angina, arrhythmias including supraventricular tachycardia, atrial fibrillation rate control, acute myocardial infarction (MI), migraine headaches, anxiety, essential tremor, and heart failure (HF) because of systolic dysfunction.

Drug name	Brand name	Drug name	Brand name
Acebutolol[a,b]	Sectral	Nadolol[c]	Corgard
Atenolol[a]	Tenormin	Nebivolol[a]	Bystolic
Betaxolol[a]	Kerlone	Pindolol[b,c]	Visken
Bisoprolol[a]	Zebeta	Propranolol[c]	Inderal, Inderal LA
Carvedilol[c,d]	Coreg, Coreg CR	Sotalol[c]	Betapace
Esmolol[a]	Brevibloc	Timolol[c]	Blocadren
Labetalol[c,d]	Normodyne, Trandate		
Metoprolol succinate[a] Metoprolol tartrate[a]	Toprol XL Lopressor, Lopressor SR		

[a]Cardioselective [b]β-blockers with intrinsic sympathomimetic activity [c]Noncardioselective [d]Combined α-β-blocker

Angiotensin-Converting Enzyme Inhibitors (ACE-I)

Indications: HTN, coronary artery disease, HF caused by systolic dysfunction, diabetic nephropathy, chronic kidney disease, and cerebrovascular disease.

Drug name	Brand name	Combination ACE-I + HCTZ	ACE-I + CCB
Benazepril	Lotensin	Lotensin HCT	Lotrel (+ amlodipine)
Captopril	Capoten	Capozide	

(continued)

Selected tables from appendix A are reprinted from ACSM, 2014, *ACSM's guidelines for exercise testing and prescription,* 9th ed. (Boston, MA: Lippincott, Williams, and Wilkins), appendix a.

(continued from previous page)

Drug name	Brand name	Combination ACE-I + HCTZ	ACE-I + CCB
Enalapril	Vasotec	Vaseretic	Lexxel (+ felodipine)
Fosinopril	Monopril	Monopril HCT	
Lisinopril	Zestril, Prinivil	Prinzide, Zestoretic	
Moexipril	Univasc	Uniretic	
Perindopril	Aceon		Prestalia (+ amlodipine)
Quinapril	Accupril	Accuretic	
Ramipril	Altace		
Trandolapril	Mavik		Tarka (+ verapamil)

HCTZ = hydrochlorothiazide, a thiazide diuretic   CCB = calcium channel blocker

Angiotensin II Receptor Blockers (ARBs)

Indications: HTN, diabetic nephropathy, and HF.

Drug name	Brand name	Combination ARB + diuretic (HCTZ[a] or chlorthalidone[b])	Combination ARB + HCTZ + CCB[c]	Combination ARB + CCB[d]
Azilsartan	Edarbi	Edarbychlor[b]		
Candesartan	Atacand	Atacand HCT[a]		
Eprosartan	Teveten	Teveten HCT[a]		
Irbesartan	Avapro	Avalide[a]		
Losartan	Cozaar	Hyzaar[a]		
Olmesartan	Benicar	Benicar HCT[a]	Tribenzor	Azor
Telmisartan	Micardis	Micardis HCT[a]		Twynsta
Valsartan	Diovan	Diovan HCT[a]	Exforge HCT	Exforge

[a]ARB + HCTZ for use in HTN and HF   [b]ARB + chlorthalidone for use in HTN   [c]ARB + HCTZ + CCB for use in HTN   [d]ARB + CCB for use in HTN

Direct Renin Inhibitor (DRI)

Indication: HTN.

Drug name	Brand name	Combination DRI + HCTZ[a]
Aliskiren	Tekturna	Tekturna HCT

[a]DRI + HCTZ for use in HTN

Calcium Channel Blockers (CCB)

Dihydropyridines

Indications: HTN, isolated systolic HTN, angina pectoris, vasospastic angina, and ischemic heart disease.

Drug name	Brand name	Drug name	Brand name
Amlodipine	Norvasc	Nicardipine	Cardene, Cardene SR
Clevidipine (IV formulation only)	Cleviprex	Nifedipine[a,b]	Adalat CC[a], Afeditab CR[a], Procardia[b], Procardia XL[a]
Felodipine	Plendil	Nimodipine	Nymalize
Isradipine	DynaCirc, DynaCirc CR	Nisoldipine	Sular

[a]Long-acting   [b]Short-acting

Nondihydropyridines

Indications: Angina, HTN, paroxysmal supraventricular tachycardia, and arrhythmia.

Drug name	Brand name	Drug name	Brand name
Diltiazem	Cardizem	Verapamil	Calan, Verelan, Covera HS, Isoptin
Diltiazem, extended release	Cardizem CD or LA, Cartia XT, Dilacor XR, Dilt CD or XR, Diltia XT, Diltzac, Taztia XT, Tiazac	Verapamil, controlled and extended release	Calan SR, Covera HS, Verelan, Verelan PM
		Verapamil + trandolapril	Tarka

Diuretics

Indications: Edema, HTN, HF, and certain kidney disorders.

Drug name	Brand name	
THIAZIDES		
Bendroflumethiazide	(+ nadolol) Corzide	
Chlorothiazide	Diuril	
Hydrochlorothiazide (HCTZ)	Microzide, Oretic	
Methylclothiazide	None	
Polythiazide	Renese	
THIAZIDE-LIKE		
Chlorthalidone	Thalitone, (+ atenolol) Tenoretic	
Indapamide	Lozol	
Metolazone	Zaroxolyn	
LOOP DIURETICS		
Bumetanide	Bumex	
Ethacrynic acid	Edecrin	
Furosemide	Lasix	
Torsemide	Demadex	
POTASSIUM-SPARING DIURETICS		**COMBINED WITH HCTZ**
Amiloride	Midamor	Moduretic, Hydro-Ride
Triamterene	Dyrenium	Dyazide, Maxzide
MINERALOCORTICOID (ALDOSTERONE) RECEPTOR BLOCKERS		**COMBINED WITH HCTZ**
Eplerenone	Inspra	
Spironolactone	Aldactone	Aldactazide

Vasodilating Agents

Nitrates and Nitrites

Indications: Angina, acute MI, HF, low cardiac output syndromes, and HTN.

Drug name	Brand name	Drug name	Brand name
Amyl nitrite (inhaled)	Amyl Nitrite	Isosorbide dinitrate + hydralazine HCl	BiDil
Isosorbide mononitrate	Monoket		
Isosorbide dinitrate	Dilatrate SR, Isordil	Nitric oxide (inhaled)	INOmax

(continued)

(continued from previous page)

Drug name	Brand name	Drug name	Brand name
Nitroglycerin, capsules ER	Nitro-Time, Nitroglycerin Slocaps	Nitroglycerin, topical ointment	Nitro-Bid
Nitroglycerin, lingual (spray)	Nitrolingual Pumpspray, Nitromist	Nitroglycerin, transdermal	Minitran, Nitro-Dur, Nitrek, Deponit
Nitroglycerin, sublingual	Nitrostat	Nitroglycerin, transmucosal (buccal)	Nitrogard

α-Blockers

Indications: HTN and benign prostatic hyperplasia.

Drug name	Brand name	Drug name	Brand name
Doxazosin	Cardura, Cardura XL	Tamsulosin	Flomax
Prazosin	Minipress	Terazosin	Hytrin

Central α-Agonists

Indication: HTN.

Drug name	Brand name	Drug name	Brand name
Clonidine	Catapres, Catapres-TTS (patch), Duraclon (injection form), Kapvay	Guanfacine	Intuniv, Tenex
Guanabenz	Wytensin	Methyldopa	Aldoril

Direct Vasodilators

Indications: HTN, hair loss, and HF.

Drug name	Brand name	Drug name	Brand name
Diazoxide	Hyperstat	Minoxidil	Loniten Topical: Rogaine, Theroxidil
Hydralazine	(+ HCTZ) Hydra-Zide, (+ isosorbide dinitrate) Bidil	Sodium nitroprusside	Nipride, Nitropress

Peripheral Adrenergic Inhibitors

Indications: HTN and psychotic disorder.

Drug name	Brand name
Reserpine	Hyserpin

Others

Cardiac Glycosides

Indications: Acute, decompensated HF in the setting of dilated cardiomyopathy, and need to increase atrioventricular (AV) block to slow ventricular response with atrial fibrillation.

Drug name	Brand name
Amrinone (inamrinone)	Inocor
Digoxin	Lanoxin, Lanoxicaps, Digitek
Milrinone	Primacor

Cardiotonic Agent

Indication: Symptomatic management of stable angina pectoris in HF. Specifically for heart rate reduction in patients with systolic dysfunction when in sinus rhythm with a resting heart rate ≥70 beats/min and either currently prescribed a maximally tolerated dose of β-blockers or with a contraindication to β-blocker use.

Drug name	Brand name
Ivabradine	Corlanor

Antiarrhythmic Agents

Indications: Specific for individual drugs but generally include suppression of atrial fibrillation and maintenance of normal sinus rhythm, serious ventricular arrhythmias in certain clinical settings, and increase in AV nodal block to slow ventricular response in atrial fibrillation.

Drug name	Brand name
CLASS IA	
Disopyramide	Norpace CR
Procainamide	Procanbid
Quinidine	Quinora, Quinidex, Quinaglute, Quinalan, Carioquin
CLASS IB	
Lidocaine	Xylocaine
Mexiletine	Mexitil
Phenytoin	Dilantin, Phenytek
CLASS IC	
Flecainide	Tambocor
Propafenone	Rythmol SR
CLASS II	
β-Blockers	
Atenolol	Tenormin
Bisoprolol	Zebeta
Esmolol	Brevibloc
Metoprolol	Lopressor, Lopressor SR, Toprol XL
Propranolol	Inderal, Inderal LA
Timolol	Blocadren
CLASS III	
Amiodarone	Cordarone, Nexterone (IV), Pacerone
Dofetilide	Tikosyn
Dronedarone	Multaq
Ibutilide	Covert (IV)
Sotalol	Betapace, Betapace AF, Sorine
CLASS IV	
Diltiazem	Cardizem CD or LA, Cartia XT, Dilacor XR, Dilt CD or XR, Diltia XT, Diltzac, Tiazac, Taztia XT
Verapamil	Calan, Calan SR, Covera HS, Verelan, Verelan PM

Antianginal Agents

Indications: Adjunctive therapy in the management of chronic stable angina pectoris. May be used in combination with β-blockers, calcium channel blockers, nitrates, angiotensin-converting enzyme inhibitors, angiotensin-receptor blockers, lipid-lowering therapy, or some combination of these.

Drug name	Brand name
Ranolazine	Ranexa

Antilipemic Agents

Indications: Elevated total blood cholesterol, low-density lipoproteins (LDL), and triglycerides; low high-density lipoproteins (HDL); and metabolic syndrome.

Drug name	Brand name
BILE ACID SEQUESTRANTS	
Cholestyramine	Prevalite
Colesevelam	Welchol
Colestipol	Colestid
FIBRIC ACID SEQUESTRANTS	
Fenofibrate	Antara, Fenoglide, Lipofen, Lofibra, Tricor, Triglide, Trilipix
Gemfibrozil	Lopid
HYDROXYMETHYLGLUTARYL CO-ENZYME A REDUCTASE INHIBITORS (STATINS)	
Atorvastatin	Lipitor
Fluvastatin	Lescol XL
Lovastatin	Mevacor, Altoprev
Lovastatin + niacin	Advicor
Pitavastatin	Livalo
Pravastatin	Pravachol
Rosuvastatin	Crestor
Simvastatin	Zocor
Simvastatin + niacin	Simcor
STATIN + CCB	
Atorvastatin + amlodipine	Caduet
NICOTINIC ACID	
Niacin (vitamin B_6)	Niaspan, Nicobid, Slo-Niacin
OMEGA-3 FATTY ACID ETHYL ESTERS	
Omega-3-carboxylic acids (EPA and DHA)	Epanova
Icosapent ethyl (EPA)	Vascepa
Omega-3 fatty acid ethyl esters (EPA and DHA)	Lovaza
CHOLESTEROL ABSORPTION INHIBITOR	
Ezetimibe	Zetia, (+ simvastatin) Vytorin

CCB = calcium channel blocker   DHA = docosahexaenoic acid   EPA = eicosapentaenoic acid

Blood Modifiers

Anticoagulants

Indications: Treatment and prophylaxis of thromboembolic disorders. To prevent blood clots, heart attack, stroke, and intermittent claudication; or vascular death in patients with established nonvalvular atrial fibrillation, deep venous thrombosis, pulmonary embolism, heparin-induced thrombocytopenia, peripheral arterial disease, or acute ST-segment elevation with MI.

Drug name	Brand name
Warfarin (vitamin K antagonist)	Coumadin, Jantoven
Apixaban (selective inhibitor of factor Xa)	Eliquis
Argatroban (direct thrombin inhibitor)	Acova
Bivalirudin (direct thrombin inhibitor)	Angiomax
Dabigatran (direct thrombin inhibitor)	Pradaxa

Drug name	Brand name
Edoxaban (selective inhibitor of factor Xa)	Savaysa
Dalteparin (LMWH)	Fragmin
Enoxaparin (LMWH)	Lovenox
Apixaban (selective inhibitor of factor Xa)	Eliquis
Fondaparinux (LMWH)	Arixtra
Rivaroxaban (selective inhibitor of factor Xa)	Xarelto

LMWH = low molecular weight heparin   factor Xa = serine endopeptidase (aka prothrombinase, thrombokinase, or thromboplastin)

Antiplatelet Agents

Indications: Antiplatelet drugs reduce platelet aggregation and are used to prevent further thromboembolic events in patients who have suffered MI, ischemic stroke, transient ischemic attacks, or unstable angina; and for primary prevention for patients at risk of a thromboembolic event. Some are also used for the prevention of reocclusion or restenosis following percutaneous coronary interventions and bypass procedures.

Drug name	Brand name
Aspirin (COX-I inhibitor)	None
Cilostazol (PDE inhibitor)	Pletal
Clopidogrel (ADP-R inhibitor)	Plavix
Dipyridamole (adenosine reuptake inhibitor)	Persantine, (+ aspirin) Aggrenox
Pentoxifylline	Trental

Drug name	Brand name
Prasugrel (ADP-R inhibitor)	Effient
Ticagrelor (ADP-R inhibitor)	Brilinta
Ticlopidine (ADP-R inhibitor)	Ticlid
Vorapaxar	Zontivity

ADP-R = adenosine diphosphate-ribose   COX-I = cyclooxygenase inhibitor   PDE = phosphodiesterase.

Respiratory

Inhaled Corticosteroids

Indications: Asthma, nasal polyp, and rhinitis.

Drug name	Brand name
Beclomethasone	Beclovent, Qvar, Vanceril
Budesonide	Pulmicort
Ciclesonide	Alvesco
Flunisolide	AeroBid

Drug name	Brand name
Fluticasone	Flovent
Mometasone	Asmanex
Triamcinolone	Azmacort

Bronchodilators

Anticholinergics (Acetylcholine Receptor Antagonists)

Indications: Anticholinergic or antimuscarinic medications are used for the management of obstructive pulmonary disease and acute asthma exacerbations. They prevent wheezing, shortness of breath, and troubled breathing caused by asthma, chronic bronchitis, emphysema, and other lung diseases.

Drug name	Brand name	Combined with sympathomimetic (β_2-receptor agonists)
Glycopyrrolate	Robinul	
Ipratropium	Atrovent	(+ albuterol) Combivent
Tiotropium	Spiriva	

Sympathomimetics (β_2-Receptor Agonists)

Indications: Relief of asthma symptoms and in the management of chronic obstructive pulmonary disease. They prevent wheezing, shortness of breath, and troubled breathing caused by asthma, chronic bronchitis, emphysema, and other lung diseases.

Drug name	Brand name	Combined with steroid
Albuterol	ProAir, Proventil, Ventolin	
Formoterol (LA)	Foradil	(+ budesonide) Symbicort (+ mometasone) Dulera
Indacaterol	Arcapta	

Drug name	Brand name	Combined with steroid
Levalbuterol	Xopenex	
Metaproterenol	Alupent, Metaprel	
Pirbuterol	Maxair	
Salmeterol (LA)	Serevent	(+ fluticasone) Advair
Terbutaline	Brethine, Brethaire, Bricanyl	

Xanthine Derivatives

Indications: Combination therapy in asthma and chronic obstructive pulmonary disease.

Drug name	Brand name
Aminophylline	Phyllocontin, Truphylline
Caffeine	None

Drug name	Brand name
Theophylline	Theo-24, Uniphyl

Leukotriene Inhibitors and Antagonists

Indications: Asthma, exercise-induced asthma, and rhinitis.

Drug name	Brand name
Montelukast	Singulair
Zafirlukast	Accolate

Drug name	Brand name
Zileuton	Zyflo CR

Mast Cell Stabilizers

Indications: To prevent wheezing, shortness of breath, and troubled breathing caused by asthma, chronic bronchitis, emphysema, and other lung diseases.

Drug name	Brand name
Cromolyn (inhaled)	Intal

Cough and Cold Products

First-Generation Antihistamines

Indications: Allergy, anaphylaxis (adjunctive), insomnia, motion sickness, pruritis of skin, rhinitis, sedation, and urticaria (hives).

Drug name	Brand name
Brompheniramine	Lodrane, Bidhist; combinations available with pseudoephedrine and phenylephrine
Carbinoxamine	Arbinoxa, Palgic
Chlorpheniramine	Aller-Chlor, Chlor-Trimeton; combinations available with pseudoephedrine and phenylephrine
Clemastine	Dayhist, Tavist
Cyproheptadine	Cyproheptadine

Drug name	Brand name
Diphenhydramine	Benadryl, Nytol; combinations available with acetaminophen (APAP), pseudoephedrine, and phenylephrine
Doxylamine	Aldex, Unisom SleepTabs, Good-Sense Sleep Aid
Promethazine	Phenergan, Prometh VC Syrup (with pseudoephedrine)
Triprolidine	Zymine, Zymine-D (with pseudoephedrine), Allerfrim (with pseudoephedrine), Aprodine (with pseudoephedrine)

Second-Generation Antihistamines

Indications: Allergic rhinitis and urticaria (hives).

Drug name	Brand name
Acrivastine	Semprex-D (with pseudoephedrine)
Cetirizine	Zyrtec, Zyrtec-D (with pseudoephedrine)
Desloratadine	Clarinex, Clarinex-D (with pseudoephedrine)

Drug name	Brand name
Fexofenadine	Allegra, Allegra-D (with pseudoephedrine)
Levocetirizine	Xyzal
Loratadine	Claritin, Claritin-D (with pseudoephedrine), Alavert, Alavert-D (with pseudoephedrine)

Sympathomimetic–Adrenergic Agonists

Indications: Allergic rhinitis and nasal congestion.

Drug name	Brand name
Phenylephrine	Sudafed PE

Drug name	Brand name
Pseudoephedrine	Sudafed; many combinations

Expectorant

Indication: Abnormal sputum (thick secretions or mucus).

Drug name	Brand name
Guaifenesin	Robitussin, Guiatuss, Mucinex (many combinations)

Antitussives

Indications: Cough and pain.

Drug name	Brand name
Benzonatate	Tessalon
Codeine	Codeine; many combinations
Dextro-methorphan	Robitussin Cough Gels, Robitussin Pediatric Cough Suppressant; many combinations

Drug name	Brand name
Hydro-codone	Many combinations

Hormonal

Human Growth Hormone

Indications: Cachexia associated with acquired immunodeficiency syndrome (AIDS), growth hormone deficiency, and short bowel syndrome.

Drug name	Brand name
Somatropin	Genotropin, Norditropin, Nutropin, Humatrope, Omnitrope

Drug name	Brand name
Mecasermin (IV)	Increlex
Tesamorelin (IV)	Egrifta

Corticosteroids

Indications: Adrenocortical insufficiency, adrenogenital syndrome, hypercalcemia, thyroiditis, rheumatic disorders, collagen diseases, dermatologic diseases, allergic conditions, ocular disorders, respiratory diseases (e.g., asthma, chronic obstructive pulmonary disorders), hematologic disorders, gastrointestinal diseases (e.g., ulcerative colitis, Crohn's disease), and liver disease among others.

Drug name	Brand name
Beclomethasone	QVAR, Beclovent
Betamethasone	Celestone, Celestone Soluspan (injectable)
Budesonide	Entocort EC, Pulmicort
Ciclesonide	Alvesco
Cortisone	Cortisone
Dexamethasone	Decadron, Dexpak
Fludrocortisone	Florinef
Flunisolide	Aerospan

Drug name	Brand name
Fluticasone	Flovent; with salmeterol: Advair
Hydrocortisone	Cortef, Hydrocortone
Methylprednisolone	Medrol, Meprolone, Solu-Medrol, Depo-Medrol, A-Methapred
Mometasone	Asmanex
Prednisolone	Orapred ODT, Prelone, Pediapred
Prednisone	Sterapred, Prednisone Intensol
Triamcinolone	Aristospan, Tac, Kenalog, Azmacort

Androgenic-Anabolic

Indications: Hypogonadism in males, catabolic and wasting disorders, endometriosis, hereditary angio-edema, fibrocystic breast disease, and precocious puberty.

Drug name	Brand name
Danazol	Danocrine
Fluoxymesterone	Halotestin, Androxy
Methyltestosterone	Android, Testred

Drug name	Brand name
Oxandrolone	Oxandrin
Testosterone	Striant, AndroGel, Andro-derm, Natesto, Testim, Delatestryl (injectable)

Estrogens

Indications: Menopause and perimenopause in women, osteoporosis, moderate to severe vasomotor symptoms, corticosteroid-induced hypogonadism, metastatic breast carcinoma, prostate carcinoma, Alzheimer's disease.

Drug name	Brand name	Combinations
Estradiol	Elestrin, EstroGel, Evamist, Menostar, Alora, Climara, Vivelle, Vivelle-Dot, Estraderm, Estrace, Estrasorb	(+ norgestimate) Prefest; (+ norethidone acetate) Activella; (+ drospirenone) Angeliq
Estradiol (acetate)	Femtrace	
Estradiol (cypionate)	Depo-Estradiol	with testosterone: Depo-Testadiol

Drug name	Brand name	Combinations
Estradiol (valerate)	Delestrogen	
Estradiol (ethinyl)		with norethindrone: Femhrt
Estrogens (conjugated)	Premarin, Cenestin (synthetic), Enjuva (synthetic)	with medroxy-progesterone: Prem-pro, Premphase
Estrogens (esterified)	Ogen, Ortho-Est, Menest	with methyltestos-terone: Covaryx

Contraceptives

Indications: Contraception, menstrual bleeding regulation, premenstrual mood disorders, acne.

Drug name	Brand name
Estrogen–progestin combinations	Oral: Beyaz, Yaz, Alesse, Lybrel, Lessina, Aviane, LoSeasonique, Loestrin, Yasmin, Microgestin, Sprintec, Ortho-Cyclen, Ortho Tri-Cyclen
Transdermal	Ortho Evra
Vaginal ring	NuvaRing

Drug name	Brand name
Intrauterine	Mirena
Progestins: etonogestrel	Subdermal implant: Implanon, Nexplanon
Progestins: levonorgestrel	Oral: Next Choice, Plan B One-Step
Progestins: norethindrone	Oral: Micronor, Nor-QD

Thyroid Agents

Indications: Hypothyroidism and pituitary thyroid-stimulating hormone suppression.

Drug name	Brand name
Levothyroxine	Levothroid, Synthroid, Levoxyl, Unithroid
Liothyronine	Cytomel

Drug name	Brand name
Liotrix	Thyrolar
Thyroid	Armour

Antidiabetic

Indication: Management of type 2 diabetes mellitus.

CLASS: α-Glucosidase Inhibitors (slow absorption of carbohydrate in the gastrointestinal tract)

Drug name	Brand name
Acarbose	Precose

Drug name	Brand name
Miglitol	Glyset

CLASS: Amylin Analogue (mimics amylin, a hormone secreted with insulin to inhibit glucose, for postprandial glycemic control)

Drug name	Brand name
Pramlintide	Symlin

CLASS: Biguanides (decrease sugar production by liver and decrease insulin resistance)

Drug name	Brand name	Combination
Metformin	Glucophage, Fortamet, Glumetza	(+ glypizide) Metaglip (+ glyburide) Glucovance

CLASS: Sodium–Glucose Cotransporter 2 (SGLT2) Inhibitors

Drug name	Brand name	Combination
Canagliflozin	Invokana	(+ metformin) Invokamet
Dapagliflozin	Farxiga	(+ metformin) Xigduo XR
Empagliflozin	Jardiance	

CLASS: Dipeptidylpeptidase-4 Inhibitors (enhance insulin release by preventing breakdown of glucagon-like peptide 1 [GLP-1] that is a potent antihyperglycemic hormone)

Drug name	Brand name	Combination	Drug name	Brand name	Combination
Alogliptin	Nesina	(+ metformin) Kazano (+ pioglitazone) Oseni	Saxagliptin	Onglyza	(+ metformin) Kombiglyze
Linagliptin	Tradjenta		Sitagliptin	Januvia	(+ metformin) Janumet

CLASS: Glucagon-Like Peptide 1 Receptor Agonists (activate GLP-1, a potent antihyperglycemic hormone that stimulates insulin release)

Drug name	Brand name	Drug name	Brand name
Exenatide	Byetta	Liraglutide	Victoza

CLASS: Meglitinides (short-acting stimulation of β-cells to produce more insulin)

Drug name	Brand name	Combination	Drug name	Brand name	Combination
Nateglinide	Starlix		Repaglinide	Prandin	(+ metformin) Prandimet

CLASS: Sulfonylureas (stimulate β-cells to produce more insulin)

Drug name	Brand name	Combination
Chlorpropamide, 1st gen.	Diabinese	
Glimepiride	Amaryl	(+ pioglitazone) Duetact (+ rosiglitazone) Avandaryl
Glipizide	Glucotrol	(+ metformin) Metaglip
Glyburide	DiaBeta, Glynase, Micronase	(+ metformin) Glucovance
Tolazamide, 1st gen.	Tolinase	
Tolbutamide, 1st gen.	Orinase	

CLASS: Thiazolidinediones (improve sensitivity of insulin receptors in muscle, liver, and fat cells)

Drug name	Brand name	Combination	Drug name	Brand name	Combination
Pioglitazone	Actos	(+ metformin) Actoplus Met XR (+ glimepiride) Duetact	Rosiglitazone	Avandia	(+ metformin) Avandamet (+ glimepiride) Avandaryl

CLASS: Insulin

Rapid-acting	Intermediate-acting	Intermediate- and rapid-acting combination	Long-acting
Humalog	Humulin L	Humalog Mix	Humulin U
Humulin R	Humulin N	Humalog 50/50	Lantus injection
Novolog	Novolin L	Humalog 75/25	Levemir
Novolin R	Novolin N	Novolin 70/30	
Iletin II R		Novolog 70/30	
Apidra			

Central Nervous System

Antidepressants

Indication: Depression.

Drug name	Brand name
Amitriptyline (TCA)	(+ chlordiazepoxide) Limbitrol, Limbitrol DS
Amoxapine (TCA)	Asendin
Bupropion	Wellbutrin (SR and XL), Zyban
Citalopram (SSRI)	Celexa
Clomipramine (TCA)	Anafranil
Desipramine (TCA)	Norpramin
Desvenlafaxine (SNRI)	Pristiq
Doxepin (TCA)	Adapin, Sinequan
Duloxetine (SNRI)	Cymbalta
Escitalopram (SSRI)	Lexapro
Fluoxetine (SSRI)	Prozac, Sarafem, (+ olanzepine) Symbyax
Fluvoxamine (SSRI)	Luvox CR
Imipramine (TCA)	Tofranil, Tofranil-PM
Isocarboxazid (MAO-I)	Marplan
Levomilnacipran (SNRI)	Fetzima

Drug name	Brand name
Maprotiline (TeCA)	Ludiomil
Milnacipran (SNRI)	Savella
Mirtazapine (TeCA)	Remeron
Nefazodone	
Nortriptyline (TCA)	Pamelor
Paroxetine (SSRI)	Paxil CR, Pexeva
Phenelzine (MAO-I)	Nardil
Protriptyline (TCA)	Vivactil
Selegiline (MAO-I)	Anipryl
Sertraline (SSRI)	Zoloft
Tranylcypromine (MAO-I)	Parnate
Trazodone (SARI)	Desyrel, Dividose
Trimipramine (TCA)	Surmontil
Venlafaxine (SNRI)	Effexor XR
Vilazodone (SARI)	Viibryd

MAO-I = monoamine oxidase inhibitor   SARI = serotonin antagonist reuptake inhibitor   SNRI = serotonin–norepinephrine reuptake inhibitor   SSRI = selective serotonin reuptake inhibitor   TCA = tricyclic antidepressant   TeCA = tetracyclic antidepressant

Antipsychotics

Indications: Bipolar disorder, Gilles de la Tourette syndrome, hyperactive behavior, psychotic disorder, and schizophrenia.

Drug name	Brand name
Aripiprazole (atypical)	Abilify
Asenapine (atypical)	Saphris
Chlorpromazine (typical)	Thorazine
Clozapine (atypical)	Clozaril, FazaClo
Fluphenazine (typical)	Permitil, Prolixin
Haloperidol (typical)	Haldol

Drug name	Brand name
Iloperidone (atypical)	Fanapt
Lithium	Eskalith CR, Lithobid
Loxapine (typical)	Loxitane
Lurasidone (atypical)	Latuda
Mesoridazine (phenothiazine)	Serentil
Molindone (typical)	Moban

(continued)

(continued from previous page)

Drug name	Brand name
Olanzapine (atypical)	Zyprexa, (+ fluoxetine) Symbyax
Paliperidone (atypical)	Invega
Perphenazine (typical)	Perphenazine
Prochlorperazine (typical)	Compazine
Pimozide	Orap
Promazine	Sparine

Drug name	Brand name
Quetiapine (atypical)	Seroquel
Risperidone (atypical)	Risperdal
Thioridazine (typical)	Mellaril
Thiothixene (typical)	Navane
Triflupromazine	Vesprin
Valproic acid	Depakote ER, Depakene
Ziprasidone (atypical)	Geodon

Antianxiety

Indications: Anxiety and panic disorder.

Drug name	Brand name
Alprazolam	Xanax XR, Niravam
Buspirone	Buspar
Chlordiazepoxide	Limbitrol DS, Librium, (+ clidinium) Librax
Clonazepam	Klonopin

Drug name	Brand name
Clorazepate	Tranxene
Diazepam	Valium
Lorazepam	Ativan
Meprobamate	Equanil, Miltown, Meprospan
Oxazepam	Serax

Sedative-Hypnotics

Indications: General anesthesia, insomnia, and sedation.

Drug name	Brand name
Amobarbital	Amytal
Butabarbital	Butisol
Chloral hydrate	Somnote, Aquachloral, Supprettes
Dexmedetomidine	Precedex
Estazolam	ProSom
Eszopiclone	Lunesta
Flurazepam	Dalmane
Fospropofol	Lusedra
Mephobarbital	Mebaral

Drug name	Brand name
Promethazine	Phenergan, Phenadoz, Prometh
Propofol	Diprivan
Quazepam	Doral, Dormalin
Ramelteon	Rozerem
Secobarbital	Seconal
Temazepam	Restoril
Triazolam	Halcion
Zaleplon	Sonata
Zolpidem	Ambien CR, Intermezzo, Edluar

Stimulants

Indications: Attention deficit hyperactivity disorder, narcolepsy, obstructive sleep apnea, and shift work sleep disorder.

Drug name	Brand name
Amphetamine salts	Adderall XR
Armodafinil	Nuvigil
Caffeine	No Doz, Vivarin
Dexmethylphenidate	Focalin XR
Dextroamphetamine	Dexedrine, Dextrostat

Drug name	Brand name
Lisdexamfetamine	Vyvanse
Methamphetamine	Desoxyn
Methylphenidate	Concerta, Metadate CD or ER; Ritalin LA, SR; Methylin ER
Modafinil	Provigil

Nicotine Replacement Therapy

Indication: Smoking cessation assistance.

Drug name	Brand name
Nicotine	Solution: Nicotrol NS Inhalant: Nicotrol Inhaler Transdermal: Nicotrol Step 1,2,3; NicoDerm CQ Step 1,2,3

Drug name	Brand name
Nicotine polacrilex	Lozenges: Commit Chewing gum: Nicorette, Nicorette DS

Nonsteroidal Anti-Inflammatory Drugs (NSAIDs)

Indications: Fever, headache, juvenile rheumatoid arthritis, migraine, osteoarthritis, pain, primary dysmenorrhea, and rheumatoid arthritis.

Drug name	Brand name
Celecoxib	Celebrex
Diclofenac	Athrotec, Cataflam, Voltaren
Diflunisal	Dolobid
Etodolac	Lodine
Fenoprofen	Nalfon
Flurbiprofen	Ansaid
Ibuprofen	Advil, Ibu-Tab, Menadol, Midol, Motrin, Nuprin, Genpril, Haltran
Indomethacin	Indocin
Ketoprofen	Actron, Orudis, Oruvail

Drug name	Brand name
Ketorolac	Toradol
Meclofenamate	Meclomen
Mefenamic acid	Ponstel
Meloxicam	Mobic
Nabumetone	Relafen
Naproxen	Aleve, Anaprox, Naprelan, Naprosyn
Oxaprozin	Daypro, Daypro Alta
Piroxicam	Feldene
Sulindac	Clinoril
Tolmetin	Tolectin

Opioids

Opiate Agonists

Indications: Pain, chronic nonmalignant pain, MI, delirium, acute pulmonary edema, preoperative sedation, cough, and opiate dependence.

Drug name	Brand name
Codeine	Codeine; with acetaminophen (APAP), pseudoephedrine, and phenylephrine: Tylenol with Codeine no. 3 and no. 4
Fentanyl	Actiq, Fentora, Duragesic (topical)
Hydrocodone	Bancap HC, Ceta-Plus, Lorcet, Hydrocet; with APAP: Lortab, Vicodin, Anexsia, Co-Gesic, Zydone, (+ ibuprofen) Vicoprofen, Reprexain
Hydromorphone	Dilaudid, Exalgo
Levorphanol	Levo-Dromoran
Meperidine	Demerol
Methadone	Dolophine, Intensol, Methadose

Drug name	Brand name
Morphine	Avinza, MS Contin, Oramorph SR, Kadian
Opium	None
Oxycodone	OxyIR, OxyContin, Endocodone, Percolone, Roxicodone; (+ APAP) Percocet, Tylox, Endocet, Roxicet; (+ aspirin [ASA]) Percodan, Endodan, Roxiprin
Oxymorphone	Opana ER
Remifentanil	Ultiva (IV)
Sufentanil	Sufenta (IV)
Tapentadol	Nucynta
Tramadol	Ultram ER, (+ APAP) Ultracet

Opiate Partial Agonists (Pain and Opiate Dependence)

Indications: General anesthesia (adjunctive) and pain.

Drug name	Brand name
Buprenorphine	Butrans (topical), Suboxone (sublingual strip), Subutex (sublingual tablet), Buprenex (injectable)
Butorphanol	Stadol (injectable), Stadol NS (nasal spray)

Drug name	Brand name
Nalbuphine	Nubain (injectable)
Pentazocine	(+ naloxone) Talwin Nx, (+ acetaminophen, pseudoephedrine, and phenylephrine) Talacen, Talwin (injectable)

Analgesics and Antipyretics

Indications: Dysmenorrhea, fever, headache, and pain.

Drug name	Brand name
Acetaminophen	Tylenol, many combinations

Drug name	Brand name
Salicylamide	BC Powder, BC Tablets

Unclassified

Antigout

Indication: To treat or prevent gout or treat hyperuricemia (excess uric acid in the blood).

Drug name	Brand name
Allopurinol	Zyloprim
Colchicine	Colcrys
Febuxostat	Uloric

Drug name	Brand name
Probenecid	(+ colchicine) Col-Benecid
Sulfinpyrazone	Anturane

References

American Society of Health-System Pharmacists. AHFS Drug Information 2014. Bethesda: American Society of Health-System Pharmacists; 2014.

Mainenti MRM, Teixeira PFS, Oliveira FP, Vaisman M. Effect of hormone replacement on exercise cardiopulmonary reserve and recovery performance in subclinical hypothyroidism. *Braz J Med Biol Res.* 2010;43(11):1095–1101.

Reents S. *Sport and Exercise Pharmacology.* Champaign, IL: Human Kinetics; 2000.

Somani SM, ed. *Pharmacology in Exercise and Sports.* Boca Raton, FL: CRC Press; 1996.

Additional Resources

The American Hospital Formulary Service (AHFS) Drug Information, www.ahfsdruginformation.com [Accessed September 14, 2015].

MICROMEDEX 2.0 (unbiased, referenced information about medications), www.micromedex.com [Accessed September 14, 2015].

U.S. Food and Drug Administration, U.S. Department of Health and Human Services, www.accessdata.fda.gov/scripts/cder/drugsatfda/index.cfm?fuseaction=Search.Search_Drug_Name [Accessed September 14, 2015].

Basic *CDD4* Recommendations

Mode	Frequency	Duration	Intensity	Progression
Aerobic • Large-muscle easily accessible activities such as walking as the basic program • Aquatics recommended for those with musculoskeletal problems during weight-bearing activity • Other fun-to-do large-muscle activities such as cycling or gardening are alternatives	4-5 days/week	Start at any duration, as tolerated Goal of 40 min/session 20 min if a combined session with strength training exercises	• Start at self-selected walking speed, at an intensity meeting the talk test • Gradually increase to an RPE of 3-5/10	• From self-selected pace, over 4 weeks gradually increase time to 40 min each session, increasing intensity as tolerated. • Persons interested in higher-intensity exercise should obtain guidance before doing it.
Strength • Functional gravity-based exercises recommended as the basic program • Weight training an alternative for those who are interested and motivated to do it	2-3 days/week	For body weight exercises: Functional exercises (see chapter 4), one set during TV commercials For weights: 1 set of 8-12 reps to fatigue	• Sit to stand: 8 reps • Alternative: stair steps (standard is 10 steps) • Arm curls: 8 reps with ~4 kg (can use plastic milk jug filled with water) • 50-70% of 1RM	• Build gradually to as many sets a day as tolerated. • For curls and weight training: Increase to 2 sets over ~8 weeks.
Flexibility Hips, knees, shoulders, and neck	3 days/week	20 s/stretch	Maintain stretch below discomfort point	Discomfort point should occur at a ROM that does not cause instability. This discomfort point will vary between people and with different joints in each person.
Warm-up and cool-down	Before and after each session	10-15 min	Easy RPE <3/10	Should be maintained as transition phase, especially for those doing higher-intensity physical activity.

RPE = rating of perceived exertion
1 RM = 1 repetition maximum, the amount of weight that can be lifted just 1 time
ROM = range of motion

Index

The italicized *f* or *t* following a page number indicates a figure or table on that page.

About the ACSM

The American College of Sports Medicine (ACSM), founded in 1954, is the largest sports medicine and exercise science organization in the world. With more than 50,000 members and certified professionals worldwide, ACSM is dedicated to improving health through science, education, and medicine. ACSM members work in a wide range of medical specialties, allied health professions, and scientific disciplines. Members are committed to the diagnosis, treatment, and prevention of sport-related injuries and the advancement of the science of exercise.

The ACSM promotes and integrates scientific research, education, and practical applications of sports medicine and exercise science to maintain and enhance physical performance, fitness, health, and quality of life.

Contributors

Senior Editors

Geoffrey E. Moore, MD, FACSM
Healthy Living & Exercise
Medicine Associates

J. Larry Durstine, PhD, FACSM
University of South Carolina

Patricia L. Painter, PhD, FACSM
University of Utah

Part Editors

Tony Babb, PhD
University of Texas

Peter H. Brubaker, PhD, FACSM
Wake Forest University

Bradley D. Hatfield, PhD
University of Maryland

Richard Macko, MD
University of Maryland

Jonathan N. Myers, PhD, FACSM
Palo Alto VA Medical Center

David C. Nieman, DrPH, FACSM
Appalachian State University

Elizabeth J. Protas, PhD, PT, FACSM
University of Texas Medical
Branch

Contributors

J. Edwin Atwood, MD
Palo Alto VA Medical Center

Abraham Samuel Babu, MPT, FCR
Manipal University, India

Stephen P. Bailey, PhD, PT, FACSM
Elon University

Bart Bartels, PT, MSc
HU University of Applied
Sciences, University Medical
Center Utrecht, Netherlands

Bart C. Bongers, PhD
Wilhelmina Children's
Hospital, University Medical
Center, Utrecht, Netherlands

Tanvi Bhatt, PT, PhD
University of Illinois at
Chicago

Clinton A. Brawner, PhD, RCEP,
FACSM
Henry Ford Hospital, Detroit,
Michigan

Kelly Chin, MD
University of Texas South-
western Medical Center

Gerson Cipriano Jr., PhD, PT
University of Brasilia, DF,
Brazil

Christopher B. Cooper, MD
David Geffen School
of Medicine, UCLA

Michael Costello, PT, DSc
Cayuga Medical Center

Janke de Groot, PT, PhD
HU University of Applied
Sciences, University Medical
Center Utrecht

Brett A. Dolezal, PhD
David Geffen School
of Medicine, UCLA

Wesley D. Dudgeon, PhD
College of Charleston

Holly Fonda, MS
Palo Alto VA Medical Center

Benjamin T. Gordon, PhD, RCEP
University of South Carolina

Gregory A. Hand, PhD, FACSM
West Virginia University

Heather Hayes, DPT, NCS, PhD
University of Utah

Connie C. W. Hsia, MD
University of Texas South-
western Medical Center

Erik Hulzebos, PT, RCEP, PhD
Child Development
& Exercise Center, Wilhelmi-
na Children's Hospital

William E. Kraus, MD
Duke University

Barry Lewis, DO, FACC
Henry Ford Hospital, Detroit,
Michigan

Michael Lockard, MA
Willamette University

G. William Lyerly, PhD
Coastal Carolina University

Sangeetha Madhavan, PT, PhD
University of Illinois at
Chicago

Désirée B. Maltais, PhD, PT
Université Laval

Jessica S. Oldham, MS
Arapahoe Community
College

Matthew W. Parker, MD, FACC
University of Massachusetts
Memorial Medical Center

Tara Patterson, MEd, PhD
University of Texas Medical
Branch

Shane A. Phillips, PT, PhD
University of Illinois at
Chicago

Sherry O. Pinkstaff, PhD, PT
University of North Florida

Kenneth W. Rundell, PhD, FACSM
The Commonwealth Medical
College

Kathryn Schmitz, PhD
Penn State University at
Hershey

Jill Seale, PT, PhD, NCS
University of Texas Medical
Branch

Audrey B. Silva, PhD, PT
Federal University of São
Carlos, SP, Brazil

Rhonda K. Stanley, PhD, PT
New Mexico Department
of Health

Tim Takken, PhD
Child Development &
Exercise Center, Wilhelmina
Children's Hospital

Maarten S. Werkman, PT, PhD
Child Development &
Exercise Center, Wilhelmina
Children's Hospital

Jo B. Zimmerman, MS
University of Maryland,
College Park